Computational Music Science

Series editors

Guerino Mazzola
Moreno Andreatta

Guerino Mazzola

The Topos of Music
IV: Roots

Appendices

Second Edition

 Springer

Guerino Mazzola
School of Music
University of Minnesota
Minneapolis, MN, USA

This is the fourth of four volumes in the set "The Topos of Music" in the series Computational Music Science. Volume I is ISBN 978-3-319-64363-2, Volume II is ISBN 978-3-319-64443-1, Volume III is ISBN 978-3-319-64479-0. The first edition of The Topos of Music appeared in one volume in the Birkhäuser program ISBN 978-3-7643-5731-3.

ISSN 1868-0305   ISSN 1868-0313   (electronic)
Computational Music Science
ISBN 978-3-030-09721-9   ISBN 978-3-319-64495-0   (eBook)
https://doi.org/10.1007/978-3-319-64495-0

Preface to the Second Edition

Comprendre, c'est
attraper le geste
et pouvoir continuer
Jean Cavaillès [181, p. 186]

A major reason for a second edition of *The Topos of Music*—besides the simple fact that the first edition is now sold out—goes back to spring 2002, when I was completing its first edition, published in fall 2002. I was asked to give a talk in the MaMuX seminar of the IRCAM in Paris, to explain how I applied the mathematics of *The Topos of Music* to my free jazz improvisations.

While preparing my talk I realized that despite the presence of mathematical music theory the decisive generator of my instant compositions was the gestural deployment of formulas, the "action painting" of musical thoughts, not the abstract formulas in their static facticity. First and foremost this was a shocking insight in view of the forthcoming publication of the formulaic setup in *The Topos of Music*.

Fortunately, I knew from Hermann Hesse that "every end is a beginning"[1], which meant in my case that the end of a scientific development as traced in the book's first edition initiated the next step: a music theory of gestures. It goes without saying that this new phase would not destroy the previous research, but incorporate it as the stratum of facticity in an extended ontology of embodiment, where facts are the output of processes and their gestural generators.

In the sequel, I discovered that I was far from being the first scholar and artist to discover the crucial role of gestures in music. For instance, free jazz pianist Cecil Tayor, music philosopher Theodor Wiesengrund Adorno, or lateral thinker Paul Valéry had clearly stressed the dancing essence of art, an insight that I had embodied in my own pianist's art, but never understood on an intellectual level.

Of course, I could not be satisfied by the very existence of gesture philosophy or gestural practice, just as I could not accept traditional music and performance theory when I started my enterprise of mathematical music theory in 1978. The gesturally colored thoughts and actions needed a rigorous conceptualization in the same vein as my efforts before the first edition of *The Topos of Music*.

In 2002, I was in the privileged position to work in the multimedia division of Peter Stucki at the Institut für Informatik of the University of Zurich. I had excellent PhD students, and we could, with one of them, Stefan Müller, realize a first experimental software for the gestural representation of a pianist's hand, a work presented at the ICMC conference in 2003 [772].

This experimental preliminary work was then taken as a point of departure for a mathematical theory of musical gestures. I presented this theory in a course in spring 2005 at École normale supérieure in Paris, a course that was later in 2007 taken as the material basis of my French book *La vérité de beau dans la musique* [718]. The first publication of a formally more evolved mathematical theory of musical gestures was

---

[1] Actually, he says that "Jeder Anfang ist ein Ende," but the reverse is immediate.

written with co-author Moreno Andreatta in 2007 [720]. This date could be called the birthday of a valid mathematical theory of musical gestures.

Until the publication of this second edition of *The Topos of Music*, several important conceptual extensions of the mathematical theory of musical gestures, models of musical gestural processes, as well as a number of theorems have been published. The decade from date of birth to the presence proved that the mathematical theory of musical gestures is an important added value to the theory described in the first edition of *The Topos of Music*.

We can however not state that this theory of gestures is in a complete state, quite the opposite is true: The coming years will reveal important news of theoretical as well as practical nature. So why did we make the decision to publish the present state of the art? The first argument is that the present state is rich enough to define concrete new directions, be it in music theory, such as harmony or counterpoint, be it in performance theory, or be it in the understanding of embodiment in the making of music. The second argument is that the present material, roughly 500 pages of new material, is ample enough to present a book's stature. And the third and very important argument is that we would like to communicate the state of the art in the spirit of Cavaillès: *Understanding is catching the gesture and being able to continue.* The co-authors of the gesture theory part, René Guitart, Jocelyn Ho, Alex Lubet, Maria Mannone, Matt Rahaim, and Florian Thalmann, are a wonderful confirmation of this philosophy. So let us continue!

Here is a summary of the new material and its authorship. Whenever I don't mention the author, it is my own contribution, all others are mentioned explicitly.

Until Part XIV, nearly everything is as in the book's first edition, refer to the preface of that edition (also included in this edition) for detailed summaries. The only new content—besides errata corrections—is Chapter 45 in Part XI, which is a shortened version of a paper [110] on a statistical analysis of Chopin's Prélude op. 28, No. 4, written with Jan Beran, Robert Goswitz, and Patrizio Mazzola.

Gesture theory starts with Part XV: Gesture Philosophy for Music. Chapter 57 gives an overview of philosophical aspects of gestures, including works by Jean-Claude Schmitt, Vilém Flusser, Michel Guérin, Adam Kendon, David McNeill, Juhani Pallasmaa, André Chastel, Émile Benveniste, Marie-Dominique Popelard, and Anthony Wall. In Chapter 58, we discuss the presemiotic approch to gestures in the French perspective of Maurice Merleau-Ponty, Gilles Deleuze, Jean Cavaillès, Charles Alunni, and Gilles Châtelet. Paul Valéry is also referenced in Section 59.4.

Chapter 59 deals with gestural aspects in cognitive science. After a review of Embodied AI and anthropology, Alex Lubet in Section 59.5 introduces us to gestural disability studies, focusing on two famous disabled pianists: Horace Parlan and Oscar Peterson (in his last years). Then in Section 59.6 Lubet reflects on perception of musical gesture as being inherently synaesthetic.

Chapter 60 concludes this part with a review of musical models of gesturality as proposed by Wolfgang Graeser, Theodor W. Adorno, Neil P. McAngus Todd, David Lewin, Robert Hatten, Marcelo Wanderley, Claude Cadoz, and Marc Leman.

Part XVI introduces the mathematics of gestures. Chapter 61 presents the mathematical concept of a gesture in a topological space and states the Diamond Conjecture, which deals with a hypothetical big space that unites algebraic and topological categories. Chapter 62 extends the theory from gestures in topological spaces to gestures in topological categories and introduces functorial gestures, i.e., functors on topological categories with values in the category of gestures, similar to functorial compositions in the previous theory.

Chapter 63 presents a generalized singular homology, where cubes are replaced by general hypergestures. Hypergesture homology applies to a gestural model of counterpoint and to a gestural refinement of performance stemma theory.

Chapter 64 presents—similar to Chapter 63—Stokes' Theorem for hypergestures. This theorem applies to problems in gestural modulation theory.

Chapters 65 and 66 discuss categories of local and global compositions, processes/networks, and gestures, together with their functorial relationships. This triple typology composition/process/gesture corresponds to the ontological dimension of embodiment with its three coordinates facts/processes/gestures.

In Sections 67.1–67.7 of Chapter 67, René Guitart develops a fascinating and demanding mathematical model of mathematical creativity, where thought is viewed as an algebra of gestures.

In Section 61.14, we, Maria Mannone and Guerino Mazzola, present a group-theoretical model of Georg Wilhelm Friedrich Hegel's initial discourse in his *Wissenschaft der Logik*, a model that applies to the Yoneda concept of creativity [726, Chapter 19.2]. We illustrate the method with an experimental composition by Mannone. This discussion extends over Sections 67.8–67.15 and completes Chapter 67.

Part XVII deals with concept architectures and software for musical gesture theory. Chapter 68 explains the denotator formalism for gestures over topological categories. Chapter 69 is a summary of the Java-based RUBATO® Composer software [739], written to sketch the framework of Chapters 70–74, where Florian Thalmann presents his gesture-oriented software component, the BigBang rubette. This discourse again follows the coordinates of the dimension of embodiment: facts, processes, gestures, which in this situation specify to: visualization and sonification of denotators (Chapter 71), BigBang's operation graph (Chapter 72), and gestural interaction and gesturalization (Chapter 73). In the final Chapter 74 of this part, Thalmann discusses musical examples.

Part XVIII is entitled *The Multiverse Perspective* because it opens up the relationship of gesture theory with string theory in theoretical physics. After a critical review of Hermann Hesse's *Glasperlenspiel* with regard to its gestural deficiencies, we, Mazzola and Mannone, develop the Euler-Lagrange formalism of world-sheets for musical gestures. This theory extends to functorial global gestures over global topological categories.

Part XIX is dedicated to applications of gesture theory to a number of musical themes.

Chapter 79 deals with singular gesture homology being applied to counterpoint.

Chapter 80 introduces a gestural restatement of modulation theory, applying in particular Stokes' Theorem for hypergestures.

Chapter 81 applies gesture theory to a gestural performance stemma theory.

Chapter 82 is written by Jocelyn Ho as a creative presentation of composition and analysis as embodied gestures in an inter-corporeal world. She presents two compositions, Toru Takemitsu's *Rain Tree Sketch II* and her own composition *Sheng* for piano, smartphones, and fixed playback.

Chapter 83 is Mannone's analysis and classification of a conductor's movements from the viewpoint of gesture theory.

Chapter 84 is a review of gestural aspects that were developed in *Flow, Gesture, and Spaces in Free Jazz* [721].

Chapter 85 is written by Matt Rahaim and presents the gestural approach to understanding Hindustani music in its vocal gesturality.

Chapter 86 is a first approach, written by Mannone, to a future theory of vocal gestures. The short addendum was written by Mazzola.

The Appendix has been enriched by additional complements on mathematics (Chapter J) plus complements on physics (Chapter K).

The Leitfaden III has been added to the original Leitfaden I & II for the gestural chapters.

The ToM_CD has been updated, containing now the present book's pdf ToposOfMusic.pdf. However the original CD is no longer added to the book, instead the ToM_CD can be downloaded from

www.encyclospace.org/special/ToM_CD.zip.

Concerning the division of the now very large book into parts, this is the split:

- Volume I: *Theory*, Prefaces and Table of Contents, Parts I to VII
- Volume II: *Performance*, Parts VIII to XIV
- Volume III: *Gestures*, Parts XV to XIX
- Volume IV: *Roots*, Appendices

My sincere acknowledgments go to my co-authors and to Springer's Ronan Nugent and Frank Holzwarth as well as to Birkhäuser's Thomas Hempfling.

Minneapolis, October 2017   Guerino Mazzola

Preface

*Man kann
einen jeden Begriff,
einen jeden Titel,
darunter viele Erkenntnisse gehören,
einen logischen Ort nennen.*
Immanuel Kant [519, p. B 324]

This book's title subject, *The Topos of Music*, has been chosen to communicate a double message: First, the Greek word "topos" ($\tau \acute{o}\pi o\varsigma$ = location, site) alludes to the logical and transcendental location of the concept of music in the sense of Aristotle's [40, 1154] and Kant's [519, p. B 324] topic. This view deals with the question of *where music is situated **as** a concept*—and hence with the underlying ontological problem: *What is the type of being and existence of music?* The second message is a more technical understanding insofar as the system of musical signs can be associated with the mathematical theory of *topoi*, which realizes a powerful synthesis of geometric and logical theories. It laid the foundation of a thorough geometrization of logic and has been successful in central issues of algebraic geometry (Grothendieck, Deligne), independence proofs and intuitionistic logic (Cohen, Lawvere, Kripke).

But this second message is intimately entwined with the first since the present concept framework of the musical sign system is technically based on topos theory, so the topos of music receives its topos-theoretic foundation. In this perspective, the double message of the book's title in fact condenses to a unified intention: to unite philosophical insight with mathematical explicitness.

According to Birkhäuser's initial plan in 1996, this book was first conceived as an English translation of my former book *Geometrie der Töne* [682], since the German original had suffered from its restricted access to the international public. However, the scientific progress since 1989, when it was written, has been considerable in theory and technology. We have known new subjects, such as the denotator concept framework, performance theory, and new software platforms for composition, analysis, and performance, such as RUBATO® or OpenMusic. Modeling concepts via the denotator approach in fact results from an intense collaboration of mathematicians and computer scientists in the object-oriented programming paradigm and supported by several international research grants.

Also, the scientific acceptance of mathematical music theory has grown since its beginnings in the late 1970s. As the first acceptance of mathematical music theory was testified to by von Karajan's legendary Ostersymposium "Musik und Mathematik" in 1984 in Salzburg [383], so is the significantly improved present status of acceptance testified to by the Fourth Diderot Forum on Mathematics and Music [711] in Paris, Vienna, and Lisbon 1999, which was organized by the European Mathematical Society. The corresponding extension of collaborative efforts in particular entail the inclusion of works by other research groups in this book, such as the "American Set Theory", the Swedish school of performance research at Stockholm's KTH, or the research on computer-aided composition at the IRCAM in Paris.

Therefore, as a result of these revised conditions, *The Topos of Music* appears as a vastly extended English update of the original work. The extension is visibly traced in the following parts which are new with respect to [682]: Part II exposes the theory of denotators and forms, part V introduces the topological theories of rhythms and motives, part VIII introduces the structure theory of performance, part IX deals with the expressive semantics of performance in the language of performance operators and stemmata (genealogical trees of successively refined performance), part X is devoted to the description of the RUBATO® software platform for representation, analysis, composition, and performance, part XI presents a statistical analysis of musical analysis, part XII concludes the subject of performance with an inverse performance theory, in fact a first formalization of the problem of music criticism.

This does however not mean that the other parts are just translations of the German text. Considerable progress has been made in most fields, except the last part XIV which reproduces the status quo in [682]. In particular, the local and global theories have been thoroughly functorialized and thereby introduce an ontological depth and variability of concepts, techniques, and results, which by far transcend the semiotically naive geometric approach in [682]. The present theory is as different from the traditional geometric conceptualization as is Grothendieck's topos theoretic algebraic geometry from classical algebraic geometry in the spirit of Segre, van der Waerden, or Zariski.

Beyond this topos-theoretic generalization, the denotator language also introduces a fairly exceptional technique of circular concept constructions. This more precisely is rooted in Finsler's pioneering work in foundations of set theory [322], a thread which has been rediscovered in modern theoretical computer sciences [5]. The present state of denotator theory rightly could be termed a Galois theory of concepts in the sense that circular definitions of concepts play the role of conceptual equations (corresponding to algebraic equations in algebraic Galois theory), the solutions of which are concepts instead of algebraic numbers.

Accordingly, the mathematical apparatus has been vastly extended, not only in the field of topos theory and its intuitionistic logic, but also with regard to general and algebraic topology, ordinary and partial differential equations, Pólya theory, statistics, multiaffine algebra and functorial algebraic geometry. It is mandatory that these technicalities had to be placed in a more elaborate semiotic perspective. However, this book does not cover the full range of music semiotics, for which the reader is referred to [703]. Of course, such an extension on the technical level has consequences for the readability of the theory. In view of the present volume of over 1300 pages, we could however not even make the attempt to approach a non-technical presentation. This subject is left to subsequent efforts. The critical reader may put the question whether music is really that complex. The answer is yes, and the reason is straightforward: We cannot pretend that Bach, Haydn, Mozart, or Beethoven, just to name some of the most prominent composers, are outstanding geniuses and have elaborated masterworks of eternal value, without trying to understand such singular creations with adequate tools, and this means: of adequate depth and power. After all, understanding God's 'composition', the material universe, cannot be approached without the most sophisticated tools as they have been elaborated in physics, chemistry, and molecular biology.

So who is recommended to read this book? A first category of readers is evidently the working scientist in the fields of mathematical music theory, the soft- and hardware engineer in music informatics, but also the mathematician who is interested in new applications from the above fields of pure mathematics. A second category are those theoretical mathematicians or computer scientists interested in the Galois theory of concepts; they may discover interesting unsolved problems. A third category of potential readers are all those who really want to get an idea of what music is about, of how one may conceptualize and turn into language the "ineffable" in music for the common language. Those who insist on the dogma that precision and beauty contradict each other, and that mathematics only produces tautologies and therefore must fail when aiming at substantial knowledge, should not read such a book.

Despite the technical character of *The Topos of Music*, there are at least four different approaches to its reading. To begin with, one may read it as a philosophical text, concentrating on the qualitative passages, surfing over technical portions and leaving those paragraphs to others. One may also take the book as a dictionary for computational musicology, including its concept framework and the lists of musical objects and processes (such as modulation degrees, contrapuntal steps) in the appendices. Observe however, that not all existing important lists have been included. For example, the list of all-interval series and the list of self-addressed chords are omitted, the reader may find these lists in other publications. Thirdly, the working scientist will have to read the full-fledged technicalities. And last, but not least, one may take the book as a source for ideas of how to go on with the whole subject of music. The GPL (General Public License[2]) software sources in the appended CD-ROM may support further development.

The prerequisites to a more in-depth reading of this book are these. Generally speaking, a good acquaintance with formal reasoning as mathematics (including formal logic) preconizes, is a conditio sine qua non. As to musicology and music theory, the familiarity with elementary concepts, like chords, motives, rhythm, and also musical notation, as well as a real interest in understanding music and not simply (ab)using it, are recommended. For the more computer-oriented passages, familiarity with the paradigm of object-oriented programming is profitable. We have not included the appendix on mathematical basics because it should help the reader get familiar with mathematics, but as an orientation in fields where the specialized mathematician possibly needs a specification of concepts and notation. The appendix was also included to expose the spectrum of mathematics which is needed to tackle the formal problems of computational musicology. It is by no means an overkill of mathematization: We have even omitted some non-trivial fields, such as statistics or Lambda calculus, for which we have to apologize.

There are different supporting instances to facilitate orientation in this book. To begin with, the table of contents and an extensive subject and name index may help find one's key-words. Further, following the list of contents, a leitfaden (on page xlv) is included for a generic navigation. Each chapter and section is headed by a summary that offers a first orientation about specific contents. Finally, the book is also available as a file `ToposOfMusic.pdf` with bookmarks and active cross-references in the appended CD-ROM (see page xlvii for its contents). This version is also attractive because the figures' colors are visible only in this version.

In order to obtain a consistent first reading, we recommend chapters 1 to 5, and then appendix A: Common Parameter Spaces (appendix B is not mandatory here, though it gives a good and not so technical overview of auditory physiology). After that, the reader may go on with chapter 6 on denotators and then follow the outline of the leitfaden (see page xlv).

This book could not have been realized without the engaged support of nineteen collaborators and contributors. Above all, my PhD students Stefan Göller and Stefan Müller at the MultiMedia Laboratory of the Department of Information Technology at the University of Zurich have collaborated in the production of this book on the levels of the LATEX installation, the final production of hundreds of figures, and the contributions sections 20.2 through 20.5 (Göller) and sections 47.3 through 47.3.6.2 (Müller). My special gratitude goes to their truly collaborative spirit.

Contributions to this book have been delivered by (in alphabetic order): By Carlos Agon, and Gérard Assayag (both IRCAM) with their precious Lambda-calculus-oriented presentation of the object-oriented programming principles in the composition software OpenMusic described in chapter 52, Moreno Andreatta (IRCAM) with an elucidating discourse on the American Set Theory in section 11.5.2 and section 16.3, Jan Beran (Universität Konstanz) with his contribution to the compositional strategies in his original composition [103] in section 11.5.1.1, as well as with his inspiring work on statistics as reported in chapters 43 and 44, Chantal Buteau (Universität and ETH Zürich) with her detailed review of chapter 22, Roberto Ferretti (ETH Zürich) with his progressive contributions to the algebraic geometry of inverse performance theory in sections 39.8 and 47.2, Anja Fleischer (Technische Universität Berlin) with her short but critical preliminaries in chapter 23, Harald Fripertinger (Universität Graz) with his 'killer' formulas concerning enumeration of finite local and global compositions in sections 11.4, 16.2.2 and appendix C.3.6, Jörg Garbers (Technische Universität Berlin) with his portation of the RUBATO® application to Mac OS X, as documented in the screenshots in chapters 40, 41, Werner Hemmert (Infineon) with a very up-to-date presentation of room acoustics in section A.1.1.1 and auditory physiology in appendix B.1 (we would have loved to include more of his knowledge), Michael Leyton (DIMACS, Rutgers University) with a formidable cover figure entitled "Dark Theory", a beautiful subtitle to this book, as well as with innumerable discussions around time and its reduction to symmetries as presented in chapter 48, Emilio Lluis Puebla (UNAM, Mexico City)

---

[2] A legal matter file is contained in the book's CD-ROM, see page xlvii.

with his unique and engaged promotion and dissipation of mathematical music theory on the American continent, especially also in the preparation and critical review of this book, Mariana Montiel Hernandez (UNAM, Mexico City) with her critical review of the theory of circular forms and denotators in section 6.5 and appendix G.2.2.1, Thomas Noll (Technische Universität Berlin) with his substantial contributions to the functorial theory of compositions, and for his revolutionary rebuilding of Riemann's harmony and its relations to counterpoint, Joachim Stange-Elbe (Universität Osnabrück) with a very clear and innovative description of his outstanding RUBATO® performance of Bach's contrapunctus III in the *Art of Fugue* in sections 42.2 through 42.4.3, Hans Straub with his adventurous extensions of classical cadence theory in section 26.2.2 and his classification of four-element motives in appendix O.4, and, last but not least, Oliver Zahorka (Out Media Design), my former collaborator and chief programmer of the NeXT RUBATO® application, which has contributed so much to the success of the Zürich school of performance theory. To all of them, I owe my deepest gratitude and recognition for their sweat and tears.

My sincere acknowledgments go to Alexander Grothendieck, whose encouraging letters and, no doubt, awe inspiring revolution in mathematical thinking has given me so much in isolated phases of this enterprise. My acknowledgments also go to my engaged mentor Peter Stucki, director of the MultiMedia Laboratory of the Department of Information Technology at the University of Zurich; without his support, this book would have seen its birthday years later, if ever. My thanks also go to my brother Silvio, who once again (he did it already for my first book [670]) supported the final review efforts by an ideal environment in his villa in Vulpera. My thanks also go to the unbureaucratic management of the book's production by Birkhäuser's lector Thomas Hempfling and the very patient copy editor Edwin Beschler. All these beautiful supports would have failed without my wife Christina's infinite understanding and vital environment—if this book is a trace of humanity, it is also, and strongly, hers.

Vulpera, June 2002   Guerino Mazzola

Volume IV Contents

Part XX Appendix: Sound

A Common Parameter Spaces 1335
   A.1 Physical Spaces 1335
      A.1.1 Neutral Data 1336
      A.1.2 Sound Analysis and Synthesis 1338
   A.2 Mathematical and Symbolic Spaces 1347
      A.2.1 Onset and Duration 1347
      A.2.2 Amplitude and Crescendo 1348
      A.2.3 Frequency and Glissando 1349

B Auditory Physiology and Psychology 1353
   B.1 Physiology: From the Auricle to Heschl's Gyri 1353
      B.1.1 Outer Ear 1354
      B.1.2 Middle Ear 1354
      B.1.3 Inner Ear (Cochlea) 1355
      B.1.4 Cochlear Hydrodynamics: The Travelling Wave 1357
      B.1.5 Active Amplification of the Traveling Wave Motion 1358
      B.1.6 Neural Processing 1360
   B.2 Discriminating Tones: Werner Meyer-Eppler's Valence Theory 1362
   B.3 Aspects of Consonance and Dissonance 1364
      B.3.1 Euler's Gradus Function 1364
      B.3.2 von Helmholtz' Beat Model 1365
      B.3.3 Psychometric Investigations by Plomp and Levelt 1367
      B.3.4 Counterpoint 1367
      B.3.5 Consonance and Dissonance: A Conceptual Field 1367

Part XXI Appendix: Mathematical Basics

C Sets, Relations, Monoids, Groups 1371
   C.1 Sets 1371
      C.1.1 Examples of Sets 1371
   C.2 Relations 1372
      C.2.1 Universal Constructions 1374
      C.2.2 Graphs and Quivers 1375
      C.2.3 Monoids 1376

C.3   Groups... 1378
     C.3.1   Homomorphisms of Groups 1378
     C.3.2   Direct, Semi-direct, and Wreath Products 1380
     C.3.3   Sylow Theorems on p-groups............................. 1380
     C.3.4   Classification of Groups 1381
     C.3.5   General Affine Groups 1382
     C.3.6   Permutation Groups 1382

D   Rings and Algebras... 1385
   D.1   Basic Definitions and Constructions................................ 1385
     D.1.1   Universal Constructions 1386
   D.2   Prime Factorization... 1389
   D.3   Euclidean Algorithm.. 1389
   D.4   Approximation of Real Numbers by Fractions 1389
   D.5   Some Special Issues 1390
     D.5.1   Integers, Rationals, and Real Numbers 1390

E   Modules, Linear, and Affine Transformations 1391
   E.1   Modules and Linear Transformations............................... 1391
     E.1.1   Examples 1391
   E.2   Module Classification 1392
     E.2.1   Dimension.. 1392
     E.2.2   Endomorphisms on Dual Numbers......................... 1394
     E.2.3   Semi-simple Modules 1394
     E.2.4   Jacobson Radical and Socle............................... 1395
     E.2.5   Theorem of Krull-Remak-Schmidt 1396
   E.3   Categories of Modules and Affine Transformations 1396
     E.3.1   Direct Sums 1397
     E.3.2   Affine Forms and Tensors 1397
     E.3.3   Biaffine Maps 1399
     E.3.4   Symmetries of the Affine Plane........................... 1402
     E.3.5   Symmetries on $\mathbb{Z}^2$ 1402
     E.3.6   Symmetries on $\mathbb{Z}^n$ 1403
     E.3.7   Complements on the Module of a Local Composition 1403
     E.3.8   Fiber Products and Fiber Sums in **Mod** 1404
   E.4   Complements of Commutative Algebra 1405
     E.4.1   Localization 1406
     E.4.2   Projective Modules 1406
     E.4.3   Injective Modules 1407
     E.4.4   Lie Algebras 1408

F   Algebraic Geometry 1411
   F.1   Locally Ringed Spaces 1411
   F.2   Spectra of Commutative Rings 1412
     F.2.1   Sober Spaces 1413
   F.3   Schemes and Functors 1414
   F.4   Algebraic and Geometric Structures on Schemes 1415
     F.4.1   The Zariski Tangent Space 1415
   F.5   Grassmannians.. 1416
   F.6   Quotients 1417

G   Categories, Topoi, and Logic 1419
   G.1   Categories Instead of Sets 1419
       G.1.1   Examples 1420
       G.1.2   Functors 1421
       G.1.3   Natural Transformations 1422
   G.2   The Yoneda Lemma 1423
       G.2.1   Universal Constructions: Adjoints, Limits, and Colimits 1423
       G.2.2   Limit and Colimit Characterizations 1425
   G.3   Topoi 1427
       G.3.1   Subobject Classifiers 1428
       G.3.2   Exponentiation 1429
       G.3.3   Definition of Topoi 1429
   G.4   Grothendieck Topologies 1430
       G.4.1   Sheaves 1431
   G.5   Formal Logic 1432
       G.5.1   Propositional Calculus 1432
       G.5.2   Predicate Logic 1435
       G.5.3   A Formal Setup for Consistent Domains of Forms 1437

H   Complements on General and Algebraic Topology 1443
   H.1   Topology 1443
       H.1.1   General 1443
       H.1.2   The Category of Topological Spaces 1444
       H.1.3   Uniform Spaces 1444
       H.1.4   Special Issues 1445
   H.2   Algebraic Topology 1445
       H.2.1   Simplicial Complexes 1445
       H.2.2   Geometric Realization of a Simplicial Complex 1446
       H.2.3   Contiguity 1447
   H.3   Simplicial Coefficient Systems 1447
       H.3.1   Cohomology 1447

I   Complements on Calculus 1449
   I.1   Abstract on Calculus 1449
       I.1.1   Norms and Metrics 1449
       I.1.2   Completeness 1450
       I.1.3   Differentiation 1451
   I.2   Ordinary Differential Equations (ODEs) 1451
       I.2.1   The Fundamental Theorem: Local Case 1452
       I.2.2   The Fundamental Theorem: Global Case 1453
       I.2.3   Flows and Differential Equations 1455
       I.2.4   Vector Fields and Derivations 1455
   I.3   Partial Differential Equations 1455

J   More Complements on Mathematics 1457
   J.1   Directed Graphs 1457
       J.1.1   The Category of Directed Graphs (Digraphs) 1457
       J.1.2   Two Standard Constructions in Graph Theory 1459
       J.1.3   The Topos of Digraphs 1459
   J.2   Galois Theory 1460

J.3   Splines 1461
    J.3.1   Some Simplex Constructions for Splines 1461
    J.3.2   Definition of General Splines 1462
J.4   Topology and Topological Categories 1463
    J.4.1   Topology 1463
    J.4.2   Topological Categories 1464
J.5   Complex Analysis 1465
J.6   Differentiable Manifolds 1466
    J.6.1   Manifolds with Boundary 1467
    J.6.2   The Tangent Manifold 1467
J.7   Tensor Fields 1468
    J.7.1   Alternating Tensors 1468
    J.7.2   Tangent Tensors 1468
J.8   Stokes' Theorem 1469
J.9   Calculus of Variations 1471
J.10   Partial Differential Equations 1471
    J.10.1   Explicit Calculation 1472
J.11   Algebraic Topology 1474
    J.11.1   Homotopy Theory 1474
    J.11.2   The Fundamental Group(oid) 1475
J.12   Homology 1475
    J.12.1   Singular Homology 1476
J.13   Cohomology 1477

Part XXII Appendix: Complements in Physics

K   Complements on Physics 1481
    K.1   Hamilton's Variational Principle 1481
        K.1.1   Euler-Lagrange Equations for a Non-relativistic Particle 1482
    K.2   String Theory 1482
    K.3   Duality and Supersymmetry 1484
    K.4   Quantum Mechanics 1485
        K.4.1   Banach and Hilbert Spaces 1486
        K.4.2   Geometry on Hilbert Spaces 1490
        K.4.3   Axioms for Quantum Mechanics 1492
        K.4.4   The Spectral Theorem 1493

Part XXIII Appendix: Tables

L   Euler's Gradus Function 1497

M   Just and Well-Tempered Tuning 1499

N   Chord and Third Chain Classes 1501
    N.1   Chord Classes 1501
    N.2   Third Chain Classes 1506

O   Two, Three, and Four Tone Motif Classes 1513
    O.1   Two Tone Motifs in $OnPiMod_{12,12}$ 1513
    O.2   Two Tone Motifs in $OnPiMod_{5,12}$ 1513

O.3   Three Tone Motifs in $OnPiMod_{12,12}$ 1514
O.4   Four Tone Motifs in $OnPiMod_{12,12}$ 1517
O.5   Three Tone Motifs in $OnPiMod_{5,12}$ 1523

P   **Well-Tempered and Just Modulation Steps** 1525
      P.1   12-Tempered Modulation Steps 1525
              P.1.1   Scale Orbits and Number of Quantized Modulations 1525
              P.1.2   Quanta and Pivots for the Modulations Between Diatonic Major Scales (No.38.1) . 1527
              P.1.3   Quanta and Pivots for the Modulations Between Melodic Minor Scales (No.47.1) .. 1528
              P.1.4   Quanta and Pivots for the Modulations Between Harmonic Minor Scales (No.54.1) 1530
              P.1.5   Examples of 12-Tempered Modulations for All Fourth Relations 1530
      P.2   2-3-5-Just Modulation Steps 1531
              P.2.1   Modulation Steps Between Just Major Scales 1531
              P.2.2   Modulation Steps Between Natural Minor Scales 1532
              P.2.3   Modulation Steps from Natural Minor to Major Scales 1532
              P.2.4   Modulation Steps from Major to Natural Minor Scales 1533
              P.2.5   Modulation Steps Between Harmonic Minor Scales 1533
              P.2.6   Modulation Steps Between Melodic Minor Scales 1534
              P.2.7   General Modulation Behaviour for 32 Altered Scales 1535

Q   **Counterpoint Steps** 1537
      Q.1   Contrapuntal Symmetries 1537
              Q.1.1   Class No. 64 1537
              Q.1.2   Class No. 68 1538
              Q.1.3   Class No. 71 1539
              Q.1.4   Class No. 75 1540
              Q.1.5   Class No. 78 1541
              Q.1.6   Class No. 82 1542
      Q.2   Permitted Successors for the Major Scale 1543

Part XXIV   References and Index

References R.1

Index R.33

Book Set Contents

Part I Introduction and Orientation

1   What Is Music About? 3
   1.1   Fundamental Activities 3
   1.2   Fundamental Scientific Domains 5

2   Topography 9
   2.1   Layers of Reality 10
      2.1.1   Physical Reality 10
      2.1.2   Mental Reality 11
      2.1.3   Psychological Reality 11
   2.2   Molino's Communication Stream 11
      2.2.1   Creator and Poietic Level 12
      2.2.2   Work and Neutral Level 13
      2.2.3   Listener and Esthesic Level 13
   2.3   Semiosis 14
      2.3.1   Expressions 15
      2.3.2   Content 15
      2.3.3   The Process of Signification 15
      2.3.4   A Short Overview of Music Semiotics 15
   2.4   The Cube of Local Topography 17
   2.5   Topographical Navigation 19

3   Musical Ontology 21
   3.1   Where Is Music? 21
   3.2   Depth and Complexity 23

4   Models and Experiments in Musicology 27
   4.1   Interior and Exterior Nature 29
   4.2   What Is a Musicological Experiment? 30
   4.3   Questions—Experiments of the Mind 31
   4.4   New Scientific Paradigms and Collaboratories 32

Part II Navigation on Concept Spaces

5   Navigation 35
   5.1   Music in the EncycloSpace 36
   5.2   Receptive Navigation 39
   5.3   Productive Navigation 39

6   Denotators 41
   6.1   Universal Concept Formats 42
      6.1.1   First Naive Approach to Denotators 43
      6.1.2   Interpretations and Comments 48
      6.1.3   Ordering Denotators and 'Concept Leafing' 50
   6.2   Forms 52
      6.2.1   Variable Addresses 53
      6.2.2   Formal Definition 54
      6.2.3   Discussion of the Form Typology 56
   6.3   Denotators 57
      6.3.1   Formal Definition of a Denotator 57
   6.4   Anchoring Forms in Modules 59
      6.4.1   First Examples and Comments on Modules in Music 60
   6.5   Regular and Circular Forms 64
   6.6   Regular Denotators 66
   6.7   Circular Denotators 72
   6.8   Ordering on Forms and Denotators 75
      6.8.1   Concretizations and Applications 78
   6.9   Concept Surgery and Denotator Semantics 83

Part III Local Theory

7   Local Compositions 89
   7.1   The Objects of Local Theory 89
   7.2   First Local Music Objects 92
      7.2.1   Chords and Scales 92
      7.2.2   Local Meters and Local Rhythms 96
      7.2.3   Motives 99
   7.3   Functorial Local Compositions 101
   7.4   First Elements of Local Theory 103
   7.5   Alterations Are Tangents 107
      7.5.1   The Theorem of Mason-Mazzola 108

8   Symmetries and Morphisms 113
   8.1   Symmetries in Music 114
      8.1.1   Elementary Examples 116
   8.2   Morphisms of Local Compositions 128
   8.3   Categories of Local Compositions 132
      8.3.1   Commenting on the Concatenation Principle 134
      8.3.2   Embedding and Addressed Adjointness 136
      8.3.3   Universal Constructions on Local Compositions 138
      8.3.4   The Address Question 140
      8.3.5   Categories of Commutative Local Compositions 142

9   Yoneda Perspectives 145
    9.1   Morphisms Are Points 147
    9.2   Yoneda's Fundamental Lemma 150
    9.3   The Yoneda Philosophy 152
    9.4   Understanding Fine and Other Arts 153
            9.4.1   Painting and Music 153
            9.4.2   The Art of Object-Oriented Programming 155

10   Paradigmatic Classification 157
    10.1   Paradigmata in Musicology, Linguistics, and Mathematics 158
    10.2   Transformation 162
    10.3   Similarity 163
    10.4   Fuzzy Concepts in the Humanities 164

11   Orbits 167
    11.1   Gestalt and Symmetry Groups 167
    11.2   The Framework for Local Classification 168
    11.3   Orbits of Elementary Structures 168
            11.3.1   Classification Techniques 169
            11.3.2   The Local Classification Theorem 170
            11.3.3   The Finite Case 177
            11.3.4   Dimension 178
            11.3.5   Chords 180
            11.3.6   Empirical Harmonic Vocabularies 181
            11.3.7   Self-addressed Chords 185
            11.3.8   Motives 187
    11.4   Enumeration Theory 190
            11.4.1   Pólya and de Bruijn Theory 190
            11.4.2   Big Science for Big Numbers 196
    11.5   Group-Theoretical Methods in Composition and Theory 198
            11.5.1   Aspects of Serialism 199
            11.5.2   The American Tradition 202
    11.6   Esthetic Implications of Classification 211
            11.6.1   Jakobson's Poetic Function 212
            11.6.2   Motivic Analysis: Schubert/Stolberg "Lied auf dem Wasser zu singen..." 214
            11.6.3   Composition: Mazzola/Baudelaire "La mort des artistes" 218
    11.7   Mathematical Reflections on Historicity in Music 220
            11.7.1   Jean-Jacques Nattiez' Paradigmatic Theme 221
            11.7.2   Groups as a Parameter of Historicity 223

12   Topological Specialization 225
    12.1   What Ehrenfels Neglected 225
    12.2   Topology 226
            12.2.1   Metrical Comparison 228
            12.2.2   Specialization Morphisms of Local Compositions 230
    12.3   The Problem of Sound Classification 232
            12.3.1   Topographic Determinants of Sound Descriptions 232
            12.3.2   Varieties of Sounds 238
            12.3.3   Semiotics of Sound Classification 240
    12.4   Making the Vague Precise 241

Part IV Global Theory

13 Global Compositions 245
   13.1 The Local-Global Dichotomy in Music 246
      13.1.1 Musical and Mathematical Manifolds 251
   13.2 What Are Global Compositions? 252
      13.2.1 The Nerve of an Objective Global Composition 253
   13.3 Functorial Global Compositions 256
   13.4 Interpretations and the Vocabulary of Global Concepts 258
      13.4.1 Iterated Interpretations 258
      13.4.2 The Pitch Domain: Chains of Thirds, Ecclesiastical Modes, Triadic and Quaternary Degrees 259
      13.4.3 Interpreting Time: Global Meters and Rhythms 266
      13.4.4 Motivic Interpretations: Melodies and Themes 270

14 Global Perspectives 273
   14.1 Musical Motivation 273
   14.2 Global Morphisms 274
   14.3 Local Domains 280
   14.4 Nerves 281
   14.5 Simplicial Weights 283
   14.6 Categories of Commutative Global Compositions 285

15 Global Classification 287
   15.1 Module Complexes 287
      15.1.1 Global Affine Functions 288
      15.1.2 Bilinear and Exterior Forms 290
      15.1.3 Deviation: Compositions vs. "Molecules" 291
   15.2 The Resolution of a Global Composition 292
      15.2.1 Global Standard Compositions 293
      15.2.2 Compositions from Module Complexes 294
   15.3 Orbits of Module Complexes Are Classifying 298
      15.3.1 Combinatorial Group Actions 299
      15.3.2 Classifying Spaces 300

16 Classifying Interpretations 303
   16.1 Characterization of Interpretable Compositions 303
      16.1.1 Automorphism Groups of Interpretable Compositions 306
      16.1.2 A Cohomological Criterion 307
   16.2 Global Enumeration Theory 309
      16.2.1 Tesselation 309
      16.2.2 Mosaics 310
      16.2.3 Classifying Rational Rhythms and Canons 312
   16.3 Global American Set Theory 314
   16.4 Interpretable "Molecules" 316

17 Esthetics and Classification 319
   17.1 Understanding by Resolution: An Illustrative Example 319
   17.2 Varèse's Program and Yoneda's Lemma 323

18 Predicates 327
  18.1 What Is the Case: The Existence Problem 327
    18.1.1 Merging Systematic and Historical Musicology 328
  18.2 Textual and Paratextual Semiosis 329
    18.2.1 Textual and Paratextual Signification 330
  18.3 Textuality 331
    18.3.1 The Category of Denotators 331
    18.3.2 Textual Semiosis 334
    18.3.3 Atomic Predicates 339
    18.3.4 Logical and Geometric Motivation 345
  18.4 Paratextuality 349

19 Topoi of Music 351
  19.1 The Grothendieck Topology 351
    19.1.1 Cohomology 354
    19.1.2 Marginalia on Presheaves 356
  19.2 The Topos of Music: An Overview 357

20 Visualization Principles 361
  20.1 Problems 361
  20.2 Folding Dimensions 363
    20.2.1 $\mathbb{R}^2 \to \mathbb{R}$ 363
    20.2.2 $\mathbb{R}^n \to \mathbb{R}$ 364
    20.2.3 An Explicit Construction of μ with Special Values. 365
  20.3 Folding Denotators 366
    20.3.1 Folding Limits 366
    20.3.2 Folding Colimits 367
    20.3.3 Folding Powersets 368
    20.3.4 Folding Circular Denotators 369
  20.4 Compound Parametrized Objects 370
  20.5 Examples 371

Part V Topologies for Rhythm and Motives

21 Metrics and Rhythmics 375
  21.1 Review of Riemann and Jackendoff-Lerdahl Theories 375
    21.1.1 Riemann's Weights 375
    21.1.2 Jackendoff-Lerdahl: Intrinsic Versus Extrinsic Time Structures ... 376
  21.2 Topologies of Global Meters and Associated Weights 378
  21.3 Macro-events in the Time Domain 380

22 Motif Gestalts 383
  22.1 Motivic Interpretation 384
  22.2 Shape Types 385
    22.2.1 Examples of Shape Types 386
  22.3 Metrical Similarity 388
    22.3.1 Examples of Distance Functions 389
  22.4 Paradigmatic Groups 390
    22.4.1 Examples of Paradigmatic Groups 391
  22.5 Pseudo-metrics on Orbits 393
  22.6 Topologies on Gestalts 394

      22.6.1 The Inheritance Property.. 395
      22.6.2 Cognitive Aspects of Inheritance 396
      22.6.3 Epsilon Topologies... 397
    22.7 First Properties of the Epsilon Topologies 399
      22.7.1 Toroidal Topologies... 401
    22.8 Rudolph Reti's Motivic Analysis Revisited 404
      22.8.1 Review of Concepts.. 404
      22.8.2 Reconstruction 406
    22.9 Motivic Weights.. 408

Part VI Harmony

23 Critical Preliminaries... 413
    23.1 Hugo Riemann 414
    23.2 Paul Hindemith 414
    23.3 Heinrich Schenker and Friedrich Salzer 414

24 Harmonic Topology... 417
    24.1 Chord Perspectives 418
      24.1.1 Euler Perspectives.. 418
      24.1.2 12-Tempered Perspectives 422
      24.1.3 Enharmonic Projection.. 425
    24.2 Chord Topologies... 427
      24.2.1 Extension and Intension 427
      24.2.2 Extension and Intension Topologies 429
      24.2.3 Faithful Addresses.. 431
      24.2.4 The Saturation Sheaf 434

25 Harmonic Semantics... 435
    25.1 Harmonic Signs—Overview 436
    25.2 Degree Theory 437
      25.2.1 Chains of Thirds 437
      25.2.2 American Jazz Theory 439
      25.2.3 Hans Straub: General Degrees in General Scales 442
    25.3 Function Theory 442
      25.3.1 Canonical Morphemes for European Harmony............................ 444
      25.3.2 Riemann Matrices 447
      25.3.3 Chains of Thirds 447
      25.3.4 Tonal Functions from Absorbing Addresses 449

26 Cadence... 453
    26.1 Making the Concept Precise 454
    26.2 Classical Cadences Relating to 12-Tempered Intonation 454
      26.2.1 Cadences in Triadic Interpretations of Diatonic Scales................... 455
      26.2.2 Cadences in More General Interpretations 456
    26.3 Cadences in Self-addressed Tonalities of Morphology 457
    26.4 Self-addressed Cadences by Symmetries and Morphisms 459
    26.5 Cadences for Just Intonation... 460
      26.5.1 Tonalities in Third-Fifth Intonation...................................... 460
      26.5.2 Tonalities in Pythagorean Intonation..................................... 461

27 Modulation 463
   27.1 Modeling Modulation by Particle Interaction 464
     27.1.1 Models and the Anthropic Principle 464
     27.1.2 Classical Motivation and Heuristics 465
     27.1.3 The General Background 467
     27.1.4 The Well-Tempered Case 469
     27.1.5 Reconstructing the Diatonic Scale from Modulation 471
     27.1.6 The Case of Just Tuning 473
     27.1.7 Quantized Modulations and Modulation Domains for Selected Scales ... 477
   27.2 Harmonic Tension 481
     27.2.1 The Riemann Algebra 481
     27.2.2 Weights on the Riemann Algebra 482
     27.2.3 Harmonic Tensions from Classical Harmony? 484
     27.2.4 Optimizing Harmonic Paths 485

28 Applications 487
   28.1 First Examples 488
     28.1.1 Johann Sebastian Bach: Choral from "Himmelfahrtsoratorium" ... 488
     28.1.2 Wolfgang Amadeus Mozart: "Zauberflöte", Choir of Priests 490
     28.1.3 Claude Debussy: "Préludes", Livre 1, No.4 492
   28.2 Modulation in Beethoven's Sonata op.106, 1st Movement 495
     28.2.1 Introduction 495
     28.2.2 The Fundamental Theses of Erwin Ratz and Jürgen Uhde 497
     28.2.3 Overview of the Modulation Structure 498
     28.2.4 Modulation $B_\flat \rightsquigarrow G$ via e^{-3} in W 499
     28.2.5 Modulation $G \rightsquigarrow E_\flat$ via U_g in W 499
     28.2.6 Modulation $E_\flat \rightsquigarrow D/b$ from W to W^* 499
     28.2.7 Modulation $D/b \rightsquigarrow B$ via $U_{d/d_\sharp} = U_{g_\sharp/a}$ within W^* 500
     28.2.8 Modulation $B \rightsquigarrow B_\flat$ from W^* to W 500
     28.2.9 Modulation $B_\flat \rightsquigarrow G_\flat$ via $U_{b_\flat}$ within W 501
     28.2.10 Modulation $G_\flat \rightsquigarrow G$ via $U_{a_\flat/a}$ within W 501
   28.3 Rhythmical Modulation in "Synthesis" 501
     28.3.1 Rhythmic Modes 502
     28.3.2 Composition for Percussion Ensemble 503

Part VII Counterpoint

29 Melodic Variation by Arrows 507
   29.1 Arrows and Alterations 507
   29.2 The Contrapuntal Interval Concept 508
   29.3 The Algebra of Intervals 509
     29.3.1 The Third Torus 510
   29.4 Musical Interpretation of the Interval Ring 511
   29.5 Self-addressed Arrows 514
   29.6 Change of Orientation 515

30 Interval Dichotomies as a Contrast 517
   30.1 Dichotomies and Polarity 517
   30.2 The Consonance and Dissonance Dichotomy 520
        30.2.1 Fux and Riemann Consonances Are Isomorphic 521
        30.2.2 Induced Polarities 523
        30.2.3 Empirical Evidence for the Polarity Function 523
        30.2.4 Music and the Hippocampal Gate Function 527

31 Modeling Counterpoint by Local Symmetries 531
   31.1 Deformations of the Strong Dichotomies 531
   31.2 Contrapuntal Symmetries Are Local 533
   31.3 The Counterpoint Theorem 534
        31.3.1 Some Preliminary Calculations 534
        31.3.2 Two Lemmata on Cardinalities of Intersections 536
        31.3.3 An Algorithm for Exhibiting the Contrapuntal Symmetries 536
        31.3.4 Transfer of the Counterpoint Rules to General Representatives of Strong Dichotomies ... 540
   31.4 The Classical Case: Consonances and Dissonances 540
        31.4.1 Discussion of the Counterpoint Theorem in the Light of Reduced Strict Style 541
        31.4.2 The Major Dichotomy—A Cultural Antipode? 542
        31.4.3 Software for Counterpoint and Theoretical Extentions 543

Part VIII Structure Theory of Performance

32 Local and Global Performance Transformations 547
   32.1 Performance as a Reality Switch 548
   32.2 Why Do We Need Infinite Performance of the Same Piece? 549
   32.3 Local Structure 550
        32.3.1 The Coherence of Local Performance Transformations 550
        32.3.2 Differential Morphisms of Local Compositions 551
   32.4 Global Structure 554
        32.4.1 Modeling Performance Syntax 556
        32.4.2 The Formal Setup 557
        32.4.3 Performance qua Interpretation of Interpretation 560

33 Performance Fields 561
   33.1 Classics: Tempo, Intonation, and Dynamics 561
        33.1.1 Tempo 561
        33.1.2 Intonation 563
        33.1.3 Dynamics 564
   33.2 Genesis of the General Formalism 565
        33.2.1 The Question of Articulation 565
        33.2.2 The Formalism of Performance Fields 568
   33.3 What Performance Fields Signify 568
        33.3.1 Th.W. Adorno, W. Benjamin, and D. Raffman 569
        33.3.2 Towards Composition of Performance 571

34 Initial Sets and Initial Performances 573
   34.1 Taking Off with a Shifter 573
   34.2 Anchoring Onset 574
   34.3 The Concert Pitch 576
   34.4 Dynamical Anchors 578

34.5   Initializing Articulation 578
34.6   Hit Point Theory 580
    34.6.1   Distances 580
    34.6.2   Flow Interpolation... 582

35   Hierarchies and Performance Scores 585
35.1   Performance Cells 585
35.2   The Category of Performance Cells 586
35.3   Hierarchies 588
    35.3.1   Operations on Hierarchies 590
    35.3.2   Classification Issues 591
    35.3.3   Example: The Piano and Violin Hierarchies..................... 593
35.4   Local Performance Scores 594
35.5   Global Performance Scores 598
    35.5.1   Instrumental Fibers 599

Part IX Expressive Semantics

36   Taxonomy of Expressive Performance...................................... 603
36.1   Feelings: Emotional Semantics 604
36.2   Motion: Gestural Semantics... 606
36.3   Understanding: Rational Semantics 609
36.4   Cross-semantical Relations 612

37   Performance Grammars 615
37.1   Rule-Based Grammars 615
    37.1.1   The KTH School 617
    37.1.2   Neil P. McAngus Todd 618
    37.1.3   The Zurich School 619
37.2   Remarks on Learning Grammars 619

38   Stemma Theory 621
38.1   Motivation from Practising and Rehearsing 622
    38.1.1   Does Reproducibility of Performances Help Understanding?...... 622
38.2   Tempo Curves Are Inadequate 623
38.3   The Stemma Concept 626
    38.3.1   The General Setup of Matrilineal Sexual Propagation 628
    38.3.2   The Primary Mother—Taking Off 629
    38.3.3   Mono- and Polygamy—Local and Global Actions............... 632
    38.3.4   Family Life—Cross-correlations 634

39   Operator Theory 637
39.1   Why Weights? 638
    39.1.1   Discrete and Continuous Weights............................... 638
    39.1.2   Weight Recombination 639
39.2   Primavista Weights 640
    39.2.1   Dynamics 640
    39.2.2   Agogics 643
    39.2.3   Tuning and Intonation 644
    39.2.4   Articulation 645
    39.2.5   Ornaments 645

39.3   Analytical Weights 646
39.4   Taxonomy of Operators 648
    39.4.1   Splitting Operators 649
    39.4.2   Symbolic Operators 650
    39.4.3   Physical Operators 651
    39.4.4   Field Operators 652
39.5   Tempo Operator 653
39.6   Scalar Operator 654
39.7   The Theory of Basis—Pianola Operators 655
    39.7.1   Basis Specialization 656
    39.7.2   Pianola Specialization 659
39.8   Locally Linear Grammars 659

Part X   RUBATO®

40   Architecture 665
40.1   The Overall Modularity 666
40.2   Frame and Modules 668
40.3   Postscriptum: The Rubato Composer Environment 668

41   The RUBETTE® Family 669
41.1   MetroRUBETTE® 669
41.2   MeloRUBETTE® 672
41.3   HarmoRUBETTE® 674
    41.3.1   A Set of New Harmonic Analysis Rubettes on RUBATO® Composer 678
41.4   PerformanceRUBETTE® 679
41.5   PrimavistaRUBETTE® 685

42   Performance Experiments 687
42.1   A Preliminary Experiment: Robert Schumann's "Kuriose Geschichte" 687
42.2   Full Experiment: J.S. Bach's "Kunst der Fuge" 688
42.3   Analysis 688
    42.3.1   Metric Analysis 688
    42.3.2   Motif Analysis 692
    42.3.3   Omission of Harmonic Analysis 693
42.4   Stemma Constructions 694
    42.4.1   Performance Setup 694
    42.4.2   Instrumental Setup 700
    42.4.3   Global Discussion 701

Part XI   Statistics of Analysis and Performance

43   Analysis of Analysis 707
43.1   Hierarchical Decomposition 707
    43.1.1   General Motivation 707
    43.1.2   Hierarchical Smoothing 708
    43.1.3   Hierarchical Decomposition 710
43.2   Comparing Analyses of Bach, Schumann, and Webern 711

44 Differential Operators and Regression 719
  44.0.1 Analytical Data 720
  44.1 The Beran Operator 721
    44.1.1 The Concept 721
    44.1.2 The Formalism 723
  44.2 The Method of Regression Analysis 726
    44.2.1 The Full Model 726
    44.2.2 Step Forward Selection 727
  44.3 The Results of Regression Analysis 727
    44.3.1 Relations Between Tempo and Analysis 727
    44.3.2 Complex Relationships 729
    44.3.3 Commonalities and Diversities 729
    44.3.4 Overview of Statistical Results 738

**45 Relating Tempo to Metric, Melodic and Harmonic Analyses in Chopin's Prélude op.
    28, No. 4** 743
  45.1 Introduction 743
  45.2 Data 745
    45.2.1 Analytical Data 745
    45.2.2 Tempo Data 745
  45.3 Short Summary of the Results 747
  45.4 Some Philosophical Comments 748

Part XII Inverse Performance Theory

46 Principles of Music Critique 751
  46.1 Boiling Down Infinity—Is Feuilletonism Inevitable? 751
  46.2 "Political Correctness" in Performance—Reviewing Gould 752
  46.3 Transversal Ethnomusicology 754

47 Critical Fibers 755
  47.1 The Stemma Model of Critique 755
  47.2 Fibers for Locally Linear Grammars 756
  47.3 Algorithmic Extraction of Performance Fields 759
    47.3.1 The Infinitesimal View on Expression 759
    47.3.2 Real-Time Processing of Expressive Performance 760
    47.3.3 Score-Performance Matching 761
    47.3.4 Performance Field Calculation 762
    47.3.5 Visualization 763
    47.3.6 The EspressoRUBETTE®: An Interactive Tool for Expression Extraction .. 764
  47.4 Local Sections 766
    47.4.1 Comparing Argerich and Horowitz 768

Part XIII Operationalization of Poiesis

48 Unfolding Geometry and Logic in Time 773
  48.1 Performance of Logic and Geometry 774
  48.2 Constructing Time from Geometry 775
  48.3 Discourse and Insight 776

49 Local and Global Strategies in Composition 779
   49.1 Local Paradigmatic Instances 780
      49.1.1 Transformations 780
      49.1.2 Variations 780
   49.2 Global Poetical Syntax 781
      49.2.1 Roman Jakobson's Horizontal Function 781
      49.2.2 Roland Posner's Vertical Function 781
   49.3 Structure and Process 782

50 The Paradigmatic Discourse on *presto*® 783
   50.1 The *presto*® Functional Scheme 783
   50.2 Modular Affine Transformations 786
   50.3 Ornaments and Variations 786
   50.4 Problems of Abstraction 789

51 Case Study I: "Synthesis" by Guerino Mazzola 791
   51.1 The Overall Organization 791
      51.1.1 The Material: 26 Classes of Three-Element Motives 792
      51.1.2 Principles of the Four Movements and Instrumentation 793
   51.2 1st Movement: Sonata Form 794
   51.3 2nd Movement: Variations 794
   51.4 3rd Movement: Scherzo 798
   51.5 4th Movement: Fractal Syntax 799

52 Object-Oriented Programming in OpenMusic 801
   52.1 Object-Oriented Language 802
      52.1.1 Patches 802
      52.1.2 Objects 803
      52.1.3 Classes 803
      52.1.4 Methods 803
      52.1.5 Generic Functions 804
      52.1.6 Message Passing 804
      52.1.7 Inheritance 805
      52.1.8 Boxes and Evaluation 805
      52.1.9 Instantiation 806
   52.2 Musical Object Framework 806
      52.2.1 Internal Representation 806
      52.2.2 Interface 808
   52.3 Maquettes: Objects in Time 811
   52.4 Meta-object Protocol 813
      52.4.1 Reification of Temporal Boxes 815
   52.5 A Musical Example 817

Part XIV String Quartet Theory

53 Historical and Theoretical Prerequisites 825
   53.1 History 825
   53.2 Theory of the String Quartet Following Ludwig Finscher 826
      53.2.1 Four Part Texture 826
      53.2.2 The Topos of Conversation Among Four Humanists 827
      53.2.3 The Family of Violins 828

54   Estimation of Resolution Parameters 831
54.1   Parameter Spaces for Violins 831
54.2   Estimation 833

55   The Case of Counterpoint and Harmony 839
55.1   Counterpoint 839
55.2   Harmony 840
55.3   Effective Selection 840

Part XV   Gesture Philosophy for Music

56   The Topos of Gestures 843

57   Gesture Philosophy: Phenomenology, Ontology, and Semiotics 845
57.1   A Short Recapitulation of Musical Ontology 845
    57.1.1   Ontology: Where, Why, and How 845
    57.1.2   Oniontology: Facts, Processes, and Gestures 846
57.2   Jean-Claude Schmitt's Historiographic and Philosophical Treatise "La raison des gestes
    dans l'Occident médiéval" 846
    57.2.1   Comments 847
57.3   Vilém Flusser's *Gesten: Versuch einer Phänomenologie* 848
    57.3.1   A Short Introduction to Flusser's Essay 848
    57.3.2   The Semiotic Neurosis 848
57.4   Michel Guérin's *philosophie des gestes* 850
    57.4.1   The Essay's Structure 850
    57.4.2   Gestural Ontology and Four Elementary Gestures 850
57.5   Flusser and Guérin: Some Consequences 851
57.6   A Program 852
    57.6.1   Circularity 852
57.7   The Semiotic Gesture Concept of Adam Kendon and David McNeill 853
    57.7.1   Comments 855
57.8   Juhani Pallasmaa and André Chastel: The Thinking Hand in Architecture and the Arts ... 855
57.9   Benveniste, Popelard, Wall 856

58   The French Presemiotic Approach 859
58.1   Maurice Merleau-Ponty 860
58.2   Francis Bacon and Gilles Deleuze 860
58.3   Jean Cavaillès and Charles Alunni 861
58.4   Gilles Châtelet 862

59   Cognitive Science 867
59.1   Embodiment 867
    59.1.1   Embodiment Science 868
59.2   Neuroscience 871
    59.2.1   Embodied AI 873
59.3   Anthropology 874
59.4   Dance 874
59.5   Disabled Gestures Versus Gestures Disabled: Parlan's Versus Peterson's Pianism 875
    59.5.1   Performative Gestures: Disabled Jazz Pianists 876
    59.5.2   Horace Parlan: Disabled Gestures 877
    59.5.3   Parlan with Bass (and Drums) 877

59.5.4   Parlan with Rhythm Section 878
59.5.5   Parlan as Soloist 879
59.5.6   Parlan's Duets with Archie Shepp 880
59.5.7   Disabled Gestures 881
59.5.8   Gestures Disabled: Oscar Peterson 882
59.5.9   Conclusion 884
59.6   Aristotle, Blind Lemon Jefferson, and Vilayanur S. Ramachandran Walk into a Bar 885
59.6.1   Introduction 885
59.6.2   Division by (Almost) Zero: Many Blind Bluesmen but Few Blind Blues 885
59.6.3   Seeing Blind Blues: Gesture, Flow, Circuitry, and Amplification 886
59.6.4   Epilogue: Puns as Gestures 887

60   Models from Music 889
60.1   Wolfgang Graeser 890
60.2   Adorno, Wieland, Sessions, Clynes 890
60.2.1   Theodor Wiesengrund Adorno 891
60.2.2   Renate Wieland 893
60.2.3   Roger Sessions 894
60.2.4   Manfred Clynes 895
60.3   Johan Sundberg and Neil P. McAngus Todd 896
60.4   David Lewin and Robert S. Hatten 898
60.5   Marcelo Wanderley and Claude Cadoz, Rolf Inge Godøy and Marc Leman 901

Part XVI Mathematics of Gestures

61   Fundamental Concepts and Associated Categories 907
61.1   Introduction 907
61.2   Towards a Musical String Theory 909
61.3   Initial Investigations: Diagrams of Curves 910
61.4   Modeling a Pianist's Hand 912
61.4.1   The Hand's Model 912
61.4.2   Transforming Abstract Note Symbols into Symbolic Gestures 912
61.4.3   From Symbolic Hand Gestures to Physical Gestures 913
61.5   The Mathematical Definition of Gestures 914
61.6   Hypergestures 915
61.6.1   Spatial Hypergestures 917
61.7   Categorically Natural Gestures 918
61.8   Connecting to Algebraic Topology: Hypergestures Generalize Homotopy 919
61.9   Gestoids 922
61.9.1   The Fundamental Group, Klumpenhouwer Networks, and Fourier Representation . . 924
61.10   Gabriel's Spectroids and Natural Formulas 925
61.10.1   Solutions of Representations of Natural Formulas by Local Networks 927
61.11   The Tangent Category 927
61.12   The Diamond Conjecture 929
61.13   Topos Logic for Gestures 930
61.14   The Escher Theorem for Hypergestures 931
61.14.1   Hypergestures and the Escher Theorem for Fux Counterpoint 931
61.14.2   Rebecca Lazier's Vanish: Lawvere, Escher, Schoenberg 933

62   Categories of Gestures over Topological Categories 937
   62.1   Gestures over Topological Categories 939
      62.1.1   The Categorical Digraph of a Topological Category 940
      62.1.2   Gestures with Body in a Topological Category 940
      62.1.3   Varying the Underlying Topological Category 942
   62.2   From Morphisms to Gestures 942
      62.2.1   Diagrams as Gestures 944
      62.2.2   Gestures in Factorization Categories 944
      62.2.3   Extensions from Homological Algebra Are Gestures 945
      62.2.4   The Bicategory of Gestures 945
      62.2.5   Entering the Diamond Space 946
   62.3   Diagrams in Topological Groups for Gestures 947
   62.4   Gestural Interpretation of Modulations in Beethoven's op.106/Allegro 950
      62.4.1   Recapitulation of the Results from Section 28.2 951
      62.4.2   The Modulation $B_\flat$-major $\leadsto$ G-major Between Measure 31 and Measure 44 952
      62.4.3   Lewin's Characteristic Gestures Identified? 955
      62.4.4   Modulation $E_\flat$-major $\leadsto$ D-major/B-minor from W to W^* 957
      62.4.5   The Fanfare 958
   62.5   Conclusion for the Categorical Gesture Approach 960
   62.6   Functorial Gestures: General Addresses 961
   62.7   Yoneda's Lemma for Gestures 962
   62.8   Examples from Music 963
      62.8.1   Collections of Acoustical Waves 963
      62.8.2   Collections of Spectral Music Data 964
      62.8.3   MIDI-Type ON-OFF Transformations 964

63   Singular Homology of Hypergestures 965
   63.1   An Introductory Example 965
   63.2   Chain Modules for Singular Hypergestural Homology 967
   63.3   The Boundary Homomorphism 968

64   Stokes' Theorem for Hypergestures 973
   64.1   The Need for Stokes' Theorem for Hypergestures 973
   64.2   Almost Regular Manifolds, Differential Forms, and Integration for Hypergestures 973
      64.2.1   Locally Almost Regular Manifolds 974
      64.2.2   Differential Forms 975
      64.2.3   Integration 975
   64.3   Stokes' Theorem 976

65   Local Facts, Processes, and Gestures 979
   65.1   Categories of Local Compositions 979
   65.2   Categories of Local Networks 980
   65.3   Categories of Local Gestures 982
      65.3.1   Local Gestures on Topological Categories of Points 982
   65.4   Connecting Functors 984
   65.5   Hypernetworks and Hypergestures 985
      65.5.1   Escher Theorems 985
   65.6   Singular Homology of Hypernetworks and Hypergestures 985

66   Global Categories 987
   66.1   Categories of Global Compositions 987
       66.1.1 Simplicial Methods 988
   66.2   Classification of Global Compositions 989
   66.3   Non-interpretable Global Compositions 990
   66.4   Categories of Global Networks 990
       66.4.1 Non-interpretable Global Networks 991
   66.5   Categories of Global Gestures 993
   66.6   Globalizing Topological Categories: Categorical Manifolds 993
   66.7   Globalizing Skeleta 996
   66.8   Functorial Global Gestures 998

67   Mathematical Models of Creativity 1001
   67.1   Forewarning: Invention of Gestures in Mathematics 1001
       67.1.1 Thinking Exactness, Like a Rolling Mind 1001
       67.1.2 Thought as an Algebra of Gestures 1002
   67.2   Method and Objects, Summarily Explained: I—Preamble 1003
       67.2.1 Prelude to a Discourse of a Method: "Caminos", "Aletheia", Irreverence 1003
       67.2.2 Our Posture 1015
   67.3   Method and Objects, Summarily Explained: II—Data 1021
       67.3.1 Simple Objects, Structures and Invariants in Mathematics 1021
       67.3.2 Complete Frameworks, Computations and Representations 1029
   67.4   Creativity in Mathematics: Gestures in Historical Contexts 1031
       67.4.1 Creativity: Phenomenology, Psychology and Skills, and Life 1031
       67.4.2 Determination of Mathematics as a History of Its Gestures 1036
       67.4.3 Invention in the Art of Mathematics 1044
   67.5   On the Mathematical Invention of Coordinations 1046
       67.5.1 Emergence of Coordinations 1047
       67.5.2 Arrows 1050
       67.5.3 Bodies, Implicit Surfaces, Abstract Relations 1055
       67.5.4 Sketches 1056
   67.6   Pulsation in the Living Process of Invention Among Shapes 1058
       67.6.1 Productions: Objects and Relations, Problems, Pulsation 1058
       67.6.2 Creativity in the Mathematical World Seen as a Living System of Shapes, in a Categorical Framework 1060
   67.7   Conclusion: Categorical Presentation of Pulsations 1064
   67.8   The Hegel Group Action on a Critical Concept's Walls 1066
   67.9   Introduction 1066
   67.10 The Hegel Concept Group $\mathcal{G}$ 1067
       67.10.1 Hegel's Initial Thought Movement in *Wissenschaft der Logik* ... 1067
       67.10.2 The Implicit Group Structure 1070
       67.10.3 The Conceptual Box Structure 1072
   67.11 The $\mathcal{G}$ Action on the Yoneda Model of Creativity 1073
   67.12 The Hegel Body $\mathcal{B}$ in the Concept Architecture of Forms and Denotators 1073
   67.13 The Usage of $\mathcal{G}$ for the Dynamics of Creativity 1074
       67.13.1 Two Preliminary Examples 1074
       67.13.2 The Challenge: Creating a Spectrum of Conceptual Extensions ... 1075
       67.13.3 Escher's Theorem for Beethoven's Fanfare in the *"Hammerklavier" Sonata op. 106* 1075
       67.13.4 The Rotation $S@N$ as a Driving Creative Force in the Incipit of Liszt's *Mephisto Walzer* No.1 1076
   67.14 An Experimental Composition 1078
   67.15 Still More Symmetries? Future Developments 1080

Part XVII Concept Architectures and Software for Gesture Theory

68 Forms and Denotators over Topological Categories 1085
   68.1 The General Topos—Theoretical Framework 1085
        68.1.1 The category **TopCat** of Small Topological Categories 1085
   68.2 Forms and Denotators 1086
   68.3 Mathematics of Objects, Structures, and Concepts............. 1087
   68.4 Galois Theory of Concepts.................................. 1087
        68.4.1 Introduction 1088
        68.4.2 Form Semiotics 1089
        68.4.3 The Category of Form Semiotics 1092
        68.4.4 Galois Correspondence of Form Semiotics 1093

69 The Rubato Composer Architecture 1095
   69.1 The Software Architecture 1096
   69.2 The Rubette World.. 1097
        69.2.1 Rubettes for Counterpoint........................... 1097
        69.2.2 Rubettes for Harmony 1098
        69.2.3 MetroRubettes..................................... 1099

70 The BigBang Rubette and the Ontological Dimension of Embodiment ... 1101

71 Facts: Denotators and Their Visualization and Sonification 1103
   71.1 Some Earlier Visualizations of Denotators 1103
        71.1.1 Göller's PrimaVista Browser 1103
        71.1.2 Milmeister's ScorePlay and Select2D Rubettes 1105
   71.2 An Early Score-Based Version of BigBang 1106
        71.2.1 The Early BigBang Rubette's View Configurations 1107
        71.2.2 Navigating Denotators 1111
        71.2.3 Sonifying Score-Based Denotators 1111
   71.3 BigBangObjects and Visualization of Arbitrary $Mod^@$ Denotators ... 1111
        71.3.1 A Look at Potential Visual Characteristics of Form Types .. 1112
        71.3.2 From a General View Concept to BigBang Objects 1114
        71.3.3 New Visual Dimensions 1115
   71.4 The Sonification of BigBangObjects 1116
   71.5 Examples of Forms and the Visualization of Their Denotators..... 1117
        71.5.1 Some Set-Theoretical Structures 1117
        71.5.2 Tonal and Transformational Theory 1119
        71.5.3 Synthesizers and Sound Design....................... 1121

72 Processes: BigBang's Operation Graph 1127
   72.1 Temporal BigBangObjects, Object Selection, and Layers 1128
        72.1.1 Selecting None and Lewin's Transformation Graphs...... 1128
        72.1.2 The Temporal Existence of BigBangObjects 1129
        72.1.3 BigBang Layers 1131
   72.2 Operations and Transformations in BigBang 1132
        72.2.1 Non-transformational Operations..................... 1132
        72.2.2 Transformations.................................... 1136
   72.3 BigBang's Process View 1138
        72.3.1 Visualization of Processes 1138
        72.3.2 Selecting States and Modifying Operations 1139

72.3.3   Alternative and Parallel Processes 1139
72.3.4   Structurally Modifying the Graph 1141
72.3.5   Undo/Redo.. 1141

73   Gestures: Gestural Interaction and Gesturalization 1143
73.1   Formalizing: From Gestures to Operations 1143
73.1.1   Modes, Gestural Operations, and the Mouse 1144
73.1.2   Affine Transformations and Multi-touch 1148
73.1.3   Dynamic Motives, Sound Synthesis, and Leap Motion 1149
73.1.4   Recording, Modifying Operations and MIDI Controllers 1151
73.2   Gesturalizing and the Real BigBang: Animated Composition History 1152
73.2.1   Gesturalizing Transformations 1152
73.2.2   Gesturalizing Other Operations 1154
73.2.3   Using Gesturalization as a Compositional Tool 1154

74   Musical Examples 1157
74.1   Some Example Compositions 1157
74.1.1   Transforming an Existing Composition 1157
74.1.2   Gesturalizating and Looping with a Simple Graph 1158
74.1.3   Drawing UPIC-like Motives and Transforming 1159
74.1.4   Drawing Time-Slices 1160
74.1.5   Converting Forms, Tricks for Gesturalizing 1160
74.1.6   Gesturalizing a Spectrum.. 1162
74.1.7   Using Wallpapers to Create Rhythmical Structures 1163
74.2   Improvisation and Performance with BigBang................................ 1163
74.2.1   Improvising by Selecting States and Modifying Transformations 1164
74.2.2   Playing Sounds with a MIDI Keyboard and Modifying Them 1164
74.2.3   Playing a MIDI Grand Piano with Leap Motion 1165
74.2.4   Playing a MIDI Grand Piano with the Ableton Push................. 1166
74.2.5   Improvising with 12-Tone Rows 1167

Part XVIII The Multiverse Perspective

75   Gesture Theory and String Theory 1173

76   Physical and Musical Multiverses....................................... 1175

77   Hesse's Melting Beads: A Multiverse Game with Strings and Gestures 1177
77.1   Review of Hesse's Glass Bead Game 1177
77.2   Frozen Glass Beads of Facticity 1178
77.3   The Revolution of Functors 1178
77.4   Gestures in Philosophy and Science 1180
77.5   Gesture Theory in Music 1181
77.6   A Remark on Gestural Creativity 1183
77.7   Gestures and Strings... 1183
77.8   Playing the Multiversed Game in a Pre-semiotic Ontology 1184

78   Euler-Lagrange Equations for Hypergestures 1185
    78.1   The Problem in Performance Theory with the Physical Nambu-Goto Lagrangian 1185
        78.1.1   Complex Time and Descartes' Dualistic Ontology 1186
    78.2   Lagrangian Density for Complex Time 1187
        78.2.1   The Lagrangian Action for Performance 1188
        78.2.2   The World-Sheet of Complex Time 1190
        78.2.3   The Space for a Hand's Gestures 1192
        78.2.4   The World-Sheet for a Simple Case 1192
        78.2.5   The Elementary Gesture of a Pianist 1192
        78.2.6   The Overarching Framework Between Note Performance and Gesture Performance 1195
        78.2.7   Examples of Functional Relations Between Potential and Physical Gesture 1196
        78.2.8   Calculus of Variations for the Physical Gesture 1202
        78.2.9   A First Solution. World-Sheet Potentials Determine a Pianist's Gesture: Calculus
                  of Variations and Fourier Analysis 1203
        78.2.10 The Calculus with Vanishing Potential 1204
        78.2.11 The Calculus with General Potential 1209
        78.2.12 Solution of the Differential Equation Using 2D Fourier Series 1212
        78.2.13 Parallels Between Performance Operators for Scores and for Gestures 1216
        78.2.14 Complex Time and the Artistic Effort 1218
        78.2.15 Opening the Aesthetic Question that Is Quantified in Lagrange Potentials 1218
        78.2.16 A Musical Composition by Maria Mannone Realized Using These Ideas 1219
    78.3   Global Performance Hypergestures 1223
        78.3.1   The Musical Situation: An Intuitive Introduction 1223
    78.4   Categorical Gestures and Global Performance Hypergestures 1224
        78.4.1   Categorical Gestures: The Case of Potentials 1224
        78.4.2   The Mathematics of Global Performance Hypergestures 1226
    78.5   World-Sheet Hypergestures for General Skeleta 1227
    78.6   A Global Variational Principle for the Lagrange Formalism 1231

Part XIX   Gestures in Music and Performance Theory, and in Ethnomusicology

79   Gesture Homology for Counterpoint 1235
    79.1   Summary of Mathematical Theory of Counterpoint: What It Is About and What Is Missing ... 1235
    79.2   Hypergestural Singular Homology 1236
    79.3   A Classical Example of a Topological Category from Counterpoint 1237
        79.3.1   Generators of $H_1(^G X)$ for a Groupoid $^G X$ Defined by a Group Action 1238
    79.4   The Meaning of H_1 for Counterpoint 1240
    79.5   Concluding Comments 1241

80   Modulation Theory and Lie Brackets of Vector Fields 1243
    80.1   Introduction 1243
        80.1.1   Short Recapitulation of the Classical Model's Structure 1244
    80.2   Hypergestures Between Triadic Degrees That Are Parallel to Vector Fields 1245
    80.3   Lie Brackets Generate Vector Fields That Connect Symmetry-Related Degrees 1245
    80.4   Selecting Parallel Hypergestures That Are Admissible for Modulation 1247
    80.5   The Other Direct Modulations 1249

81 Hypergestures for Performance Stemmata 1253
   81.1 Motivation, Terminology, and Previous Results 1253
       81.1.1 Performance Stemmata and Performance Gestures of Locally Compact Points 1254
   81.2 Gestures with Lie Operators in Stemma Theory 1255
   81.3 Connecting Stemmatic Gestures for Weights and Performance Fields 1256
   81.4 Homology of Weight Parameter Stemmata 1258
   81.5 A Concrete Example 1260
   81.6 A Final Comment 1261

82 Composing and Analyzing with the Performing Body 1263
   82.1 Gesture: A Sign or a Totality? 1264
   82.2 A Gesture-Based Structural Reading in *Rain Tree Sketch II* by Toru Takemitsu 1266
       82.2.1 Process I: Synergy of Mirroring and Parallel Gestures 1267
       82.2.2 Process II: Towards Relaxation, Balance, and Weightfulness 1270
   82.3 The Last Leg of a Bodily Journey 1274
       82.3.1 *Sheng* for Piano, Smartphones, and Fixed Playback 1278
       82.3.2 Cross-modality of Gestures 1279
       82.3.3 Learning the Smartphone Instrument 1279
       82.3.4 Kinesthetic Awareness and Modes of Listening 1281
   82.4 Conclusion: Foregrounding the Performer's Body 1283

83 Gestural Analysis and Classification of a Conductor's Movements 1285
   83.1 Gestures and Communication in Orchestral Conducting: A Case Study 1285
       83.1.1 Problematics and Solving Methods 1286
       83.1.2 Results, Consequences, Applications 1288
       83.1.3 Some Remarks 1289
   83.2 Hints for a Mathematical Description 1289
   83.3 Data Analysis 1290
   83.4 Conclusion 1292
   83.5 Addendum 1293

84 Reviewing Flow, Gesture, and Spaces in Free Jazz 1295
   84.1 Improvisation: Defining Time 1295
   84.2 Flow, Gestures, Imaginary Time and Spaces in the Music Movie *Imaginary Time* 1295
       84.2.1 The Compositional Character of the Pieces 1297
       84.2.2 Large Forms 1298
       84.2.3 Precision of Attacks 1298
       84.2.4 Co-presence of Different Time Layers 1299
       84.2.5 The Reality of Imaginary Time 1300
       84.2.6 Measuring Flow 1300
       84.2.7 Explicit Perception of Gestures 1300

85 Gesture and Vocalization 1301
   85.1 Vocal Gesture 1301
   85.2 Vocal and Manual Motion 1303
   85.3 Gait 1304
   85.4 Hindustani Vocal Music 1305
   85.5 Notic Models and Kinetic Models 1306
   85.6 The Realist Pitfall 1307
   85.7 The Subjectivist Pitfall 1308
   85.8 Speech Gesture 1311

86 Elements of a Future Vocal Gesture Theory 1313
  86.1 Why a Theory of Vocal Gestures? 1313
      86.1.1 Studying the Voice Without the Singer? 1314
      86.1.2 Parts of the Phonatory System and Their Functions 1314
      86.1.3 Imaginary Gestures in Real Time? 1315
      86.1.4 Space of Voice Parameters Gestures 1315
      86.1.5 About the Importance of Breathing and of Laryngeal Movements ... 1316
      86.1.6 Mathematical Description of Vocal Gestures 1317
      86.1.7 Gestures Thought by Singers 1321
  86.2 A Powerful Tool from the Past 1322
      86.2.1 Gestures in Gregorian Chant Didactics 1323
      86.2.2 Concept of Rhythm and Time 1323
      86.2.3 The Neumes 1327
  86.3 Connecting Physiology, Gestures and Notation. Toward New Neumes? ... 1329
      86.3.1 New Neumes 1330

Part XX Appendix: Sound

A Common Parameter Spaces 1335
  A.1 Physical Spaces 1335
      A.1.1 Neutral Data 1336
      A.1.2 Sound Analysis and Synthesis 1338
  A.2 Mathematical and Symbolic Spaces 1347
      A.2.1 Onset and Duration 1347
      A.2.2 Amplitude and Crescendo 1348
      A.2.3 Frequency and Glissando 1349

B Auditory Physiology and Psychology 1353
  B.1 Physiology: From the Auricle to Heschl's Gyri 1353
      B.1.1 Outer Ear 1354
      B.1.2 Middle Ear 1354
      B.1.3 Inner Ear (Cochlea) 1355
      B.1.4 Cochlear Hydrodynamics: The Travelling Wave 1357
      B.1.5 Active Amplification of the Traveling Wave Motion 1358
      B.1.6 Neural Processing 1360
  B.2 Discriminating Tones: Werner Meyer-Eppler's Valence Theory 1362
  B.3 Aspects of Consonance and Dissonance 1364
      B.3.1 Euler's Gradus Function 1364
      B.3.2 von Helmholtz' Beat Model 1365
      B.3.3 Psychometric Investigations by Plomp and Levelt 1367
      B.3.4 Counterpoint 1367
      B.3.5 Consonance and Dissonance: A Conceptual Field 1367

Part XXI Appendix: Mathematical Basics

C Sets, Relations, Monoids, Groups 1371
  C.1 Sets 1371
      C.1.1 Examples of Sets 1371

C.2   Relations 1372
    C.2.1   Universal Constructions 1374
    C.2.2   Graphs and Quivers 1375
    C.2.3   Monoids 1376
C.3   Groups 1378
    C.3.1   Homomorphisms of Groups 1378
    C.3.2   Direct, Semi-direct, and Wreath Products 1380
    C.3.3   Sylow Theorems on p-groups 1380
    C.3.4   Classification of Groups 1381
    C.3.5   General Affine Groups 1382
    C.3.6   Permutation Groups 1382

D   **Rings and Algebras** 1385
D.1   Basic Definitions and Constructions 1385
    D.1.1   Universal Constructions 1386
D.2   Prime Factorization 1389
D.3   Euclidean Algorithm 1389
D.4   Approximation of Real Numbers by Fractions 1389
D.5   Some Special Issues 1390
    D.5.1   Integers, Rationals, and Real Numbers 1390

E   **Modules, Linear, and Affine Transformations** 1391
E.1   Modules and Linear Transformations 1391
    E.1.1   Examples 1391
E.2   Module Classification 1392
    E.2.1   Dimension 1392
    E.2.2   Endomorphisms on Dual Numbers 1394
    E.2.3   Semi-simple Modules 1394
    E.2.4   Jacobson Radical and Socle 1395
    E.2.5   Theorem of Krull-Remak-Schmidt 1396
E.3   Categories of Modules and Affine Transformations 1396
    E.3.1   Direct Sums 1397
    E.3.2   Affine Forms and Tensors 1397
    E.3.3   Biaffine Maps 1399
    E.3.4   Symmetries of the Affine Plane 1402
    E.3.5   Symmetries on $\mathbb{Z}^2$ 1402
    E.3.6   Symmetries on $\mathbb{Z}^n$ 1403
    E.3.7   Complements on the Module of a Local Composition ... 1403
    E.3.8   Fiber Products and Fiber Sums in **Mod** 1404
E.4   Complements of Commutative Algebra 1405
    E.4.1   Localization 1406
    E.4.2   Projective Modules 1406
    E.4.3   Injective Modules 1407
    E.4.4   Lie Algebras 1408

F   **Algebraic Geometry** 1411
F.1   Locally Ringed Spaces 1411
F.2   Spectra of Commutative Rings 1412
    F.2.1   Sober Spaces 1413

F.3   Schemes and Functors 1414
F.4   Algebraic and Geometric Structures on Schemes........................... 1415
       F.4.1   The Zariski Tangent Space 1415
F.5   Grassmannians 1416
F.6   Quotients 1417

G   Categories, Topoi, and Logic 1419
G.1   Categories Instead of Sets 1419
       G.1.1   Examples 1420
       G.1.2   Functors 1421
       G.1.3   Natural Transformations 1422
G.2   The Yoneda Lemma 1423
       G.2.1   Universal Constructions: Adjoints, Limits, and Colimits 1423
       G.2.2   Limit and Colimit Characterizations 1425
G.3   Topoi 1427
       G.3.1   Subobject Classifiers.. 1428
       G.3.2   Exponentiation 1429
       G.3.3   Definition of Topoi 1429
G.4   Grothendieck Topologies 1430
       G.4.1   Sheaves 1431
G.5   Formal Logic 1432
       G.5.1   Propositional Calculus 1432
       G.5.2   Predicate Logic 1435
       G.5.3   A Formal Setup for Consistent Domains of Forms 1437

H   Complements on General and Algebraic Topology 1443
H.1   Topology 1443
       H.1.1   General 1443
       H.1.2   The Category of Topological Spaces 1444
       H.1.3   Uniform Spaces 1444
       H.1.4   Special Issues 1445
H.2   Algebraic Topology 1445
       H.2.1   Simplicial Complexes 1445
       H.2.2   Geometric Realization of a Simplicial Complex 1446
       H.2.3   Contiguity... 1447
H.3   Simplicial Coefficient Systems 1447
       H.3.1   Cohomology 1447

I   Complements on Calculus 1449
I.1   Abstract on Calculus 1449
       I.1.1   Norms and Metrics 1449
       I.1.2   Completeness 1450
       I.1.3   Differentiation 1451
I.2   Ordinary Differential Equations (ODEs)..................................... 1451
       I.2.1   The Fundamental Theorem: Local Case 1452
       I.2.2   The Fundamental Theorem: Global Case 1453
       I.2.3   Flows and Differential Equations 1455
       I.2.4   Vector Fields and Derivations 1455
I.3   Partial Differential Equations 1455

J   More Complements on Mathematics 1457
   J.1   Directed Graphs 1457
      J.1.1   The Category of Directed Graphs (Digraphs) 1457
      J.1.2   Two Standard Constructions in Graph Theory 1459
      J.1.3   The Topos of Digraphs 1459
   J.2   Galois Theory 1460
   J.3   Splines 1461
      J.3.1   Some Simplex Constructions for Splines 1461
      J.3.2   Definition of General Splines 1462
   J.4   Topology and Topological Categories 1463
      J.4.1   Topology 1463
      J.4.2   Topological Categories 1464
   J.5   Complex Analysis 1465
   J.6   Differentiable Manifolds 1466
      J.6.1   Manifolds with Boundary 1467
      J.6.2   The Tangent Manifold 1467
   J.7   Tensor Fields 1468
      J.7.1   Alternating Tensors 1468
      J.7.2   Tangent Tensors 1468
   J.8   Stokes' Theorem 1469
   J.9   Calculus of Variations 1471
   J.10   Partial Differential Equations 1471
      J.10.1   Explicit Calculation 1472
   J.11   Algebraic Topology 1474
      J.11.1   Homotopy Theory 1474
      J.11.2   The Fundamental Group(oid) 1475
   J.12   Homology 1475
      J.12.1   Singular Homology 1476
   J.13   Cohomology 1477

Part XXII   Appendix: Complements in Physics

K   Complements on Physics 1481
   K.1   Hamilton's Variational Principle 1481
      K.1.1   Euler-Lagrange Equations for a Non-relativistic Particle 1482
   K.2   String Theory 1482
   K.3   Duality and Supersymmetry 1484
   K.4   Quantum Mechanics 1485
      K.4.1   Banach and Hilbert Spaces 1486
      K.4.2   Geometry on Hilbert Spaces 1490
      K.4.3   Axioms for Quantum Mechanics 1492
      K.4.4   The Spectral Theorem 1493

Part XXIII   Appendix: Tables

L   Euler's Gradus Function 1497

M   Just and Well-Tempered Tuning 1499

N   Chord and Third Chain Classes 1501
    N.1   Chord Classes 1501
    N.2   Third Chain Classes 1506

O   Two, Three, and Four Tone Motif Classes 1513
    O.1   Two Tone Motifs in $OnPiMod_{12,12}$ 1513
    O.2   Two Tone Motifs in $OnPiMod_{5,12}$ 1513
    O.3   Three Tone Motifs in $OnPiMod_{12,12}$ 1514
    O.4   Four Tone Motifs in $OnPiMod_{12,12}$ 1517
    O.5   Three Tone Motifs in $OnPiMod_{5,12}$ 1523

P   Well-Tempered and Just Modulation Steps 1525
    P.1   12-Tempered Modulation Steps 1525
        P.1.1   Scale Orbits and Number of Quantized Modulations 1525
        P.1.2   Quanta and Pivots for the Modulations Between Diatonic Major Scales (No.38.1) . 1527
        P.1.3   Quanta and Pivots for the Modulations Between Melodic Minor Scales (No.47.1) . . 1528
        P.1.4   Quanta and Pivots for the Modulations Between Harmonic Minor Scales (No.54.1) 1530
        P.1.5   Examples of 12-Tempered Modulations for All Fourth Relations 1530
    P.2   2-3-5-Just Modulation Steps 1531
        P.2.1   Modulation Steps Between Just Major Scales 1531
        P.2.2   Modulation Steps Between Natural Minor Scales 1532
        P.2.3   Modulation Steps from Natural Minor to Major Scales 1532
        P.2.4   Modulation Steps from Major to Natural Minor Scales 1533
        P.2.5   Modulation Steps Between Harmonic Minor Scales 1533
        P.2.6   Modulation Steps Between Melodic Minor Scales 1534
        P.2.7   General Modulation Behaviour for 32 Altered Scales 1535

Q   Counterpoint Steps 1537
    Q.1   Contrapuntal Symmetries 1537
        Q.1.1   Class No. 64 1537
        Q.1.2   Class No. 68 1538
        Q.1.3   Class No. 71 1539
        Q.1.4   Class No. 75 1540
        Q.1.5   Class No. 78 1541
        Q.1.6   Class No. 82 1542
    Q.2   Permitted Successors for the Major Scale 1543

Part XXIV   References and Index

References R.1

Index R.33

Leitfaden

Leitfaden I & II

Leitfaden III

XIX. Gestures in Music and Performance Theory, and in Ethnomusicology

XV. Gesture Philosophy for Music

XVI. Mathematics of Gestures

XVIII. The Multiverse Perspective

XVII. Concept Architectures and Software for Gesture Theory

XX. Appendix:
Sound

XXI. Appendix:
Mathematical Basics

XXII. Appendix:
Complements in Physics

XXIII. Appendix:
Tables

XXIV. References
Bibliography
Index

ToM_CD

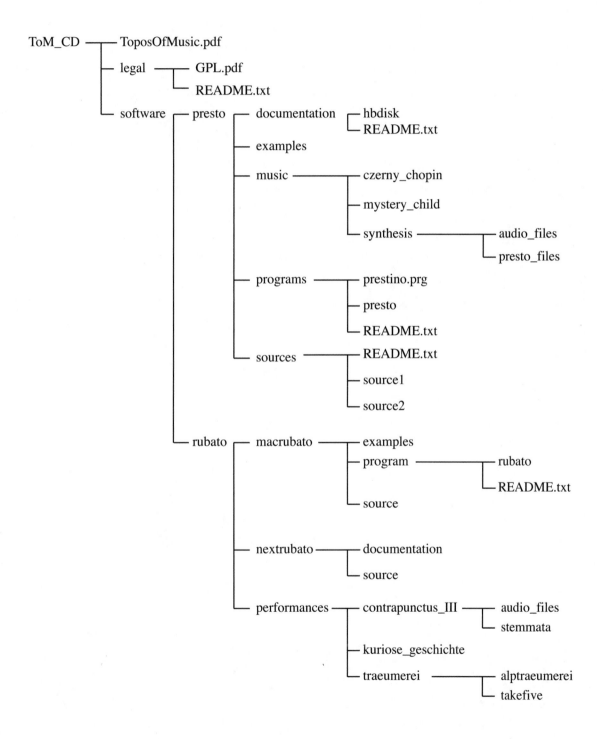

Part XX

Appendix: Sound

A

Common Parameter Spaces

This appendix chapter is an overview, not an exhaustive treatise of spaces which parametrize sound objects. These spaces where sounds are positioned always define an aspect, never the totality of music thinking, and every attempt to define a preferred space will narrow the music thinking, not the music. The best that can occur is that we offer an encompassing or at least a representative ensemble of parameter spaces which are interrelated by a precise relation. To this end, it is recommended to distinguish topographic positions, above all in their realities and communicative perspectives. This will also entail the corresponding mathematics.

We start by the physical descriptions, turn over to more mathematical abstractions and the describe more symbolic viewpoints which we call interpretative since they are not just a new mathematical rephrasing of a priori equivalent physical description, but express abstraction with some mental background constructions.

A.1 Physical Spaces

As a physical object, a *sound*[1] is a more or less regular variation of normal air pressure[2] as a function of time. Starting at a determined *onset time e sec*, it starts from a source Q at position $q = q_Q$ m in the ordinary physical space and expands as a wave. At a location x and time t, the *pressure variation* (the difference from the normal pressure) $p_x^Q(t)$ Nm^{-2} is perceived as a longitudinal air wave, i.e., with a pressure front perpendicular to the waves expanding direction, see Figure A.1.

For a punctual sound source, however, the wave front at $x \neq q$ is a spheric surface; we can write

$$p_x^Q(t) = |x - q|^{-1} p_q(t - |x - q|/v), \qquad (A.1)$$

where v is the expansion velocity of the wave[3]. It is calculated by the formula $v = C.\sqrt{T}$, where T K is the absolute temperature in Kelvin degrees, and C is a constant with value $C = 20.1$ for the normal pressure. For normal conditions, we have $v \approx 343$ $msec^{-1}$. But with complex sound sources and room-specific reflection and refractions, the pressure variation of a sound may be an overlapping of different spheric wave components. If several sound sources $Q_1, \ldots Q_s$ are given, the resulting pressure variation sums up to $p_x^Q(t) = \sum_{1 \leq i \leq s} p_x^{Q_i}(t)$, and it is the integral $\int dp_x^{dQ}(t)$ of a family of infinitesimal point sources dQ. In general, from the knowledge of $p_x^Q(t)$, one cannot infer the original source functions. This is like with painting where in general one perspectivic image does not allow us to infer the original object configuration: The esthesic position is not sufficient to reconstruct poiesis[4].

---

[1] German: "Klang".

[2] At the zero height above sea and zero degrees Celsius, this is $\approx 1.1013.10^5 Nm^{-2}$.

[3] It is known [914] that the square of the pression variation is proportional to the intensity, i.e., energy flow per surface and time unit, and the latter, by energy conservation, decreases proportionally to the square of the distance $|x - q|$, whence the formula.

[4] See chapter 2 for the concepts of music topography, such as "esthesic", "neutral", or "poietic".

© Springer International Publishing AG, part of Springer Nature 2017
G. Mazzola, *The Topos of Music IV: Roots*, Computational Music Science,
https://doi.org/10.1007/978-3-319-64495-0_1

Fig. A.1. The prototypical punctual sound source and the spherical sound wave.

The fundamental problem of musical acoustics is that the neutral data, the objectively measurable pressure values, are far from what is intended by musicians, i.e., the neutral data is not what is interesting and what is the message. So one of the most important tasks of musical acoustics is the interpretation of the neutral data, in other words, of what is behind the data, what could be the hidden parameters of the audible phenomenena. And it is one of the worst tragedies of traditional music-acoustics and psycho-acoustics that the fundamental difference of neutral and poietic levels is ignored and disregarded.

A.1.1 Neutral Data

In order to describe the neutral sound data, let us first concentrate on the information at the source location $q = q_Q$ of a point source Q, in an idealized model of a single instrument. We shall come back to room acoustics in the next Section A.1.1.1. The source sound (variation) event $p_q(t)$ is usually a finite event, starting at time e, and ending after the duration d. The variation between these time limits is also limited by the maximal amplitude Am of the total pressure variation. So the function $p_q(t)$ is the affine image of a normalized function $p_q^0(t)$ which starts at time $e = 0$, has duration $d = 1$, and amplitude $A = 1$. More precisely:

$$p_q(t) = Support(A, e, d)(p_q^0)(t) = A.p_q^0((t - e)/d). \tag{A.2}$$

The operator $Support(A, e, d)$ reduces the unknown sound event to a normalized event. What happens between the normalized unit supports is however completely arbitrary. It may be a percussive sound or the sound of a Stradivari violin. The normalization by the support operator is completely harmless, but not much more than this data can be traced on the neutral level. The theory of all the rest is far from neutral; we are going to deal with this in Section A.1.2.

A.1.1.1 Room Acoustics

The room is an important part in the information chain from the information source (instrument, speaker, public address system) to the receiver (listener, director, artist). Room sizes vary from small living rooms to huge cathedrals or concert halls. In this section, we want to tackle only basic features, reverberation time and acoustical power. For a large, irregular room, we can visualize the acoustical conditions by imagining a wave traveling inside the room. This wave travels in a straight line until it strikes a surface. It is reflected off the surface at an angle equal to the angle of incidence and travels in this direction until it strikes another surface. Because sound travels about $343 \ msec^{-1}$, many reflections will occur within a small time span.

Absorption. After a wave has undergone a reflection from a wall that is absorbing, its intensity will be less during its next traverse of the room. In a large, irregular room, the number of waves traveling are so numerous that at each surface all directions of incident flow are equally probable. The sound absorption coefficient α is therefore taken to be averaged for all angles of incidence. All materials have absorption coefficients that are different at different frequencies[5] [111]. In the frequency range of interest between $250 \ Hz$ and $4 \ kHz$, plain walls and floors, as well as closed windows, have absorption indices below 0.2. Higher absorption can be achieved with acoustic tiles and, at least in the upper frequency range, with thick carpets and draperies. The total absorption A of a room is the sum of the product of surfaces $A_i \ m^2$ and absorption coefficients α_i: $A = \Sigma_i \alpha_i A_i$.

If there is an open window in the room, all the energy incident on its area will pass outdoors and none will be reflected.[6] The absorption of an area of acoustical material in a room can therefore be expressed in terms of the equivalent area of an open window. For this reason, the total absorption A of a room can be characterized by its equivalent "open window surface".

Critical Distance. The sound field which builds up in a room is fundamentally different from the free field situation. Let us first assume a sound source in a free field, i.e., a loudspeaker on a high post emitting sound in all directions. If the sound radiation is unidirectional and no reflections occur, the source will emit a spherical wave and sound pressure will decrease inversely proportional with the distance from the source.

For loss-less reflecting walls on all sides around the source, the sound waves will be reflected over and over. In the case of no absorption in the air, the sound pressure will now rise and rise, as no energy loss occurs. Small absorption will cause an equilibrium. However, the sound field in the room will no longer be a directional spherical field because the reflected waves will by far dominate over the direct sound wave. The sound field will be diffuse, reflections will arrive from all directions with equal probability. Only in the close vicinity of the small sound source is the sound field directional, because there the sound pressure of the unreflected direct sound wave will dominate over the diffuse sound field. In summary, the sound field in a room will be directional only close to the sound source Q. There, as we know from the above, at point x, the sound pressure falls with $1/|x - q|$. The sound field at a large distance from the source is diffuse. Sound pressure is almost constant and much higher compared to a free sound field. The distance from the source, where the transition between these distinct regions, occurs is called *critical distance* or *diffuse field distance* r_H. If the average absorption $\bar{\alpha} \leqslant 0.4$, the critical distance for the absorption A can be calculated with a precision of about 10% [111]:

$$r_H = \sqrt{A/16\pi} \approx 0.14\sqrt{A}. \tag{A.3}$$

By the inverse distance law, the sound *pressure decrease factor* ρ_D in the diffuse field of a room can be estimated by the formula:

$$\rho_D = 1/r_H \approx \frac{7.1}{\sqrt{A}}. \tag{A.4}$$

Reverberation Time. Temporal effects during the onset and the decay of sound are essential features in rooms and are not present in a free sound field. In large rooms, such as a cathedral, long decay times are apparent. The human auditory system can distinguish whether a sound source is located in a large room, a small room or even whether it is not in a room at all, but outside. The human ear is obviously able to extract information about the room size from the temporal structure of sound. We will therefore discuss the onset and decay of sound.

---

[5] See Section A.1.2 for the discussion of the frequency concept.

[6] This statement is strictly true only if the window is several wavelengths wide and high, otherwise diffraction will occur.

When the sound leaves its source, the direct sound reaches the receiver first, its delay is determined by the distance sound has to travel divided by the velocity of sound. With further delay, reflected sound waves with only a slightly longer travel distance arrive, and later, also reflections with longer routes and multiple reflections arrive. The sound pressure is increasing until it reaches its steady state value. Only the direct sound gives information about the location of the source, which is exploited by our hearing system for sound localization. The build-up of the reverberation is only perceived in highly reverberant rooms. The reverberation is much more perceptible after the source is muted, and the echo may be still noticeable after seconds. The direct sound ceases after the propagation time from the source to the receiver, however, all the reflected sound waves still arrive. Their intensity will be reduced by factor $1 - \bar{\alpha}$ after each reflection on a wall. Therefore, the sound pressure will decrease exponentially.

The *reverberation time* T is defined as the time required for the sound to decay by a sound pressure level[7] of 60 dB. Let α_L m^{-1} be the absorption in air (a highly frequency- and humidity-dependent variable), and let V m^3 be the volume of the room. Then the reverberation time T can be estimated with *Sabine's* formula [1158]:

$$T \approx \frac{0.161V}{A + 0.46\alpha_L V} \; sec. \tag{A.5}$$

In the frequency range below 4 kHz, sound absorption in air is usually neglected, whereas for frequencies above 4 kHz, the reverberation time T is mainly determined by the absorption in air α_L.

The reverberation time is the most important variable to describe the acoustics of rooms. In rooms with a long reverberation time, sources with a relatively low level yield a high sound intensity, however, speech intelligibility is decreased due to increased temporal masking. As a compromise, the reverberation time must be "appropriate" to the room size. For speech, reverberation time should be between 0.5 sec and 1 sec (increasing with room size), for music presentations 1 sec to 2 sec are acceptable.

Beyond the global description of a room by the reverberation time, the temporal fine-structure of the reverberation, the temporal incidence of the reflections, is of interest. From the first reflections, the human hearing system is able to extract information about the size of the room. If the first reflections occur very early (1 $msec$ to 10 $msec$ after the direct sound wave), the sound color—especially for music recordings—is altered. Reflections in the time span from 10 $msec$ to 50 $msec$ increase the perceived loudness. Single echoes, arriving with a delay of more than about 100 $msec$, are perceived as echoes. Very disturbing are periodic echoes, which are generated between parallel walls, for example. For good room acoustics, the reflections should be homogeneous and the intensity should decrease with time. Single echoes should not be larger than 5 dB compared with their temporal vicinity. In concert halls, the reverberation time is measured as a function of frequency and additional sound absorbers and reflectors are placed to improve the acoustics.

Remark 29 This sketchy discussion shows that room acoustics is a complex topic, therefore acoustic experts should be consulted already in the planning of music rooms. This need is documented by plenty of examples, where the acoustics of rooms built for the purpose of audio presentations is so bad that speech intelligibility is severely hampered. Thoroughly planned room acoustics is essential not only for concert and lecture halls, but also for most other rooms such as offices and even hallways and production areas, to keep noise levels down and achieve an environment which is pleasing to the ear.

Remark 30 We have also included this discussion since it makes plausible that there is no chance to integrate a poor theory of room acoustics in a valid music-theoretic framework, and that great efforts should be made to lift this status in order to give the compositions with room acoustical specifications a firm background.

A.1.2 Sound Analysis and Synthesis

This central subject of acoustics relates to the poietic (synthesis) and esthesic (analysis) aspects of neutral sound data, more specifically, of the sound pressure variation function $p(t) = p_q^0(t)$ at a source location q

---

[7] See Section A.2.2 for the definition of loudness.

and cast in a determined standard support (see Section A.1.1). More precisely, we are considering a time function p which is defined for all times, vanishes outside a finite time interval, e.g., $[0, 1]$, and has a finite absolute amplitude supremum $sup(|p|)$, e.g., $sup(|p|) = 1$.

Sound synthesis means that we have exhibited an operator σ which for each sequence of its finite or infinite number of numeric (mostly real-valued) parameters $x_1, x_2, \ldots$ yields a function $p = \sigma(x_1, x_2, \ldots)$ of the described type. Analysis then means that we are given p and the operator σ and would like to determine a sequence $x_1, x_2, \ldots$ of arguments such that $p = \sigma(x_1, x_2, \ldots)$. The general map $\sigma : (x_1, x_2, \ldots) \mapsto \sigma(x_1, x_2, \ldots)$ will be neither surjective nor injective. So synthesis is neither a synthesis of any imaginable p, nor is analysis unambiguous, i.e., the fiber of σ could be a large set of parameters. Moreover, one may have a synthesis operator σ_1 and an analysis operator σ_2 such that the analysis of a tame synthesis may become pathological.

In the following sections, we shall discuss four such operators: Fourier, frequency modulation, wavelets, and physical modeling. We shall however not deal with mixed synthesis/analysis problems which, mathematically speaking, are wild ones—let alone the associated technological problems.

A.1.2.1 Fourier

The Fourier approach deals with periodic functions, in our case periodic pressure variations $p(t)$ as functions of time t. This however is not the direct approach to produce a pressure function which is inserted into the support operator, since such a p has an infinite support. In order to turn a periodic function f into one with a standard support $Support(1, 0, 1)$, say, it is usually multiplied by an *envelope* function H, see Figure A.2. This is a continuous, piecewise differentiable[8] non-negative function on $\mathbb{R}$ which fits in the standard support

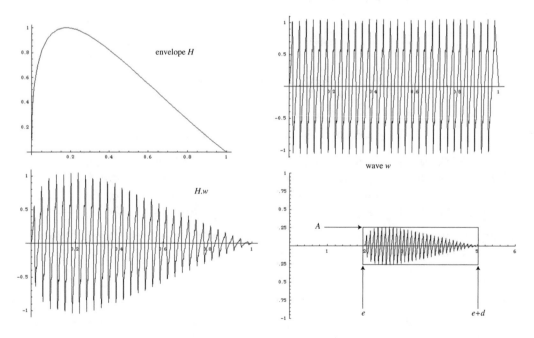

Fig. A.2. The envelope H (top left), the wave w (top right), and its combination $H.w$ (bottom left), as well as the affine deformation by the support operator $Support(A, e, d)$ (bottom right).

$Support(1, 0, 1)$, i.e., $H(t) = 0$ outside $[0, 1]$ and $\|H\|_\infty = Max(H) = 1$. In most technological applications, H is even a spline function (often even a linear, i.e., polygonal spline) modeling the attack and decay of a sound event. Then, the pressure function is given by $p = w.H$. Observe that this is already a source of

---

[8] Differentiable except for a finite set of points

poietic ambiguities: neither w (not even its frequency), nor H are uniquely determined by p. We then say that the standardized pressure function p is defined by the envelope H and the *wave w*.

Fourier's theorem deals with periodic wave functions w which are *piecewise smooth*[9] waves. For any piecewise smooth function $w : \mathbb{R} \to \mathbb{R}$, the additive group $\mathbb{R}$ acts by translations: $(e^P.w)(t) = w(t + P)$. The group of periods $Periods_w$ of w is the isotropy group of w under this action. For any non-zero period P, the inverse $f_P = 1/P$ is called a *frequency of w*, its unit is *Hertz, Hz*. If $P_w = inf(\{P \in Periods_w, 0 < P\}) = 0$, w is evidently constant, otherwise, $Periods_w = \langle P_w \rangle$ is the discrete group generated by the smallest positive period P_w. To avoid ambiguities in the periods or frequencies of a wave, one addresses this smallest period P_w or frequency $f_w = 1/P_w$ if one speaks about the "fundamental period" or the "fundamental frequency" of w (otherwise, not even the period or frequency of a wave would be uniquely determined). Fourier's theorem is this (for a proof, see [551]):

Theorem 45 *If w is a piecewise smooth wave function and $P \in Periods_w$ is a positive period, with $f = 1/P$ the corresponding frequency, then there are two sequences $(A_n, Ph_{n+1})_{n=0,1,2,...}$ of real numbers such that*

$$w(t) = A_o + \sum_{n=1,2,3,...} A_n \sin(2\pi n f t + Ph_n), \tag{A.6}$$

i.e., the infinite series converges and represents the wave for every time t for all points where the function is continuous. For the given period, the coefficients A_n and Ph_n are uniquely determined and can be calculated as follows:

$$A_0 = f \int_{-P/2}^{P/2} w(t)dt,$$

$$a_n = 2f \int_{-P/2}^{P/2} w(t) \cos(2\pi n f t)dt, 0 < n,$$

$$b_n = 2f \int_{-P/2}^{P/2} w(t) \sin(2\pi n f t)dt, 0 < n,$$

$$A_n = \sqrt{a_n^2 + b_n^2}, 0 < n,$$

$$Ph_n = \arcsin(a_n/A_n) \text{ if } A_n \neq 0 \text{ and } Ph_n = 0 \text{ else.}$$

The A_n is called the nth amplitude, *whereas Ph_n is called the nth* phase *of the wave with respect to the selected period. The sequence $(A_n)_n$ is called the* amplitude spectrum, $(A_n^2)_n$ *is called the* energy spectrum *since the energy of a wave is proportional to the square of the amplitude, and the sequence $(Ph_n)_n$ is called the* phase spectrum. *If the period/frequency is the fundamental period/frequency, one omits these specifications.*

An equivalent representation (with coefficients a_n, b_n, unique for a given period) is obtained for the explication of the sinoidal components via the goniometric formula

$$\sin(a + b) = \sin(a) \cos(b) + \cos(a) \sin(b)$$

and yields

$$w(t) = A_o + \sum_{n=1,2,3,...} a_n \cos(2\pi n f t) + b_n \sin(2\pi n f t). \tag{A.7}$$

Remark 31 It is well known *[617, Thm. 6.7.2]*, that the function sequences

$$(\sin(2\pi n f t))_{n=1,2,3,...}, (\cos(2\pi n f t))_{n=0,1,2,...}$$

---

[9] Continuous, except for a finite set of discontinuities (it need not be defined in these points, but the left and right limits of the functions exist in these points), with a continuous derivative, except for a finite set of points (where the derivative is not continuous or even not defined, but the left and right limits exist in all these singularities). Many examples are plain C^1 functions, but the saw-tooth function is not.

form an orthogonal basis of the pre-Hilbert space[10] $C^0[-P/2, P/2]$ of the continuous functions on $[-P/2, P/2]$ for the 2-norm (see Appendix I.1.2), where $f = 1/P$. This follows in particular from the trigonometric orthogonality relations of the defining scalar product $(f, g) = \int_{-P/2}^{P/2} f(t)g(t)dt$ of functions $f, g \in C^0[-P/2, P/2]$, i.e.,

$$(\sin(2\pi nft), \cos(2\pi mft)) = (\sin(2\pi nft), \sin(2\pi nft)) = (\cos(2\pi nft), \cos(2\pi mft)) = 0 \qquad (A.8)$$

for $n \neq m$. There is an infinity of such orthogonal bases for $C^0[-P/2, P/2]$, and mathematically, nothing distinguishes the sinoidal basis chosen by Fourier from the other orthogonal bases. Moreover, sinoidal functions are all but elementary. Mathematically, they are very complex, as is evident from Euler's identity $\cos(x) + i.\sin(x) = e^{ix}$. A justification for using sinoidal waves lies in the fact that simple mechanical differential equations, such as the spring equation $m.\ddot{x} = -k.x$, have sinoidal functions as their solutions. But this is a physical argument which must be coupled with a dynamical system of this equational type in order to give these functions any preference.

In order to meet the requirement for a unit amplitude, the coefficients of the Fourier representation can be dilated by a common factor, and we are done with the periodic wave. A common generalization of the Fourier representation (A.7) is defined if the frequency and coefficients are also functions of time: $f = f(t), A_n = A_n(t), Ph_n = Ph_n(t)$, a situation which is also needed to represent sounds of physical instruments with glissandi, crescendi, and their natural damping effects.

This construction is the poietic perspective. The esthesic one deals with the problem of constructing an envelope H, a periodic wave w and its Fourier representation (A.7) for a given sound function p. As was already mentioned above, the wave and the envelope cannot be reconstructed unambiguously in general. Even if the wave is known, the envelope is not reconstructible, although a number of obvious candidates can be calculated, e.g., a polygonal envelope defined by the local maxima and minima of the enveloped wave. As to the wave, one candidate for such can be guessed by the analysis of a time window $[t_1, t_2]$ of p within the supporting duration, such that the local maxima of p are relatively constant (neither at the initial, nor at the decay phase of the sound). One can then take a multiple of the period as a time window, and calculate the Fourier representation of this time window which is interpreted as a finite interval of a really periodic function, i.e., prolongation of this window to infinity. Although this period will not be the fundamental period of the wave, the Fourier representation will yield the right coefficients modulo a multiple of the fundamental frequency. If the fundamental period is small relatively to the total duration of the sound, there is a chance to calculate the underlying wave. In general, this is a highly ambiguous situation.

Once the wave is reconstructed, the Fourier coefficients are uniquely determined by the Fourier theorem, and we are done. But this is only the last phase of a highly ambiguous situation. It is of course always possible to find an underlying wave, it suffices to take the total duration of the support and to set it to the wave's period, i.e., prolongation to infinity of the sound by adding copies of itself to the left and to the right of the sound support. For the relation of these reconstructions to what is heard, see below Appendix B.

Remark 32 A final remark on the terminology of sound frequencies in music. Usually, when we deal with "the frequency of a sound", we do not mean that this frequency is a neutral property of the sound (although it could happen that the sound really has a fundamental frequency), but the fundamental frequency of a wave that is used in the standard representation of the sound as a product of its (periodic) wave and a deformed envelope. *This is a poietic definition, and this is what we will use because a neutral definition does not exist for general sounds.* In this setup, we have the sound function $p = Support(A, e, d)(H)w_{A.,Ph.}(f)$. Here, we may assume that $w_{A.,Ph.}$ is the formal representation of the trigonometric sum by the amplitude and phase spectra, and such that the total amplitude maximum is 1; the frequency is given as an additional argument, and the envelope H is given in its standard support. We call the 4-vector (e, f, d, A) the *geometric coordinates* of the sound, and the pair $(w_{A.,Ph.}, H)$ the *color coordinates* since they are responsible for the sound color (timbre) in this representation. For other poieses of sound which may also use the frequency coordinate (for example those to be discussed in the following sections), this one would also be referred to, but there is no neutral access to this concept.

---

[10] A normed real vector space whose norm is defined by a positive definite symmetric bilinear form.

A word of caution: The way humans "detect" sound frequencies is not neutral, this is an esthesic psycho-physiological system whose function is far from understood, so do never mix up neutral facts with poietic or esthesic facts when dealing with sound attributes!

A.1.2.2 Frequency Modulation

In this section, we have a similar decomposition as discussed before: the sound p is written as a product $p = H.w$, where H is the envelope, and w is a "wave" function. However, this time we do not use the additive combination of sinoidal functions of Fourier synthesis to build w. The combination is rather a functional concatenation of such functions, i.e., sinoidal functions have in their arguments other sinoidal functions, and so on. This synthesis operator was introduced by John Chowning [198] and implemented first in the legendary Yamaha's DX7 synthesizers. The formal definition of frequency modulation (FM) functions in terms of circular denotators is given in Section 6.7, example 3, and yields this type of expressions:

$$FMsound(myFMObject)(t) = \sum_i^n A_i \sin(2\pi F_i t + Ph_i + FMsound(myModulator_i)(t))$$

where $myModulator_i$ is the FM-Object factor of the limit type denotator $Knot_i$. In the terminology of FM synthesis, the (respective) interior functions $FMsound(myModulator_i)(t)$ are called the *modulators* with respect to the (respective) exterior sinoidal functions $A_i \sin(2\pi F_i t + Ph_i)$ which are named the *carriers*. We symbolize this relation by an arrow $myModulator_i \Rightarrow myFMObject$. So the FM functions start as a sum of carriers where modulators are inserted, and these modulators are again of this nature, etc., until no modulator appears and the recursion terminates. In FM synthesis it is also allowed to have circular denotators in the sense that a modulator can be a carrier in one and the same function! Observe that the existence of a denotator $myFMObject$ describing such a "self-modulating" function does not imply the existence of the corresponding function $FMsound(myFMObject)$. However in digital sound synthesis, one often resolves this problem by taking the time argument of the modulator one digital step before the time argument of the carrier, i.e.,

$$FMsound(myFMObject)(t_n) = \sum_i^n A_i \sin(2\pi F_i t_n + Ph_i + FMsound(myModulator_i)(t_{n-1})).$$

In the DX7 implementation, the FM recursion scheme is presented in the graphical block diagram format of a so-called *algorithm* (not an adequate wording, though); Figure A.3 shows Yamaha's 32 algorithms.

Each sinoidal carrier component is written as a block, whereas the modulators of a given carrier are those blocks which are above the carrier and are connected by a line with this carrier. Each of the building blocks can be specified in the respective parameters.

The point of the FM synthesis is that it needs a small number of sinoidal functions—in fact only six sinoidal oscillators are needed for the DX7 algorithms—to simulate complex instrumental sounds, which is a theoretical and technological advantage.

But the elegant FM synthesis has also drawbacks concerning the uniqueness question: How many different denotators $myFMObject$ do yield the same function $FMsound(myFMObject)$? A general answer seems difficult, we have two partial solutions. The first regards a tower of modulators:

Lemma 65 *Let $G : \mathbb{R} \to \mathbb{R}$ be a bounded C^1 function. Then G is determined by its value $G(0)$ and by any function $F : \mathbb{R} \to \mathbb{R}$ of the shape $F(t) = a + b \sin(ct + G(t))$, with $b \neq 0$.*

Proof. Evidently, by the boundedness of G, $a+b = max(F), a-b = min(F)$, so $a = (max(F)+min(F))/2, b = (max(F) - min(F))/2$, are determined by F. Moreover, the continuous function $G' = \dfrac{bF'}{\sqrt{b^2-(F-a)^2}} - c$ is a function of F, and the value $G(0)$, together with the integral of G' which is a function of F, completely determine G, QED.

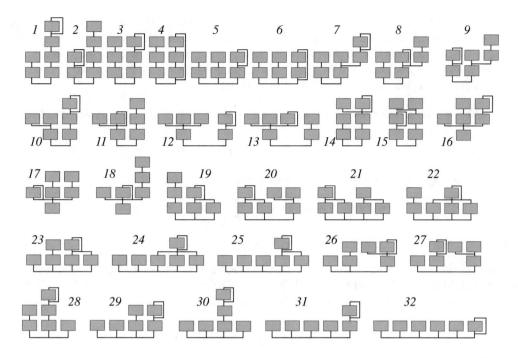

Fig. A.3. Yamaha's 32 algorithms for FM synthesis. Only six sinoidal oscillators are required to generate a variety of more or less natural sounds.

Proposition 73 *If $M = (M_n \Rightarrow M_{n-1} \Rightarrow M_{n-2} \Rightarrow \ldots M_0)$ is an FM denotator defined by a sequence of relative modulators for the functions $M_i = A_i \sin(2\pi f_i t + ?)$, then the resulting sound function $FMsound(M)$ determines all the modulators.*

Proof. This follows from lemma 65 by recursion on modulators, starting at total function $FMsound(M)(t) = A_0 \sin(2\pi f_0 t + FMsound(M_1))$, where M_1 is the denotator starting from M_1 instead of M_0. In the lemma, take $F = FMsound(M), G = FMsound(M_1)$, and observe that $G(0) = 0$, QED.

The second solution regards a flat sequence of unrelated carriers:

Proposition 74 *Let $F(t) = \sum_{i=1,\ldots k} a_i \sin(b_i t)$ with $0 < b_1 < \ldots b_k, 0 < a_i$ be the function associated with a flat FM denotator. Then the coefficients, and therefore the denotator, are all uniquely determined by F.*

Proof. We have $(-1)^n F^{2n+1}(0) = \sum_{i=1,\ldots k} a_i b_i^{2n+1}$ for all $n \geq 1$. Consider the function $H(a.,b.)(x) = \sum_{i=1,\ldots k} a_i b_i^x$ of the real variable x. Suppose we have sequences $(a_i)_{i=1,\ldots k}, (b_i)_{i=1,\ldots k}$ and $(a'_i)_{i=1,\ldots k}, (b'_i)_{i=1,\ldots k}$ such that they yield the same function F. Then we have $H(a.,b.)(x) = H(a'.,b'.)(x)$ for all $x = 2n+1, n \geq 1$. Write $b_i = e^{\beta_i}, b'_i = e^{\beta'_i}$, and suppose $b_k < b'_k$. Then

$$\frac{H(a.,b.)(x)}{H(a'.,b'.)(x)} = \frac{\sum_{i=1,\ldots k} a_i e^{\beta_i x}}{\sum_{i=1,\ldots k} a'_i e^{\beta'_i x}} = \frac{\sum_{i=1,\ldots k} a_i e^{(\beta_i - \beta_k)x}}{\sum_{i=1,\ldots k} a'_i e^{(\beta'_i - \beta_k)x}} = \frac{\sum_{i=1,\ldots k-1} a_i e^{(\beta_i - \beta_k)x} + a_k}{\sum_{i=1,\ldots k} a'_i e^{(\beta'_i - \beta_k)x}}.$$

But for $x \to \infty$, the denominator goes to 0 whereas the numerator goes to a_k, contradicting the fact that this quotient is 1 for all $x = 2n+1, n \geq 1$. Therefore $b_k = b'_k$, so in the above quotient, the limit for $x \to \infty$ is a_k/a'_k which is also 1, and we have equal coefficients for the index k. Therefore, we may proceed by induction to $k-1$ and we are done, QED.

These very special results show that given the tower or the flat FM schemes, the functions determine their coefficients (the background denotators) uniquely, but for general FM schemes, there is no such a result. Moreover, if the FM scheme is not known, we have no idea of how the scheme should be determined from the function. This is a drawback compared to the Fourier operator, where the coefficients are always uniquely determined once the fundamental frequency is fixed. In other words, FM synthesis is much more efficient than Fourier synthesis, but one has to pay for this when turning to the respective analyses.

The switch between Fourier and FM operators is essentially managed by Bessel functions. These are defined directly from a core situation from FM synthesis, i.e.,

Definition 125 *Let z be a real number. Then the Fourier expansion of the 2π-periodic function $\sin(z\sin(t))$ of t is*

$$\sin(z\sin(t)) = 2 \sum_{n=0,1,2,\dots} J_{2n+1}(z)\sin((2n+1)t), \tag{A.9}$$

whereas the Fourier expansion of the 2π-periodic function $\cos(z\sin(t))$ of t is

$$\cos(z\sin(t)) = J_0(z) + 2 \sum_{n=1,2,3,\dots} J_{2n}(z)\cos(2nt). \tag{A.10}$$

The functions $J_m, m = 0,1,2,\dots$ are called the mth Bessel functions.

The above definition relies on the Fourier representation of the respective functions and on their properties as odd or even functions. An alternative definition of Bessel functions is $J_m(z) = \frac{1}{\pi}\int_0^\pi \cos(mt - z\sin(t))dt$. From Definition 125, one obtains the following fundamental equations for Bessel functions:

$$\sin(r + z\sin(s)) = \sum_{-\infty}^{\infty} J_n(z)\sin(r + ns),$$

$$\cos(r + z\sin(s)) = \sum_{-\infty}^{\infty} J_n(z)\cos(r + ns).$$

In particular, with $s = 2\pi ft, z = I, t = 2\pi gt$, we have

$$\sin(2\pi ft + I\sin(2\pi gt)) = \sum_{-\infty}^{\infty} J_n(I)\sin(2\pi(f + n.2\pi g)t),$$

which is a Fourier type linearization, however, it is not a proper Fourier representation since the so-called "nth side band" frequencies $f + n.2\pi g$ are not a multiple of a fundamental frequency in general. This is rather a reduction of a FM concatenation to a flat FM configuration. Conversely, every (finite) Fourier decomposition is evidently a flat FM configuration.

A.1.2.3 Wavelets

Although the FM operator is much more efficient than the Fourier operator, it is still an operator which produces functions with an infinite support, a property which no real sound shares, and we have in fact added an envelope to cope with this requirement for Fourier and FM operators. From this point of view, wavelets are fundamentally better suited for handling finite sound objects without any envelope casting. Refer to [622] and [559] for the wavelet theory and its applications to sound and music. Let f be a square integrable function (element of $L^2(\mathbb{R})$, i.e., $\int_{\mathbb{R}} |f(x)|^2 dx < \infty$). Then its Fourier transform is defined and is the function

$$\hat{f}(\omega) = (2\pi)^{-1}/2 \int_{\mathbb{R}} f(x)e^{-ix\omega}dx. \tag{A.11}$$

Definition 126 *A square integrable function ψ is called a* wavelet *if $0 < c_\psi = \int_{\mathbb{R}} \frac{|\hat{\psi}|^2}{|\psi|}d\psi < \infty$. For a wavelet ψ, the* wavelet-transformed *of a square integrable function f is the function (of two variables a, b)*

$$L_\psi f(a,b) = c_\psi^{-1/2}|a|^{-1/2}\int_{\mathbb{R}} f(t)\psi(\frac{t-b}{a})dt, \tag{A.12}$$

with $a \in \mathbb{R} - \{0\}, b \in \mathbb{R}$.

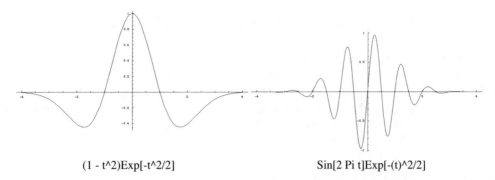

(1 - t^2)Exp[-t^2/2]   Sin[2 Pi t]Exp[-(t)^2/2]

Fig. A.4. Two wavelets: Murenzi's Mexican hat [776] (left) and Morlet's wavelet (imaginary part) [385] deduced from the sinoidal function (right).

Figure A.4 shows two typical examples of wavelets. The wavelet transform is a function of two variables defined on every couple $(a, b) \in (\mathbb{R} - \{0\}) \times \mathbb{R}$. The point of this representation is that it is a kind of system of coefficients $(L_\psi f(a, b))_{(a,b) \in \mathbb{R} - \{0\} \times \mathbb{R}}$ which is parametrized by two real numbers, corresponding to a scalar product

$$c_\psi^{-1/2}(f, Support(|a|^{-1/2}, b, a)(\psi)) \tag{A.13}$$

of the function f with an affinely deformed version[11] $Support(|a|^{-1/2}, b, a)(\psi)$ of the "mother" wavelet ψ. This deformation $Support(|a|^{-1/2}, b, a)$ is an isometry on the square integrable functions, see also Figure A.5 for some deformed wavelets. By the following formula, the wavelet transformations $L_\psi f(a, b)$ redetermine

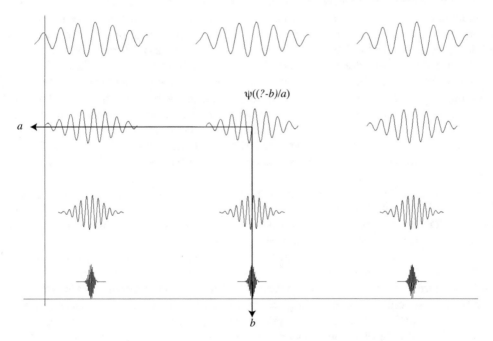

Fig. A.5. Various affine deformations of the mother wavelet ψ.

the original function f:

---

[11] Observe that the operator $Support(|a|^{-1/2}, b, a)$ defines a linear action $Support : \overrightarrow{GL}(\mathbb{R}) \to GL(L^2(\mathbb{R})) : e^b a \mapsto Support(|a|^{-1/2}, b, a)$ of the affine group $\overrightarrow{GL}(\mathbb{R})$ on the space $L^2(\mathbb{R})$ of square integrable functions.

$$f(t) = c_\psi^{-1/2} \int_{\mathbb{R}^2} L_\psi f(a,b) Support(|a|^{-1/2}, b, a)(\psi)(t) \frac{da\, db}{a^2}. \qquad (A.14)$$

For selected wavelets, it is possible to generate an orthonormal basis of $L^2(\mathbb{R})$ which is defined as a so-called *frame*. For $a_0 > 1, b_0 > 0$, and for the Meyer wavelet[12] ψ we consider the frame $(Support(a_0^{-m/2}, nb_0 a_0^m, a_0^m)(\psi))_{m,n\in\mathbb{Z}}$ of deformed versions of ψ. Then this is an orthonormal basis of $L^2(\mathbb{R})$ for $a_0 = 2, b_0 = 1$. This is an analogous situation as encountered for Fourier series and their sinoidal bases with the known trigonometric orthogonality relations in formulas (A.8).

For a comparison of Fourier and wavelet analysis, see [1105].

A.1.2.4 Some Remarks on Physical Modeling

The previous operators were directly acting on the production of a sound function. Their poietic nature was a mathematical one, just to construct a time function $p(t)$ by a mathematical procedure from a certain type of "atomic", i.e., basis functions. In contrast, physical modeling is one step more poietic in that it does not directly deal with sound, but with a physical system that produces sound. On the one hand, this is a strong restriction since sound does not care for the physical system that evokes that sound. It seems however that this drawback is compensated by the fact that musical expressivity is largely determined by the physical device which the artist manipulates when interpreting or improvising music. It is also an important argument that the simulation of the physical instrument and then a possible canonical extension could yield more interesting sounds than just "abstract nonsense" procedures. For a general survey on physical modeling, see [911, Chapter 7], we restrict our discussion of this extensive topic to some systematic remarks.

The idea is that one considers a physical model of a sound production device and then implements this model as a software which—if sufficient calculation power is available—calculates the physical output on the level of the sound wave that is emitted by the modeled instrument. At present, there are three methodologies for such a modeling: mass-spring, modal synthesis, and waveguide. The mass-spring paradigm just models a physical instrument (a string, a drum) by a finite space configuration of point masses that are related by springs and damping effects [469]. The modeling is built upon the classical mechanics of Newton's law and the corresponding dynamic behavior that eventually terminates in the air's vibration.

The modal synthesis paradigm [171] reduces the vibrating physical system to a system of vibrating substructures, usually very small in number compared with the mass-spring components. These substructures are characterized by their frequencies, damping coefficients and parameters for the vibrating mode's shape. This adds up to a sum of modal vibrations. Whereas in simple configurations these data can be obtained from classical literature in equations for vibrating systems, the complex data must be extracted from experimental results. A prototypical implementation of this paradigm is MOSAIC, developed by Jean-Marie Adrien and Joseph Morrison, see [911, p.276 ff.].

The waveguide paradigm has been implemented in commercial physical modeling synthesizers by YAMAHA and KORG , see [911, p.282 ff.]. It has mainly been developed by Julius O Smith III and collaborators [989, 990]. The waveguide model implements the traveling wave along a medium, such as a tube or a string. See [991] for an update of physical modeling strategies.

Although physical modeling is a successful approach in the simulation of musical instrumental sounds, it is a step back from the neutral sound objects to their generators. This has not only been a technological requirement for performance theory (where physical modeling is a core approach), it is also a consequence of the failure in the understanding of the topological semantics of sound objects as exposed in Section 12.3. In particular, it is an open problem to relate the physical modeling theory and technology to the other operators, such as Fourier, FM, and wavelets. The deeper question here is whether the neutral sound objects are really the most relevant ingredients of musical performance, i.e., how strongly the gestural components influence and characterize the sounding reality. It is not clear how deeply a sound conveys the generating gestures in its autonomous structure. To our knowledge, sound classification has not been directed towards a gestural coordinate in the neutral sound description, except, perhaps, in the straightforward envelope component.

---

[12] See [622, 2.1.25].

A.2 Mathematical and Symbolic Spaces

The physical description of sounds is not what can be used for music theories. This is based on (1) the way humans perceive sounds, (2) the shape of music thinking, and (3) available instrumental technologies. Therefore physical parameter spaces must be transformed into spaces which essentially encode the same information, but do so in a way which is more adapted to music. Basically, we shall present mathematical structures where physical parameters are represented. Based upon these spaces, we shall explain a derived set of representations which encode different music-topographic aspects. Our discussion regards the geometric coordinate pairs (basis coordinate plus associated pianola coordinate) onset and duration (A.2.1), amplitude and crescendo (A.2.2), and frequency and glissando (A.2.3).

A.2.1 Onset and Duration

In music, one speaks about tempo, metronome, quavers, semiquavers, triplets, 3/4 meter, etc. The relation to the physical time parameters is as follows. To begin with, the *physical* onset time e *sec* and duration d *sec* are opposed to a *musical* onset time E *note* and duration D *note*, usually also in real values, and in units such as "*note*", meaning a whole note. However, in many practical contexts, the rational number field $\mathbb{Q}$ will do. In this latter context, we have the ratios of integer numbers $E = w/n, D = z/n, n > 0$, and the denominator is of the form $n = 2^r 3^s 5^t 7^u$ with natural exponents r, s, t, u, whereas the numerators w, z are integers, i.e., we are working in the ring localization[13] $\mathbb{Z}[1/2, 1/3, 1/5, 1/7]$ at the primes $2, 3, 5, 7$. Musically, this means that one is allowed to add, subtract, multiply such numbers at will without leaving the domain, and also division of such numbers by $2, 3, 5, 7$ leaves the domain invariant, i.e., the construction of duplets, triplets, quintuplets, septuplets is possible without any restriction.

The musical time shares a mental reality and should not be confused with physical time. Genetically, musical time is an abstraction from physical time, but they are by no means equivalent. This abstraction is also a creation of an autonomous time quality where mental constructions such as a score can be positioned. The relation between these two time qualities is defined by the tempo, usually encoded as a metronomic indication of x quarters per minute according to Mälzel, meaning a difference quotient (velocity) $\Delta E/\Delta e$ of musical time per physical time. This presupposes that we are given a one-to-one performance mapping $E \mapsto e(E)$ from musical time to physical time. *It is common in mathematical music theory to write physical parameters in lower case letters, whereas musical parameters are written in upper case letters.*

Let x, y be positive integers. Then every onset E can be written uniquely as $E = \delta + \tau.x/y$, where $0 \leqslant \delta < x/y$ is in $\mathbb{Q}$, and $\tau \in \mathbb{Z}$. If $\delta = q/(ny)$, we say that *in x/y time*[14], *E is on the $(q + 1)$st ny-tuplet in bar $\tau + 1$.* The additive group $\mathbb{Z}.1/y$ is called the *meter* of the x/y time. Evidently, this initializes a score at time $E = 0$, but this is pure convention. Pay attention *not* to view the symbol x/y as a plain fraction, but as a pair x, y giving rise to a fraction. In other words, in music notation, the symbol x/y is the mathematical fraction x/y *plus* the meter.

Example 91 Let $x = 3, y = 4, E = 15.375$. Then we have $E = 3/8 + 20.3/4$, i.e., E is on the fourth quaver of bar 21.

If we work in $\mathbb{Z}[1/2, 1/3, 1/5, 1/7]$, then with E and $1/y$, the remainder δ and the meter are automatically in this domain. If we are given a duration D, then durations of form D/n are also called n-tuplets (with respect to D).

The reason why the concept of onset time is not common in music lies in the fact that a note's onset can be deduced from its position in the score. The time signature, bar-lines, and simultaneous or preceding notes help establish the time context of each note. This helps calculate onsets algorithmically by recursion from the first score onset. This algorithm is derived from the linear syntagm of written language, but it sometimes leads to ambiguities.

---

[13] Make the integers $2, 3, 5, 7$ invertible, i.e., admitting fractions $1/2, 1/3, 1/5, 1/7$. But see Appendix E.4.1 for a formal definition.

[14] We stick to the continental terminology since "meter" will be reserved for a different concept.

A.2.2 Amplitude and Crescendo

The amplitude A of a sound relates to the loudness sensation, i.e., to musical dynamics. However, the chain of transformations to the physiologically relevant measures is quite complex [914]. The first member of this chain is the transformation which associates A with the *sound pressure level* (SPL)

$$l(A) = 20. \log_{10}(A/A_{threshold}) \; dB. \tag{A.15}$$

Here, $A_{threshold} = 2.10^5 \; Nm^{-2}$ is close to the amplitude of the threshold pressure variation for a 1 kHz sound in a young, normal hearing human subject. The unit dB for l is *Dezibel*. Of course, these constants are pure convention. From the auditory physiology, the shape $l(A) = a \ln(A) + b$ is essential, i.e., $l(A)$ is a conventionally normed linear function of the logarithm of the amplitude. The physio-psychological motivation for this approach lies in the Weber-Fechner law according to which the sensation of a difference of sensory stimuli is proportional to the stimuli [494], yielding the logarithmic representation as an adequate encoding of this sensorial modality. Figure A.6 shows some environmental SPL values (the musical units *ppp*, *pp*, etc. will be discussed below). For example, the SPL $\approx 120 \; dB$ of an air jet is one million times the threshold SPL.

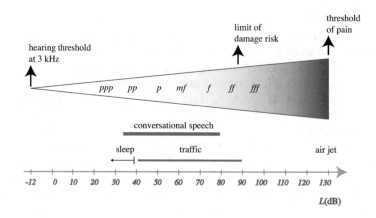

Fig. A.6. Some environmental sound pressure level values.

Just as duration is the time interval from the onset time to the "offset" time of a sound event, the *crescendo* parameter c is the difference between the onset SPL and the offset SPL. In normal piano sounds, this vanishes, but for violins, trombones, etc., this is a relevant quantity. In this context, one supposes that the sound representation $p = Support(A, e, d)(w_{A.,Ph.}(f).H)$ has a time-dependent amplitude $A = A(t)$. At the offset time $e + d$, the amplitude has changed by the amount $A(e + d)/A(e)$, or in terms of loudness, $c = l(A(e + d)) - l(A(e))$. As with onset and duration, it is not specified what happens in the sound process between the onset and the offset, it's just the difference that matters. This difference c is termed the (physical) *crescendo*.

In the musical abstraction L for loudness, one also works in the field $\mathbb{R}$ of real numbers, but the common score notation is not well defined for all real values. More precisely, the first still physically motivated codification is a conventional calibration such as, for example, $l : \mathbb{Z} \to \mathbb{R}$ with $l(L) = 10 \; dB.L + 60 \; dB$ such that the values

$$0 \mapsto mf$$
$$1 \mapsto f \qquad \qquad -1 \mapsto p$$
$$2 \mapsto ff \qquad \qquad -2 \mapsto pp$$
$$3 \mapsto fff \qquad \qquad -3 \mapsto ppp$$

are arranged symmetrically around the *mezzoforte* sign. This looks like an identification of the score symbols with precise physical values. But this is wrong, it is a transformation from mental to physical reality. In fact, the affine transformation $l(L)$ is only a "default" assignment of a mental quantity symbolized by a dynamic symbol (which is codified by an integer). This fact is more evident if one recalls the velocity parameter in the MIDI code. This quantity is an integer in the interval $[0, 127]$, but the physical meaning of a velocity value depends on the assignment by specific technological calibrations, in particular by the output chain where the loudspeakers can take values that have nothing to do with the velocities. Units in the mental level of loudness are not standard, but we could, for example, take *vel* for MIDI's velocity parameter.

The extension of integer values to real or rational values (depending on the specific usage) is a reasonable procedure, for instance preconized by MIDI velocity codification. It is also useful for finer loudness data management with relative loudness signs such as "crescendo", "diminuendo" which make sense in a mental crescendo parameter which we shall denote by C. See also our discussion of performance transformations and primavista weights in Section 39.2.

A.2.3 Frequency and Glissando

Sounds which share a frequency parameter f are musically important, at least in the European tradition. However, the corresponding mathematics is a bit more involved than for time and amplitude. This is an expression of an intense discussion of traditional harmonies and interval theories with an evolving instrumental technology.

To begin with, f behaves like amplitude: For reasons of auditory physiology (see also Appendix B), f is transformed via the logarithmic formula $h(f) = u \ln(f) + v$, yielding the *physical pitch* h of frequency f. The common unit of pitch is the *Cent Ct*, it corresponds to the logarithm of a relative frequency increase by the factor $2^{1/1200}$, i.e., one percent of a well-tempered semitone (see below for tuning types); this entails $u = 1200/\ln(2)$, but the constants u, v are purely conventional. Presently, and in the Western framework, the relevant frequencies have the shape $f = 132\ Hz.2^p.3^s.5^t$, where $p, s, r \in \mathbb{Q}$. It is based on the *chamber pitch* $440\ Hz$ of the one-line a, as fixed in London 1939. Instead, we have chosen the (unlined) c with the frequency $132\ Hz = 440\ Hz.2^{-1}.3.5^{-1}$ as a starting frequency in order to relate the examples with ease to c. Mathematically, the restriction to the first three primes $2, 3, 5$ is not essential. One could as well take any sequence $p_1, p_2, \ldots$ of mutually prime natural numbers (larger than 1), the rational powers of which are multiplied by $132\ Hz$.

The natural logarithm of f is

$$\ln(f) = p.\ln(2) + s.\ln(3) + r.\ln(5) + \ln(132).$$

But music theory is rather interested in the relative pitch, i.e.,

$$\ln(f) - \ln(132) = p.\ln(2) + s.\ln(3) + r.\ln(5).$$

Moreover, the passage to another logarithmic basis b could be desirable, i.e.,

$$\ln(b)^{-1}.\ln(f) - \log_b(132) = p.\log_b(2) + s.\log_b(3) + r.\log_b(5),$$

so that we are given the linear function

$$u.X + v = \ln(b)^{-1}.X - \log_b(132)$$

for $X = ln(f)$. For music theory, the restriction to rational exponents is essential, since this hypothesis enables an unambiguous representation, see Appendix E.2.1. Every frequency f can be replaced by a point $x = (p, s, r) \in \mathbb{Q}^3$. Such a point represents the frequency $f(x) = 132\ Hz.2^p.3^s.5^r$, and from $f(x) = f(y)$, we conclude $x = y$. With this interpretation, $\mathbb{Q}^3$ is called the *Euler space* and a point $x = (p, s, r) \in \mathbb{Q}^3$ is called an *Euler point*. This means that a real number $p.\log_b(2) + s.\log_b(3) + r.\log_b(5)$ is viewed as a vector which is a rational linear combination of linearly independent vectors $\log_b(2), \log_b(3), \log_b(5)$.

The choice of the three first primes stems from the tradition of just tuning, where for two frequencies f, g, we have:

$(O) f/g = 2/1 : f$ is the *octave* (frequency) for g.

$(Q) f/g = 3/2 : f$ is the (just) *fifth* (frequency) for g.

$(T) f/g = 5/4 : f$ is the (just) *major third* (frequency) for g.

This is why 2 is associated with the octave, 3 with the fifth, and 5 with the major third. We therefore call p the *octave coordinate*, r the *fifth coordinate*, and s the *third coordinate* (of the sound with frequency f or of the corresponding Euler point. The Euler point $o = (1, 0, 0)$ is called the *octave point*, the point $q = (0, 1, 0)$ is called the *fifth point*, and $t = (0, 0, 1)$ is called the *third point*. Observe that fifth and third points are not the fifth and octave. The structural meaning of these points is explained in Section 6.4.1. The minor just third frequency 6/5 does not add a new prime number.

For a fixed chamber pitch, the *2-3-5 just tuning* (short: just tuning) is the set of frequencies which are represented by integer coordinates, i.e., the subgroup $\mathbb{Z}^3 \subset \mathbb{Q}^3$ of the Euler space. This is the three-dimensional grid which was introduced by Leonhard Euler [307]. The group of those grid points whose third coordinate vanishes is the *Pythagorean tuning*.

The Euler space is derived from the logarithmic representation of pitch, but the coefficients are beyond physical reality, this is why we view this space as a mental space. The same is valid if we consider the codomain space of pitch h, but interpreted with rational coefficients (written as $\mathbb{R}_{[\mathbb{Q}]}$ instead of the usual real line $\mathbb{R}$, the one-dimensional real vector space[15]). More precisely, one should name H the pitch on the mental pitch space *Pitch*, and h the pitch on the physical pitch space *PhysPitch*. Whereas the unit for physical pitch is *Cent*, the unit for mental pitch could be *Semitone* or MIDI's key(number), but no standard exists here. In performance, we have a transformation $\wp : Pitch \to PhysPitch$. Both spaces have the real numbers as underlying sets, but the meaning of the spaces is different. The mental pitch space encodes pitch as it is symbolized on the score or as a key number in MIDI code. Ideally, performance transforms this abstract data into mathematical pitch $Math(H) = (p, s, r)$ in the Euler space, and this is transformed to the pitch $\ln(f) = p. \ln(2) + s. \ln(3) + r. \ln(5) + \ln(132)$ in $\mathbb{R}_{[\mathbb{Q}]}$. Only after forgetting about the coefficients, we are in *PhysPitch*. And this is by no means a formal play: It is a dramatic change of reality if one views the reals as an infinite-dimensional space with linearly independent octave, fifth, and third logarithm vectors, or as a line, where everything shares the same direction! In analogy to onset and loudness, one introduces physical glissando g and its symbolic counterpart G.

In the common visualization of the Euler space, all grid cells are shown as cubes. This could suggest that angles and distances are relevant to this space. So far, this has however no musicological reason, and it is nonsense to argue with angles and distances as Arthur von Oettingen [806], Carl Eitz[297], and Martin Vogel [1089] have done.

Evidently, the mathematical structure of just tuning is independent of the historical choice of the sequence 2,3,5. One could as well take any pairwise prime positive numbers $1 < p_1 < p_2 < p_3$ and would get a p_1-p_2-p_3 just tuning.

If the pitch range in Euler space is described by non-integer coefficients, one speaks of *tempered tunings*. The most well known are defined by a uniform construction mode. For a natural number $w > 1$ one considers all pitches whose octave coordinate is a fraction of shape $p = x/w$, whereas the other coordinates vanish. This is called the *w-tempered tuning*. This defines a grid $\mathbb{Z}.1/w.o \xrightarrow{\sim} \mathbb{Z}$ with step width $1/w$ of the octave point. By the same recipe, tempered tunings in fifth and third direction can be defined. The 12-tempered tuning is the famous "well-tempered" tuning. The 1200-tempered tuning is less interesting for conventional composers than for measurement techniques, where the unit step is the Ct step defined above.

More generally, one may define tunings which consist of tempered and just components. The procedure runs as follows. Take three positive integers w_1, w_2, w_3 and consider the grid

$$\mathbb{Z}.\frac{1}{w_1}.o + \mathbb{Z}.\frac{1}{w_2}.q + \mathbb{Z}.\frac{1}{w_3}.t$$

---

[15] See Appendix E.2.1.

which specializes to the 2-3-5 just tuning as well as the tempered tunings as defined above. Call this construction the w_1-w_2-w_3 *just-tempered tuning*. Historically relevant is the *mediante* tuning, which is the 1-1-2 just-tempered tuning, and which includes the tempered whole-tone step in the major third.

With respect to the auditory psycho-physiology (see Appendix B), we should consider the distribution of the just tuning grid vectors $x = (p, s, r)$ in $\mathbb{Z}^3$ with respect to the mathematical pitch $H(x) = p.\log_b(2) + s.\log_b(3) + r.\log_b(5)$. More generally, take any vector $x \in \mathbb{R}^3$ and denote by $H_{prime} = (\log_b(2), \log_b(3), \log_b(5))$ the *prime vector*. This means that pitch is the usual scalar product

$$H(x) = (H_{prime}, x) \tag{A.16}$$

of the prime vector with the generalized Euler point x. Therefore,

Proposition 75 *With the definition* (A.16) *two vectors* $x, x' \in \mathbb{R}^3$ *have the same pitch* $H(x) = H(x')$ *iff their difference is orthogonal to the prime vector* H_{prime}.

Call $E = H_{prime}^{\perp}$ the plane orthogonal to the prime vector. Then the proposition means that for any generalized Euler point x, $x + E$ is the set of points with same pitch as x. Now, according to what we know, E lies so skew in $\mathbb{R}^3$ that for a point $x \in \mathbb{Q}^3$,

$$(x + E) \cap \mathbb{Q}^3 = \{x\}.$$

Nonetheless, every real number ϕ can be approached within any given error by points of the just tuning grid: For any positive bound δ, there is $x \in \mathbb{Z}^3$ such that $|H(x) - \phi| < \delta$, see Appendix D.4 for a proof. In particular, we have the following proposition which has dramatic consequences for theories of hearing of just tuned pitch (see Section B.2):

Proposition 76 *If* $\phi = H(x_0)$ *is a mathematical pitch of a grid point* x_0, *then for any positive bound* δ, *there is an infinity of grid points* x *such that* $|H(x) - H(x_0)| < \delta$.

B

Auditory Physiology and Psychology

"Music listening" is a metonymy of understanding music: For all participants, the ear functions as an interface for perceiving music between physical, psychological, and mathematical reality. But a metonymy is not the matter as such. This is what deaf Beethoven teaches us impressively: his innermost ear was an organ of imagination that was uncoupled from the material ear. And the physiology of the hearing process teaches us that the neural coupling of the ear to the respective cortical regions is extremely complex and still hardly understood. Section B.1 is written to give an overview on auditory physiology.

Beyond receptive processes hearing means also an active shaping according to templates of esthesis and poiesis. One of the difficult basic problems concerning the activity of hearing deals with the compatibility of these templates and the physical input, and in particular the notorious "straightened out hearing" (German: "Zurechthören") which keeps alive a lot of wishful thinking in music theory. Basically, the problem is that we do not perceive physical sounds, but classes of indistinguishable sounds, what Werner Meyer-Eppler coined "valences" in [735]. Valence theory, which is sketched in Section B.2, has dramatic consequences for the relation between mathematical theories and their semantic potential for music.

Historically and materially, and in view of the genealogy of a mathematical theory, the subject pairing of "consonance-dissonance" is an excellent illustration of auditory physiology and psychology. We deal with the formal and some physiological aspects of this subject in part VII. In the following Section B.3, we want to expose the stratification of the phenomenon of consonance and dissonance. Hereby, the problem setup as well as the approaches to its solution demonstrate a strong dependency of the addressed reality layer. Methodologically, Section B.3 is important since it makes evident that known approaches of the mathematical argumentation in musicology turn out to be too narrow with respect to the existing music, and too dogmatic and scientifically unbased with respect to music thinking.

B.1 Physiology: From the Auricle to Heschl's Gyri

Two fundamentally different regions of sound processing in the auditory system can be distinguished. In the peripheral region, mechanical preprocessing takes place, especially in the fluid-filled inner ear. The sensory cells encode the preprocessed mechanical oscillations into electrical nerve action potentials. In the second region of the hearing system, neural processing of the sound information is conducted in ascending nuclei, which finally leads to auditory sensation. The hearing system, especially in mammals, has pushed its bandwidth up in a frequency range, where the limits of neural processing in the time-domain are exceeded by far. Instead of processing high frequencies in the time domain, evolution has developed the so-called frequency-place principle. In the inner ear, the frequency contents of sound signals are separated in the spatial domain. The sensitivity of our hearing system is thereby remarkable, in its most sensitive region, the threshold of the hearing system is limited only by thermal noise.

© Springer International Publishing AG, part of Springer Nature 2017
G. Mazzola, *The Topos of Music IV: Roots*, Computational Music Science,
https://doi.org/10.1007/978-3-319-64495-0_2

B.1.1 Outer Ear

Sound energy is collected by the outer ear and transmitted through the outer ear canal to the ear drum (Figure B.1).

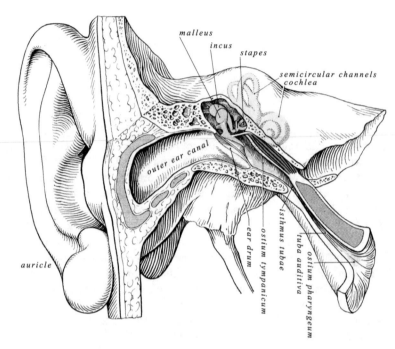

Fig. B.1. Schematic view of the outer, middle and inner ear, modified from [1157], © Springer-Verlag Berlin Heidelberg 1999. With permission of Springer.

The delicate structures of our hearing system are well protected inside the skull. For the sound transmission, the outer ear canal acts like an open pipe with a length of about 20 to 30 mm. Its quarter-wave resonance is responsible for the high sensitivity of our hearing organ in this frequency range, indicated by the dip of the threshold in quiet around 4 kHz. This high sensitivity is however also the reason for high susceptibility to noise-induced damage in the region around 4 kHz.

B.1.2 Middle Ear

The fluids of the inner ear must be excited by the sound-induced vibrations of the air particles in front of the ear drum. The light but sturdy funnel-shaped ear drum (tympanic membrane) operates over a wide frequency range as a pressure receiver. It is firmly attached to the long arm of the hammer (malleus) (Figure B.1). The motions of the eardrum are so transmitted via the anvil (incus) to the stirrup (stapes). The stapes foot plate, together with a ring-shaped membrane called the oval window, forms the entrance to the inner ear. The middle ear optimizes the energy flow from air-borne sound in front of the ear drum to fluid motion in the inner ear by a mechanism called impedance transformation. One part of the impedance transformation is based on the lever ratio of about 1.5:1 produced by the different lengths of the arms of malleus and incus [162]. The lever ratio transforms oscillations of the ear drum with small forces into motions of the fluid with large forces. The law of energy conservation implies also that the tiny displacements of the ear drum are transformed into still smaller oscillations in fluid. An even larger transformation of the pressure is due to the ratio of the large ear drum to that of the small oval window. This ratio is about 17 [97]. Through the lever and area ratios, an almost perfect impedance match is reached in man in the middle frequency range between 1 and 4 kHz. It allows optimization of the energy flow into the inner ear, which otherwise would be reflected.

The middle ear operates normally when it is filled with air at atmospheric pressure. The Eustachian tube, which connects the middle ear cavity to the upper throat, normally opens and closes periodically, thereby insuring that the static pressure in the middle ear will remain the same as atmospheric pressure. We experience a pressure difference when the Eustachian tube fails to open during ascent or descent in an elevator. For an elevation of $8\ m$, the change in atmospheric pressure is $100\ Nm^{-2}$, corresponding to a sound pressure of $130\ dB$ relative to the $20\ \mu Nm^{-2}$ reference! This pressure causes a static deflection of the ear drum and increases the stiffness of the middle ear transmission system and sound transmission is attenuated.

B.1.3 Inner Ear (Cochlea)

The shape of the cochlea resembles that of a snail shell with two and one-half turns (in humans) and hence its name (Figure B.2). The central conical bony core of the cochlea is called the *modiolus*. The auditory nerve fibers run in this bone and exit the cochlea at its base. The outer wall of the modiolus forms the inner wall of a $30\ mm$ long canal which spirals the full two and one-half turns around the central core.

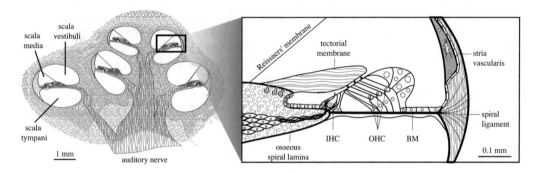

Fig. B.2. Section through the human cochlea (left) and magnified view of the organ of Corti (right). IHC: inner hair cell, OHC: outer hair cell, BM: basilar membrane. [1157], © Springer-Verlag Berlin Heidelberg 1999. With permission of Springer.

This canal is separated into three partitions called scales: Scala tympani is separated by the so-called cochlear partition, which is formed by a thin shelf of bone projecting from the modiolus (the ossesus spiral lamina), which is connected by the basilar membrane and the spiral ligament to the outer wall of the cochlea. The sensory organ of hearing, the organ of Corti, is located on top of the basilar membrane. As can be seen in the cross section of the cochlear spiral, the cochlear scalae become smaller and smaller in cross-sectional area as the apex is approached. Directly opposed, the basilar membrane becomes progressively wider towards the apex (Figure B.2). This is because the osseous spiral lamina is broadest at the cochlear base where the basilar membrane is only about $0.16\ mm$ wide (in humans); at the apex the basilar membrane has broadened to about $0.52\ mm$.

Scala media and scala vestibuli are separated by a thin membrane, called Reissner's membrane (Figure B.2). At the apical end of the cochlea, scala vestibuli and scala tympani are connected by an opening in the cochlear partition, called helicotrema. Scala tympani and scala vestibuli are filled with perilymph, which resembles in its chemical composition other extracellular fluids. Perilymph is characterized by high sodium (Na^+) concentration of about $140 - 150\ mM$ and its low potassium (K^+) content of only around $5\ mM$. Scala media is filled with endolymph, which is unlike any other extracellular fluid found in the body. From its chemical composition, it resembles intracellular fluids. Its predominant cation is potassium with a concentration of about $157\ mM$; sodium is very low ($1.3\ mM$). In addition to its special chemical composition, the endolymphatic space exhibits a considerable positive electrical potential within scala media of about $+80\ mV$ relative to scala tympani and scala vestibuli, called the endocochlear potential. The chemical composition and the electrical potential of the endolymphatic space is sustained by active ion transport provided by the cellular layers of stria vascularis.

The organ of Corti lies just between the endolymphatic and perilymphatic spaces (Figure B.2). Its surface is sealed by tight junctions to keep the fluids separate. The basilar membrane, on which the organ of Corti rests, is composed mainly of extracellular matrix material with embedded fibers. In contrast to the tight surface of the organ of Corti, the basilar membrane is thought to be permeable to perilymph. In the organ of Corti, two types of sensory cells, one row of inner hair cells and three to four rows of outer hair cells are embedded. In humans, there are approximately 3,500 inner- and 12,000 outer hair cells. The membrane potential of inner hair cells is about $-40\ mV$, of outer hair cells even as low as $-70\ mV$. Both types display "hair bundles" or stereocilia, which project into the endolymphatic space. The stereocilia are arranged in several rows, which are graded in size. Stereocilia from different rows are connected by a fine filament, called tip-link. The current theory of transduction assumes that hair bundle deflection pulls on the tip-links, opening transducer channels which are close to the attachment points. This concept also explains that the hair bundle is only sensitive to mechanical stimulation in the direction of the tallest stereocilia. The transducer channel is a nonselective cation channel which is very impermeable to anions. Therefore, the transduction current is mainly carried by potassium (K^+) and calcium (Ca^{2+}) cations, driven by the large electrical potential between the endolymphatic space and the receptor cells. The driving potential sums up to $120\ mV$ for inner hair cells and to $150\ mV$ for outer hair cells. Potassium leaves the sensory cells via potassium channels present in the basolateral cell membrane and diffuses into scala tympani. It is interesting to notice that the intracellular concentration of potassium is as high as that of endolymph. Potassium is driven into the cells by the electrical potential and because of the concentration gradient, it can diffuse into scala tympani without requiring energy from the sensory cells.

The inner and outer pillar cells, the phalangeal processes of the Deiters' cells and the cylindrical bodies of the outer hair cells build a complex, three-dimensional truss (see Figure B.3). One peculiarity is that outer

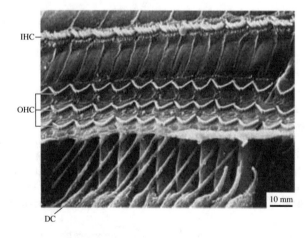

Fig. B.3. Scanning electron micrograph of the three-dimensional arrangement of the organ of Corti. IHC: inner hair cells, OHC: outer hair cells, DC: Deiters cells [301], with kind permisson of the authors..

hair cells are not in contact with other cells along their lateral surface but immersed in extracellular fluid. From Figure B.3 also the different morphology of the hair bundles becomes apparent: Whereas the bundles of the inner hair cells are arranged in a straight line, these of the three rows of outer hair cells are W-shaped.

The hair bundles of outer hair cells are excited by a shearing motion between the surface of the organ of Corti and the so-called tectorial membrane. The tectorial membrane is a gel-like structure composed of extracellular matrix material and it is in direct contact with the longest row of outer hair cell stereocilia. In contrast, the hair bundles of the inner hair cells seem not to be in direct contact with the tectorial membrane. Their bundles are probably driven by fluid forces. The transduction current flowing through the stereocilia is converted into a receptor potential in the cell body of the hair cell. At low frequencies, the receptor potential follows the stimulus cycle-by-cycle. Upon mechanical stimulation of the hair bundle in excitatory direction, tip-links are stretched and the transduction channels open. As positive K^+ ions are driven into the cell, the potential inside the cell becomes more positive. If the bundle is stimulated in the other direction, tip links relax and transduction channels close. However, as for inner hair cells at rest only 20% of transduction channels are open, the receptor potential is highly asymmetric when stimulated with a tone. On the basal pole of the inner hair cells about 10 to 30 afferent synapses are located, and upon depolarization of the cell membrane, voltage-sensitive Ca^{2+}-channels located in the basal pole of the cell membrane open, the increased Ca^{2+} level causes vesicles filled with transmitter to fuse with the lateral cell membrane and release transmitter into the synaptic cleft. This transmitter release triggers an action potential in the afferent nerve, and these electrical spikes transmitted to the brain finally lead to the hearing sensations.

These mechanisms work well at low frequencies, where the events of neural processing can easily follow the sound stimulus. Because of its refractory period, a single auditory nerve fiber can not respond to each successive cycle of a high-frequency sound. This problem can be partially overcome by the fact that more than one nerve fiber contacts a single inner hair cell. Each fiber is incapable of responding to every cycle of the stimulus, but collectively, they can do so. The temporal structure of the sound is conserved, as each fiber responds in the depolarized state of the inner hair cell. We call this behavior "phase locking". Even if the firing rate of a nerve fiber is too slow to follow the stimulus, basic features of its temporal structure—like the phase—are still coded in the neural pulse train. Still, the effect of phase locking is limited by several factors. The lateral membrane has a time constant of about 1 kHz, and above this frequency, the AC-amplitude of the receptor potential decreases. Also the synaptic processes like vesicle release and the generation of the postsynaptic potential are limited in speed and accuracy. This leads to a gradual loss of phase locking starting above 1 kHz, and above 3 kHz, phase locking is completely lost in humans.

In the frequency region, where the receptor potential can no longer follow the stimulus on a cycle-by-cycle basis, we see a depolarization of the inner hair cells membrane potential while stimulated. This so-called DC-component of the receptor potential follows the stimulus *envelope*. This is because of the asymmetry of the hair-bundle transduction. The depolarizing currents dominate upon sinusoidal stimulation and the receptor potential, low-pass filtered by the cell membrane, cause a depolarization of the cell membrane. All the information about the sound's frequency is lost in this signal, therefore, a different way to code high frequencies had to be developed by evolution.

This task was achieved by "sorting" sounds by frequency. Sound signals are mechanically preprocessed in a way that they are separated spatially. This concept is well-known as the frequency-place principle where high frequencies are located in the basal part of the cochlea only and low frequencies in the apical part. Different sets of auditory nerve fibers elicit different auditory sensations by virtue of their central connections. We will examine the mechanical frequency separation process in the sec Section B.1.4 and to do so, we will have to focus on the mechanical properties of the inner ear.

It is however not clear how specific excitations converge to yield well-defined pitch in general. There is increasing evidence that pitch is extracted in the time domain (periodicity analysis [580]). Moreover, pitch was found to be independent of the frequency-place mapping of the components of complex tones [581].

B.1.4 Cochlear Hydrodynamics: The Travelling Wave

The cochlea consists of three fluid-filled scalae, but from a mechanical point of view, the elastic properties of the thin Reissner membrane can be neglected compared to the stiffness of the basilar membrane. We can therefore simplify our mechanical investigations to a fluid model with two chambers separated by the cochlear partition (compare Figure B.4). The basilar membrane contributes a large part of the elastic properties of the partition. Its width is increasing from base to apex from about 0.16 μm to 0.52 μm (in humans). Its thickness, on the other hand, decreases along the cochlea. Thus, its stiffness decreases greatly along the length of the cochlea.

At the basal end of the cochlea there are two openings to the cochlear ducts, one on each side of the cochlear partition, that are covered by membranes. One is called the oval window and, as we already mentioned, it is in contact with the stapes foot plate. The other, the round window, is just below the oval window. Because we can assume the cochlear fluids as incompressible, the round window has to move out of phase if the oval window is driven by stapes motion.

If we consider very slow stapes motions, the fluid is pushed along the entire length of scala vestibuli, through the helicotrema and back along scala tympani. Thus, the helicotrema provides a low-frequency shunt for extremely low frequencies. When the stapes moves into the cochlea, pressure builds up in the fluid, which deflects the cochlear partition. Pushing fluid requires overcoming inertial forces generated by the fluid mass. The elastic properties of the basilar membrane in combination with the mass of the surrounding fluid constitutes second-order resonators. As the stiffness of the basilar membrane is very high at the basal end of the cochlea and much lower at its apical extreme, the resonant frequency of the cochlear partition monotonically decreases from base to apex. Inversely, the time constant of each resonator increases from base to apex. However, the basilar membrane is not under tension, and it does not respond like a series of *independent* resonators, like the strings of a harp. Instead, each part of the cochlear partition is coupled to the next by the cochlear fluids, and due to the large inherent friction, they are highly damped. For a periodical motion of the stapes, the cochlear partition is first set into motion at the basal extreme, where the mechanical time-constant is smallest. Because the stiffness is very high, the deflection of the basilar membrane is fairly small. The deflection propagates in the form of a wave in apical direction. As the time-constants of the partition increase and the stiffness decreases, the response will be more and more delayed but its amplitude will grow. At the location of resonance, the wave will reach its maximum and lose its energy, its amplitude will drop very rapidly. The location of "cochlear resonance", the place where the displacement—and therefore the excitation—of the cochlear partition reaches its maximum, depends on the stimulus frequency. Low frequencies will travel along the basilar membrane and reach a maximum close to the cochlear apex, high frequency sounds will exhibit their maximum response close to the cochlea base and fade out.

The exact form of the vibration response of the cochlear partition was investigated by Georg von Békésy in human cadaver ears [97], a feat which earned him the Nobel Prize. He found that the deformation of the basilar membrane is a traveling wave. The wave starts at the cochlear base, where the basilar membrane is stiffest. It propagates toward the apex with a time delay that depends upon its own mechanical properties and the properties of the surrounding fluid. Its vibration amplitude is increasing until it reaches a maximum, close to the location of cochlear resonance, and from then on, the wave diminishes rapidly. Because the stiffness gradient of the basilar membrane is approximately logarithmic, the peaks of the excitation patterns of sinoidal tones are located on the basilar membrane with a logarithmic frequency spacing. Figure B.4 schematically illustrates traveling waves elicited by a stimulus composed of three frequencies, together with the envelope of the peak displacement. The peaks of the three waves are clearly separated along the cochlea, however, there is also considerable overlap between the waves and recordings from the auditory nerve have been found to be much more frequency selective, especially close to threshold, than the mechanical responses observed in cadaver ears. Therefore, a second mechanism is required to boost the frequency selectivity of the vibration responses.

B.1.5 Active Amplification of the Traveling Wave Motion

We have so far neglected the function of the outer hair cells, the second group of receptor cells within the organ of Corti. Despite the at about three to four times higher number of outer hair cells, only 5% of the afferent fibers innervate outer hair cells. The fibers are so-called type II fibers, they are highly branched and each fiber innervates dozens of outer hair cells. Little is known about these fibers because they are small and unmyelinated, making it difficult to record their activity. They are expected to be less sharply tuned, since they innervate a broad region of the cochlea, and, because of their lack of myelination, their conduction

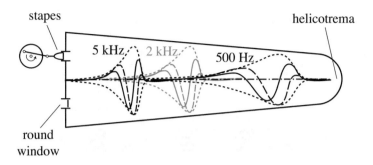

Fig. B.4. Schematic illustration of the traveling waves elicited by three pure tones with frequencies of 500 Hz, 2 kHz and 5 kHz. The displacement of the basilar membrane is shown at the instant T_0 (solid line) and a quarter of a cycle later (dashed line). The dotted line indicates the envelope of the wave.

velocity is likely to be very slow. Most of the frequency selective and time-critical auditory information must therefore be carried by the afferent fibers originating from the inner hair cells. Whereas inner hair cells are not directly innervated by the efferent system, myelinated fibers from the medial to the medial superior olivary nuclei make direct synaptic contact with the outer hair cells. Outer hair cells seem to be under neural control, much more like muscles than sensory cells. A final indication of their "active" role was established when their ability to change their length upon electrical stimulation was detected. Modern research assumes that the outer hair cells indeed provide an active, mechanical amplification of the cochlear traveling wave. As neural processing is by far too slow to keep up with high frequency hearing, current concepts assume that amplification relies on a local mechanical feed-back process [239]:

> Outer hair cells sense motions of the cochlear partition by converting shearing motion between the surface of the organ of Corti and the tectorial membrane into an electrical receptor potential. Upon depolarization, the outer hair cell reacts with a contractile force, which is fed back into the motion of the basilar membrane.

From theoretical calculations derived from measurements in the inner ear, it is required that energy is pumped into the vibration of the cochlear partition in a region starting before the traveling wave reaches its maximum up to the place of cochlear resonance. The function of this amplification process is still unclear in its details, but from the observations of active cochlear mechanics it has wide reaching consequences. Measurements in the basal part of the cochlea indicate, that the amplification boosts cochlear sensitivity up to thousand-fold. The amplification is limited to a narrow frequency range, covering only about half an octave. The active traveling-wave response becomes very sharp in the region of cochlear resonance. The amplification of the vibration response is required to achieve the extraordinary sensitivity of the hearing system.

The second hallmark of amplification is non-linearity. The amplification process boosts only weak sounds, it saturates at increasing levels. This non-linearity greatly compresses the dynamic range of the mechanical responses. This is important, because the dynamic range of the inner hair cell receptor potential is limited to a range certainly not exceeding 60 dB. This non-linearity also has unwanted side-effects: If two sinoidal tones are presented, the non-linearities of the hearing-organ generate so-called distortion products, additional tones which we perceive under certain circumstances. In general, however, by virtue of its construction, artifacts due to the non-linearities of the inner ear are surprisingly small.

Figure B.5 shows the excitation pattern expected on the human basilar membrane when stimulated with a 3 kHz sinoidal tone. For high sound levels (i.e., 100 dB), the feed-back amplifier is saturated and the traveling-wave is almost purely passive. It is highly damped and its envelope shows the characteristic shallow

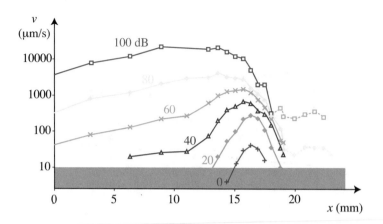

Fig. B.5. Reconstruction of the excitation pattern in the human cochlea for a 3 kHz sinoidal tone. Original measurements were recorded in a chinchilla cochlea at a distance of 3.5 mm from its most basal extreme [781]. The characteristic frequency of this location was 9.5 kHz. Data has been converted assuming a frequency-place map of 8 mm/octave to illustrate the excitation pattern in a human cochlea. The shaded area indicates excitation below neural threshold, which is expected between 50 $\mu msec^{-1}$ and 1 nm.

increase from base to apex and a sharp decay after the maximum is reached.[1] The threshold of the auditory nerve is somewhere between a minimal basilar-membrane velocity of 50 $\mu msec^{-1}$ and a displacement of 1 nm, exact values are still unknown. The broad traveling wave at high levels indicates that a large number of nerve fibers are stimulated, especially in the basal part of the cochlea. For faint sounds, the traveling-wave response becomes sharper and sharper and its envelope is almost symmetrical for levels below 40 dB. The location of maximum amplitude for a sinoidal tone at low levels is called its characteristic place. Only nerve fibers originating from a very narrow region around the characteristic place of the cochlea are stimulated. Note that for increasing levels, the maximum of the traveling wave shifts considerably in the basal direction. If we analyze the level-dependence of the responses at various locations of the cochlea, we clearly see the effects of the non-linearities of the amplification. At the characteristic place (about 16 mm in Figure B.5), the velocity amplitude increases from a value of 42 $\mu msec^{-1}$ at 0 dB to 5 $msec^{-1}$ at 100 dB. Without amplification, the amplitude of the traveling-wave response would be expected to drop by a factor of 10^5, or from 5 $mmsec^{-1}$ to 50 $nmsec^{-1}$ from 100 dB to 0 dB! The amplification therefore is almost a factor of 1000 or 60 dB. In addition to enabling the detection of weak signals, the amplification therefore also compresses the dynamic range, again by almost 60 dB. The inner hair cells, at the location of the characteristic place, have to cope only with a stimulus ratio of a little bit more than 40 dB to cover a 100 dB-level change of the sound stimulus.

B.1.6 Neural Processing

The difficulty to understand neural processing of sound strongly stems from the extremely complex innervation from the auditory nerve to primary auditory cortex (Heschl's gyri) in the temporal lobe (see Figure B.6). We have to stress that the image in this figure is considerably simplified, and that in particular, there are also connections from the cochlea to the ipsilateral auditory cortex, as well as efferent nerves from the auditory cortex down to the cochlea.

---

[1] The data in Figure B.5 (dashed lines) indicates that further towards the apex, the wave does not die out completely in this experiment. It is still under debate, whether this remaining response is also present in the intact human cochlea.

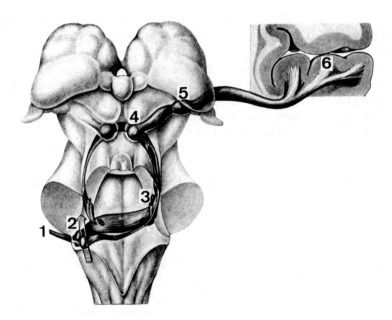

Fig. B.6. (Private archive H. G. Wieser and G. Mazzola) This simplified image shows the six relevant relais stations of the auditory path from the cochlea to the Heschl gyri: (1) Nervus cochlearis, (2) nucleus cochlearis, (3) nuclei superiores olivae, (4) colliculus inferior, (5) corpus geniculatum mediale, (6) gyri Heschl.

The auditory system includes at least five "relays stations", whence it is clear that any particular functional decomposition (like Fourier's) will not be transferred unchanged to the auditory cortex. For example, the tonotopy of the latter is a multiply distributed one, see Figure B.7.

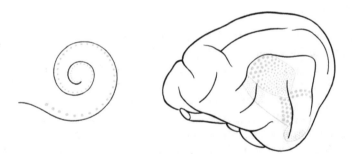

Fig. B.7. (With kind permission of the Thieme-Verlag, [516]) Left: tonotopy in the spiral of the cochlea: high pitches are thick points, low pitches are thin points. Right: Corresponding multiple tonotopy in the auditory cortex of the cat.

By tonotopy, one understands the spatial distribution of excitation patterns according to specific pitch. More recent research also demonstrates that outside the auditory cortex, i.e., in the limbic system (the hippocampal formation, to be precise, which plays an important role for emotional and memory tasks), one finds a refined processing of pitch information [678, 679, 1122, 1123].

The "template fitting model" of Julius Goldstein [378] shows how far we are from understanding the neural pitch processing. In this model, the mathematical principle of a neural "central pitch detector" is proposed, from which the fundamental frequency of a periodic wave can be extracted if its Fourier components are known. The central pitch detector is however charged with the solution of local minima problems for functions in two variables—a rather heroic task for a small neural population. It seems hopeless to identify within the neural network the physical realization of a dynamic system that solves these differential conditions. We presently have no chance to verify the model physiologically, since the human ethics excludes adequate experiments in humans.

In view of these facts, it is not only logically erroneous and experimentally very delicate to infer the higher sound processing from the superficial auditory physiology of the ears. And ethically, such an attempt is problematic since one runs the risk to "justify" discrimination of "degenerate music" against the acceptance of so-called "commonly accepted" music. It is also not clear how much such investigations reveal grown and trained configurations instead of biological inheritance.

B.2 Discriminating Tones: Werner Meyer-Eppler's Valence Theory

From our everyday experience and from specific experiments it follows that we do not really hear the single tone events or chords, motives or rhythms as they have been parametrized on the level of physical or mathematical description. In fact, despite the very sensitive physiology of Corti's organ and its hair cells, we cannot distinguish all physically or mathematically distinct sound objects. For instance, sounds with frequency above $20,000\ Hz$ or below $0\ dB$ are indistinguishable since you cannot hear them. Or two sounds p, q with a phase shift, i.e., $p(t + \Delta) = q(t)$ are indistinguishable. More important is that even under ideal condition we cannot distinguish sounds with arbitrary precision. Every singer or violinist who has to adapt his/her pitch to a prescribed context knows this. Investigations on variations of instrumental intonation show a remarkable bandwidth [117]. The same is valid for listening to time values, loudness degrees and instrumental colors.

It is mandatory but not easy to take into account these phenomena. Werner Meyer-Eppler [735, 736] has attempted to solve the problem by use of the concept of a "valence". According to this approach,

Definition 127 *In a specific context, given two sound objects s_1, s_2, s_1 is* metamere *to s_2 with regard to a given predicate P (short: P-metamere, in symbols: $s_1 \sim_P s_2$) iff the s_1 cannot be distinguished from s_2 with respect to P by human listeners. The set $\sim_P s$ of sound objects which are P-metamere to a given sound s are called the* valence *of s.*

Since usually the relation $\sim_P$ is symmetric, one calls two sound objects s_1, s_2 metamere if $s_1 \sim_P s_2$. If a sound object s is defined by a sequence $P_1, \ldots P_k$ of predicates, one defines its valence as being the sequence of valences $\sim_{P_1}, \ldots \sim_{P_k}$, and we may then reduce the total valence to those components which do not include all sound objects since these predicates are not relevant to the distinction of sound objects. The union $\bigcup_{\text{supporting valences}} \sim_{P_{i_j}}$ of supporting valences is in fact the valence of the conjunction predicate $\bigwedge_i P_i$. Predicates which are relevant in this sense are called *valence supporting* by Meyer-Eppler.

In practice, if we are given a sound whose sound color is described by partials with frequencies up to the limit frequency $f_0\ Hz$, and which last $d\ sec$, Meyer-Eppler deduces a maximal number of numerical predicates (dimensions) that are relevant to the valence, i.e., of valence supporting numerical predicates. This limit is called the *maximal structure content*, and its value is $K = 2.d.f_0$. It is however an open fundamental question of sound color theory, which and how many valence supporting predicates must be chosen in order to yield a differentiated perception of sound colors. Probably, these valences define a multiply connected topological space in the physical parameter space, but see also Section 12.3.

Although Meyer-Eppler's conceptualization is plausible, it hides two delicate problems. The first concerns the context where valences take place, as related to the predicate in question. If the context's specification is neglected, the valence concept loses its meaning. Let us make two representative examples concerning pitch and onset. To begin with, we have to agree on who is accepted as a listener. In the sense of a statistical approach, Meyer-Eppler proposes that two pitches should be called metamere if at least 90% of the test subjects cannot distinguish them [736]. Moreover, one has to agree on the parameter(relations) of the test sounds. The result will depend essentially on the choice of instruments, flute, brass etc., and conditions upon the duration and the onset distances of test sounds. For example, the simultaneous presentation of sounds of several seconds duration will yield smaller valences because of beat effects, as compared to a comparison of non-overlapping sound events.

Besides the parametric conditions, the context can also depend on the chosen music. Let us discuss this on the onset parameter, for example. In [736], the duration valence density is indicated by $50 - 60$ per second. This means that sounds which are less than $1/50 - 1/60$ *sec* apart are perceived as being simultaneous. However, the musical context of such a claim is relevant. Within a very slow piece with a small number of instruments, the temporal variation of $1/8$ *sec* will scarcely be noticed, whereas in a rhythmically very dense and fast piece, $1/60$ *sec* is known to define a quite coarse grid.

The second problem is important for the theoretical significance of the valence concept. It relates to the fact that metamery is not an equivalence relation[2] in general. More precisely, any of the wanted properties: reflexivity, symmetry, and transitivity, can be violated. Reflexivity can be violated if the comparison of two sounds is temporally so separated, by several hours, say, that the human memory fails to recognize one and the same sound. Symmetry is violated if the order of appearance of sound objects is relevant, for example, a very loud sound, immediately followed by a very soft one, can mask the latter's properties. The most dramatic failure is the absence of transitivity: $s \sim_P t, t \sim_P u$ does not always imply $s \sim_P u$. For example, if three pitches are such that the first is perceived as being equal to the second, and the second being equal to the third pitch, this is not entailed by equality of first and third pitch! Therefore, pitch valences are not equivalence classes, they may overlap. This means that the attempt to define pitch by an esthesic position in music psychology must fail. The perceptional concept of pitch is a non-transitive relation among tones, and therefore is not an attribute of tones. You hear that two tones have the same pitch, but you do not hear the pitch. Therefore corresponding attempts such as [784] must fail.

This has important consequences for musical practice and for theoretical aspects. In practice it is desirable to select grids of sounds such that their valences would not overlap. This needn't be a grid which is fixed once for ever, it might be a time-dependent construction, but it must yield a locally disjoint valence set.

With the common notation, such a grid is realized as an orientation device as well as an acoustic and performative scheme. The continuum of onsets, durations, pitches, and sound pressure values is quantized in the well-known way such that Boulez' "notched tone space" can be taken as a grid behind the *realiter* played or heard.

The reality layer of a grid is a mental or psychological one. The blurredness of hearing, as it is expressed in valences, has to be subjected to a cognitive interpretation. Semiologically speaking, the valence is the significant, the expression for a meaning which relates to our understanding of music. By use of the grid which is superposed to the valence perception, it is possible to associate the valences of perceived sound to objects of our imagination. The psychological quality of pitch then results from this mapping as a grid object in our imagination. The semiotic power of this signification depends upon the definition of the actual grid.

A music-theoretically fundamental grid is the selection of the pitch arsenal, the tuning, wherein tones may be played. Let us first look at the chromatic w-tempered scales (see Section 7.2.1.3). If w is not too large, valences of neighboring tones can be separated. From our experience with microtonal music, $w = 36$, i.e., tempered sixth-tone intervals, is not too large. So the postulate of an adjustment of pitch by inner grids is acceptable for w-tempered scales with $w \leqslant 36$.

The situation of just intonation (see Section 7.2.1.4 and Appendix A.2.3) is much more delicate. For each pitch $H(x_0)$ of a point x_0 in the Euler grid $\mathbb{Z}^3$ of just intonation, and for any positive real number ϵ, there are infinitely many points x such that $|H(x - x_0)| < \epsilon$, see Proposition 76 in Appendix A.2.3. In particular, infinitely many pitches fall into the valence of x_0. *In the valence semiology, just intonation is infinitely homonymic.* This problem can be solved by a restriction of the context, where just intonation music is played. As soon as the local context is a small region of the Euler space, for example, a small neighborhood of a determined finite portion of the chromatic scale, valences can be used to distinguish just grid points.

A second difficulty for just intonation relates to the perception of pitch differences. A classical argument of just music theory is Euler's substitution theory [308], according to which intervals are heard in a way such that the frequency relations of their tones form fractions a/b (in reduced representation) with minimal

---

[2] See Appendix C.2.

numerator a and denominator b. This is the basis of the classical consonance-dissonance theory which we address in the following section. In the grid of just intonation this means that the pitch difference of the interval is corrected/adapted to an interval which is realized by two points of the just intonation grid under the constraint that their (Euclidean or 1-norm) distance is minimal. Of course, this correction has to happen within a valence in order to be a reasonable process of human hearing. However, it is easy to see that in general, there are several solutions $a/b, a'/b', a''/b'', \ldots$ with minimal distance. So Euler's substitution theory would have to impose a contextual restriction for single tones as well as for intervals.

These considerations should be taken for nothing more than they are: esthesic aspects of hearing. Nothing prevents us from doing music theory on neutral and poietic layers without bothering about valences. But then, one has to be conscious of the fact that highly differentiated mathematical structures may be blurred by the semiology of valences from auditory psychology and physiology.

B.3 Symbolic, Physiological, and Psychological Aspects of Consonance and Dissonance

For a long time, mathematical reflections in music were centered around the problematic concept couple of consonance and dissonance. This is based on the ancient Greek Pythagorean tradition where consonance and dissonance of intervals was laid in the involved frequency relation. Perfect consonant intervals corresponded to ratios like 2/1 for the octave, 3/2 for the fifth, and 4/3 for the fourth. This simple arithmetic corresponded to the philosophy of the metaphysical tetractys. See [672, 784] for a historical discussion of these roots.

Here, we simply want to recall that a unified mathematical foundation of musical thinking in the paradigm of simple consonant frequency ratios could not survive the differentiated development of theories in the contrapuntal setting [924], the psychological foundation of musical relations as introduced by René Descartes [262], and the discovery of physical partials by Marin Mersenne [630].

According to these more recent positions, the problem of consonances and dissonances changes as a function of the layer of reality where it is investigated—and on each layer it is not a minor one. In the present shorthand presentation, this result seems to be a provocative one since the conceptual unity seems violated. We want to make clear that we really are dealing with three different meanings of the sonance concept—Euler's gradus suavitatis on the mental layer, Helmholtz' beat model on the physical layer, and Plomp-Levelt's psychometrics on the psychological layer. In itself, each of these approaches is consistent, the problem only arises if one attempts to reduce one reality to another one. Following the knowledge about the neural processing of sounds (see the previous discussion in this chapter), it is hardly astonishing that psychological and physiological layers are not congruent: what the ear (in Helmholtz' model) does not "like" can very well be "agreeable" for the limbic system or the auditory cortex.

B.3.1 Euler's Gradus Function

Being a number theorist, Euler was interested in prime numbers. A priori, his gradus function Γ [307] is a purely number-theoretic function, it is defined as follows: According to the prime factorization of integers (see Appendix D.2), a positive integer a is the unique product $a = p_1^{e_1}.p_2^{e_2}.\ldots.p_n^{e_n}$ of positive powers of primes $p_1 < p_2 < \ldots p_n$ (with the singular case of zero factors and a product $= 1$ for $a = 1$). Euler's formula, the *gradus suavitatis*, is $\Gamma(a) = 1 + \sum_{1 \leqslant k \leqslant n} e_k(p_k - 1)$, and more generally $\Gamma(x/y) = \Gamma(x.y)$ for a reduced fraction x/y.

In just intonation—where Euler's approach belongs—only intervals of tones are considered whose frequency ratio is a positive rational number x/y. The gradus function is defined for such intervals. Each interval evaluates to a positive integer. The frequency ratios with Γ-values $\leqslant 10$ are listed in table Appendix L, see also Figure B.8.

In just intonation, intervals can be read as differences $\Delta = x - y = (e, f, g)$ of Euler points x and y in the grid $\mathbb{Z}^3$. For the gradus function, we may restrict to non-negative coordinates e, f, g. We then have

$$\Gamma(\Delta) = \Gamma(2^e.3^f.5^g) = 1 = 1 + (2-1)e + (3-1)f + (5-1)g \tag{B.1}$$

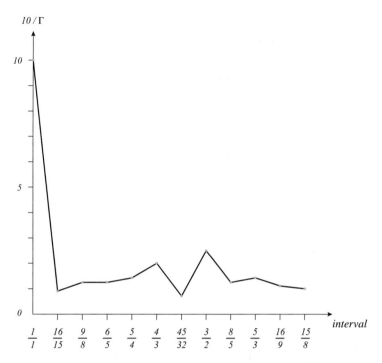

Fig. B.8. The agreeableness of intervals within the octave in just intonation according to Vogel's chromatic [1089], see Section 7.2.1.4, is represented. It is reasonable to represent the reciprocal values $10/\Gamma$ instead of Γ since the latter is rather a 'gradus dissuavitatis': small Γ values are taken for the octave and the fifth, large ones for the second and tritone. The factor 10 is only a scaling constant. The order of Euler's valuation is this: Prime, fifth, fourth, major third/major sixth, minor third/minor sixth/major second, minor seventh, major seventh, minor second, tritone.

or as a scalar product

$$\Gamma(\Delta) = 1 + (\Phi, \Delta) \text{ with } \Phi = (1, 2, 4).\tag{B.2}$$

If we compare this formula with the pitch formula

$$H(\Delta) = (H_{prime}, \Delta)$$

where H_{prime} is the prime vector from Section A.2.3, then we observe a similar construction. The gradus function is something like a pitch function, but the 'direction' is Φ instead of H_{prime}, see Figure B.9. The ranking of intervals by the gradus function was already criticized by Euler's contemporaries Mattheson, Mitzler, and Rameau [161]. But it is the merit of Euler to have defined a ranking by a linear expression which, together with pitch and octave coordinate defines a coordinate system for the Euler space on the one hand, and considers the consonances and dissonance, on the other. The disadvantage of Euler's approach is that it is based on the valence-theoretically invalid substitution hypothesis (see Section B.2).

B.3.2 von Helmholtz' Beat Model

Hermann von Helmholtz proceeds from the hypothesis that beats between partials of two tones is responsible for sonance phenomena. The fact that he uses partials is bound to Ohm's postulate that we have a cochlear Fourier analysis [457]. Hence Helmholtz' approach only regards the cochlear basis of music perception and not the higher limbic and cortical auditory processing. A *beat* is the periodic amplitude variation which results from the superposition of two sinoidal waves which have a frequency difference $\Delta = f - g$, the *beat frequency*[3] which is small with respect to their frequencies f, g.

---

[3] One uses the trigonometric equation $\sin(x) + \sin(y) = 2\sin(\frac{x+y}{2})\cos(\frac{x-y}{2})$, see Figure B.10.

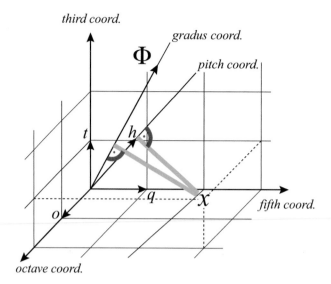

Fig. B.9. The gradus function is something like a pitch function, but the 'direction' is Φ instead of H_{prime}, Both values are obtained from a scalar product of an Euler point x (of a difference point Δ, respectively) with H_{prime} (with Φ, respectively). However, the point is not uniquely determined by the 'gradus coordinate', in contrast to the 'pitch coordinate'.

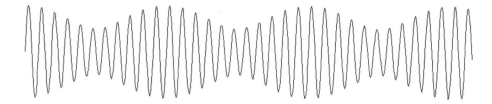

Fig. B.10. A beat between two superposed sinoidal waves is characterized by a periodic amplitude variation whose frequency is twice (!) the difference of the given frequencies.

Helmholtz calculates the *roughness*, i.e., the degree of dissonance of an interval which consists of two tones p, q as the sum of the *beat intensities* $I_{n,m}$, which are associated with the nth partial of p and the mth partial of q, where $I_{n,m}$ is supposed to have a strong maximum for beat frequency $\Delta_{n,m} = 33\ Hz$.

Therefore Helmholtz' dissonance concept depends on the pitches and the involved sound colors. On the example of the violin, Helmholtz obtained good coincidences with Euler's gradus function. This model is impressive since it explains the experience according to which consonance is a function of the instrument and the absolute pitch of the interval tones. Its experimental verification is somewhat problematic. It essentially depends on the measurability of the beat intensities in the cochlea. Since non-linear distortions on the sound's way to the cochlea change spectra, one would have to perform ethically problematic invasive cochlear measurements. Moreover, individual statistical variations of the non-linear distortions would make the experiments even less robust.

A fundamental doubt on the model's validity results from binaural experiments in hearing by Heinrich Husmann [494], where the interval tones are presented on separate left and right headphone inputs. In this case, no beats can intervene in the cochlea. Nonetheless, the experiments also revealed consonance as "happy moments of within the general (interval) disaster". The hypothesis that in the binaural experiments,

Helmholtz beats must occur in the relays station of the corpus geniculatum mediale (see Figure B.6) is speculative and demonstrates the limit of physiological models.

B.3.3 Psychometric Investigations by Plomp and Levelt

In the psychological reality of interval perception, the judgment of interval qualities looks quite different. In their investigation of "pleasantness" of intervals, Reiner Plomp and Wilhelm Levelt [850] have presented pairs of sinoidal tones and asked for pleasantness as a function of the given interval. The experiment was intendedly performed with musically untrained individuals in order to avoid judgments as a function of musical knowledge. Figure B.11 shows the resulting valuation curve. It is quite different from Euler's function as shown in Figure B.9.

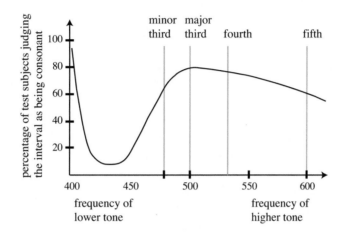

Fig. B.11. (From [914], with permission of Springer-Verlag) The psychometric investigation [850] of Plomp and Levelt yields a valuation which differs significantly from Euler's Γ function and is based on sinoidal tones shown in Figure B.9.

Using this curve, Plomp and Levelt have tried to infer a description of Helmholtz' beat intensities, a procedure which was already recognized as being problematic in Appendix B.3.2.

B.3.4 Counterpoint

There is however another—quite remarkable—point of view of the consonance-dissonance phenomenon, which has been poorly recognized within the psychoacoustic discussion, namely the prominent meaning of the concept pairing in the contrapuntal tradition, which was elaborated in the High and Late and Middle Ages, and which was encoded in an exemplary way by Johann Joseph Fux' *Gradus ad Parnassum* [349]. Carl Dahlhaus [230] has rightly pointed out that the textural function of the contrapuntal consonance concept is not yet fully understood. Interestingly, in the framework of the core theory of counterpoint, the interval of the fourth is dissonant, in contradiction to the other theories. It is inconceivable that the mathematical and physiological theories of counterpoint never included this perspective. A mathematical model of counterpoint is discussed in Chapters 29 through 31.

B.3.5 Consonance and Dissonance: A Conceptual Field

Even on the mental level, the concept of consonance and dissonance is multiply explained, without necessarily leading to contradictions. In fact, Euler's approach was a neutral mental (number-theoretic) one, whereas the contrapuntal approach is poietic. If we look at the status quo of the consonance-dissonance discussion and the fight for a valid final semantics, we are confronted with a disaster. The mathematically arbitrary ornaments which composers such as Klarenz Barlow [82] or neo-Pythagoreans such as Martin Vogel [1091] add to Euler's formula does not interest psychoacousticians such as Ernst Terhardt [1050], who would extrapolate cochlear findings into the auditory cortex, and the latter approach cannot shed light onto the esthetics of music.

What is common to all these positions adds to a concept of consonance and dissonance which is a conceptual field within the topography of music, a field with a quite ubiquitous presence. As a musical thought it results from a fundamental linear dynamic between polar extremals. It should be a main task of mathematical music theory to elaborate reliable and semantically reasonable, but esthetically undogmatic, models for such a way of thinking music.

Part XXI

Appendix: Mathematical Basics

C

Sets, Relations, Monoids, Groups

C.1 Sets

The language of sets describes mathematical facts in a classical way. An alternative foundation to sets is the language of categories, see Appendix G.

A *set* M is an object which is defined as a collection of uniquely determined objects which are also sets. These objects are called the *elements* or *points* of M. Two sets are equal iff they have the same elements. Whenever we say that M is a set, we mean that it is defined in a consistent way, i.e., without causing any contradiction. Existence of a mathematical object means that the object's definition causes no contradiction in classical logic (A is identical to A, (exclusive) either A or non A, and exclusion of a third). Then, for any set m it is either an element of M or it is not. One writes $m \in M$ for "m is an element of M", or also "m is a point of M". In order to define a set M by its elements $m, m' \ldots$, one also writes $M = \{m, m', \ldots\}$. Observe that multiple enumeration does not change the set, for example $\{x, x, x, y\} = \{x, y\}$.

A set N whose elements are all elements of M is called a subset of M, in signs: $N \subseteq M$, also sometimes $N \subset M$ iff $N \subseteq M$ and $N \neq M$. Two sets are equal iff they are mutually subsets of each other. The *empty set* $\varnothing$ is defined as having no elements. It is a subset of any set. A set is called *finite* if is empty or its elements can be indexed[1] by a sequence[2] $0, 1, 2, 3, \ldots n$ of natural numbers. Otherwise it is called *infinite*.

C.1.1 Examples of Sets

Example 92 $\mathbb{Z} = \{0, \pm 1, \pm 2, \ldots \pm n, \ldots\}$, the set of *integers*; $\mathbb{N} = \{n \in \mathbb{Z}, n \geqslant 0\}$, the set of *natural numbers*; the set $\mathbb{Q} = \{p/q, p, q \in \mathbb{Z}, q \neq 0\}$ of *rational numbers*; the set $\mathbb{R}$ of *decimal or real numbers*, e.g., $x = -741.76, \pi = 3.1415926\ldots$. We have $\mathbb{N} \subset \mathbb{Z} \subset \mathbb{Q} \subset \mathbb{R}$, where integers p are identified with rational numbers of form $p/1$, whereas rational numbers are identified with periodic real numbers.

Example 93 If M, N are sets, their *difference* $M - N$ or *the complement of N in M* is the set of points of M which are not in N. If $V = (M_i)_i$ is a family of sets, we denote by $\bigcup V$ the *union of V*, whose elements

---

[1] A mathematically correct definition of finiteness is this: A set if finite iff it is not in bijection with any proper subset.

[2] Recall that in this book, we make the logical, though not very common usage of the ellipsis symbol "$\ldots$": it means that one has started with a sequence of symbol combinations which follows an evident law, such as $1, 2, \ldots n$, or $a_1 + a_2 + \ldots a_n$. The evidence is built upon the starting unit, such as "1," or "a_1+" in our examples, and then the following unit, such as "2," or "a_2+", and then inducing the following units to be denoted, such as "3," or "a_3+", "4," or "a_4+", etc., until the sequence is terminated by the last symbol, such as "n" or "a_n" in our examples. The ellipsis means that the building law is repeated, and as such, it is a meta-sign referring to the inductive offset. Therefore the more common notation $1, 2, \ldots, n$, or $a_1 + a_2 + \ldots + a_n$ is not correct. In the limit, for $n = 3$, it would imply a notation such as $1, 2, , 3$ or $a_1 + a_2 + + a_3$, which is nonsense. Moreover, in complicated indexing situation, the common notation would be overloaded.

© Springer International Publishing AG, part of Springer Nature 2017
G. Mazzola, *The Topos of Music IV: Roots*, Computational Music Science,
https://doi.org/10.1007/978-3-319-64495-0_3

are precisely the elements collected from any of the M_i, for finite families $V = M_1, \ldots M_n$, one also writes $\bigcup V = M_1 \cup \ldots M_n$. In particular, $\bigcup \emptyset = \emptyset$. A covering of a set X is a family V such that $\bigcup V = X$. The *intersection of a family* V is the set $\bigcap V$ consisting exactly of those points which are points in any of the family's member M_i. For finite families $V = M_1, \ldots M_n$, one also writes $\bigcap V = M_1 \cap \ldots M_n$. In particular, $\bigcap \emptyset = AllSet$, the set whose elements are all existing sets. In less universal contexts, one only takes the intersection with regard to a large superset of the family's members. Observe that there is no reason why *AllSet* should not exist, however, not any of its subcollections defined by predicates equally exists. For example, the subset of all sets not containing themselves as elements does not.

A *partition of a non-empty set* X is a covering such that any two of its members are non-empty and *disjoint*, i.e., intersect in the empty set.

C.2 Relations

Definition 128 *If x, y are two sets, the* ordered pair (x, y) *is defined to be the set*

$$(x, y) = \begin{cases} \{\{x\}\} & \text{if } x = y, \\ \{\{x\}, \{x, y\}\} & \text{else.} \end{cases} \tag{C.1}$$

Lemma 66 *For any four sets a, b, c, d, we have $(a, b) = (c, d)$ iff $a = c, b = d$.*

A triple (a, b, c) is a pair of form $((a, b), c)$. Clearly, $(a, b, c) = (a', b', c')$ iff $a = a', b = b', c = c'$.

Definition 129 *Given two sets A, B their* Cartesian product *is the set*

$$A \times B = \{(a, b) | a \in A, b \in B\}$$

consisting of all ordered pairs (a, b), with the first *coordinate a an element in the* first *factor A, and the* second *coordinate b an element in the* second *factor B. A* relation from A to B *is a subset $R \subseteq A \times B$. One writes aRb for $(a, b) \in R$, and, if the relation is clear, more simply $a \sim b$. The* inverse *of a relation is the set $R^{-1} = \{(b, a) | aRb\}$. If $A = B$, one also speaks of a relation on A.*

Definition 130 *A* graph $f : A \to B$ *from A to B is a triple (A, B, f) with f a relation from A to B; it is called*

(*i*) total *iff every $a \in A$ is the first factor of a pair in f;*
(*ii*) functional *iff $(a, b), (a, b') \in f$ implies $b = b'$; one writes $f(a)$ instead of the uniquely determined b, or also $f : a \mapsto b$.*
(*iii*) *a* function *or* map *iff it is a total functional graph. The set A is the* domain *of f, whereas B is the function's* codomain.

If $f : A \to B$ and $g : B \to C$ are two functions, their *composition* or *concatenation* function $g \circ f : A \to C$ is defined by $g \circ f(a) = g(f(a))$. The function $f : A \to A$ with $f(a) = a$ for all $a \in A$ is called the identity on A and is denoted by Id_A.

Exercise 83 *Verify that $g \circ f$ is indeed a function. If $h : C \to D$ is a third function, we have $(h \circ g) \circ f = h \circ (g \circ f)$ and therefore we write $h \circ g \circ f$. Show that $Id_B \circ f = f \circ Id_A = f$.*

Definition 131 *A function f is called*

(*i*) injective *if $f(a) = f(a')$ always implies $a = a'$;*
(*ii*) surjective, *iff for every $b \in B$, there is $a \in A$ such that $f(a) = b$;*
(*iii*) bijective, *iff f is injective and surjective.*

Lemma 67 *For a function $f : A \to B$, the following conditions are equivalent:*

(i) *There is a function $g : B \to A$ such that $g \circ f = Id_A$ and $f \circ g = Id_B$.*
(ii) *The function f is bijective.*

The g in this lemma is uniquely determined by f and is called the *inverse function of f*, it is denoted by f^{-1}. Bijections $f : A \to A$ are also called *permutations of A*. If there is a bijection $f : A \to B$ between two sets A, B, we say that *they have the same cardinality*, and write $card(A) = card(B)$. On *AllSet*, the cardinality relation is an equivalence relation, and one may define the *cardinality $card(A)$ of a set A* as the equivalence class $[A]$ under this relation. A finite set is one whose cardinality is that of a set of form $\{1, 2, 3, \ldots n\}$, with a natural number $0 \leqslant n$, where we take the empty set for $n = 0$.

Definition 132 *A binary relation $\leqslant$ on a set S is said to be*

(i) reflexive *iff $x \leqslant x$ for all $x \in S$;*
(ii) transitive *iff $x \leqslant y$ and $y \leqslant z$ implies $x \leqslant z$ for all $x, y, z \in S$;*
(iii) symmetric *iff $x \leqslant y$ implies $y \leqslant x$ for all $x, y \in S$;*
(iv) antisymmetric *iff $x \leqslant y$ and $x \neq y$ excludes $y \leqslant x$ for all $x, y \in S$;*
(v) total *iff $x \leqslant y$ or $y \leqslant x$ for all $x, y \in S$.*

Definition 133 *A binary relation $\leqslant$ on a set S is called an* equivalence relation *iff it is reflexive, transitive, and symmetric. In this case, the relation is usually denoted by "$\sim$" instead of "$\leqslant$".*

Lemma 68 *Let $\sim$ be an equivalence relation on S. Then the subsets $[s] = \{t \mid s \sim t\}$ are called* equivalence classes *of $\sim$. The set of equivalence classes is denoted by $S/\sim$. It defines a partition of S, i.e., it covers S, and for any two elements $s, t \in S$, either $[s] = [t]$ or $[s] \cap [t] = \varnothing$.*

Definition 134 *A binary relation $\leqslant$ on a set S is called a* partial ordering *iff it is reflexive, transitive, and antisymmetric. A partial ordering is called* linear *iff it is total. A linear ordering is called* well-ordered *iff every non-empty subset $T \subset S$ contains a minimal element.*

Lemma 69 *Let $\leqslant$ be a binary relation on a set S. Denoting $x < y$ iff $x \leqslant y$ and $x \neq y$, the following two statements are equivalent:*

(i) *The relation $\leqslant$ is a partial ordering.*
(ii) *The relation $\leqslant$ is reflexive, the relation $<$ is transitive, and for all $x, y \in S$, $x < y$ excludes $y < x$.*

If these equivalent properties hold, we have $x \leqslant y$ iff $x = y$ or else $x < y$. In particular, if we are given $<$ with the properties (ii), and if we define $x \leqslant y$ by the preceding condition, then the latter relation is a partial ordering.

Proof. (i) $\Rightarrow$ (ii) Clearly $\leqslant$ is reflexive. If $x < y$ and $y < z$, then $x \leqslant z$. If we had $x = z$, then we were in the second statement of (ii), and it suffices to prove this one. But $x < y$ and $y < x$ implies $x = y$ by the asymmetry of $\leqslant$, a contradiction. (ii) $\Rightarrow$ (i) Transitivity: If $x \leqslant y$ and $y \leqslant z$, and either $x = y$ or $y = z$, we are done. Otherwise $x < y$ and $y < z$, whence $x < z$, therefore $x \leqslant z$. Asymmetry: If $x \leqslant y$ and $y \leqslant x$, then $x \neq y$ is excluded by (ii) whence the claim. The last statement is clear, QED.

Example 94 Suppose that $(I, <)$ is a linearly ordered "index set", and that we are given a family $((T_i, <_i, s_i))_I$ of linearly ordered sets $(T_i, <_i)$, each having a distinguished "constant" element s_i. Consider the set W of all nearly constant "words" $(t_i)_i \in \prod_i T_i$, i.e., $t_i = s_i$ except for a finite set of indexes. Then W has a canonical linear ordering, the *lexicographic ordering*, defined as follows. If we are given two different words, consider the smallest index where they differ. The smaller letter at that index defines the smaller word. Prove that this is a linear ordering.

In the usual lexicographic situation, the sets T_i are all an alphabet with its alphabetic linear ordering—as expressed, for example, by the *ASCII* or *UNICODE* encoding by natural numbers—, and the empty space as constant element. The indexes are natural, $i = 0, 1, 2, \ldots$.

Axiom 6 (Axiom of choice) *Let V be a partition of a set X. Then there is a subset $Y \subseteq X$ such that for every member M_i of the family V, the intersection $M_i \cap Y = \{m_i\}$ is a one-element set (a singleton).*

This is one of the many equivalent versions of the axiom of choice [561]. A particularly important variant is this:

Theorem 46 (Zermelo) *If the axiom of choice holds there is a well-ordered relation for every set. Conversely, if every set can be well ordered, the axiom of choice holds.*

For a proof, see [561, p.261].

Proposition 77 *Let $\leqslant$ be a partial ordering on a set S. Let $\leq$ be the following binary relation on the set $Fin(S)$ of finite subsets of S. For $A, B \subset S$, let*

$$A \leq B \begin{cases} A = B \text{ or} \\ A \neq B \text{ and for all } x \in A - B, \text{ there is } y \in B - A \text{ with } x < y. \end{cases} \tag{C.2}$$

Then $\leq$ is a partial ordering. If $\leqslant$ is linear, so is $\leq$.

Proof. We set $A < B$ as in Lemma 69. If $\leqslant$ is total, then for any $A \neq B$ in $Fin(S)$, if both disjoint difference sets $A - B$ and $B - A$ are non-empty, they contain different maximal elements, and we are done. Next, we verify statement (ii) about $<$. Suppose that we have $A < B$ and $B < A$ for $A \neq B$. Hence, we may suppose $A - B \neq \varnothing$. Take a maximal $x \in A - B$. There is $y \in B - A$ with $x < y$. Then there is $z \in A - B$ with $y < z$, but by transitivity of $<$, $x < z$, a contradiction to maximality of x. To show transitivity, take $A \neq B \neq C$ in $Fin(S)$. Since $A - C = \varnothing$ is trivial, suppose that x is a maximal element of $A - C$.

Suppose first that $x \in B$. Then $x \in B - C$, and there is $x < z$ with $z \in C - B$, take a maximal element z of this type. If $z \in A$, then $z \in A - B$, hence there is $z < w$, $w \in B - A$. If $w \in C$, $w \in C - A$, and $x < z < w$ which was required. Else, $w \notin C$ whence $w \in B - C$, and there is $u \in C - B$ with $w < u$, but $z < w < u$ contradicts maximality of z in $C - B$. Now if $z \notin A$, we have $z \in C - A$, and we are also done.

Suppose now that $x \notin B$, and therefore $x \in A - B$. There is $z \in B - A$ with $x < z$, take a maximal such element in $B - A$. If $z \in C$, then $z \in C - A$ and there is $w \in C - B$ with $z < w$. If $w \notin A$, then $w \in C - A$ and $x < z < w$ which was required. If $w \in A$ we have $w \in A - B$, and there is $u \in B - A$ with $u < w$. But then $z < w < u$ contradicts maximality of z in $B - A$, QED.

Definition 135 *For a finite partial ordering Rel on a set X, the* Hasse diagram *$Rel(X)$ is the relation whose pairs are the uniquely determined minimal set of generating relations of Rel.*

Proof of uniqueness. Suppose that two minimal generating sets A, B are different. Then any relation of $A - B$ points at a decomposition with a factor in $B - A$, and so forth, vice versa, such that we obtain an infinite chain of relations, contradicting the finiteness of X.

For a finite partial ordering Rel on X, the level function $lev : X \to \mathbb{N}$ is defined as follows. For minimal elements x in the Hasse diagram $Rel(X)$, put $lev(x) = 0$, for any element $x \in X$, set $pre(x) = \{y| \ y < x \text{ in } Rel(X)\}$. Then we put $lev(x) = 1 + Max(lev(y), y \in pre(x))$.

C.2.1 Universal Constructions

If $V = (M_i)_{i \in I}$ is a family of sets, the *product* set $\prod_I M_i$ is the set whose elements are the functions $f : I \to \bigcup V$ such that $f(i) \in M_i$ for all $i \in I$. Evidently, pairs, triples are special cases of such functions, but they are basic to the definition of functions, and therefore are treated separately. For each $j \in I$, we have the projection function $p_j : \prod_I M_i \to M_j$. Here is the *universal*[3] *property* of the product[4] set:

---

[3] This is a terminology from category theory, stating (in its most generic form) that a certain object in a category is final, but see Appendix G.

[4] This is a special case of a limit set, see Appendix G.2.1.

Lemma 70 *With the above notation, if $(f_i : X \to M_i)_i$ is a family of functions, then there is exactly one function $f : X \to \prod_I M_i$ such that $f_i = p_i \circ f$ for all $i \in I$, i.e., $f(x)(i) = f_i(x)$.*

For the given family V, the *coproduct* set $\coprod_I M_i$ is the union of the family $V' = (\{i\} \times M_i)_i$. For each index j, we have a function $\iota_j : M_j \to \coprod_I M_i$ defined by $\iota_j(m) = (j, m)$. The *universal property* of the coproduct[5] is this:

Lemma 71 *For every family $(f_i : M_i \to X)_i$ of functions, there is exactly one function $f : \coprod_I M_i \to X$ such that $f \circ \iota_i = f_i$, i.e., $f((i, m)) = f_i(m)$.*

Lemma 72 *If $f : A \to B$ is a function, then the* image *$Im(f) = \{b | b \in B,$ there exists $a \in A$ such that $b = f(a)\}$, together with the inclusion function $i : Im(f) \to B$ and the surjective function $f' : A \to Im(f) : a \mapsto f(a)$ has the following* universal property*: We have $f = i \circ f'$, and for every factorization $f = u \circ v$, there is a unique factorization $u = i \circ h$, and $f' = h \circ v$.*

Lemma 73 *Let A^B be a* powerset*, i.e., the set of functions $f : B \to A$. Then there is a natural bijection of* adjunction *$ad : C^{A \times B} \to (C^B)^A$, defined by $ad(g)(a)(b) = g(a, b)$.*

For a function $f : A \to B$ and a subset $C \subseteq B$, we define by $f^{-1}(C) = \{x | x \in A, f(x) \in C\}$ the *inverse image of C* under f; if $C = \{c\}$ is a singleton, we also write $f^{-1}(c)$ instead and call the set the *fiber of c*.

Lemma 74 *Let $2 = \{0, 1\}, 0 = \varnothing, 1 = \{0\}$, and let $Sub(X)$ be the set of subsets of X (also called the* powerset *of X). Then there is a natural bijection $\chi : Sub(X) \to 2^X$ defined by $\chi(Y)(z) = 0$ iff $z \in Y$, and $\chi(Y)(z) = 1$ iff $z \notin Y$. The function $\chi(Y)$ is called the* characteristic function *of Y. The inverse function maps $c : X \to 2$ to $Y_c = c^{-1}(0)$.*

C.2.2 Graphs and Quivers

Definition 136 *For a set X, denote by $P_2(X)$ the set of subsets of cardinality one or two in X, i.e., the set of singletons and unordered pairs in X. Then a* multigraph *is a triple $(L, V, G : L \to P_2(V))$. The elements of L are called the* lines*, the elements of V are called the* vertexes *of the multigraph. Often, we identify the multigraph with its map G. If G is injective, the multigraph is called a* graph*. A multigraph is* finite*, iff all involved sets are finite.*

Definition 137 *A* quiver *is a pair $G = (\text{head}, \text{tail} : A \rightrightarrows V)$ of set maps. The elements of V are called* vertexes*, the elements of A are called* arrows*. If every pair of vertexes is head and tail of at most one arrow, the quiver is called a* directed graph*.*

A path *p in G is either a vertex $p = v$ (a 'lazy' path) or else a sequence $p = (a_0, a_1, \ldots a_n)$ of arrows with $\text{head}(a_i) = \text{tail}(a_{i+1})$ for each index $i = 0, 1, \ldots n - 1$. The* length *$l(p)$ of a path is 0 for a lazy path, and n for a general path. A* closed path *is a path with $\text{head}(a_n) = \text{tail}(a_0)$. A lazy path is also a closed path of length 0. A closed path of length 1 is also called a* loop*. A* cycle *is an equivalence class of closed paths which differ from each other by their start/end point.*

We also use the symbols $\text{head}(p), \text{tail}(p)$ for the head, or tail, respectively, of the last, or first arrow of p, respectively (or just the unique vertex v for the lazy path).

Example 95 *For a set V, a* complete *quiver is a quiver $G = (\text{head}, \text{tail} : A \rightrightarrows V)$ such that the map*

$$f : A \to V^2 : a \mapsto (\text{tail}(a), \text{head}(a))$$

is a bijection.

For two paths p, q with $\text{head}(q) = \text{tail}(p)$, we have a composed path $p.q$ which is the evident juxtaposition of arrow sequences or the respectively other path if one path is lazy. As with functions, the composition of paths is associative if it is defined.

---

[5] This is a special case of a colimit set, see Appendix G.2.1.

C.2.3 Monoids

Definition 138 *A* semigroup *is a couple* (M, μ) *consisting of a set M and an associative binary operation* $\mu : M \times M \to M : (m, n) \mapsto \mu(m, n) = m.n$. *The semigroup M is a* monoid *if there exists a neutral element* e, *i.e.,* $e.m = m.e = m$ *for all $m \in M$. Since e is uniquely determined by this property, it is called the* neutral element. *For a given subset $S \subset M$ of a semigroup, the semigroup $\langle S \rangle$ generated by S is the smallest sub-semigroup of M containing S. If M is a monoid, the submonoid with neutral element e generated by S is denoted by $\langle S \rangle_e$. If $m.n = n.m$ for all $m, n \in M$, the semigroup is called* commutative. *An element $x \in M$ such that there is y with $x.y = e$ is called* invertible, *its set $M^\star$ is a submonoid. A monoid (M, μ) is* finite, *iff the underlying set M is so.*

Given two monoids M, N a monoid homomorphism *is a set map $f : M \to N$ such that $f(e_M) = e_N$ and $f(m.m') = f(m).f(m')$ for all $m, m' \in M$. A monoid* isomorphism *is a bijective monoid homomorphism, its inverse set map is then automatically a monoid isomorphism.*

Exercise 84 *Show that the map $N \mapsto N_e = N \cup \{e\}$ defines a projection from the set of sub-semigroups of M onto the set of sub-monoids of M with $N_e = N$ iff N is a submonoid and $N_e = N'_e$ iff $N = N'$ for sub-semigroups N, N' which are not sub-monoids.*

Example 96 For a set[6] *Alphabet*, there is a monoid $FM(Alphabet)$ and an injection

$$i : Alphabet \to FM(Alphabet)$$

such that for any set map $f : Alphabet \to M$ into a monoid M, there is exactly one monoid homomorphism $FM(f) : FM(Alphabet) \to M$ such that $FM(f) \circ i = f$. One calls the monoid $FM(Alphabet)$ the *free monoid over Alphabet* and the above property is the universal property that characterizes the free monoid up to isomorphism. The free monoid consists of all "word" expressions $b_1 \ldots b_k$ for $b_i \in Alphabet$, $k > 0$, and the empty word (). The product is defined by juxtaposition of words, () is neutral. The free monoid is also called "word monoid over the alphabet *Alphabet*".

Consider the equivalence relation on $FM(Alphabet)$ with $b_1 \ldots b_k \sim b_{\pi(1)} \ldots b_{\pi(k)}$ for any permutation π, then the quotient space $FCM(Alphabet) = FM(Alphabet)/\sim$ with the induced multiplication is a commutative monoid and is called the *free commutative monoid over Alphabet*.

Definition 139 *In a semigroup M, an* idempotent *is an element x such that $x = x.x = x^2$. We denote $Idempot(M) = \{x| \ x$ is idempotent in $M\}$. For a subset $X \subset M$, we define its* radical *by*

$$\sqrt{X} = \{y | \exists n, 0 < n, \ such \ that \ y^n \in X\}.$$

In particular, given an idempotent x, the radical $\sqrt{x}$ of x is defined by

$$\sqrt{x} = \{y | \exists n, 0 < n, \ such \ that \ x = y^n\}.$$

Since for every $y \in \sqrt{x}$, $\langle y \rangle \subset \sqrt{x}$, we also have $\sqrt{x} = \bigcup_{y \in \sqrt{x}} \langle y \rangle$.

Example 97 For example, the neutral element in a monoid is idempotent.

Example 98 If Q is a left module[7] over a ring R, the set of idempotents $Idempot(End(Q))$ of the linear endomorphism ring of Q is in bijection with the set $Dir(Q)$ of direct decompositions $Q = U \oplus V$ of Q. In fact, the bijection is set up by

$$Idempot(End(Q)) \overset{\sim}{\to} Dir(Q) : x \mapsto (Im(x), Im(Id_Q - x))$$

---

[6] E.g. *Alphabet = ASCII, UNICODE*.

[7] See Appendix E.

whose inverse defines the projection $p : Q \to U$ onto the factor U of $Q = U \oplus V$. The image of an idempotent x is a direct decomposition by the equation $Id_Q = x + (Id_Q - x)$ which yields $Q = Im(x) \oplus Im(Id_Q - x)$ with $x = Id_{Im(x)} \oplus 0_{Im(Id_Q - x)}$. Conversely, the projection $x = pr_U$ yields $U = Im(x), V = Im(Id_Q - x)$. So the above map is a bijection.

Lemma 75 *The idempotents of the monoid $Q@Q$ of affine endomorphisms*[8] *are the elements $e^q.y$ such that $y \in Idempot(End(Q))$ and $q \in Im(Id_Q - y) = Ker(y)$.*

Example 99 For example, the idempotents of $\mathbb{Z}_{12}$ are $0, 1, 4, 9$, and $Ker(0) = \mathbb{Z}_{12}, Ker(1) = 0, Ker(4) = 3.\mathbb{Z}_{12}, Ker(9) = 4.\mathbb{Z}_{12}$, so we have a total of 20 affine idempotents here.

Lemma 76 *With the previous notation, for two idempotent elements x, y of a semigroup M, $\sqrt{x} \cap \sqrt{y} \neq \varnothing$ iff $x = y$. Hence we have a partition of radicals in M, viz., $\sqrt{Idempot(M)} = \coprod_{x \in Idempot(M)} \sqrt{x}$, the idempotent components*[9] *of M.*

Clearly, if $f : M \xrightarrow{\sim} M$ is an automorphism of the semigroup M, we have $f(\sqrt{x}) = \sqrt{f(x)}$. In particular, if $M = Q@Q$ for a module Q, and if $y \in End(Q)$ is an idempotent linear endomorphism of Q, the conjugation by a translation $e^q, q \in Q$ gives the formula

$$e^q \sqrt{y} e^{-q} = \sqrt{e^{(1-y)q}.y} \tag{C.3}$$

which means that the translation exponent $(1 - y)q$ is a general element of $Ker(y)$. Therefore:

Proposition 78 *For a module Q, the orbits of idempotent components under the conjugation action of the translation group e^Q are the sets of idempotent components associated with the linear idempotent endomorphisms of Q.*

Lemma 77 *The canonical surjective linear factor projection $p : Q@Q \to End(Q)$ is compatible with the idempotent partitions, and the fiber of every idempotent component in $End(Q)$ is the orbit of the translation group action under conjugation (see Proposition 78).*

Proposition 79 *If in the above example the module Q is finite, the respective idempotent components define partitions*

$$\sqrt{Idempot(End(Q))} = End(Q)$$

and

$$\sqrt{Idempot(Q@Q)} = Q@Q.$$

Proof. Let $y \in End(Q)$. By Fitting's lemma (Appendix 88), there is a positive power $z = y^n$ and a direct decomposition $Q = U \oplus V$ such that $z|U \in GL(U), Ker(z) = V$. So by finiteness of U, there is a positive power $(z|U)^m = Id_U$. Therefore, $z^m = y^{nm}$ is an idempotent. As to the affine case, if $y = e^t.y_0 \in Q@Q$, we know by the preceding that a power $y_0^m = t$ is idempotent. So WLOG., we may suppose that y_0 is idempotent. Set $e^{t_k}.y_0 = (e^t.y_0)^k = y^k$; the positive powers of y must have recurrent values, so take $y^u = y^{u+k}$, for positive k. Then $e^{t_u}.y_0 = e^{t_u}.y_0.e^{t_k}.y_0 = e^{t_u + y_0(t_k)}.y_0$, i.e., $t_k \in Ker(y_0)$, whence y^k is an idempotent by Lemma 75, QED.

Lemma 78 *For the subgroup*[10] *$M^\star$ of invertible elements of the monoid M, we have the intersection formula*

$$M^\star \cap \sqrt{Idempot(M)} = \sqrt{1_M}.$$

---

[8] See Appendix E.
[9] Terminology of Noll, [802].
[10] See section C.3.

Proof. Clearly, $\sqrt{1_M} \subset M^\star$. On the other hand, if $x \in \sqrt{t} \subset M^\star$, we have $x^k = t = t^2 = x^{2k}$, whence $1_M = x^k = t$.

In the special case of a total partition $M = \sqrt{Idempot(M)}$, we get

$$M^\star = \sqrt{1_M}.$$

This is the case for an affine endomorphism monoid $Q@Q$ of a finite module Q.

C.3 Groups

Definition 140 *A monoid (M, μ) with the neutral element e such that every element m has a left inverse n, i.e., $n.m = e$, is called a* group. *A left inverse of m is also a right inverse, $m.n = e$, and the inverse is uniquely determined by m, it is denoted by m^{-1}. A subgroup of a group M is a submonoid N which is a group. A commutative group is also called* abelian. *If $n \in \mathbb{N}$, then we write m^n for the n-fold product $m.m. \ldots .m$; if n is a negative integer, then we write $m^n = (m^{-1})^{|n|}$, we also set $m^0 = e$. Evidently $m^n.m^l = m^{n+l}$.*

If a group is abelian, the product is usually written additively, i.e., $m + n$ instead of $m.n$, further $m - n$ instead of $m.n^{-1}$, and the neutral element is noted by 0 instead of e.

Exercise 85 *Observe that the neutral element e of a group is the only idempotent element $x.x = x$. Therefore, a subgroup has necessarily the same neutral element as the supergroup.*

C.3.1 Homomorphisms of Groups

If (G, γ), (H, η) are two groups, a set map $f : G \to H$ is called a *group homomorphism* iff $f(g.g') = f(g).f(g')$ (products in the respective groups). We have $f(e_G) = e_H$ and $f(x^{-1}) = f(x)^{-1}$. Clearly, the set-theoretic composition of two group homomorphisms is also a group homomorphism. The set of group homomorphisms $f : G \to H$ is denoted by $Hom(G, H)$. If f is a bijection, its inverse is also a group homomorphism; f is then called a *group isomorphism*. G, H are said to be *isomorphic* iff there is an isomorphism between them. Clearly, group isomorphisms are an equivalence relation, whence the term "isomorphism classes". The explicit description of isomorphism classes is the main task of group theory. For finite groups, this is essentially solved[11]. For finite commutative groups, the classification is described in Appendix C.3.4.

The set $Aut(G)$ of isomorphisms of a group G onto itself (the *automorphisms* of G) is a group under the composition of group homomorphisms, the identity Id_G being the neutral element of $Aut(G)$. The group $\mathfrak{S}_M$ of permutations of a set M is called the *symmetric group of M*.

Example 100 *A homomorphism $\mu : G \to \mathfrak{S}_M$ is called a* left action of G on M. *Let G^{opp} denote the* opposite group to G *where products are interchanged: $g.^{opp}h = h.g$. Then a right action is a homomorphism $\mu : G^{opp} \to \mathfrak{S}_M$. A left action is equivalent to a map $\mu : G \times M \to M$ (same notation for μ) with $\mu((e, m)) = m, \mu(g(\mu(h, m)) = \mu(g.h, m)$, all $g, h \in G, m, n \in M$. If the action is clear, we write $g.m$ instead of $\mu(g, m)$. We usually mean* left *actions when we speak of actions.*

A group action μ is called effective *iff it is an injective homomorphism. For an element $m \in M$, the group $G_m = \{g \in G, g.m = m\}$ is called the* stabilizer *or* isotropy *or* fixpoint group *of m. If all stabilizers are trivial, the action is called* (fixpoint) free *or* faithful.

For a group action μ, an orbit *is a set of form $G.m = \{g.m | g \in G\}$. The* orbit space $G \backslash M$ *defines a partition of M, with the canonical map $\pi : M \to G \backslash M : m \mapsto G.m$. The action is* transitive *iff the orbit space is a singleton. Clearly*

$$card(G.m).card(G_m) = card(G).$$

If $\mu_1 : G \times M \to M, \mu_1 : G \times N \to N$ are two group actions of group G, and if $h : M \to N$ is a set map, we say that h is *equivariant* iff it commutes with these group actions, i.e., if for all $m \in M, g \in G$, we have $h(\mu_1(g, m)) = \mu_2(g, h(m))$.

---

[11] The classification of all finite simple groups is one of the main results in group theory of the 20th century, see [381].

Example 101 For a set *Alphabet*, there is a group $FG(Alphabet)$ and an injection $i : Alphabet \to FG(Alphabet)$ such that for any set map $f : Alphabet \to G$ into a group G, there is exactly one group homomorphism $FG(f) : FG(Alphabet) \to G$ such that $FG(f) \circ i = f$. The group $FG(Alphabet)$ is called the *free group over Alphabet* and the above property is the universal property that characterizes the free group up to isomorphism. The free group consists of all "word" expressions $b_1^{n_1} \dots b_k^{n_k}$, where $b_i \in Alphabet, n_i \in \mathbb{Z}$ which are reduced, i.e., $b_i \neq b_{i+1}, n_i \neq 0$, and the empty word (). The product is defined by juxtaposition and cancelling of powers of adjacent letters.

Lemma 79 *For a group homomorphism $f : G \to H$, $Im(f)$ is a subgroup of H. For a subgroup $I \subseteq H$, the inverse image $f^{-1}(I)$ is a subgroup of G. The inverse image of the trivial group (the singleton consisting of the neutral element e_H) in H is called the* kernel *of f and is denoted by $Ker(f)$*

Example 102 For every element $g \in G$ of a group, we have a special group automorphism

$$Int_g : G \to G : h \mapsto g.h.g^{-1}$$

the *conjugation with g*. This yields a group homomorphism $Int : G \to Aut(G)$.

Proposition 80 *For a group G, a subgroup H and an element $g \in G$, we write $gH = \{g.h | h \in H\}, Hg = \{h.g | h \in H\}$ for the* left *and* right *cosets of H. The set of left, right cosets of G is denoted by G/H, $H \backslash G$, respectively. Denoting $(G : H) = card(G/H)$, we have the Lagrange equation[12]:*

$$card(G) = card(H).(G : H)$$

and in particular, if G is finite, any subgroup cardinality divides the order *$card(G)$ of G.*

Proposition 81 *If H is a subgroup of G, the following statements are equivalent:*

(i) *For all $g \in G$, $gH = Hg$.*
(ii) *There is a group homomorphism $f : G \to K$ with $Ker(f) = H$.*

If H has the properties of Proposition 81, it is called a *normal subgroup of G*, in symbols: $H \lhd G$. A *simple* group is one which has no normal subgroup except the trivial and the full subgroup. The group homomorphism in Proposition 81 is constructed by the *quotient group G/H* structure on G/H, where the product is defined by $gH.g'H = g.g'H$. The *canonical* homomorphism f is defined by $K = G/H$, and $f(g) = gH$.

Proposition 82 *Let $H \lhd G$ be a normal subgroup, $\pi : G \to G/H$ the canonical homomorphism, and K any group. Then*

$$Hom(G/H, K) \xrightarrow{\sim} \{f \in Hom(G, K) | H \subseteq Ker(f)\} : t \mapsto t \circ \pi$$

is a canonical bijection. If $H = Ker(f)$, then the corresponding morphism $G/H \to K$ is an isomorphism onto $Im(f)$.

Exercise 86 *If $f : G \to H$ is an isomorphism of groups, then the map*

$$Int_f : Aut(G) \to Aut(H) : t \mapsto f.t.f^{-1}$$

is an isomorphism of groups.

Here are the basic isomorphism theorems:

Theorem 47 (First isomorphism theorem) *Let G, H, N be groups with $N \lhd G, H \subseteq G$. Then:*

---

[12] Define $card(A).card(B) = card(A \times B)$.

(*i*) *The product set $H.N$ is a subgroup.*
(*ii*) *We have $N \lhd H.N$ and $N \cap H \lhd H$.*
(*iii*) *We have an isomorphism $H/N \cap H \xrightarrow{\sim} H.N/N : h(N \cap H) \mapsto hN$.*

Theorem 48 (Second isomorphism theorem) *Let $N \lhd G$ be a normal subgroup.*

(*i*) *The normal subgroups $\bar{M} \lhd G/N$ are in bijection with the subgroups M with $N \lhd M \lhd G$, and*
(*ii*) *we have an isomorphism $G/M \xrightarrow{\sim} (G/N)/(M/N) : gM \mapsto (gN).M/N$.*

C.3.2 Direct, Semi-direct, and Wreath Products

Given a family $(G_i)_{i \in I}$ of groups, the set-theoretic product $\prod_I G_i$ becomes a *(direct) product group* via the coordinate-wise product, i.e., $(g_i).(h_i) = (g_i.h_i)$.

Given two groups H, N and a group homomorphism $\phi : H \to Aut(N)$, the *semidirect product* $N \rtimes_\phi H$ is the group structure on the set $N \times H$ defined by

$$(n, g).(m, h) = (n.^g m, g.h)$$

where $^g m = \phi(g)(m)$. We have two group injections $i : N \to N \rtimes_\phi H : n \mapsto (n, e_H)$ and $j : H \to N \rtimes_\phi H : h \mapsto (e_N, h)$, and $i(N) \lhd N \rtimes_\phi H$. Further, $i(N) \cap j(H) = e$, $i(N).j(H) = N \rtimes_\phi H$, and the conjugation action identifies to the group homomorphism ϕ, i.e., $^g m = Int_g(m)$. We therefore have the short exact sequence (*i* is injective, π is surjective, and $Ker(\pi) = Im(i)$):

$$e \longrightarrow N \xrightarrow{\ i\ } N \rtimes_\phi H \underset{\pi}{\overset{j}{\rightleftarrows}} H \longrightarrow e$$

which is split, i.e., $\pi \circ j = Id_H$. Conversely, any such split sequence

$$e \longrightarrow N \xrightarrow{\ i\ } G \underset{\pi}{\overset{j}{\rightleftarrows}} H \longrightarrow e$$

identifies the middle group G to $N \rtimes_\phi H$.

Example 103 For two groups G, H, we have the direct product group $G^H = \prod_H G$ and an action $\phi : H \to Aut(G^H)$ via $\phi(h)(f)_k = f_{k.h}, k \in H$. Then the wreath product is $G \wr H = G^H \rtimes_\phi H$.

C.3.3 Sylow Theorems on p-groups

Proposition 83 (Sylow's proposition) *If G is a finite group of order n, p is prime, and $p^k | n$, then there is a subgroup $H \subseteq G$ with order p^k.*

A finite group G has the order $card(G) = p^k$ for a prime p iff all its elements have a power of p as their orders. In this case, G is called a *p-group*. A maximal p-subgroup in a group G is called a *p-Sylow group* in G.

Theorem 49 (Sylow's Theorem) *Let G be a finite group, p a prime. Then:*

(*i*) *The p-Sylow groups in G are the p-subgroups S with $p \nmid (G : S)$.*
(*ii*) *Any two p-Sylow groups are conjugate to each other.*
(*iii*) *Let σ_p be the cardinality of the set of p-Sylow groups in G. Then we have $\sigma_p | card(G)$ and $\sigma_p \equiv 1 \pmod{p}$.*

See [41] for a proof.

C.3.4 Classification of Groups

C.3.4.1 Classification of Cyclic Groups

If $S \subset G$ is a subset of a group G, the smallest subgroup in G containing S is denoted by $\langle S \rangle$ and is called the *group generated by* S. A *finitely generated* group is one that admits a finite set of generators. A cyclic group is one that is generated by one element $G = \langle s \rangle$. For such a group, the group homomorphism $s^? : \mathbb{Z} \to G : n \mapsto s^n$ is a surjection with kernel $Ker(s^?) = O(s).\mathbb{Z}$, where $s^{O(s)} = e$ is the smallest positive power of s which yields e, the *order of* s, or $O(s) = 0$ if no positive power of s vanishes, in which case the order of s is said to be infinite. This means that $\langle s \rangle \overset{\sim}{\to} \mathbb{Z}/O(s).\mathbb{Z}$, and we have classified all cyclic groups: They are isomorphic to the quotient groups $\mathbb{Z}_n = \mathbb{Z}/n.\mathbb{Z}$ with $card(\mathbb{Z}_n) = n$ for positive n, and ∞ for $\mathbb{Z}_0 = \mathbb{Z}$.

Among these groups, the groups $\mathbb{Z}_p$ of prime order p are *simple*. For abelian groups, simplicity means being cyclic of prime order.

The number of generators of a finite cyclic group $\mathbb{Z}_n$ is the number of numbers $0 < t < n$ prime to n, i.e., the *Euler function* $\phi(n) = card(\mathbb{Z}_n^\times)$, where $\mathbb{Z}_n^\times$ denotes these numbers modulo n, this is in fact the group of invertible elements of the ring $\mathbb{Z}_n$, see Appendix D.1.

Proposition 84 *If* $h = r.s$ *is a factorization by natural numbers* $1 < r, s$, *and* $(r, s) = 1$, *then* $\phi(h) = \phi(r).\phi(s)$. *If* $h = p^k$ *is a positive power of a prime number* q, *then* $\phi(p^k) = q^{k-1}(q - 1)$. *Therefore, if* $h = \prod_i q_i^{k_i}$ *is the prime decomposition of* h, *then*

$$\phi(h) = h. \prod_i (1 - 1/q_i).$$

C.3.4.2 Classification of Finitely Generated Abelian Groups

Let p be an prime integer ($p = 2, 3, 5, 7, 11, \ldots$). Take a weakly increasing sequence $u. = u_1 \leqslant u_2 \leqslant \ldots u_w$ of positive integers. We set
$$S(p, u.) = \mathbb{Z}_{p^{u_1}} \times \mathbb{Z}_{p^{u_2}} \times \ldots \mathbb{Z}_{p^{u_w}}.$$

If we are given an increasing sequence $p. = (p_1 < p_2 < \ldots p_t)$ of primes and for each such p_i a sequence $u.^i$, i.e., a sequence of sequences $u.^{\cdot}$. Then we set

$$T(p., u.^{\cdot}) = \prod_i S(p_i, u.^i). \tag{C.4}$$

Theorem 50 *For every finitely generated abelian group* G, *there is a natural number* f *and a system of primes* $p.$ *and positive, weakly increasing sequences* $u.^{\cdot}$ *such that*

$$G \overset{\sim}{\to} \mathbb{Z}^f \times T(p., u.^{\cdot}).$$

All the numbers f *(the torsion-free rank of* G), *and* $p., u.^{\cdot}$ *are uniquely determined. The length of the sequence* $u.^i$ *is called the* p_i-*rank of* G. *The image* $T(G)$ *of the factor* $T(p., u.^{\cdot})$ *in* G *is the torsion group of* G. *The subgroups in* G *corresponding to* $S(p_i, u.^i)$ *are called the* p_i-*Sylow groups. They are the subgroups of all elements with an order equal to a power of* p_i.

For a proof, see [1078, Vol.II].

Corollary 31 *If* G *is a finite abelian group, and if* $m|card(G)$, *then there is a subgroup* H *of* G *with* $card(H) = m$.

C.3.5 General Affine Groups

For a commutative ring R, we denote by $GL_n(R)$ the group $\mathbb{M}_{n,n}(R)^\times$ of invertible $n \times n$-matrices over R, see Appendix D.1; we also denote $GL(n,p) = GL_n(\mathbb{Z}_p)$ and $\overrightarrow{GL}(n,p) = e^{\mathbb{Z}_p} \cdot GL(n,p)$, the affine automorphism group of $\mathbb{Z}_p^n$.

Theorem 51 (Minkowski) *Suppose that G is a finite subgroup of $GL_n(\mathbb{Z})$, and let $q \in \mathbb{N}$. Consider the canonical projection homomorphism*

$$\Phi : G \to GL_n(\mathbb{Z}_q).$$

If $3 \le q$, then $Ker(\Phi)$ is trivial. If $q = 2$, then there is $U \in GL_n(\mathbb{Z})$ with

$$UNU^{-1} = Diag(\epsilon_1, \dots \epsilon_n) \text{ diagonal matrix,}$$

where $\epsilon_i = \pm 1$ for all $N \in Ker(\Phi)$. In particular, $card(Ker(\Phi))|w^n$.

For a proof, see [536, Satz 5.1].
For a prime number p and a power n, we have the following cardinalities:

$$card(GL(n,p)) = (p^n - 1)(p^{n-1} - 1) \dots (p-1)p^{n(n-1)/2}$$

and

$$card(\overrightarrow{GL}(n,p)) = p^n card(GL(n,p)).$$

C.3.6 Permutation Groups

Proposition 85 (Cayley) *Every group G is isomorphic to a subgroup of permutations.*

Proof. In fact, the left regular representation map $l_? : G \to \mathfrak{S}_G$ with $l_?(g) = l_g : h \mapsto g \cdot h$ is such an embedding, QED.
A permutation group $G \subseteq \mathfrak{S}_P$ by definition acts on the underlying set P. Suppose that P is finite. For $g \in G$, the orbits of $\langle g \rangle$ are called the cycles of g. Then these finite sets are arranged as sequences $C = (x, g.x, g^2.x, \dots g^{k(x)}.x)$ of pairwise different elements, i.e., $g^{k(x)+1}$ is the generator of the stabilizer of x. The permutation g can be represented as sequence $(C_1, C_2, \dots C_r)$ of orbits in cycle representation.

Definition 141 *If $G \subseteq \mathfrak{S}_P$ is a permutation group on a finite set P of cardinality p, the cycle index of an element $g \in G$ is the polynomial*

$$X^{cyc(g)} = X_1^{c_1} \cdot X_2^{c_2} \cdot \dots X_p^{c_p}$$

with the cycle type of g $cyc(g) = (c_1, c_2, \dots c_p)$, and $c_i = card(\{C = \text{ cycle of } g, card(g) = i\})$.

Here is Fripertinger's cycle index formula [343]

$$Z(\overrightarrow{GL}(\mathbb{Z}_{12}^2)(1 + x, 1 + x^2, \dots 1 + x^{144})$$

for orbits of zero-addressed local compositions in $\mathbb{Z}_{12}^2$:

$x^{144} + x^{143} + 5x^{142} + 26x^{141} + 216x^{140} + 2\,024x^{139} + 27\,806x^{138} + 417\,209x^{137} + 6\,345\,735x^{136} + 90\,590\,713x^{135} +$
$1\,190\,322\,956x^{134} + 14\,303\,835\,837x^{133} + 157\,430\,569\,051x^{132} + 1\,592\,645\,620\,686x^{131} + 14\,873\,235\,105\,552x^{130} +$
$128\,762\,751\,824\,308x^{129} + 1\,037\,532\,923\,086\,353x^{128} + 7\,809\,413\,514\,931\,644x^{127} + 55\,089\,365\,597\,956\,206x^{126} +$
$365\,290\,003\,947\,963\,446x^{125} + 2\,282\,919\,558\,918\,081\,919x^{124} + 13\,479\,601\,808\,118\,798\,229x^{123} +$
$75\,361\,590\,622\,423\,713\,249x^{122} + 399\,738\,890\,367\,674\,230\,448x^{121} + 2\,015\,334\,387\,723\,540\,077\,262x^{120} +$
$9\,673\,558\,570\,858\,327\,142\,094x^{119} + 44\,275\,002\,111\,552\,677\,715\,575x^{118} + 193\,497\,799\,414\,541\,699\,555\,587x^{117} +$
$808\,543\,433\,959\,017\,353\,438\,195x^{116} + 3\,234\,171\,338\,137\,153\,259\,094\,292x^{115} +$
$12\,397\,650\,890\,304\,440\,505\,241\,198x^{114} + 45\,591\,347\,244\,850\,943\,472\,027\,532x^{113} +$

$160\,994\,412\,344\,908\,368\,725\,437\,163x^{112} + 546\,405\,205\,018\,625\,434\,948\,486\,100x^{111} +$

$1\,783\,852\,127\,215\,514\,388\,216\,575\,524x^{110} + 5\,606\,392\,061\,138\,587\,678\,507\,139\,578x^{109} +$

$16\,974\,908\,597\,922\,176\,404\,758\,662\,419x^{108} + 49\,548\,380\,452\,249\,950\,392\,015\,617\,673x^{107} +$

$139\,517\,805\,378\,058\,810\,895\,892\,716\,876x^{106} + 379\,202\,235\,047\,824\,659\,955\,968\,634\,895x^{105} +$

$995\,405\,857\,334\,028\,240\,446\,249\,995\,969x^{104} + 2\,524\,931\,913\,311\,378\,421\,460\,541\,875\,013x^{103} +$

$6\,192\,094\,899\,403\,308\,142\,319\,324\,646\,830x^{102} + 14\,688\,225\,057\,065\,816\,000\,841\,247\,153\,422x^{101} +$

$33\,716\,152\,882\,551\,682\,431\,054\,950\,635\,828x^{100} + 74\,924\,784\,036\,765\,597\,482\,162\,224\,697\,378x^{99} +$

$161\,251\,165\,409\,134\,463\,248\,992\,354\,275\,261x^{98} + 336\,225\,833\,888\,858\,733\,322\,982\,932\,904\,265x^{97} +$

$679\,456\,372\,086\,288\,422\,448\,712\,466\,252\,503x^{96} + 1\,331\,179\,830\,182\,151\,403\,666\,404\,596\,530\,852x^{95} +$

$2\,529\,241\,676\,111\,626\,447\,928\,668\,220\,456\,264x^{94} + 4\,661\,739\,558\,127\,027\,290\,220\,867\,616\,981\,880x^{93} +$

$8\,337\,341\,899\,567\,786\,249\,391\,103\,289\,453\,916x^{92} + 14\,472\,367\,067\,576\,451\,752\,984\,797\,361\,008\,304x^{91} +$

$24\,388\,618\,572\,337\,747\,341\,932\,969\,998\,362\,288x^{90} + 39\,908\,648\,567\,034\,355\,259\,311\,114\,115\,744\,392x^{89} +$

$63\,426\,245\,036\,529\,210\,051\,949\,169\,850\,308\,102x^{88} + 97\,921\,220\,397\,909\,924\,969\,018\,620\,386\,852\,352x^{87} +$

$146\,881\,830\,585\,458\,073\,270\,850\,321\,720\,445\,928x^{86} + 214\,098\,939\,483\,879\,341\,610\,433\,150\,629\,060\,274x^{85} +$

$303\,306\,830\,919\,747\,863\,651\,620\,555\,026\,700\,930x^{84} + 417\,668\,422\,888\,061\,171\,460\,770\,548\,484\,103\,836x^{83} +$

$559\,136\,759\,653\,084\,522\,330\,064\,385\,877\,590\,780x^{82} + 727\,765\,306\,194\,069\,123\,565\,702\,210\,626\,823\,392x^{81} +$

$921\,077\,965\,629\,957\,077\,012\,552\,741\,715\,036\,692x^{80} + 1\,133\,634\,419\,214\,796\,834\,928\,853\,170\,296\,724\,314x^{79} +$

$1\,356\,926\,047\,220\,511\,677\,349\,073\,201\,120\,481\,570x^{78} + 1\,579\,704\,950\,475\,555\,411\,914\,967\,237\,903\,930\,342x^{77} +$

$1\,788\,783\,546\,844\,376\,088\,722\,000\,995\,922\,467\,990x^{76} + 1\,970\,254\,341\,437\,213\,013\,502\,048\,964\,983\,877\,090x^{75} +$

$2\,110\,986\,794\,386\,177\,596\,749\,436\,553\,816\,924\,660x^{74} + 2\,200\,183\,419\,494\,435\,885\,449\,671\,402\,432\,366\,956x^{73} +$

$2\,230\,741\,522\,540\,743\,033\,415\,296\,821\,609\,381\,912x^{72} + 2\,200\,183\,419\,494\,435\,885\,449\,671\,402\,432\,366\,956x^{71} +$

$2\,110\,986\,794\,386\,177\,596\,749\,436\,553\,816\,924\,660x^{70} + 1\,970\,254\,341\,437\,213\,013\,502\,048\,964\,983\,877\,090x^{69} +$

$1\,788\,783\,546\,844\,376\,088\,722\,000\,995\,922\,467\,990x^{68} + 1\,579\,704\,950\,475\,555\,411\,914\,967\,237\,903\,930\,342x^{67} +$

$1\,356\,926\,047\,220\,511\,677\,349\,073\,201\,120\,481\,570x^{66} + 1\,133\,634\,419\,214\,796\,834\,928\,853\,170\,296\,724\,314x^{65} +$

$921\,077\,965\,629\,957\,077\,012\,552\,741\,715\,036\,692x^{64} + 727\,765\,306\,194\,069\,123\,565\,702\,210\,626\,823\,392x^{63} +$

$559\,136\,759\,653\,084\,522\,330\,064\,385\,877\,590\,780x^{62} + 417\,668\,422\,888\,061\,171\,460\,770\,548\,484\,103\,836x^{61} +$

$303\,306\,830\,919\,747\,863\,651\,620\,555\,026\,700\,930x^{60} + 214\,098\,939\,483\,879\,341\,610\,433\,150\,629\,060\,274x^{59} +$

$146\,881\,830\,585\,458\,073\,270\,850\,321\,720\,445\,928x^{58} + 97\,921\,220\,397\,909\,924\,969\,018\,620\,386\,852\,352x^{57} +$

$63\,426\,245\,036\,529\,210\,051\,949\,169\,850\,308\,102x^{56} + 39\,908\,648\,567\,034\,355\,259\,311\,114\,115\,744\,392x^{55} +$

$24\,388\,618\,572\,337\,747\,341\,932\,969\,998\,362\,288x^{54} + 14\,472\,367\,067\,576\,451\,752\,984\,797\,361\,008\,304x^{53} +$

$8\,337\,341\,899\,567\,786\,249\,391\,103\,289\,453\,916x^{52} + 4\,661\,739\,558\,127\,027\,290\,220\,867\,616\,981\,880x^{51} +$

$2\,529\,241\,676\,111\,626\,447\,928\,668\,220\,456\,264x^{50} + 1\,331\,179\,830\,182\,151\,403\,666\,404\,596\,530\,852x^{49} +$

$679\,456\,372\,086\,288\,422\,448\,712\,466\,252\,503x^{48} + 336\,225\,833\,888\,858\,733\,322\,982\,932\,904\,265x^{47} +$

$161\,251\,165\,409\,134\,463\,248\,992\,354\,275\,261x^{46} + 74\,924\,784\,036\,765\,597\,482\,162\,224\,697\,378x^{45} +$

$33\,716\,152\,882\,551\,682\,431\,054\,950\,635\,828x^{44} + 14\,688\,225\,057\,065\,816\,000\,841\,247\,153\,422x^{43} +$

$6\,192\,094\,899\,403\,308\,142\,319\,324\,646\,830x^{42} + 2\,524\,931\,913\,311\,378\,421\,460\,541\,875\,013x^{41} +$

$995\,405\,857\,334\,028\,240\,446\,249\,995\,969x^{40} + 379\,202\,235\,047\,824\,659\,955\,968\,634\,895x^{39} +$

$139\,517\,805\,378\,058\,810\,895\,892\,716\,876x^{38} + 49\,548\,380\,452\,249\,950\,392\,015\,617\,673x^{37} +$

$16\,974\,908\,597\,922\,176\,404\,758\,662\,419x^{36} + 5\,606\,392\,061\,138\,587\,678\,507\,139\,578x^{35} +$

$1\,783\,852\,127\,215\,514\,388\,216\,575\,524x^{34} + 546\,405\,205\,018\,625\,434\,948\,486\,100x^{33} +$

$160\,994\,412\,344\,908\,368\,725\,437\,163x^{32} + 45\,591\,347\,244\,850\,943\,472\,027\,532x^{31} +$

$12\,397\,650\,890\,304\,440\,505\,241\,198x^{30} + 3\,234\,171\,338\,137\,153\,259\,094\,292x^{29} +$

$808\,543\,433\,959\,017\,353\,438\,195x^{28} + 193\,497\,799\,414\,541\,699\,555\,587x^{27} + 44\,275\,002\,111\,552\,677\,715\,575x^{26} +$

$9\,673\,558\,570\,858\,327\,142\,094x^{25} + 2\,015\,334\,387\,723\,540\,077\,262x^{24} + 399\,738\,890\,367\,674\,230\,448x^{23} +$

$75\,361\,590\,622\,423\,713\,249x^{22} + 13\,479\,601\,808\,118\,798\,229x^{21} + 2\,282\,919\,558\,918\,081\,919x^{20} +$

$365\,290\,003\,947\,963\,446x^{19} + 55\,089\,365\,597\,956\,206x^{18} + 7\,809\,413\,514\,931\,644x^{17} + 1\,037\,532\,923\,086\,353x^{16} +$

$128\,762\,751\,824\,308x^{15} + 14\,873\,235\,105\,552x^{14} + 1\,592\,645\,620\,686x^{13} + 157\,430\,569\,051x^{12} + 14\,303\,835\,837x^{11} +$

$1\,190\,322\,956x^{10} + 90\,590\,713x^9 + 6\,345\,735x^8 + 417\,209x^7 + 27\,806x^6 + 2\,024x^5 + 216x^4 + 26x^3 + 5x^2 + x + 1.$

D

Rings and Algebras

D.1 Basic Definitions and Constructions

Definition 142 *A* (unitary) ring *is a triple* (R, α, μ) *where* (R, α) *is an abelian group whose operation* α *is written additively* $(\alpha(r, s) = r + s)$ *with neutral element* 0_R*, and* (R, μ) *is monoid, written multiplicatively* $(\mu(r, s) = r \cdot s)$ *with multiplicative neutral element* 1_R *such that these operations are coupled by* distributivity*, i.e.,* $(r + s) \cdot t = r \cdot t + s \cdot t, t \cdot (r + s) = t \cdot r + t \cdot s$ *for all* $r, s, t \in R$*. A ring is* commutative *iff its multiplicative monoid is commutative.*

A set map $f : R \to S$ *of rings* R, S *is a* ring homomorphism *iff it is a homomorphism of the underlying additive groups and a homomorphism of the underlying multiplicative monoids. The set of ring homomorphisms from* R *to* S *is again denoted by* $Hom(R, S)$ *if no confusion is likely.*

An element x *of a non-zero ring* R *is called* invertible*, iff there is a multiplicative (left) inverse* y*, i.e.,* $y \cdot x = 1_R$*. The subset* $R^\times$ *of invertible elements is a multiplicative group. A* skew field *is a ring such that* $R^\times = R - \{0_R\}$*. A* (commutative) field *is a commutative ring which is a skew field.*

The subring $Z(R)$ *of all elements in a ring* R *which commute with all of* R *is called the* center *of* R*. For a commutative* R*, a ring homomorphism* $\varphi : R \to S$ *whose image is in* $Z(S)$ *is called an* R-algebra*. If* φ *is clear, one says that "S is an R-algebra", and* φ *is called the "structural homomorphism". If* $\varphi : R \to S, \psi : R \to T$ *are two R-algebras, a ring homomorphism* $f : S \to T$ *is an* R-algebra homomorphism*, iff* $\psi = f \circ \varphi$*, i.e., iff the 'R-elements are conserved under f'. One als often writes* r *instead of* $\varphi(r)$ *if the algebra structure is clear.*

The set-theoretic composition $g \circ f$ of two ring homomorphisms $f : R \to S, g : S \to T$ is again a ring homomorphism. A bijective ring homomorphism automatically has an *inverse* ring homomorphism, i.e., this is a ring *isomorphism*, and a ring *endomorphism* is a homomorphism with domain equal to its codomain, whereas an *automorphism* is an endomorphism which is an isomorphism. The corresponding concepts for R-algebras are evident: The homomorphisms have to be algebra homomorphisms, so, for example, an R-algebra automorphism is an automorphism which conserves the structural homomorphism.

Example 104 Classical examples of rings are the rings $\mathbb{Z}, \mathbb{Q}, \mathbb{R}, \mathbb{C}, \mathbb{H}$ of integers, rational numbers, real numbers, complex numbers and Hamilton quaternions. Except the integers, these rings are also skew fields, and $\mathbb{H}$ is not commutative. The conjugation $z = a + i.b \mapsto \bar{z} = a - i.b$ is an automorphism of the $\mathbb{R}$-algebra $\mathbb{C}$.

To every ring R, one has the opposite ring R^{opp} which is the same additive group, but multiplication is defined by reversing to given multiplication, i.e., $r \cdot^{opp} s = s \cdot r$. An anti-homomorphism of rings is a homomorphism into the opposite codomain ring.

Every ring R is a $\mathbb{Z}$-algebra in a unique way by $\varphi(z) = z.1_R$; the latter means that $z.1_R = 1_R + 1_R + \ldots 1_R$ if $z > 0$, it is 0_R if $z = 0$, and it is $-(-z).1_R$ if $z < 0$.

By the natural inclusions, $\mathbb{R}$ is a $\mathbb{Q}$-algebra, $\mathbb{C}$ is an $\mathbb{R}$-algebra, and $\mathbb{H}$ is an $\mathbb{R}$-algebra, but not a $\mathbb{C}$-algebra.

© Springer International Publishing AG, part of Springer Nature 2017
G. Mazzola, *The Topos of Music IV: Roots*, Computational Music Science,
https://doi.org/10.1007/978-3-319-64495-0_4

The 2×2 matrices over a commutative ring R with coefficient-wise addition and usual matrix multiplication form a non-commutative ring. They are an R-algebra by the diagonal map $r \mapsto \left(\begin{smallmatrix} r & 0 \\ 0 & r \end{smallmatrix} \right)$.

Proposition 86 *If I is a subgroup of a ring R, the following statements are equivalent:*

(i) *For all $r \in R$, $r \cdot I \subseteq I$, i.e., I is a* left *ideal in R, and $I \cdot r \subseteq I$, i.e., I is a* right *ideal in R.*
(ii) *There is a ring homomorphism $f : R \to S$ with $Ker(f) = I$.*

If I has the equivalent properties of Proposition 86, we call it an *ideal* or a *two-sided ideal* in R. So ideals correspond to normal subgroups. The *quotient ring* R/I (and the associated projection $\pi : R \to R/I$) is just the quotient group (with respect to the additive structure), together with the well-defined multiplication $(r+I)(s+I) = rs+I$, and we have $Ker(\pi) = I$. If G is any subset of a ring R, the smallest ideal containing G is denoted by (G) and is called the *ideal generated by G*.

Proposition 87 *Let I be an ideal in R, $\pi : R \to R/I$ the canonical homomorphism, and S any ring. Then*

$$Hom(R/I, S) \xrightarrow{\sim} \{f \in Hom(R,S) | I \subseteq Ker(f)\} : t \mapsto t \circ \pi$$

is a canonical bijection. If $I = Ker(f)$, then the corresponding morphism $R/I \to I$ is an isomorphism onto $Im(f)$.

Example 105 By the Euclidean algorithm (D.3), every subgroup of $\mathbb{Z}$ is of the form $\langle n \rangle = n.\mathbb{Z}$ for a uniquely determined non-negative n. Such a subgroup is also an ideal. The quotient ring $\mathbb{Z}_n = \mathbb{Z}/n.\mathbb{Z}$ is the *ring of integers modulo n*. It is the so-called "prime field of characteristic p" iff the Euler function $\phi(n) = n-1$, i.e., $n = p$ is a prime (see also Proposition 84 in Appendix C.3.4.1). For any skew field F, we have the unique $\mathbb{Z}$-algebra structure $p : \mathbb{Z} \to F$ whose kernel must be either (0) or a prime ideal $p.\mathbb{Z}$. In the latter case, the image $Im(p)$ is the smallest subfield in F; in the former case, the smallest subfield of F is evidently isomorphic to $\mathbb{Q}$. The prime field of a skew field is called its *prime field*, whereas the generator (zero or a positive prime) of $Ker(p)$ is called the characteristic $char(F)$ of the skew field.

A skew field F has only two ideals: $(0), F$, and conversely, a ring which has only these ideals is called *simple*. The ring $\mathbb{M}_{n,n}(F)$ of $n \times n$-matrices (see E.1.1) over a skew field F is simple (but not a skew field for $1 < n$).

D.1.1 Universal Constructions

If M is a multiplicative monoid and R is a commutative ring, we have a ring $R\langle M \rangle$, the *monoid algebra* by the following construction. The underlying set is the set $R^{(M)}$ of functions $f : M \to R$ with $f(m) = 0$ except of a finite number of arguments. This is an additive group under the addition $(f+g)(m) = f(m)+g(m)$. The product is defined by $(f \cdot g)(m) = \sum_{n \cdot n' = m} f(n) \cdot g(n')$, which is reasonable since the summands which do not vanish are finite in number. The multiplicative neutral element $1_{R\langle M \rangle}$ is the function $1_{R\langle M \rangle}(e_M) = 1_R$ and zero else. The elements $f : M \to R$ of the monoid algebra are also written as a formal sum $\sum_{f(m) \neq 0} f(m)m$. The algebra structure is given by $r \mapsto re_M$. Here is the *universal property* of the monoid algebra:

Proposition 88 *Every monoid homomorphism $\varphi : M \to (S, \mu)$ into the multiplicative monoid of an R-algebra S, can be extended in a unique way to a homomorphism $\Phi : S\langle M \rangle \to S$ of R-algebras. If $\sum f_m.m \in S\langle M \rangle$, its image is $\Phi(\sum f_m.m) = \sum f_m \cdot \varphi(m)$.*

A special case of a monoid algebra is for the free monoid $FM\langle Alphabet \rangle$ over *Alphabet*, see Example 96 in Appendix C.2.3. One usually writes $S\langle Alphabet \rangle$ instead of $S\langle FM(Alphabet) \rangle$. For example, if $Alphabet = \{X_1, X_2, \ldots X_n, \ldots\}$, we have the *algebra of non-commutative polynomials in the indeterminates $X_1, X_2, \ldots X_n, \ldots$*. If instead we take the free commutative monoid $FCM(Alphabet)$, we get the monoid algebra $S[Alphabet] = S\langle FCM(Alphabet) \rangle$ of commutative polynomials in the indeterminates $\{X_1, X_2, \ldots X_n, \ldots\}$ of *Alphabet*.

Example 106 For the R-algebra of polynomials in one variable $R[X]$, the ideal (X^2) defines the *R-algebra of dual numbers* $R[\varepsilon] = R[X]/(X^2)$ with $\varepsilon = X + (X^2)$ the class of X. An element of $R[\varepsilon]$ is uniquely described as a linear polynomial $a + \varepsilon.b$ in ε, and the multiplication is $(a + \varepsilon.b)(c + \varepsilon.d) = ac + \varepsilon.(ad + bc)$. The group $R[\varepsilon]^\times$ consists of those elements $a + \varepsilon.b$ with $a \in R^\times$.

Definition 143 *Let $(R_i)_{i \in I}$ be a family of rings, then the* product $\prod_{i \in I} R_i$ *of this family is the following ring: As a set, it is the product of the underlying sets, addition and multiplication are defined coordinate-wise[1], i.e.,*

$$(x_i) + (y_i) = (x_i + y_i) \text{ and}$$
$$(x_i) \cdot (y_i) = (x_i \cdot y_i),$$

and the unity is the family (1_i) of unities 1_i in the respective rings R_i. For each index $j \in I$, we have a canonical projection ring homomorphism

$$p_j : \prod_{i \in I} R_i \to R_j$$

defined by $p_j((x_i)) = x_j$.

The product and its canonical projections shares the universal property of a product in the category of rings (see Appendix G.2.1 for the definition of a product in a category):

Lemma 80 *Let S be a ring, and let $(R_i)_{i \in I}$ be a family of rings. For each family $(f_i : S \to R_i)_{i \in I}$ of ring homomorphisms, there is a unique ring homomorphism $f : S \to \prod_{i \in I} R_i$ such that $f_i = p_i \cdot f$ for all $i \in I$.*

Clearly, the map $f(s) = (f_i(x))$ solves the problem.

More generally, consider a diagram[2] $\mathbf{D}$ of rings with vertex set I. Then we have a subring $lim(\mathbf{D})$ of the product $\prod_{i \in I} R_i$ consisting of all families (x_i) such that

$$f(x_j) = x_k \text{ for any homomorphism } f : R_j \to R_k$$

corresponding to a $\mathbf{D}$-arrow from vertex j to vertex k.

Proposition 89 *With the above notation, the ring of $lim(\mathbf{D})$, together with the induced canonical projections $p_i : lim(\mathbf{D}) \to R_i$, is the* limit *of the diagram $\mathbf{D}$ of rings, i.e., for any family $(f_i : S \to R_i)_{i \in I}$ of ring homomorphisms such that $f \cdot f_j = f_k$ whenever $f : R_j \to R_k$ is a homomorphism corresponding to a $\mathbf{D}$-arrow from vertex j to vertex k, there is a unique ring homomorphism $g : S \to lim(\mathbf{D})$ such that $f_i = p_i \cdot g$, all $i \in I$.*

In particular, if we have a fiber product[3] diagram $A \xrightarrow{f} C \xleftarrow{g} B$ of rings, there is a limit, the fiber product of this diagram, usually denoted by $A \times_C B$ and inserted in the commutative "pullback" square

$$
\begin{array}{ccc}
A \times_C B & \xrightarrow{p_2} & B \\
{\scriptstyle p_1} \downarrow & & \downarrow {\scriptstyle g} \\
A & \xrightarrow{f} & C
\end{array}
\tag{D.1}
$$

of ring homomorphisms.

For non-commutative rings, fiber sums do not exist in general. However, if the diagram's homomorphisms are algebras over a commutative ring, we have the following well-known result [145, ch.III, No.2]:

---

[1] Addition and multiplication are taken in the respective rings.
[2] See definition 165, Appendix G.1.2, for a formal definition of a diagram in a category.
[3] See Appendix G.2.1 for the concept of a fiber product and the dual one of a fiber sum.

Theorem 52 *Let $A \xleftarrow{f} C \xrightarrow{g} B$ be a fiber sum diagram of algebras over the commutative ring C. Then the tensor product $A \otimes_C B$ defines a fiber sum*

$$
\begin{array}{ccc}
C & \xrightarrow{\ f\ } & A \\
{\scriptstyle g}\big\downarrow & & \big\downarrow{\scriptstyle i_1} \\
B & \xrightarrow{\ i_2\ } & A \otimes_C B
\end{array}
\tag{D.2}
$$

of C-algebras with $i_1(a) = a \otimes 1_B$ and $i_2(b) = 1_A \otimes b$.

D.1.1.1 Quiver Algebras

Definition 144 *Suppose that we are given a quiver $Q = (\mathrm{head}, \mathrm{tail} : A \rightrightarrows V)$ (see Appendix C.2.2, definition 137). The path category of Q is the set $P(Q)$ of paths of Q (the morphisms of the category, the lazy paths being the objects of the category), together with the composition $p.q$ of two paths p, q, defined if $\mathrm{head}(p) = \mathrm{tail}(q)$. In that case, $p.q$ is just the evident composed path of length $l(p.q) = l(p) + l(q)$.*

Definition 145 *Suppose that we are given a finite quiver $Q = (\mathrm{head}, \mathrm{tail} : A \rightrightarrows V)$ and a commutative ring R. The quiver algebra $R\langle Q\rangle$ of Q with coefficients in R is the free left R-module whose basis is the path set $P(Q)$, together with R-bilinear extension of the composition in the path category, i.e., the product of two paths is their composition, if possible, and zero else. This means:*

$$
\left(\sum_i r_i.p_i\right)\left(\sum_j s_j.q_j\right) = \sum_{i,j}(r_i.s_j)p_i.q_j.
$$

The unity

$$
1_{R\langle Q\rangle} = \sum_{v \in V} v
$$

is the sum of all vertexes.

Definition 146 *Let Q be a quiver, and $R\langle Q\rangle$ its quiver algebra over the coefficient ring R. A sub-path of a path p in $R\langle Q\rangle$, is a triple (u, w, v) of paths such that $p = u.w.v$. If the external factors u, v are clear, the sub-path is identified with the middle member w, and we write $w \sqsubset p$.*

For example, if $l(p) = 0$ (a vertex), the only sub-path of p is p, in fact, $p = p.p.p$ is the only factorization of this kind. If $p = \circlearrowleft x$ is a loop, the sub-paths of p are $x.x.p$, $p.x.x$, and $x.p.x$. So the middle member x appears two times since it is in different positions in the factorization.

If $p = u_1.w_1.v_1 = u_2.w_2.v_2$ are two subpaths of p with $l(w_1) = l(w_2)$, we write $(u_1, w_1, v_1) < (u_2, w_2, v_2)$ iff $l(u_1) < l(u_2)$. Clearly this relation is transitive and antisymmetric. Among all subpaths (u, w, v) of p with fixed middle length $l(w) = const.$, the relation corresponding to relation $\leqslant$ is an linear ordering. In fact, if $(u_1, w_1, v_1), (u_2, w_2, v_2)$ are two subpaths of p such that $l(u_1) = l(u_2)$ and $l(w_1) = l(w_2)$, then they are identical, so the ordering relation is total. We shall use this total ordering to define linear endomorphisms, so-called *sub-path operators* of the quiver algebras as follows.

We shall define an endomorphism $\sqsubset_i^{\lambda^i}$ for each natural index $i = 0, 1, 2, \ldots$ and for each system λ^i of coefficients in the following sense. We set

$$
\lambda^i = (\lambda^i_{lj})_{i \leqslant l, 1 \leqslant j \leqslant l - i + 1}
$$

with $\lambda^i_{lj} \in R$. The endomorphisms run as follows. Let w be a path of length $l(w) = l$. Then

$$
\sqsubset_i^{\lambda^i}(w) = \begin{cases} 0 & \text{if } l < i, \\ \sum_{v_j \sqsubset w, l(v_j) = i} \lambda^i_{lj} v_j & \text{else,} \end{cases}
$$

with the indexes of the subpaths of w referring to their total order defined above. In the musical applications, we shall encounter linear combinations $\Phi = \sum_i \mu_i \sqsubset_i^{\lambda^i}$ of such weighted sub-path operators.

D.2 Prime Factorization

An integer $1 < p$ is called *prime* iff $p = u \cdot v$ with positive factors implies $u = 1$ or $v = 1$.

Theorem 53 *Every non-zero integer x is a product*

$$x = \pm 1 \cdot p_1^{n_1} \cdot p_2^{n_2} \cdot \ldots p_k^{n_k}$$

with an increasing sequence of primes $p_1 < p_2 < \ldots p_k$ and positive exponents n_i which are all uniquely determined.

Corollary 32 *Let $p_1 < p_2 < \ldots p_k$ be an increasing sequence of primes. For two sequences of rational numbers $q_1 < q_2 < \ldots q_k$, $r_1 < r_2 < \ldots r_k$, the equation*

$$p_1^{q_1} \cdot p_2^{q_2} \cdot \ldots p_k^{q_k} = p_1^{r_1} \cdot p_2^{r_2} \cdot \ldots p_k^{r_k}$$

implies the equality of the sequences of rational numbers.

This implies that the logarithms $ln(p)$ of primes are linearly independent (see E.2.1) in the rational vector space $\mathbb{R}_{[\mathbb{Q}]}$ (see E.1.1), a central fact for the construction of the Euler module of pitch systems (see Section A.2.3).

D.3 Euclidean Algorithm

Proposition 90 *Euclidean algorithm: Given a non-zero integer d, every integer x has a unique representation $x = a.d + b, 0 \leqslant b < d$.*

Lemma 81 *Let $n > 1$ be an integer. Then every positive integer x has a representation*

$$x = \sum_{i=0}^{t} x_i . n^i$$

with $0 \leqslant x_i < n$, and $x_t \neq 0$. The x_i are uniquely determined. The representation is usually written as

$$x = x_t x_{t-1} \ldots x_1 x_0$$

and known as the n-adic representation.

D.4 Approximation of Real Numbers by Fractions

Lemma 82 *Let $L = \log_2(3)$. Then for every real number $\delta > 0$, there is a pair n, m of integers such that*

$$0 < n + m \cdot L < \delta.$$

Proof. We construct a sequence $(n_1, m_1), (n_2, m_2), \ldots (n_s, m_s), \ldots$ of pairs such that $0 < n_s + m_s \cdot L < 1/2^s$. We may start by $(n_1, m_1) = (2, -1)$ since $L \approx 1.58$. Suppose that we have found (n_s, m_s) such that $0 < n_s + m_s \cdot L < 1/2^s$. There is a maximal positive integer k such that $k \cdot (n_s + m_s \cdot L) < 1$. Then $(k + 1) \cdot (n_s + m_s \cdot L) > 1$ since L is not rational by Corollary 32. For the same reason, either

$$(k + 1) \cdot (n_s + m_s \cdot L) - 1 < 1/2 \cdot 1/2^s = 1/2^{s+1}$$

or

$$1 - k \cdot (n_s + m_s \cdot L) < 1/2^{s+1}.$$

Then either

$$(n_{s+1}, m_{s+1}) = ((k + 1) \cdot n_s - 1, (k + 1) \cdot m_s)$$

or

$$(n_{s+1}, m_{s+1}) = (1 - k \cdot n_s, -k \cdot m_s)$$

solves the problem, QED.

D.5 Some Special Issues

D.5.1 Integers, Rationals, and Real Numbers

Definition 147 *The index function index* $: \mathbb{R} \to \mathbb{Z}$ *is defined by*

$$
index(x) = \begin{cases} 1 & \text{if } 0 < x, \\ -1 & \text{if } 0 > x, \\ 0 & \text{if } 0 = x \end{cases} \tag{D.3}
$$

for $x \in \mathbb{R}$.

A real number x has unique representation $x = bottom(x) + x_+$ with $bottom(x) \in \mathbb{Z}, 0 \leqslant x_+ < 1$; we set $top(x) = floor(x) + 1$.

Definition 148 *The rounding function round* $: \mathbb{R} \to \mathbb{Z}$ *is defined by*

$$
round(x) = \begin{cases} floor(x) & \text{if } x_+ \leqslant 0.5, \\ top(x) & \text{else} \end{cases} \tag{D.4}
$$

for $x \in \mathbb{R}$.

E

Modules, Linear, and Affine Transformations

E.1 Modules and Linear Transformations

Definition 149 *Let R be a ring, then a* (left)[1] *R-module is a triple $(R, M, \mu : R \times M \to M)$ where M is an additively written abelian group and μ is the* scalar multiplication, *usually written as $\mu(r, m) = r.m$ if μ is clear (R is also called the ring of* scalars, *and M the group of* vectors*), with the properties:*

1. *We have $1_r.m = m$ for all $m \in M$.*
2. *For all $r, s \in R$ and $m, n \in M$, we have $(r+s).m = r.m + s.m$, $r.(m+n) = r.m + r.n$, and $r.(s.m) = (r \cdot s).m$.*

If $(R, M, \mu : R \times M \to M), (R, N, \nu : R \times N \to N)$ are two R-modules, a group homomorphism $f : M \to N$ is called R-*linear* (or a module homomorphism if the rest is clear) *iff* it is "homogeneous", i.e., $f(r.m) = r.f(m)$, for all $r \in R, m \in M$, with the respective scalar multiplications. The set of R-linear homomorphisms from M to N is denoted by $Lin_R(M, N)$. It is an additive group under the pointwise addition $(f + g)(m) = f(m) + g(m)$. The set-theoretic composition $g \circ f$ of two module homomorphisms $f : M \to N, g : N \to L$ is also a module homomorphism, and we have distributivity, i.e., $(g_1 + g_2) \circ f = g_1 \circ f + g_2 \circ f$ for $g_i : N \to L, f : M \to N$ and $g \circ (f_1 + f_2) = g \circ f_1 + g \circ f_2$ for $f_i : M \to N, g : N \to L$.

By the distributivity of composition of module homomorphisms, the group $End_R(M) = Lin_R(M, M)$ is a ring, the *endomorphism ring* of M, which contains the multiplicative *automorphism group* $Aut_R(M) = End_R(M)^\times$ of M.

An R-linear homomorphism $f : M \to N$ has a group-theoretic kernel $Ker(f)$ and an image $Im(f)$ which are also submodules. For a submodule $N \subseteq M$, the quotient group M/N is also an R-module by the scalar multiplication $r.(m + N) = r.m + N$. The group-theoretic results Proposition 82 and the isomorphism Theorems 47, 48 in Appendix C, are valid literally if we replace the respective groups by modules (the normality of subgroups is automatic here).

E.1.1 Examples

Abelian groups G are canonically identified to $\mathbb{Z}$-modules by $z.g = g + g + \ldots g$, z times for $z > 0$, $(-z).g = -(z.g)$, and $0.g = 0_G$.

If M is an R-module, a submodule $N \subseteq M$ is a subgroup that is stable under scalar multiplication, i.e., $R.N = N$. If $S \subset M$ is a subset, the smallest submodule containing S is denoted by $\langle S \rangle$ and consists of all *linear combinations* $\sum_i r_i.s_i, r_i \in R, s_i \in S$. If there is a finite set S such that $\langle S \rangle = M$, M is called *finitely generated*. Finitely generated $\mathbb{Z}$-modules are completely classified, see C.3.4.2. If $(M_i)_{i \in I}$ is a family of submodules of M, we denote by $\Sigma_I M_i$ the module $\langle \bigcup_I M_i \rangle$. It consists of all finite sums $x_{i_1} + \ldots x_{i_k}, x_{i_j} \in M_{i_j}$.

Every ring R is a left R-module $_R R$ and a right R-module R_R by the given multiplication. For commutative rings these structures coincide. A left, right ideal in R identifies to a submodule of $_R R, R_R$, respectively.

---

[1] Right modules are defined in complete analogy, the scalar multiplication being written $m.r$ instead of $r.m$.

© Springer International Publishing AG, part of Springer Nature 2017
G. Mazzola, *The Topos of Music IV: Roots*, Computational Music Science,
https://doi.org/10.1007/978-3-319-64495-0_5

If $\varphi : S \to R$ is a ring homomorphism and M is an R-module, we have an S-module structure $M_{[\varphi]}$ on M via $s.m = \varphi(s).m$, the *module defined by restriction of scalars*. If φ is clear, one also writes $M_{[S]}$. For the $\mathbb{Z}$-algebra structure of every ring R, $M_{[\mathbb{Z}]}$ gives the underlying structure of an abelian group M back. In particular, an R-algebra S is an R module via $(_S S)_{[R]}$.

A *dilinear homomorphism* from an S-module M to an R-module N is a pair $(\varphi : S \to R, f : M \to N_{[\varphi]})$ consisting of a scalar restriction φ and an S-linear homomorphism f. If $(\varphi : S \to R, f : M \to N_{[\varphi]}), (\psi : T \to S, g : L \to M_{[\psi]})$ are two dilinear homomorphisms, their composition is defined by $(\varphi \circ \psi : T \to R, f \circ g : L \to N[\varphi \circ \psi])$. The set of dilinear homomorphisms from M to N is denoted by $Dil(M, N)$. If the scalar restriction is fixed by φ, we denote the corresponding set by $Dil_\varphi(M, N)$, and the special case $Dil_{Id_R}(M, N)$ is just $Lin_R(M, N)$ as above.

For any family $(M_i)_{i \in I}$ of R-modules, we have the *product module* $\prod_I M_i$. This is the product of the underlying groups, together with coordinatewise scalar multiplication $r.(m_i) = (r.m_i)$ The submodule of those (m_i) with only finitely many $m_i \neq 0$ is the *direct sum module* $\bigoplus_I M_i$. For every index j, one has the canonical (linear) projections $\pi_j : \prod_I M_i \to M_j$ and $\pi_j : \bigoplus_I M_i \to M_j$, via $\pi_j((m)_i) \mapsto m_j$, as well as the canonical (linear) injections $\iota_j : M_j \to \bigoplus_I M_i$ with $\iota_j(m)$ having zero coordinates except for coordinate index j where the value is m.

Lemma 83 (Universal limit property of direct products of modules) *For every family $(f_i : X \to M_i)_i$ of linear homomorphisms there is exactly one linear homomorphism $f : X \to \prod_I M_i$ such that $f_j = \pi_j \circ f$ for all $j \in I$.*

(Universal colimit property of direct sums of modules) *For every family $(f_i : M_i \to X)$ of linear homomorphisms there is exactly one linear homomorphism $f : \bigoplus_I M_i \to X$ such that $f_j = \circ f \circ \iota_j$ for all $j \in I$.*

A sum $\Sigma_I M_i$ of submodules M_i of a module M is called (inner) *direct*, iff the linear homomorphism $\bigoplus_I M_i \to M$ which is induced by the inclusions $M_i \subseteq M$ is an isomorphism.

For two positive integers m, n denote by $m \times n = [1, m] \times [1, n]$ the set of all pairs (i, j), $1 \leq i \leq m, 1 \leq j \leq n$. For a ring R, we have the direct sum $\mathbb{M}_{m,n}(R) = \bigoplus_{m \times n} R$ whose elements are written in the *matrix notation*

$$(r_{i,j}) = \begin{pmatrix} r_{1,1} & \cdots & r_{1,n} \\ \cdots & r_{i,j} & \cdots \\ r_{m,1} & \cdots & r_{m,n} \end{pmatrix}$$

whose *rows* or *columns* are the submatrices with constant first or second index, respectively.

If $(r_{i,j}) \in \mathbb{M}_{m,n}(R)$ and $(s_{j,k}) \in \mathbb{M}_{n,l}(R)$, we have the *matrix product* $(r_{i,j}) \cdot (s_{j,k}) = (t_{i,k}) \in \mathbb{M}_{m,l}(R)$ with $t_{i,k} = \sum_j r_{i,j} \cdot s_{j,k}$. Whenever defined, the product is associative. It is also distributive, i.e., $((r_{i,j}) + (r'_{i,j})) \cdot (s_{j,k}) = (r_{i,j}) \cdot (s_{j,k}) + (r'_{i,j}) \cdot (s_{j,k})$, and $(r_{i,j}) \cdot ((s_{j,k}) + (s'_{j,k})) = (r_{i,j}) \cdot (s_{j,k}) + (r_{i,j}) \cdot (s'_{j,k})$. For $m = n$, one has the identity matrix $E_m = (\delta_{ij})$ with the *Kronecker delta* $\delta_{ii} = 1$, $\delta_{ij} = 0$ for $i \neq j$. With this identity and the matrix addition and multiplication, $\mathbb{M}_{m,m}(R)$ is a ring. With the matrix multiplication as scalar multiplication, $\mathbb{M}_{m,n}(R)$ becomes a left $\mathbb{M}_{m,m}(R)$-module and a right $\mathbb{M}_{n,n}(R)$-module.

If R is commutative, $\mathbb{M}_{m,m}(R)$ is an R-algebra via $r \mapsto r.E_m$, i.e., the scalar multiplication of the R-module $\mathbb{M}_{m,m}(R)$ coincides with the multiplication with R-elements from the algebra embedding.

E.2 Module Classification

E.2.1 Dimension

For any set C and ring R, we have the *free R-module* $\bigoplus_C R = R^C$ of rank $card(C)$ which is the direct sum of C copies of $_R R$ (for $C = \varnothing$, we take the zero module 0_R). A *free module* M is one that is isomorphic to a free module R^C. It is well known that the rank $card(C)$ is then uniquely determined and called the *dimension* $dim(M)$ of the free module M. If $R = F$ is a skew field, an F-module is called a *vector space* (over F), and we have the main fact of linear algebra:

Theorem 54 *Every vector space M over the skew field F is free, and the dimensions are a complete system of invariants of isomorphism classes of vector spaces.*

The proof of this theorem is based on the concept of *linear (in)dependence* in a module. A family $(m_i)_i$ of elements $m_i \in M$ is called *linearly independent* iff any (finite) linear combination $0 = \sum_{j=1,\ldots k} r_j.m_{i_j}$ implies $r_j = 0$, for all j. Otherwise the family is called *linearly dependent*. A *base* of a module is a family of linearly independent elements which generates the module. The main Theorem 54 is proved by the *exchange theorem* which states that any family $(m_i)_i$ of linearly independent vectors can be inserted in a given basis by exchanging some of its elements with the $(m_i)_i$.

Example 107 $dim(\mathbb{R}_{[\mathbb{Q}]}) = card(\mathbb{R}) = 2^{\aleph_0}$, and the sequence of b-logarithms $(\log_b(p))_{p=\text{ prime}}$ is linearly independent by Corollary 32 in Appendix D.2. This means that for any finite increasing sequence $p. = (p_1, p_2, \ldots p_k)$ of primes and the corresponding sequence

$$H_{p.} = (\log_b(p_1), \log_b(p_2), \ldots \log_b(p_k)),$$

the scalar product map

$$H : \mathbb{Q}^k \to \mathbb{R}_{[\mathbb{Q}]}$$

with $H(x) = (H_{p.}, x) = \sum_i \log_b(p_i) x_i$ is a linear injection. The special case of the first three primes and $H_{prime} = (\log_b(2), \log_b(3), \log_b(5))$ was discussed in section A.2.3. Here we can prove the result needed for Proposition 76:

Lemma 84 *For any positive real bound δ and every real number ϕ, there is $x \in \mathbb{Z}^3$ such that $|H(x) - \phi| < \delta$*

Proof. WLOG, we can work with logarithm basis $b = 2$. We know from Lemma 82 in Appendix D.4 that there is $x' = (n, m, 0) \in \mathbb{Z}^3$ such that $0 < H(x') < \delta$. Clearly, there is an integer multiple $x = z.x'$ which does the job, QED.

Corollary 33 (of Theorem 54) *If $f : N \xrightarrow{\sim} R^n$ and $g : M \xrightarrow{\sim} R^m$ are two free modules of finite ranks n, m, isomorphic to the free modules R^n, R^m via isomorphisms f, g, then the linear homomorphisms are described by matrices: If $h : N \to M$ is a linear homomorphism, there is a uniquely determined matrix $H \in \mathbb{M}_{m,n}(R)$ such that for $x \in N$, we have $h(x) = g^{-1}(H \cdot f(x))$, where $f(x)$ is written as a column matrix in $\mathbb{M}_{n,1}(R)$ which canonically identifies to R^n. And conversely, each such matrix defines a linear homomorphism. In other words, there is an isomorphism*

$$Lin_R(N, M) \xrightarrow{\sim} \mathbb{M}_{m,n}(R) \tag{E.1}$$

of additive groups.

If R is a commutative ring, then an R-algebra S is also an R-module. We have an injective homomorphism of R-algebras

$$\Lambda : S \to End_R(S) \tag{E.2}$$
$$\Lambda(s)(s') = s \cdot s'$$

which is called the *left regular representation of S*. If S is a free R-module of dimension n, then the isomorphism (E.1) induces the left regular representation in matrices:

$$\lambda : S \to \mathbb{M}_{n,n}(R). \tag{E.3}$$

More generally, a *linear representation* of an R-algebra A is an algebra homomorphism $f : A \to End_R(M)$ into the R-algebra of endomorphisms of an R-module M.

Definition 150 *The points of a k-element local composition K in $\mathbb{R}^n$ is in general position iff $dim(\mathbb{R}.K) = k - 1$.*

Theorem 55 *Let $v_1, v_2, \ldots v_n$ be n vectors in a $\mathbb{Q}$-module, and take two submodules G, H of dimensions $g = dim(G), h = dim(H)$. Suppose that the module which is generated by the vectors v_i, G, and H, has dimension $n + g + h$. Then we can have at most $n + g + h$ points in general position in the union $\bigcup_i v_i + G + H$.*

Proof. WLOG, one may suppose $v_1 = 0$ after a shift. Take bases $x_1, \ldots x_g$, $y_1, \ldots y_h$ of G, H, respectively. Then, the vectors $0, v_2, \ldots v_n, v_1 + x_1, v_1 + x_2, \ldots x_g, v_1 + y_1, \ldots v_1 + y_h$ are in general position. Conversely, if the vectors $z_1, z_2, \ldots z_m$ in the union $\bigcup_i v_i + G + H$ are in general position, then $dim(\mathbb{Q}.\{z_1, z_2, \ldots z_m\}) = m - 1$. But $\langle z_1, z_2, \ldots z_m \rangle$ is contained in the module which is spanned by $v_2, \ldots v_n$, G, and H, whose dimension is $n + g + h - 1$, i.e., $m - 1 \leqslant n + g + h - 1$, QED.

E.2.2 Endomorphisms on Dual Numbers

For a commutative ring R, we have the commutative R-algebra $R[\varepsilon]$ of dual numbers (see Example 106 in Appendix D.1.1). As an R-module, it has dimension 2 and is isomorphic to R^2 under the map $a + \varepsilon.b \mapsto (a, b)$. By the above Corollary 33, the R-linear endomorphism ring of $R[\varepsilon]$ identifies to the four-dimensional matrix ring $\mathbb{M}_{2,2}(R)$. In this situation, the left regular representation of $R[\varepsilon]$ is the homomorphism of R-algebras

$$\lambda : R[\varepsilon] \to \mathbb{M}_{2,2}(R)$$

with

$$\lambda(a + \varepsilon.b) = \begin{pmatrix} a & 0 \\ b & a \end{pmatrix},$$

which represents the linear endomorphism of multiplication by $a + \varepsilon.b$. We have shown in section 29.6 that $\mathbb{M}_{2,2}(R)$ is generated by the four R-linear basis elements $\lambda(1_R), \lambda(\varepsilon), \alpha_+, \alpha_+ \cdot \lambda(\varepsilon)$, where α_+ is the sweeping orientation

$$\alpha_+ = \begin{pmatrix} 1 & 1 \\ 0 & 0 \end{pmatrix},$$

and that the R-algebra $\mathbb{M}_{2,2}(R)$ identifies to the quotient

$$R\langle \lambda(\varepsilon), \alpha_+ \rangle / (\lambda(\varepsilon)^2, \alpha_+^2 - \alpha_+, \alpha_+ \cdot \lambda(\varepsilon) + \lambda(\varepsilon) \cdot \alpha_+ - \lambda(1 + \varepsilon))$$

of the polynomial R-algebra in the two non-commuting variables $\lambda(\varepsilon), \alpha_+$.

E.2.3 Semi-simple Modules

A module $M \neq 0_R$ which has no submodules except $0_R, M$ is called *simple*. A module M is called *semi-simple* iff it has the following equivalent properties:

Lemma 85 *Let M be an R-module. The following statements are equivalent:*

(i)   Every submodule of M is a sum of simple submodules.
(ii) M is the sum of simple submodules.
(iii) M is the direct sum of simple submodules.
(iv) Every submodule N of M is a direct summand (i.e., there is a submodule N' of M such that $M = N \oplus N'$.

The following is immediate:

Lemma 86 *A linear homomorphism between simple R-modules is either an isomorphism or zero. Hence the endomorphism ring $End_R(M)$ of a simple module M is a skew field.*

For example, the left $\mathbb{M}_{m,m}(F)$-module $\mathbb{M}_{m,n}(F)$ and the right $\mathbb{M}_{n,n}(F)$-module $\mathbb{M}_{m,n}(F)$ of $m \times n$-matrices over a skew field F is semi-simple. For the left module, the n columns are the simple submodules, whereas for the right module, the m rows are the simple submodules. Moreover, the only left and right submodules of $\mathbb{M}_{m,n}(F)$ are the zero module and $\mathbb{M}_{m,n}(F)$. In particular, the only two-sided ideals in the ring $\mathbb{M}_{m,m}(F)$ are 0 and $\mathbb{M}_{m,m}(F)$, i.e., this ring is simple.

Theorem 56 (Wedderburn) *The semi-simple rings are the finite products $\prod_i \mathbb{M}_{m_i,m_i}(F_i)$ of matrix rings $\mathbb{M}_{m_i,m_i}(F_i)$ over skew fields F_i.*

If G is a finite group, and if K is a commutative field, we have defined the monoid algebra $K\langle G \rangle$ in D.1.1, it is called the group algebra in this case. Here are the semi-simple group algebras over commutative fields:

Theorem 57 (Maschke) *The group algebra $K\langle G \rangle$ is semi-simple iff $char(K) \nmid card(G)$.*

E.2.4 Jacobson Radical and Socle

Definition 151 *The intersection of all maximal submodules of an R-module $M \neq 0$ is called the Jacobson radical of M, it is denoted by $Rad(M)$.*

Sorite 13 *let M, N be R-modules. Then:*

(*i*) *For $f \in Lin_R(M, N)$, we have $f(Rad(M)) \subseteq Rad(N)$.*
(*ii*) *We have $Rad(M \oplus N) = Rad(M) \oplus Rad(N)$.*
(*iii*) *We have $Rad(M/Rad(M)) = 0$.*
(*iv*) *We have $Rad(_R R).M \subseteq Rad(M)$.*
(*v*) *If M is semi-simple, then $Rad(M) = 0$.*
(*vi*) *If a submodule N of M has M/N semi-simple, then $Rad(M) \subseteq N$.*

For a ring R, one may look at its left radical $Rad(_R R)$, or at its right radical $Rad(R_R)$. Fortunately, there is no difference in that:

Proposition 91 *The left and right radicals of a ring R coincide, $Rad(_R R) = Rad(R_R)$, and this (two-sided) ideal $Rad(R)$ is the maximal ideal I which annihilates every semi-simple module M, i.e., $I.M = 0$.*

For a ring R, we set

$$J_r = \{r \in R | 1_R \notin r \cdot R\}, J_l = \{r \in R | 1_R \notin R \cdot r\}.$$

Proposition 92 *For a ring R the following conditions are equivalent:*

(*i*) *The quotient ring $R/Rad(R)$ is a skew field.*
(*ii*) *We have $J_r = Rad(R)$.*
(*iii*) *We have $J_l = Rad(R)$.*
(*iv*) *The set J_r is additively closed.*
(*v*) *The set J_l is additively closed.*

Definition 152 *A ring with the equivalent properties of Proposition 92 is called* local. *In particular, a commutative ring R is local iff it has a unique maximal ideal.*

The *length $l(M)$ of a module* M is the maximum length l of finite chains $0 \subsetneqq N_1 \subsetneqq N_2 \subsetneqq \ldots N_l = M$ of submodules (if that maximum is ∞, we set $l(M) = \infty$). A module is called *indecomposable* iff it is not the direct sum of two proper submodules.

Lemma 87 *If M is an R-module with local endomorphism ring, then M is indecomposable. Conversely, let M be an indecomposable module of finite length l. Then $End_R(M)$ is a local ring, and the radical $S = Rad(End_R(M))$ is nilpotent, namely $S^l = 0$.*

Proposition 93 *Let $X \subset M, Y \subset N$ be two non-zero submodules of indecomposable R-modules M, N of finite lengths. If $F : M \to N, g : N \to M$ are linear maps which induce mutually inverse isomorphisms $f|X : X \xrightarrow{\sim} Y, g|Y = (f|X)^{-1} : Y \xrightarrow{\sim} X$, then f, g are isomorphisms.*

Proof. Since $f \cdot f$ restricts to the identity of X, so does any of its positive powers. So non of them can be the zero endomorphisms of M. But from Lemma 87 we know that $End(M)$ is local and that a non-nilpotent endomorphism must be invertible. A symmetric argument yields inversibility of $f \cdot g$ and therefore of both, f and g, QED.

Lemma 88 *(Fitting's lemma) Let $f : M \to M$ be a linear endomorphism of a module M of finite length. Then there is a positive power f^n and a direct decomposition $M = N \oplus Ker(F^n)$ such that f^n is an automorphism on N.*

Definition 153 *The socle $Soc(M)$ of a module M is the sum of all its simple submodules.*

Theorem 58 *For a module M of finite length, the following three statements are equivalent:*

(i) M is semi-simple.
(ii) $Rad(M) = 0$.
(iii) $Soc(M) = M$.

E.2.5 Theorem of Krull-Remak-Schmidt

Theorem 59 *(Krull-Remak-Schmidt) Let $M_1, \ldots M_k, M'_1, \ldots M'_l$ be modules with local endomorphism rings (in particular these modules are all non-zero). Suppose that the direct sums $\bigoplus_{i=1,\ldots k} M_i$ and $\bigoplus_{j=1,\ldots l} M_j$ are isomorphism of R-modules. Then $k = l$, and there is a permutation σ of the indices such that we have isomorphisms $M_i \xrightarrow{\sim} M'_{\sigma(i)}$ for all $i = 1 \ldots k$.*

Corollary 34 *A module of finite length is a direct sum of indecomposable submodules in a unique way up to permutation and isomorphisms of the summands.*

This follows from Theorem 59 in view of Lemma 87, QED.

E.3 Categories of Modules and Affine Transformations

See Appendix G for a reference to category theory.

For an additive group M and an element $m \in M$, the *translation by m* is the set map $e^m : M \to M : x \mapsto e^m(x) = m + x$. The exponential notation is chosen because $e^? : M \to \mathfrak{S}_M$ is an injective group homomorphism. We denote by e^M the group of translations on M, a group which is isomorphic to M.

Definition 154 *For two rings R, S, an R-module M and an S-module N, a diaffine homomorphism f is a map of form $e^n \cdot f_0$, where e^n is a translation on N and $f_0 \in Dil(M, N)$. The set of diaffine homomorphisms $f : M \to N$ is denoted by $M @ N$. If we fix the underlying scalar restriction $\varphi : R \to S$ and only take $f_0 \in Dil_\varphi(M, N)$, the corresponding set is denoted by $M @_\varphi N$. In particular, if $\varphi = Id_R$, we write $M @_R N$ for the set of (R-)affine homomorphisms.*

Sorite 14 *If R, S, T are rings and M, N, L are modules over these rings, respectively, we have the following facts:*

(*i*) *If $f = e^n \cdot f_0 \in M\,@N$, then $n = f(0)$, and $f_0 = e^{-n} \cdot f$. So the* translation part e^n *and the* dilinear part f_0 *are uniquely determined.*

(*ii*) *If $f = e^n \cdot f_0 \in M\,@N, g = e^l \cdot g_0 \in N\,@L$, then the set-theoretic composition $g \cdot f \in M\,@L$ and*

$$g \cdot f = e^{l+g(n)} \cdot g_0 \cdot f_0. \tag{E.4}$$

(*iii*) *The diaffine $f = e^n \cdot f_0 \in M\,@N$ is an isomorphism iff its dilinear part f_0 is so, and then the inverse is $f^{-1} = e^{-f_0^{-1}(n)} \cdot f_0^{-1}$.*

(*iv*) *If $f,g \in M\,@_\varphi N$, then so is the pointwise difference $f - g$, hence $M\,@_\varphi N$ is an additive group. If S is commutative, $M\,@_\varphi N$ is an S-module and also an R-module under φ.*

(*v*) *For $M = N, \varphi = Id_R$, the* general affine group $\overrightarrow{GL}(M)$, *i.e., the group of affine automorphisms is isomorphic to the semidirect product $M \rtimes_\phi Gl(M)$ of M with the* general linear group $GL(M)$ *of linear automorphisms of M under the group action $\phi = Id_{GL(M)}$.*

The category of modules and diaffine homomorphisms is denoted by **Mod**, whereas the subcategory of (left) R-modules and R-affine homomorphisms is denoted by **Mod**$_R$. If $f : R \to S$ is a ring homomorphism, we have the scalar extension functor

$$S \otimes_R ? : \mathbf{Mod}_R \to \mathbf{Mod}_S : M \mapsto S \otimes_R M \tag{E.5}$$

which acts by scalar extension of the linear parts of morphisms and by the canonical map $M \to S \otimes_R M :$ $x \mapsto 1 \otimes x$ on the translation part. See [145, II.5.1] for details.

In **Mod**, we have to add an additional object $\varnothing_R$ for each ring R. This is the empty set plus the unique possible scalar multiplication. This is *not* a module in the usual sense since it is not even a group! But there are important category-theoretic reasons to introduce these objects. Observe that for an S-module N, $\varnothing_R@N$ is in bijection with the set of ring homomorphisms $Hom(R, S)$, whereas $N@\varnothing_R$ is empty if N is not empty.

By **Mod**$^@$, we denote the category of set-valued presheaves on **Mod**, i.e., the contravariant set-valued functors $F : \mathbf{Mod} \to \mathbf{Ens}$. In particular, the Yoneda embedding **Mod** $\to$ **Mod**$^@$ yields the representable presheaf $@M$ for a module M, with $@M(X) = X@M$ for $X \in {}_0\mathbf{Mod}$. This is one reason why we also write $X@F$ for the evaluation $F(X)$, even if F is not representable. By $M@$ we denote the covariant functor with $M@(X) = M@X$. In the context of presheaves, we often call a module X that is an argument of such presheaves an *address*; the reasons for this wording are made explicit in the musicological Chapter 6 on forms and denotators.

E.3.1 Direct Sums

Proposition 94 *Let A be an R-module, and n a natural number. Then there is a canonical isomorphism $@A^{\oplus n} \xrightarrow{\sim} (@A)^n$, i.e., $A^{\oplus n}$ represents the n-fold product functor.*

Proof. Let X be any module. Then every affine homomorphism $f = e^t \cdot f_0 : X \to A^{\oplus n}$ projects to the n factors $f_i = p_i \cdot f$ via the respective projections $p_i : A^{\oplus n} \to A$. Also, the dilinear part f_0 projects to the n dilinear factors $f_{0,i} : X \to A$. Let t_i be the ith component of t. Then we have $f_i = e^{t_i} \cdot f_{0,i}$. This yields the desired bijection $X@A^{\oplus n} \xrightarrow{\sim} (X@A)^n$, and this is functorial in X. QED.

E.3.2 Affine Forms and Tensors

In this section we suppose that all modules have a commutative coefficient ring R, i.e., we work in the category **Mod**$_R$. Tensor products[2] are automatically taken over R. By $X^\star$ we denote the R-linear dual $Lin_R(X; R)$ of the R-module X.

---

[2] See [145] for tensor products.

For the R-module $A^\star$ of affine forms on an R-module (address) A and an R-module M, we have a canonical linear injection $M \rightarrowtail A^\star \otimes M$. In fact, there is an R-linear isomorphism $A^\star \xrightarrow{\sim} R \oplus A^\star : e^r \cdot x \mapsto r + x$, and we deduce an R-linear isomorphism $A^\star \otimes M \xrightarrow{\sim} M \oplus A^\star \otimes M$, whence the above injection; it maps $m \in M$ to $e^{1_R} . 0 \otimes m$.

With the above notation, fix an R-module A (an address). We have the subfunctor

$$A@_R : \mathbf{Mod}_R \to \mathbf{Ens} : M \mapsto A@_R M \tag{E.6}$$

of $A@$, and induced on the subcategory $\mathbf{Mod}_R$ of $\mathbf{Mod}$. We further have the functor

$$A^\star \otimes : \mathbf{Mod}_R \to \mathbf{Ens} : M \mapsto A^\star \otimes M. \tag{E.7}$$

This functor acts as follows on affine morphisms $F = e^n \cdot F_0 : M \to N$. Identify n with the canonically associated element of $A^\star \otimes N$. Then, we define

$$A^\star \otimes F = e^n \cdot A^\star \otimes F_0, \tag{E.8}$$

in other words, $(A^\star \otimes F)_0 = A^\star \otimes F_0$. For the composition

$$e^{t + G_0(n)} \cdot G_0 F_0 : M \xrightarrow{F} N \xrightarrow{G} T$$

of affine morphisms $F = e^n \cdot F_0$ and $G = e^t \cdot G_0$ this implies

$$
\begin{aligned}
A^\star \otimes GF &= e^{t + G_0(n)} \cdot A^\star \otimes G_0 F_0 \\
&= e^t \cdot e^{G_0(n)} \cdot A^\star \otimes G_0 \cdot A^\star \otimes F_0 \\
&= e^t \cdot A^\star \otimes G_0 \cdot e^n \cdot A^\star \otimes F_0 \\
&= A^\star \otimes G \cdot A^\star \otimes F,
\end{aligned}
$$

whence the claimed functoriality.

Lemma 89 *With the above notation, there is a natural transformation*

$$\theta : A^\star \otimes \to A@_R. \tag{E.9}$$

If M is an R-module, it is defined by its action on pure tensors $x \otimes m \in A^\star \otimes M$ by

$$\theta(x \otimes m) : A \to M : a \mapsto x(a)m. \tag{E.10}$$

The natural transformation θ is an isomorphism if A is a finitely generated projective module.

Proof. The formula (E.10) is an extension of a classical formula in the linear case, see [145, II.74]. In fact, write $A^\star \otimes M \xrightarrow{\sim} M \oplus A^\star \otimes M$, and then $A@M \xrightarrow{\sim} M \oplus Lin_R(A, M)$. Then the classical formula $\theta_0 : A^\star \otimes M \to Lin_R(A, M) : x \otimes m \mapsto \theta_0(x \otimes m)$ with $\theta_0(x \otimes m)(a) = x(a)m$ extends to the linear map

$$\theta'(m + \lambda) = m + \theta_0(\lambda)$$

which for special pure tensor arguments $x \otimes m = e^r \cdot x_0 \otimes m$ yields

$$\theta'(rm + x_0 \otimes m) = rm + \theta_0(x_0 \otimes m), \text{ corresponding to}$$
$$e^{rm} \cdot \theta_0(x_0 \otimes m),$$

and the latter evaluates to

$$(e^{rm} \cdot \theta_0(x_0 \otimes m))(a) = rm + x_0(a)m = x(a)m$$

which means $\theta' = \theta$. Let us then prove the naturality of θ. Let $F = e^n \cdot F_0 : M \to N$ be an affine morphism. We have to show that the diagram

$$
\begin{array}{ccc}
A^\star \otimes M & \xrightarrow{\;\theta\;} & A@M \\
{\scriptstyle A^\star \otimes F}\downarrow & & \downarrow{\scriptstyle A@F} \\
A^\star \otimes N & \xrightarrow{\;\theta\;} & A@N
\end{array}
\tag{E.11}
$$

is commutative. It suffices to verify it for pure tensors $x \otimes m \in A^\star \otimes M$. Take $a \in A$. Then

$$
\begin{aligned}
A@F(\theta(x \otimes m))(a) &= F(\theta(x \otimes m)(a)) \\
&= F(x(a)m) \\
&= n + F_0(x(a)m) \\
&= n + x(a)F_0(m) \\
&= \theta(n) + \theta(x \otimes F_0(m))(a) \\
&= \theta(n + x \otimes F_0(m))(a) \\
&= \theta((A^\star \otimes F)(x \otimes m))(a)
\end{aligned}
$$

and we are done with diagram (E.11). If A is finitely generated projective, the classical linear map θ_0 is iso, and hence so is θ. QED.

Observe that the special case $A = 0$ of zero address is included in the lemma and that in this case, θ identifies to the identity transformation on the forgetful functor.

E.3.3 Biaffine Maps

In this section we again suppose that all modules have a commutative coefficient ring R, i.e., we work in the category $\mathbf{Mod}_R$. In classical module theory, the tensor product is a universal construction relating to bilinear maps. The extension to biaffine maps runs as follows.

Definition 155 *Let U, V, W be modules in $\mathbf{Mod}_R$. A map*

$$f : U \times V \to W$$

is called biaffine *if it is affine in each variable, i.e., if*

$$
\begin{aligned}
f_u : V \to W : v \mapsto f_u(v) = f(u, v) \;\; and \\
f^v : U \to W : u \mapsto f^v(u) = f(u, v)
\end{aligned}
$$

are all affine, i.e., $f_u \in V@_R W$ and $f^v \in U@_R W$. The set of all biaffine maps f is denoted by $\mathcal{A}_2(U, V; W)$.

Lemma 90 *For R-modules U, V, W in $\mathbf{Mod}$, there is a canonical bijection*

$$\mathcal{A}_2(U, V; W) \xrightarrow{\sim} U@_R(V@_R W). \tag{E.12}$$

If these sets are given their canonical structure of R-modules, bijection E.12 is an isomorphism of R-modules.

Proof. Let $f : U \times V \to W$ be a biaffine map. Then the associated map $f_u, u \in U$ by definition stays in $V@_R W$. So we have a map $f_? : U \to V@_R W$. Let us show that this map is affine. Set $\lambda(v) = f(0, v)$, this affine map is the constant part of our candidate $f_?$, i.e., we claim that $f_u - \lambda$ is linear in u. But $f_u(v) - \lambda(v) = f^v(u) - \lambda(v) = f^v(u) - f^v(0)$ is linear in u and we are done. Conversely, each $g \in U@_R(V@_R W)$ defines $\tilde{g}(u, v) = g(u)(v)$ with $\tilde{g} \in \mathcal{A}_2(U, V; W)$, and this clearly is an inverse to the map $f \mapsto f_?$. That we have a module isomorphism is clear. QED.

Proposition 95 *Let U, V, W be R-modules in* **Mod**, *and define the* affine tensor product $U \boxtimes V = U \otimes V \oplus U \oplus V$. *Then we have a canonical bijection*

$$\boxtimes : \mathcal{A}_2(U, V; W) \xrightarrow{\sim} (U \boxtimes V) @_R W, \tag{E.13}$$

i.e., $U \boxtimes V$ is a universal object in the affine category $\mathbf{Mod}_R$ *like the tensor product is for R-linear maps. If $f \in \mathcal{A}_2(U, V; W)$, then its image $\boxtimes f$ applies a typical element $u \otimes v + r + s$ to*

$$\boxtimes f(u \otimes v + r + s) = \Box f(u \otimes v) + f^0(r) + f_0(s) - f(0, 0) \tag{E.14}$$

where $\Box f$ is the linear map associated with the bilinear map

$$f_\Box(u, v) = f(u, v) - f_0(v) - f^0(u) + f(0, 0). \tag{E.15}$$

The universal map $i : U \times V \to U \boxtimes V$ is defined by $i(u, v) = u \otimes v + u + v$.

Proof. The proposition follows directly from Lemma 90, the definition of the affine tensor product and the universal property of the linear tensor product. We then have $\boxtimes f(i(u, v)) = \boxtimes f(u \otimes v + u + v) = \Box f(u \otimes v) + f^0(u) + f_0(v) - f(0, 0) = f_\Box(u, v) + f^0(u) + f_0(v) - f(0, 0) = f(u, v) - f_0(v) - f^0(u) + f(0, 0) + f^0(u) + f_0(v) - f(0, 0) = f(u, v)$. QED.

Definition 156 *For modules U, V, W, X in* $\mathbf{Mod}_R$ *and affine maps $f : U \to W$, $g : V \to X$, the affine tensor product map*

$$f \boxtimes g : U \boxtimes V \to W \boxtimes X \tag{E.16}$$

is defined as the canonical affine map $\boxtimes h$ according to Proposition 95 which is associated with the biaffine map

$$h : U \times V \to W \boxtimes X : (u, v) \mapsto i(f(u), g(v)).$$

Sorite 15 *For modules U, V, W, X in* $\mathbf{Mod}_R$ *and with the notation of section E.3.8, we have:*

(i) $U \boxtimes V \xrightarrow{\sim} V \boxtimes U$.
(ii) $U \boxtimes 0_R \xrightarrow{\sim} 0_R \boxtimes U \xrightarrow{\sim} U$.
(iii) $U \boxtimes (V \boxtimes W) \xrightarrow{\sim} (U \boxtimes V) \boxtimes W$, *i.e., we can identify these products and write $U \boxtimes V \boxtimes W$.*
(iv) $(U \amalg V) \boxtimes W \xrightarrow{\sim} U \boxtimes W \amalg V \boxtimes W$.

For a module M in $\mathbf{Mod}_R$, the functor $@_R M : \mathbf{Mod}_R \to \mathbf{Ens} : B \mapsto B @_R M$ contains redundant structure in B since there are elements in B which are annihilated by all linear maps into M. We want to reduce B to a module where this annihilator set is the zero submodule.

Definition 157 *With the above notation, set $An(B, M) = \bigcap_{k \in B @ M} Ker(k_0)$, denote $B_{/M} = B/An(B, M)$ and write $/M : B \to B_{/M}$ for the canonical projection. The module $B_{/M}$ is called the M-reduction of B.*

The following lemma is clear:

Lemma 91 *The assignment $?_{/M} : B \mapsto B_{/M}$ defines a functor on* $\mathbf{Mod}_R$. *The projection $/M : B \to B_{/M}$ and the uniquely defined commutative diagrams*

$$\begin{array}{ccc} C & \xrightarrow{/M} & C_{/M} \\ f \downarrow & & \downarrow f_{/M} \\ B & \xrightarrow{/M} & B_{/M} \end{array} \tag{E.17}$$

which are associated with affine homomorphisms $f : C \to B$ define a natural transformation on $Id_{\mathbf{Mod}_R} \to ?_{/M}$.

Proposition 96 *Let $@_R^{red} M = @_R M \cdot ?_{/M}$ Then we have a natural isomorphism $@_R^{red} M \overset{\sim}{\to} @_R M$.*

Proof. In fact, the natural transformation $/M : B \to B_{/M}$ induces an isomorphism (of R-modules) $B@_R^{red} M \overset{\sim}{\to} B@_R M$. QED.

Corollary 35 *With the above notation, the functors $B\tilde{@}M$ and $B\tilde{@}^{red} M = B_{/M}\tilde{@}M$ are canonically isomorphic.*

Proof. If $A \in {}_0\mathbf{Mod}_R$, we have

$$
\begin{aligned}
A@B\tilde{@}^{red} M &= A@B_{/M}\tilde{@}M \\
&= A \boxtimes B_{/M}@_R M \\
&\overset{\sim}{\to} A@_R(B_{/M}@_R M) \\
&\overset{\sim}{\to} A@_R(B@_R M) \\
&\overset{\sim}{\to} A \boxtimes B@_R M \\
&= A@B\tilde{@}M.
\end{aligned}
$$

Let M, A be modules in $\mathbf{Mod}_R$. Then we have

Lemma 92 *There is an isomorphism of R-modules*

$$ Lin_R(M, A@_R R) \overset{\sim}{\to} A@_R M^\star \tag{E.18} $$

which is functorial in both, A and M.

Proof. We have these functorial isomorphisms:

$$
\begin{aligned}
Lin_R(M, A@_R R) &\overset{\sim}{\to} Lin_R(M, A^\star \oplus R) \\
&\overset{\sim}{\to} Lin_R(M, A^\star) \oplus M^\star \\
&\overset{\sim}{\to} (M \otimes A)^\star \oplus M^\star \\
&\overset{\sim}{\to} (A \otimes M)^\star \oplus M^\star \\
&\overset{\sim}{\to} Lin_R(A, M^\star) \oplus M^\star \\
&\overset{\sim}{\to} A@_R M^\star.
\end{aligned}
$$

Proposition 97 *Let M be as above, and $A = R^n$, $0 \leqslant n$, then we have canonical R-linear maps*

$$ u : A@_R M^\star \to (A@_R M)^\star, \tag{E.19} $$
$$ d : A@_R M \to (A@_R M^\star)^\star, \tag{E.20} $$

which are isomorphisms for M finitely generated and projective.

Proof. As to the first map, a linear map $v : Lin_R(A, M^\star) \to Lin_R(A, M)^\star$ is defined as follows: For $g : A \to M^\star$ and $f : A \to M$, we have the composition $f^\star \cdot g : A \to A^\star$, which is a bilinear form on A, and we may set $e(g)(f) = tr(f^\star \cdot g)$, a linear function of g, calculated in the canonical bases of A and $A^\star$. This map is canonically extended by the identity on M and we are done for u. For the second map, we have the canonical bidual linear map $l : A@_R M \to A@_R M^{\star\star}$, and we may apply the first map to the bidual of M. The statement concerning finitely generated projective modules is standard. QED

E.3.4 Symmetries of the Affine Plane

We consider symmetries, i.e., affine transformations $D = e^t \cdot H$ on $\mathbb{R}^2$. From the geometric point of view, the set $\mathbb{R}^2@\mathbb{R}^2$ of these maps is described as follows. Fix a (zero-addressed) local composition $\Delta = \{u, v, w\}$ in the real plane $\mathbb{R}^2$, with three points in general position (see Appendix E.2.1). Then we know from section 15.2.1 that the map

$$B(\Delta) : \mathbb{R}^2@\mathbb{R}^2 \to (\mathbb{R}^2)^3 : D \mapsto (D(u), D(v), D(w)) \tag{E.21}$$

is a bijection. We use this bijection to describe some special transformations:

Shearings. Let G be a straight line in $\mathbb{R}^2$ (not necessarily through the origin), and let u, v, w be in general position such that u, v lie on G, whereas w does not. A *shearing S relating to G* is a symmetry which leaves both, u, v, fixed and transforms w into $w + r.(v - u), r \in \mathbb{R}$. Then G remains fixed identically, and w is shifted in parallel motion with respect to G. The nth power S^n of S is the shearing which fixes G and transforms w into $w + nr.(v - u)$.

Dilatations. For a point $u \in \mathbb{R}^2$, and two scalars $\delta, \sigma \in \mathbb{R}$, a *dilatation D by factors δ, σ and centered in u* is defined by the prescription that $D(u) = u$, and that for two other points v, w such that u, v, w are in general position, we have $D(v) = u + \delta.(v - u), D(w) = u + \sigma.(w - u)$. A dilatation with $\delta = \sigma = -1$ is called *point reflection with center u*, it corresponds to a rotation by $180°$ around u.

Glide Reflections. Let G be a straight line, and take u, v, w as in the above paragraph E.3.4 about shearing. A *glide reflection P* is a symmetry, such that $P(u)$ lies on G, and $P(v) - P(u) = v - u, P(w) - P(u) = u - w$. Therefore, $P(v)$ lies also on G, and $P(w)$ lies on the "opposite side" of w on the line through w, u, i.e., P is a translation by $P(u) - u$, followed by a 'skew' reflection in G in the direction of the line through w and u. Especially for the diagonal $G = \mathbb{R}.(1, 1)$ and $u = P(u) = 0, w = (-1, 1)$, we obtain the exchange of coordinate axes: the *parameter exchange*.

E.3.5 Symmetries on $\mathbb{Z}^2$

Theorem 60 *Every symmetry $f \in \mathbb{Z}^2@\mathbb{Z}^2$ is the product of some of the following symmetries:*

1. *a translation $T = e^{(0,1)}$,*
2. *a shearing S which leaves the first axis $\mathbb{Z}.(1, 0)$ fixed and transforms $(0, 1)$ to $(1, 1)$,*
3. *the parameter exchange P,*
4. *the reflection K at the second axis,*
5. *the dilatations $D_m, 0 \leqslant m$ in the direction of the first axis by factor m.*

The general affine group $\overrightarrow{GL}(\mathbb{Z}^2)$ is generated by T, S, P, K.

Proof. The statement concerning the general affine group is immediate from the first part of the theorem. To show the latter, we say that a symmetry X is "good" if X can be written as a product of symmetries of the required shape. Here are the matrices for the generators:

$$S = \begin{pmatrix} 1 & 1 \\ 0 & 1 \end{pmatrix}, K = \begin{pmatrix} -1 & 0 \\ 0 & 1 \end{pmatrix}, P = \begin{pmatrix} 0 & 1 \\ 1 & 0 \end{pmatrix}, D_m = \begin{pmatrix} m & 0 \\ 0 & 1 \end{pmatrix}.$$

We first show that all 2×2-matrices, i.e., all linear maps $X = \begin{pmatrix} a & b \\ c & d \end{pmatrix}$ are products of symmetries of type S, K, D_m, P. Observe that $P^2 = K^2 = E_2$.

1. If $P \cdot X$ is good, then so is $X = P^2 \cdot X = P \cdot (P \cdot X)$. The same is valid for $X \cdot P$. Here, $P \cdot X$ is the exchange of rows in X, whereas $X \cdot P$ is the exchange of columns in X.

2. We have $S^{-1} = K \cdot S \cdot K = \begin{pmatrix} 1 & -1 \\ 0 & 1 \end{pmatrix}$. Therefore the powers $S^n = \begin{pmatrix} 1 & n \\ 0 & 1 \end{pmatrix}$ are good for integer n. Further,

$P \cdot S^{\pm 1} \cdot P = \begin{pmatrix} 1 & 0 \\ \pm 1 & 1 \end{pmatrix}$ and therefore $(P \cdot S \cdot P)^n = \begin{pmatrix} 1 & 0 \\ n & 1 \end{pmatrix}$ is good.

3. For $m \geqslant 0$, $P \cdot D_m \cdot P = \begin{pmatrix} 1 & 0 \\ 0 & m \end{pmatrix}$ and $P \cdot K \cdot D_m \cdot P = \begin{pmatrix} 1 & 0 \\ 0 & -m \end{pmatrix}$ are good.

4. If a coefficient of X vanishes, one can enforce $c = 0$ by a row or column exchange. Because of

$$\begin{pmatrix} a & b \\ 0 & d \end{pmatrix} = \begin{pmatrix} 1 & 0 \\ 0 & d \end{pmatrix} \cdot \begin{pmatrix} 1 & b \\ 0 & 1 \end{pmatrix} \cdot \begin{pmatrix} a & 0 \\ 0 & 1 \end{pmatrix}$$

and the preceding results, such an X is also good.

5. If no coefficient of X vanishes, one can apply $\pm E_2$ and exchange of columns that satisfy $c \geqslant d > 0$. Applying the Euclidean algorithm (section D.3), we can write $c = nd + r, 0 \leqslant r < d$. This yields

$$\begin{pmatrix} a & b \\ c & d \end{pmatrix} \cdot \begin{pmatrix} 1 & 0 \\ -n & 1 \end{pmatrix} = \begin{pmatrix} a - nb & b \\ c - nb & d \end{pmatrix} = \begin{pmatrix} a' & b \\ r & d \end{pmatrix}.$$

If $r = 0$, we are in case 4 above. Else, we have $d > r > 0$. Via column exchange we obtain $\begin{pmatrix} b & a' \\ d & r \end{pmatrix}$, and we may set $c' = d, d' = r$. But $d' < d$, so the algorithm leads to case 4 after a finite number of steps. This settles the linear maps.

For translations, observe the following identities:

$$e^{(1,0)} = P \cdot T \cdot P, e^{(-1,0)} = K \cdot e^{(1,0)} \cdot K, e^{(0,-1)} = P \cdot e^{(-1,0)} \cdot P.$$

Therefore all transpositions are good since we have natural numbers x, y such that $e^{(\pm x, \pm y)} = (e^{(\pm 1,0)})^x \cdot (e^{(0,\pm 1)})^y$. This settles all the cases, QED.

E.3.6 Symmetries on $\mathbb{Z}^n$

For integers $n \geqslant 2$ and $1 \leqslant i < j \leqslant n$, we have the *diagonal embedding* $\Delta_{i,j} : GL(\mathbb{Z}^2) \rightarrowtail GL(\mathbb{Z}^n)$ defined by

$$\Delta_{i,j}\left(\begin{pmatrix} a & b \\ c & d \end{pmatrix}\right) = (x_{u,v}) \text{ with}$$

$$x_{i,i} = a, x_{i,j} = b, x_{j,i} = c, x_{j,j} = d,$$
$$x_{u,u} = 1 \text{ for } u \neq i, j, \text{ and}$$
$$x_{u,v} = 0 \text{ else.}$$

Theorem 61 *For an integer $n \geqslant 2$ the group $GL(\mathbb{Z}^n)$ is generated by the diagonal embedded groups $\Delta_{1,j}GL(\mathbb{Z}^2), 1 \leqslant j \leqslant n$.*

The proof goes by induction on n and uses the Euclidean algorithm, we leave it as an exercise.

E.3.7 Complements on the Module of a Local Composition

Lemma 93 *Let A be an address module over the commutative coefficient ring R and $(K, A@M)$ a commutative local composition. Then:*

(i) $R.K \subset \langle K \rangle$.

(ii) $R.K = \langle K \rangle$ iff K is embedded.

(iii) Let $f : K \to L$ is a morphism of embedded commutative local compositions

$$(K, A@_R M), (L, A@_R N)$$

at the same address A. Then any underlying symmetry $F : M \to N$ restricts to $R.K$ and $R.L$, i.e., $F(R.K) \subset R.L$, and this restriction is uniquely determined by f. We denote this affine map by $R@f : R.K \to R.L$.

Proof. The first statement is clear since $R.K = \langle x - x_0 | \ x \in K \rangle$. If $R.K = \langle K \rangle$, then obviously $K \subset \langle K \rangle = R.K$, and K is embedded. Conversely, if $K \subset R.K$, then also $\langle K \rangle \subset R.K$ and equality follows from (i). As to (iii), observe that we have a linear application $R.f : R.K \to R.L$ which is induced by the linear part F_0 of F, and which is only a function of f, by Lemma 6 of Chapter 8. Further, if $F = e^n \cdot F_0$, and if $k \in K$, we have $n = F(k) - F_0(k) = f(k) - R.f(k)$ since $K \subset R.K$; in other words, $n = n_f$ is only a function of f, and not of the underlying F. Therefore, since both, $f(k)$ and $R.f(k)$, are elements of $R.L$, $n_f \in R.L$. This means that for $x \in R.K$, $F(x) = n_f + F_o(x) = n_f + R.f(x) = R@f(x) \in R.L$, QED.

E.3.8 Fiber Products and Fiber Sums in Mod

Theorem 62 *The category* **Mod** *of modules and diaffine transformations has arbitrary fiber products.*

Proof. We are given a fiber product diagram

$$K \underset{f}{\to} M \underset{g}{\leftarrow} L \tag{E.22}$$

of modules over the fiber product diagram

$$A \underset{u}{\to} C \underset{v}{\leftarrow} B \tag{E.23}$$

of corresponding coefficient rings. If any of these modules K, L or M is empty, or if intersection $\mathrm{Im}(f) \cap \mathrm{Im}(g)$ is empty, then the empty module over the fiber product $A \times_C B$ of coefficient rings does the job. So we may suppose that neither of these four spaces is empty.

Consider the dilinear parts f_0 and g_0 of f and g. Then we have the dilinear homomorphism $d : K \oplus L \to M : (k, l) \mapsto f_0(k) - g_0(l)$ with regard to the fiber product ring homomorphism $A \times_C B \to C$. Take any couple $(k, l) \in K \oplus L$ with $f(k) = g(l)$. Then the set-theoretic fiber product $\Delta \subset K \oplus L$ equals $Ker(d) + (k, l)$. This implies that the diaffine embedding $e^{(k,l)} : Ker(d) \overset{\sim}{\to} \Delta \subset K \oplus L$, followed by the projections to K and L defines a fiber product

$$
\begin{array}{ccc}
Ker(d) & \overset{p_2 \cdot e^{(k,l)}}{\longrightarrow} & L \\
{\scriptstyle p_1 \cdot e^{(k,l)}}\big\downarrow & & \big\downarrow{\scriptstyle g} \\
K & \overset{f}{\longrightarrow} & M
\end{array}
\tag{E.24}
$$

of modules and diaffine transformations. QED.

Theorem 63 *The category* **Mod** *has fiber sums for all pushout diagrams of modules over a fixed coefficient ring, i.e., where the scalar restrictions are the identity.*

Proof. We are given a fiber sum diagram

$$K \underset{f}{\leftarrow} M \underset{g}{\to} L \tag{E.25}$$

of modules over coefficient ring A. If M is empty we have to construct the sum $K \amalg L$, and we may suppose that both summands are nonvoid, the other cases being trivial. Consider the direct sum $S = K \oplus L \oplus A$ and the affine injections

$$i_1 : K \rightarrowtail S : k \mapsto (k, 0, 1) \text{ and}$$
$$i_2 : L \rightarrowtail S : l \mapsto (0, l, 0).$$

Suppose we are given two diaffine transformations

$$f : K \rightarrow X,$$
$$g : L \rightarrow X,$$

with factorizations $f = e^x \cdot f_0$ and $g = e^y \cdot g_0$ and scalar restriction $s : A \rightarrow B$. Define a dilinear map $h_0 : S \rightarrow X : (k, l, t) \mapsto f_0(k) + g_0(l) + s(t)(x - y)$. Then we have a diaffine transformation $h = e^y \cdot h_0$ which does the job, in fact, $h \cdot i_1 = f$, and $h \cdot i_2 = g$. Since i_2 is linear, we have $h(0) = h(i_2(0)) = g(0) = y$. Hence the affine part of h is uniquely determined. If we had two candidates h and h^* for universal arrows, they would only differ in their dilinear parts h_0 and h_0^*. But then, their difference $d = h_0 - h_0^*$ would vanish on all elements of shape $(0, l, 0)$, $l \in L$ and on all $(k, 0, 1)$, $k \in K$. The latter implies that $d(0, 0, 1) = 0$, and by dilinearity of d, $d(k, 0, 0) = d((k, 0, 1) - (0, 0, 1)) = 0$, whence the uniqueness of h.

On the other hand, if M is non-empty, so are K and L. We then have two arrows $u = i_1 \cdot f, v = i_2 \cdot g : M \rightrightarrows K \amalg L$ from the diagram

$$\begin{array}{ccc} M & \xrightarrow{\;\;f\;\;} & K \\ {\scriptstyle g}\downarrow & & \downarrow{\scriptstyle i_1} \\ L & \xrightarrow{\;\;i_2\;\;} & K \amalg L \end{array} \qquad (E.26)$$

and we are done if we can show that there is a coequalizer[3] of the couple u and v. If we have the factorizations $u = e^t \cdot u_0$ and $v = e^s \cdot v_0$, take the quotient module $E = K \amalg L / A(t - s) + \mathrm{Im}(u_0 - v_0)$. Clearly, the projection $p : K \amalg L \rightarrow E$ equalizes u and v. If $r : K \amalg L \rightarrow X$ is any diaffine transformation with scalar restriction $s : A \rightarrow B$ and equalizing the couple u and v, then r has a unique factorization through E. In fact, we may suppose without loss of generality that r is dilinear. In this case, r has the required factorization since it annihilates $A(t - s) + \mathrm{Im}(u_0 - v_0)$.

With this construction we define $K \amalg_M L = E$ and obtain a commutative diagram

$$\begin{array}{ccc} M & \xrightarrow{\;\;f\;\;} & K \\ {\scriptstyle g}\downarrow & & \downarrow{\scriptstyle p \cdot u} \\ L & \xrightarrow{\;\;p \cdot v\;\;} & K \amalg_M L \end{array} \qquad (E.27)$$

which is the required pushout diagram in **Mod**. Observe that this proof technique—build the sum and then the coequalizer—is a special case of the fact that existence of fiber sums is equivalent to existence of sums and coequalizers provided that we have an initial object, see Appendix G.2.1. QED.

Proposition 98 *A dilinear morphism $f : M \rightarrow N$ over scalar restriction $g : A \rightarrow B$ is mono iff f is diinjective, i.e., iff f, g are both injective.*

Proof. If both, f, g are injective, then clearly the dilinear morphism is mono. If f is not injective, there are two different affine morphisms $k_i : 0_{\mathbb{Z}} \rightarrow M, i = 1, 2$, which are equalized by f. If the scalar restriction g is not injective, there are two different ring homomorphisms $r_1, r_2 : \mathbb{Z}[X] \rightarrow A$ on the polynomial ring $\mathbb{Z}[X]$ with $r_1(X) \in ker(g), i = 1, 2,$, and the zero morphism $= 0_{\mathbb{Z}[X]} \rightarrow M$ for these two scalar restrictions does the job.

E.4 Complements of Commutative Algebra

In this section, all coefficient rings are commutative.

---

[3] See Appendix G.2.1.

E.4.1 Localization

See also [146, II] for concepts and facts described in this section.

Let S be a multiplicative subset of a ring A, i.e., $st \in S$ for all $s, t \in S$, and $1 \in S$. The *localization* $S^{-1}A$ is the set of equivalence classes of $A \times S$ modulo the relation $(a, s) \sim (a', s')$ iff there is $t \in S$ such that $t(as' - a's) = 0$. The equivalence class of (a, s) is denoted by the fraction a/s or $\frac{a}{s}$. It is a ring by the well-defined addition $a/s + a'/s' = (as' + a's)/ss'$ and multiplication $a/s.a'/s' = aa'/ss'$. The canonical map $i_S : A \to S^{-1}A : a \mapsto a/1$ is a ring homomorphism with the universal property that for any ring homomorphism $f : A \to B$ such that $f(S) \subset B^\times$, there is a unique ring homomorphism $j : S^{-1}A \to B$ such that $f = j \circ i_S$. The ring $S^{-1}A$ is called the *localization of A in S*. Classical example: A is a domain (no zero divisors), $S = A - \{0\}$, whence $S^{-1}A$ is the classical *field $fr(A)$ of fractions over A*.

If M is an A-module, the localization $S^{-1}M$ is the set $S^{-1}M$ of equivalence classes of $M \times S$ for the equivalence relation $(m, s) \sim (m', s')$ iff there is $t \in S$ such that $t(ms' - m's) = 0$. The addition $m/s + m'/s' = (ms' + m's)/ss'$ and the scalar multiplication $r/s.m/t = rm/st$ makes $S^{-1}M$ a $S^{-1}A$-module. One has the canonical dilinear homomorphism $i_M : M \to S^{-1}M : m \mapsto m/1$ with respect to the homomorphism i_S. It has this universal property: For every homomorphism $f : M \to N$ of A-modules, such that every dilatation $s? : N \to N : n \mapsto s.n$ is bijective, there is a unique homomorphism of A-modules $j : S^{-1}M \to N$ such that $f = j \circ i_M$. It is easily seen that the tensor product $S^{-1}A \otimes_A M$, together with the canonical homomorphism of A-modules $M \to S^{-1}A \otimes_A M : m \mapsto 1 \otimes m$ is isomorphic to the localization $S^{-1}M$.

For the multiplicative set $S_s = \{1, s, s^2, s^3, \ldots\}$, $s \in A$, one writes A_s, M_s instead of $S_s^{-1}A, S^{-1}M$. For a prime ideal $q \subseteq A$, the complement $S = A - q$ is multiplicative by definition. And one then writes A_q, M_q instead of $S^{-1}A, S^{-1}M$.

By the universal property of localization, if $f : A \to B$ is a ring homomorphism and if $S \subseteq A, T \subseteq B$ are multiplicative sets such that $f(S) \subseteq T$, then there is a canonical ring homomorphism $f_{S,T} : S^{-1}A \to T^{-1}B$ which extends f. If $g : M \to N$ is a dilinear homomorphism over f, it extends uniquely to a dilinear homomorphism $g_{S,T} : T^{-1}M \to T^{-1}N$ over $f_{S,T}$. For a multiplicative set $S \in A$ let $S' = \{t \in A | \text{there exists } a \in A, s \in S \text{ such that } s = at\}$ be the *saturation of S*. Then the canonical homomorphisms $S^{-1}A \to s'^{-1}A$ and $S^{-1}M \to s'^{-1}M$ are isomorphisms and we may identify the two corresponding localizations. In particular, if $s \in A$, we identify A_s, M_s with $S_s'^{-1}A, S_s'^{-1}M$. So if for $s, t \in A$, one has $S_s' \subseteq S_t'$, one has canonical homomorphisms $A_s \to A_t, M_s \to M_t$.

Proposition 99 *If $i_S : A \to S^{-1}A$ is the localization homomorphism, the inverse image $q \mapsto i_S^{-1}(q)$ is an order preserving bijection from the set of maximal (resp. prime) ideals in $S^{-1}A$ to the set of maximal (resp. prime) ideals in A which are disjoint from S. In particular, if $S = A - p$ for a prime ideal p, the localization A_p is a local ring with maximal ideal $m_p = p_p$ and the residue field $\kappa_p = A_p/m_p$ is isomorphic to the field of fractions $fr(A/p)$.*

E.4.2 Projective Modules

Definition 158 *An A-module P is* projective *iff it is a direct summand of a free A-module. Equivalently, it is projective if for each pair of homomorphisms $u : P \to N$, $v : M \to N$, v an epimorphism, there is a homomorphism $w : P \to M$ such that $u = v \circ w$.*

Let $U \mapsto U^{\star\star}$ be the bidual functor on A-modules. Then a direct summand $U \subset V$ is mapped into a direct summand $U^{\star\star}$ of $V^{\star\star}$. Now, if U is projective, it is a direct summand of a free module $R^{(n)}$. It is easily seen that the bidual map $R^{(n)} \to (R^{(n)})^{\star\star}$ is injective. Therefore, if U is projective, the bidual map $U \to U^{\star\star}$ is also injective. If U is finitely generated and projective, the bidual map is an isomorphism.

Let $i : N \to M$ be the inclusion of a submodule N of an R-module M. For any $x \in M$ and positive exponent r, let

$$\wedge^r x : \bigwedge\nolimits^r N \to \bigwedge\nolimits^{r+1} M : y \mapsto \bigwedge\nolimits^r i(y) \wedge x$$

the linear map defined by the rth exterior power of i and the wedge product with x in the exterior algebra of M. This defines a linear map

$$\wedge^r : M \to Lin_R(\bigwedge^r N, \bigwedge^{r+1} M) \tag{E.28}$$

which has the following property:

Lemma 94 *With the above notation, if N is a direct factor of M which is locally free of rank r, then $Ker(\wedge^r) = N$.*

Proof. Let $x \in M$. In the special case where N is free of rank r, if $\bigwedge^r N = R.u$ for a basis vector u of $\bigwedge^r N$, the claim $x \in Ker(\wedge^r)$ is equivalent to $u \wedge x = 0$. But this follows from [145, ch.III, §7, no.9, Prop.13]. In the general case, clearly the condition is sufficient. Conversely, suppose $\wedge^r x = 0$. Recall from [146, ch.II, §5 no.3, Th.2] that an R-module is projective of rank $r \in \mathbb{N}$ iff it is locally free of rank r. Notice that localizing commutes with exterior powers ([146, ch.II, §2 no.8]), so we may localize in $f \in R$ such that for the localized element $x_f \in M_f$, we have $\wedge^r x_f \in N_f$ and N_f is free of rank r, whence $x_f = 0$. As there is a cover of $\mathrm{Spec}(R)$ with basic open sets $D(f_i)$ associated with localizations R_{f_i} such that N_{f_i} is free, we deduce that $x \in N$. QED.

E.4.3 Injective Modules

Proposition 100 ([145, ch.II, §2, exercise 11]) *For an R-module M, the following properties are equivalent:*

(i) *The functor $Lin_R(?, M)$ is exact.*

(ii) *The functor $Lin_R(?, M)$ is exact on short exact sequences.*

(iii) *For every R-module E and every linear injection $F \rightarrowtail E$, every linear map $F \to M$ extends to a linear map $E \to M$.*

(iv) *For every ideal $\mathfrak{a} \subset R$ and every linear map $f : \mathfrak{a} \to M$, there is $m \in M$ such that $f(a) = a.m$ for all $a \in R$.*

(v) *M is a direct factor of every module which contains it.*

(vi) *For every R-module E which is sum of M and of a module I with one generator, M is a direct factor of E.*

Definition 159 *An R-module M is said to be* injective *iff it has the equivalent properties of Proposition 100. The ring R is said to* self-injective *iff it is injective as a (left) module over itself.*

Exercise 87 *Show the following statement: A direct sum of R-modules is injective iff each factor is.*

Example 108 Let $1 < N$ be a natural number. Then the Ring $\mathbb{Z}_N$ is self-injective. In fact, let $N = p_1^{n_1} \cdot \ldots p_r^{n_r}$ be its prime factorization (see Appendix D.2). Then $\mathbb{Z}_N \xrightarrow{\sim} \mathbb{Z}_{p_1^{n_1}} \times \ldots \mathbb{Z}_{p_r^{n_r}}$, and injectivity can be checked on each factor, so suppose $N = p^n$. We apply criterion (iv) of Proposition 100. An ideal $\mathfrak{a}$ in $\mathbb{Z}_{p^n}$ is generated by $p^m, m \leq n$, and we have the isomorphism of $\mathbb{Z}_{p^n}$-modules $\mathfrak{a} \xrightarrow{\sim} \mathbb{Z}_{p^{(n-m)}}$. Then a linear map $f : \mathbb{Z}_{p^{(n-m)}} \to \mathbb{Z}_{p^n}$ evaluates to $f(a) = f(a.1_{\mathbb{Z}_{p^{(n-m)}}}) = a.f(1_{\mathbb{Z}_{p^{(n-m)}}})$, and we are done.—In particular, every free $\mathbb{Z}_N$-module $\mathbb{Z}_N^r$ is injective.

The self-injectiveness of $\mathbb{Z}_{p^n}$ also follows from this:

Proposition 101 ([299, Proposition 21.5]) *Let R be a zero-dimensional local ring. The following are equivalent:*

(i) *R is self-injective.*

(ii) *The socle of R is simple.*

In fact, the socle of $\mathbb{Z}_{p^n}$ is isomorphic to $\mathbb{Z}_p$, the simple group of order p and it is well known that $\mathbb{Z}_{p^n}$ is zero-dimensional.

Proposition 102 *A finitely generated $\mathbb{Z}_{p^n}$-module M is injective iff it is free of finite rank.*

Proof. Clearly, by Example 108, a free module of finite rank is injective. Conversely, by the main theorem on finitely generated abelian groups E.2 we have $M \xrightarrow{\sim} \mathbb{Z}_{p^{n_1}} \times \ldots \mathbb{Z}_{p^{n_k}}, n_i \leqslant n$. By exercise 87 above, it suffices to see that $\mathbb{Z}_{p^m}$ cannot be injective if $m < n$. In fact, if the injection $\mathbb{Z}_{p^m} \leftarrow \mathbb{Z}_{p^n} : x \mapsto p^{n-m}.x$ had a left inverse h, we would have $x = h(p^{n-m}.x) = p^{n-m}.x$ which only works for $x = 0$, a contradiction. QED.

E.4.4 Lie Algebras

Definition 160 *For a module L over commutative ring R, a Lie algebra structure is an R-bilinear multiplication $[\ ] : L \times L \to L$, the Lie bracket, such that $[xx] = 0$ identically, and the Jacobi identity*

$$[x[yz]] + [y[zx]] + [z[xy]] = 0$$

holds for all $x, y, z \in L$. A homomorphism of Lie algebras $f : L_1 \to L_2$ is a linear homomorphism such that $f([xy]) = [f(x)f(y)]$ for all $x, y \in L_1$. The corresponding category of Lie algebras over R is denoted by $\mathbf{Lie}_R$.

Example 109 If $L \xrightarrow{\sim} R^n$ is free, and if (x_i) is a basis, a Lie algebra structure on L is defined by bilinearity and skew symmetry (which follows from $[xx] = 0$) of the Lie bracket if the Lie brackets $\sum_k a_{ij}^k x_k = [x_i, x_j], i < j$ are known. The condition for such a bracket to generate a Lie algebra is

$$a_{ii}^k = 0, \text{ all } i, k, \tag{E.29}$$

$$a_{ij}^k + a_{ji}^k = 0, \text{ all } i < j, k, \tag{E.30}$$

$$\sum_k a_{ij}^k a_{kl}^m + a_{jl}^k a_{ki}^m + a_{li}^k a_{kj}^m = 0, \text{ all } i, j, l, m. \tag{E.31}$$

The coefficients a_{ij}^k are called *structural constants of the Lie algebra in the given basis.*

Example 110 For any module L, the R-algebra of linear endomorphisms $End(L)$ becomes the *general linear algebra* $\mathfrak{gl}(L)$ by the bracket $[xy] = x \circ y - y \circ x$. A sub-Lie-algebra of a general linear algebra is called a *linear Lie algebra.* If $L \xrightarrow{\sim} R^n$ is free of rank n, we also write $\mathfrak{gl}(L) = \mathfrak{gl}(n, R)$. Its subalgebra of endomorphisms with vanishing trace (check that it is a sub-Lie-algebra!) is called the *special linear algebra* and denoted by $\mathfrak{sl}(L)$ or $\mathfrak{sl}(n, R)$.

Example 111 *Let L be any R-module with a bilinear product $x \cdot y$ (no other conditions required). A derivation is a linear endomorphism $D : L \to L$ such that $D(x \cdot y) = x \cdot D(y) + D(x) \cdot y$. The set $Der(L)$ is a submodule of $End(L)$, and in fact a Lie subalgebra of the general linear algebra $\mathfrak{gl}(L)$.*

In particular, if we take the Lie algebra structure on L, $Der(L)$ is another Lie algebra. Observe that for $x \in L$, the left multiplication $ad(x) = [x?] : y \mapsto [xy]$ is a derivation by the Jacobi identity. One has the Lie algebra homomorphism of adjunction

$$ad : L \to Der(L), \tag{E.32}$$

a representation of L in the general linear algebra of L. A derivation $ad(x)$ is called inner, any other is called outer derivation.

Proposition 103 *If x is a nilpotent endomorphism in a linear algebra, then its adjoint $ad(x)$ is also a nilpotent endomorphism.*

See [491, p.12] for the easy proof.

Suppose that L is a linear algebra in $End(V)$ for an R-module V, and that the conjugation $Int_e, e \in GL(V)$ leaves L invariant. Then evidently the conjugation is an automorphism of L. Suppose now that R is a $\mathbb{Q}$-algebra. If $ad(x)$ is nilpotent, then the exponential

$$exp(ad(x)) = 1 + ad(x) + ad(x)^2/2! + \ldots ad(x)^k/k! + \ldots \tag{E.33}$$

is defined. We have:

Lemma 95 *If $ad(x)$ is nilpotent, then $exp(ad(x))$ is an automorphism of the Lie algebra L. Moreover, if x is nilpotent, $exp(x)$ is defined and we have*

$$Int_{exp(x)} = exp(ad(x)).$$

See [491, p.9] for the proof.

F

Algebraic Geometry

For this chapter, we refer to [146, 259, 395, 396, 300].

F.1 Locally Ringed Spaces

Given a topological space X, its system of open sets $Open_X$ is viewed as a category with inclusions as morphisms. If $f : X \to Y$ is a continuous map, the inverse image map $U \mapsto f^{-1}U$ defines a functor $Open_f : Open_Y \to Open_X$. This defines a functor $Open_? : \mathbf{Top} \to \mathbf{Cat}$ into the category of categories and functors[1].

Let $\mathbf{C}$ be a category of sets with some additional algebraic structure, such as the categories

$$\mathbf{Mod}, \mathbf{Mon}, \mathbf{Grp}, \mathbf{Ab}, \mathbf{Rings}, \mathbf{ComRings}$$

of modules, monoids, groups, abelian groups, rings, or commutative rings, respectively. A contravariant functor $F : Open_X \to \mathbf{C}$ is called a $\mathbf{C}$-space (this is a presheaf plus the algebraic morphism conditions). For example, a ringed space is just a $\mathbf{Rings}$-space. In the present context, we always suppose that a ringed space is one with values in $\mathbf{ComRings}$, i.e., a commutatively ringed space. The set of $\mathbf{C}$-spaces on X is denoted by $\mathbf{C}_X^{spaces}$. The contravariant functor $Open_?$ induces a functor $\mathbf{C}_?^{spaces} : X \mapsto \mathbf{C}_X^{spaces}$. It maps the continuous map $f : X \to Y$ to the set map $\mathbf{C}_f^{spaces} : F \mapsto F \circ Open_f$. The image $F \circ Open_f$ is denoted by f_*F and is called the *direct image of F*. If $F : Open_X \to \mathbf{C}, G : Open_Y \to \mathbf{C}$ are $\mathbf{C}$-spaces and $f : X \to Y$ is continuous, then an *f-morphism* $h : F \to G$ is a natural transformation $h : G \to f_*F$. These morphisms define an evident category, the *category* $\mathbf{C}^{spaces}$ *of* $\mathbf{C}$-spaces.

Suppose that the category $\mathbf{C}$ has colimits for filtered diagrams[2], such as

$$\mathbf{Rings}, \mathbf{ComRings}, \mathbf{Mod}, \mathbf{Grp}, \mathbf{Ab},$$

and take a $\mathbf{C}$-space F. For each point $x \in X$, the filtered system $Open_{X,x}$ of open neighborhoods of x defines an object $F_x = colim_{U \in Open_{X,x}} F(U)$, the *stalk of F at x*. So for a (commutatively) ringed space, this is a (commutative) ring. Let $h : F \to G$ be an f-morphism for $f : X \to Y$. For $x \in X$, we have the restriction $Open_{f,x} : Open_{Y,f(x)} \to Open_{X,x}$ of $Open_f$ to the neighborhood systems $Open_{Y,f(x)}$ and $Open_{X,x}$. This induces a $\mathbf{C}$-morphism $h_x : G_{f(x)} \to F_x$.

The subcategory $\mathbf{LocRgSpaces}$ of $\mathbf{ComRings}$-spaces consists of all ringed spaces F which have a local ring F_x with maximal ideal m_x in each point $x \in X$, and of those morphisms h which induce local morphisms h_x in all stalks, i.e., $h_x^{-1}(m_x) = m_y$. It is called the *category of locally ringed spaces*. The residue

---

[1] Restricted to a universe, if the limitless collection bothers the reader, or even the category of partially ordered sets with order preserving maps, to stick to reality.

[2] Meaning that for any two objects in the diagram quiver, there are two arrows with these domains targeting at a common codomain.

© Springer International Publishing AG, part of Springer Nature 2017
G. Mazzola, *The Topos of Music IV: Roots*, Computational Music Science,
https://doi.org/10.1007/978-3-319-64495-0_6

field in a point x of such a space F is the field $\kappa(x) = F_x/m_x$. For a section $s \in F(U)$ over the open set U, we denote by $s(x)$ the canonical image of s in $\kappa(x)$.

For the category **ComMod** of modules over commutative rings and dilinear homomorphisms, a **ComMod**-space F can also be described by the underlying ringed space R and the abelian group space F (same notation), together with a scalar multiplication $R(U) \times F(U) \to F(U)$ in each open set U, and the evident dilinear transition maps. One therefore also says that F *is an R-module*. Same wording, mutatis mutandis, for an R-algebra or for an R-ideal.

If we are given a **C**-space F on X which is a sheaf, and if $\mathfrak{B}$ is a topological base for X, the restriction $F|\mathfrak{B}$ to this subcategory of $Open_X$ completely determines F. If $U \in Open_X$, $U = colim(B \in \mathfrak{B}|B \subseteq U)$, and

$$F(U) = lim(F(B), B \in \mathfrak{B}).$$

Conversely, if we are given a contravariant functor $F : \mathfrak{B} \to \mathbf{C}$, we obtain a **C**-space F' by

$$F'(U) = lim(F(B), B \in \mathfrak{B})$$

and by the universally given transition morphisms. This presheaf is a sheaf if F is a *sheaf on $\mathfrak{B}$*, i.e., if for every covering (B_i) of $B \in \mathfrak{B}$ by elements of the base, the canonical application $F(B) \to \prod_i F(B_i)$ is a bijection $x \mapsto (x|B_i)$ onto the tuples (x_i) such that for every i, j and base element $B' \subseteq B_i \cap B_j$, we have $x_i|B' = x_j|B'$.

F.2 Spectra of Commutative Rings

Definition 161 *The (prime) spectrum is a contravariant functor*

$$Spec : \mathbf{ComRings} \to \mathbf{LocRgSpaces}$$

which is defined as follows: Let A, B be commutative rings, and let $f : A \to B$ be a ring homomorphism.

1. *The topological space consists of the set $Spec(A) = \{p \text{ a prime ideal in } A\}$. The closed sets are the sets of the form $V(E) = \{p|E \subseteq p\}$ for a subset $E \subseteq A$. Equivalently, a base of open sets is given by the system $D_f = \{p|f \notin p\}, f \in A$, and we have $D_f \cap D_g = D_{fg}$.*
2. *For the base $\mathfrak{D} = \{D_f|f \in A\}$, we have a sheaf on $\mathfrak{D}$, which is defined by $D_f \mapsto A_f$, the localization at the saturated multiplicative set $S(f)$ defined by f, a well-defined setup since $D_f = D_g$ iff $S(f) = S(g)$ see [395, I.1.3.2]. This presheaf is a sheaf on $\mathfrak{D}$, and the associated sheaf on $Spec(A)$ is denoted by $\tilde{A}$ and called the ring sheaf associated with A. If $p \in Spec(A)$ is a prime ideal, we have $(\tilde{A})_p \overset{\sim}{\to} A_p$, i.e., $\tilde{A} \in \mathbf{LocRgSpaces}$.*
3. *For the homomorphism $f : A \to B$, the inverse image map on prime ideals $Spec(f) : Spec(B) \to Spec(A) : p \mapsto f^{-1}p$ is defined, and we have $Spec(f)^{-1}(D_g) = D_{f(g)}$, $Spec(f)^{-1}(V(E)) = V(f(E))$, i.e., $Spec(f)$ is continuous. Furthermore, we have a canonical map $f_g : A_g \to B_{f(g)}$ which is natural and therefore induces a morphism $Spec(f) : \tilde{A} \to \tilde{B}$ over the continuous (synonymous) map $Spec(f)$. The stalk homomorphism $f_p : \tilde{A}_p \to \tilde{B}_{Spec(f)(p)}$, colimit of the natural homomorphisms $f_g : A_g \to B_{f(g)}$, is local. One therefore has a contravariant functor Spec as announced, and one often denotes $Spec(A)$, when meaning the locally ringed space $\tilde{A}$ over $Spec(A)$.*

Theorem 64 *The functor Spec is fully faithful, the inverse* global section functor Γ *of a* **LocRgSpaces**-*morphism $u : \tilde{B} \to \tilde{A}$ is given by the ring homomorphism $u(Spec(A)) : A = \tilde{A}(Spec(A)) \to \tilde{B}(Spec(B)) = B$.*

See [395, I.1.6.3] for a proof. Since one often writes $F(U) = \Gamma(U, F)$ and calls the elements section above U, the theorem's notation is justified by the global section notation $\Gamma(\tilde{A}) = \Gamma(Spec(A), \tilde{A})$. Let **Aff** be the full subcategory of **LocRgSpaces** consisting of the objects which are isomorphic to prime spectra. These spaces are called *affine schemes*. We therefore have that the map

$$Spec : \mathbf{ComRings} \to \mathbf{Aff}$$

is an equivalence of categories.

If M is an A-module, we have a $\tilde{A}$-module $\tilde{M}$, whose sections on the base $\mathfrak{D}$ are defined by $\Gamma(D_g, \tilde{M}) = A_g \otimes M = M_g$, the localization of M at the multiplicative set $S(g)$.

Proposition 104 *The map $M \mapsto \tilde{M}$ is an exact[3] and fully faithful functor from the category of A-modules $\mathbf{Mod}_A$ to the category $\mathbf{Mod}_{\tilde{A}}$ of $\tilde{A}$-modules. It also commutes with colimits of modules, with tensor products, Hom-modules, with sums and intersections of submodules.*

The inverse to this functor is the global section functor $\tilde{M} \mapsto \Gamma(Spec(A), \tilde{M})$, which is also exact.

See [395, I.1.3] for a proof. The modules in $\mathbf{Mod}_{\tilde{A}}$, which are hit by this tilding process are the quasi-coherent ones: A module $\mathcal{M}$ over a ringed space $\mathcal{A}$ over a topological space X is *quasi-coherent* iff there is a covering X_i of X such that each restriction $\mathcal{M}_i = \mathcal{M}|X_i$ is the cokernel of a homomorphism $f_i : \mathcal{A}_i^{I_i} \to \mathcal{A}_i^{J_i}$, where $\mathcal{A}_i = \mathcal{A}|X_i$.

Theorem 65 *An $\tilde{A}$-module $\mathcal{M}$ is isomorphic to a module $\tilde{M}$ iff it is quasi-coherent.*

See [395, I.1.4.1] for a proof. This means that we have an equivalence of categories of quasi-coherent modules over $\tilde{A}$ and $\mathbf{Mod}_A$ (with linear homomorphisms).

For a ring element $f \in A$, one has $Spec(A_f) \overset{\sim}{\to} D_f$. When restricting the associated ring and module sheaves $\tilde{A}, \tilde{M}$ to basic open sets D_f, this yields ring and module sheaves which are isomorphic to $\tilde{A}_f, \tilde{M}_f$.

Theorem 66 *Let M be an A-module. The following conditions are equivalent:*

(*i*) *M is projective[4] and finitely generated.*

(*ii*) *There is a finite family (f_i) of elements of A which generate the ideal A, i.e., $Spec(A) = \bigcup_i D_{f_i}$ such that the localizations $M_{f_i} = \Gamma(D_{f_i}\tilde{M})$ are free of finite rank over A_{f_i}.*

This is why one can also define a finitely generated projective module as being a locally free module of (locally defined) finite ranks. If the locally constant rank is constant n, the module is said to be *locally free of rank n*.

F.2.1 Sober Spaces

A topological space X is *irreducible* iff every non-empty open subset is dense, or, equivalently, if any two non-empty open sets have a non-empty intersection. A subset of a topological space is called irreducible if it is so with its relative topology. A point x of an irreducible space X is said to be *generic* iff its (always irreducible) closure $\overline{\{x\}}$ is X. We say that *a point x dominates a point y, in signs $x > y$*, iff $\overline{\{y\}} \subseteq \overline{\{x\}}$. This is a partial order relation on X. An irreducible component of a space X is a maximal irreducible subset.

Sorite 16 *These are the sorite properties concerning irreducibility:*

(*i*) *A subset of a topological space is irreducible iff its closure is.*

(*ii*) *Irreducible components are closed.*

(*iii*) *Every irreducible subset is contained in an irreducible component, in particular, a topological space is the union of its irreducible components.*

(*iv*) *The image $f(E)$ of an irreducible subset $E \subseteq X$ under a continuous map $f : X \to Y$ is irreducible.*

---

[3] A sequence $K \overset{f}{\to} M \overset{g}{\to} L$ of linear homomorphisms of modules is exact in M iff $Im(f) = Ker(g)$. Such exact sequences are preserved by the functor.

[4] See definition 158 in Appendix E.4.2.

Definition 162 *A topological space X is* sober *iff each closed irreducible subset has a unique generic point. Call* **Sob** *the full subcategory of the category* **Top** *of topological spaces consisting of sober spaces.*

If A is a commutative ring, and if $E \subseteq Spec(A)$, then we denote $J(E) = \bigcap_{p \in E} p$, and $\overline{E} = V(J(E))$. This ideal is prime iff E is irreducible. In this case, $\overline{E} = \overline{\{J(E)\}}$. In fact, for two points p, q in $Spec(A)$, $p > q$ iff $p \subseteq q$. In particular, $Spec(A)$ *is a sober space.* Its irreducible components correspond to the minimal prime ideals.

Proposition 105 *The canonical injection*

$$j : \mathbf{Sob} \to \mathbf{Top}$$

has a left adjoint $?^s : \mathbf{Top} \to \mathbf{Sob}$.

Proof idea. This adjoint associates with any X a sober space X^s which is defined as follows. Its points are the irreducible closed sets in X. The open sets are the sets $V^s = \{Y \in X^s | Y \cap V \neq \varnothing\}$, where V varies over all open sets in X. Clearly, $(V \cap W)^s = V^s \cap W^s$ and $(\bigcup W_i)^s = \bigcup W_i^s$ for any family (W_i) of open sets.

On continuous maps $f : X \to Y$, the functor acts via $f^s : X^s \to Y^s : E \mapsto \overline{f(E)}$. One has a canonical continuous map $q_X : X \to X^s : x \mapsto \overline{\{x\}}$ and a commutative diagram of continuous maps:

$$
\begin{array}{ccc}
X & \xrightarrow{\;f\;} & Y \\
{\scriptstyle q_X}\downarrow & & \downarrow{\scriptstyle q_Y} \\
X^s & \xrightarrow{\;f^s\;} & Y^s
\end{array}
\qquad\qquad (\text{F.1})
$$

The map $q_X : X \to X^s$ is a homeomorphism if X is sober. The adjunction is given by the mutually reciprocal maps $\mathbf{Top}(X, j(Y)) \to \mathbf{Sob}(X^s, Y) : f \mapsto q_Y^{-1} \circ f^s$ and $\mathbf{Sob}(X^s, Y) \to \mathbf{Top}(X, j(Y)) : g \mapsto g \circ q_X$.

Lemma 96 *The canonical continuous map* $q_X : X \to X^s$ *is a quasi-homeomorphism,. i.e., the inverse image map* $2^{X^s} \to 2^X$ *is a bijection between the open sets of* X^s *and those of* X.

F.3 Schemes and Functors

A scheme $(X, \mathcal{O}_X)$ is a ringed space $\mathcal{O}_X$ on X which locally is isomorphic to a spectrum of a commutative ring, i.e., there is an open covering (X_i) of X and a family A_i of rings such that $(X_i, \mathcal{O}_X | X_i) \xrightarrow{\sim} Spec(A_i)$. The category **Schémas** of schemes is the subcategory of **LocRgSpaces** whose objects are schemes. By Yoneda, we have a fully faithful functor $Y : \mathbf{Schémas} \to \mathbf{Schémas}^@$.

Proposition 106 *The restriction*
$$Y_{\mathbf{Aff}} : \mathbf{Schémas} \to \mathbf{Aff}^@$$
is fully faithful. Equivalently, the corresponding functor

$$Y_{\mathbf{ComRings}} : \mathbf{Schémas} \to \mathbf{ComRings}_@$$

into the category $\mathbf{ComRings}_@$ *of covariant set-valued functors on* **ComRings** *is fully faithful.*

This means that we may consider schemes as special covariant functors on the category of commutative rings. The functors which correspond to schemes are characterized by a sheaf condition:

Property 3 *We are given a functor $G \in \mathbf{ComRings}_@$. For every ring $A \in \mathbf{ComRings}$, and every finite family (f_i) of elements of A which generate A as an ideal, the diagram*

$$G(A) \rightarrow \prod_i G(A_{f_i}) \rightrightarrows \prod_{i,j} G(A_{f_i f_j})$$

is exact, we say (by abuse of language, but theoretically justifiable) that G is a sheaf.

Then the full subcategory of $\mathbf{ComRings}_@$ consisting of sheaves G in the sense of property 3, together with the property that there is a family of rings R_t and morphisms[5] $a_t : R_t@ \rightarrow G$ such that for every field K, $G(K) = \bigcup_t R_t@K$, comprises the functors which are isomorphic to images of schemes under the Yoneda map $Y_{\mathbf{ComRings}}$, see [395, I.2.3.6] and [300, VI.2] for details. This means that schemes are characterized without any reference to the geometry of ringed spaces. See also [639, III.3] for the relation of this setup to the systematic topos-theoretic restatement of the schemes in terms of the *Zariski site*.

The most important universal property of the category of schemes is that it has fiber products. In the affine case, we have $Spec(A) \times_{Spec(C)} Spec(B) \xrightarrow{\sim} Spec(A \otimes_C B)$.

F.4 Algebraic and Geometric Structures on Schemes

If a scheme is viewed as a set-valued functor on rings, the sets may also be enriched by algebraic structures, such as groups, monoids, etc., to yield a category $\mathbf{C}$. We then view a scheme as a functor $S : \mathbf{ComRings} \rightarrow \mathbf{C}$, and say that S is a $\mathbf{C}$-scheme, for example, an abelian group-scheme if $\mathbf{C} = \mathbf{Ab}$.

Example 112 For $0 \leqslant n$, we have the additive group scheme $\mathbb{A}^n$ whose functor is $\mathbb{A}^n(R) = R^n$, with the canonical addition of this free module, and the canonical transitions $\mathbb{A}^n(R) \rightarrow \mathbb{A}^n(S)$ for a ring homomorphism $f : R \rightarrow S$.

Example 113 The n-dimensional linear group scheme is given by the functor $R \mapsto GL(n, R) \subset \mathbb{M}_{n,n}(R) \xrightarrow{\sim} \mathbb{A}^{n^2}(R)$, together with the canonical map $GL(n, R) \rightarrow GL(n, S)$ for a ring homomorphism $f : R \rightarrow S$. The set $GL(n, R)$ is defined as the set of $n \times n$-matrices M with invertible determinant: $det(M) \in R^\times$. The functor is represented by the affine scheme $GL_n = Spec(\mathbb{Z}[X_{ij}, 1 \leqslant i, j \leqslant n]_{det})$, where $det = Det(X_{ij})$. The group structure is the multiplication of invertible matrices.

F.4.1 The Zariski Tangent Space

For a field K, we are given a K-scheme X, i.e., a scheme $s : X \rightarrow Spec(K)$ in the comma category $\mathbf{Schémas}/Spec(K)$ (see Section G.2.1). Suppose that we have a K-rational point $x : Spec(K) \rightarrow X$, i.e., a section of s. This means that the corresponding K-algebra $\mathcal{O}_{X,x}$ has an isomorphism $K \xrightarrow{\sim} \kappa_{X,x} = \mathcal{O}_{X,x}/m_{X,x}$. The *Zariski tangent space in x* is the linear K-dual $T_{X,x} = (m_{X,x}/m_{X,x}^2)^*$. Consider the K-scheme $D_K = Spec(K[\varepsilon])$ over the dual numbers $K[\varepsilon]$, see Example 106 in Appendix D.1.1. It has the K-rational point $\varepsilon : Spec(K) \rightarrow D_K$ corresponding to the projection $K[\varepsilon] \rightarrow K : \varepsilon \mapsto 0$.

Proposition 107 *With the above hypotheses and notation, there is a bijection of the Zariski elements t of the tangent space $T_{X,x}$ and the morphisms $\tau : D_K \rightarrow X$ of K-schemes which map the K-rational point ϵ to the K-rational point x.*

See [300, VI.1.3] for a proof. In particular, if the scheme X is given by its functor on rings, this means that the tangents are special elements of the evaluation of the functor in dual numbers $X(K[\varepsilon])$. For example, if $X = \mathbb{A}_K^1 = \mathbb{A}^1 \times_{\mathbb{Z}} K$, we have tangents $x + \varepsilon.\tau, \tau \in K$, over the rational point $x \in K \xrightarrow{\sim} \mathbb{A}_K^1(K)$.

---

[5] $R_t@$ is the covariant functor on rings, i.e., $R_t@R = Hom_{\mathbf{ComRings}}(R_t, R)$.

F.5 Grassmannians

A *subfunctor* $G \to F$ in $\mathbf{ComRings}_@$ *is open* iff for every morphism $a : R@ \to F$ (corresponding to an element $a \in F(R)$ via Yoneda), the fiber product projection $G \times_a R@ \to R@$ is isomorphic to the functor of an open subscheme of $Spec(R)$. Clearly, then, if $b : X@ \to F$ is a morphism from a representable functor $X@$ of a scheme X, then the projection $G \times_b X@ \to X@$ is isomorphic to an open subscheme of X. An *open covering of a functor* F in $\mathbf{ComRings}_@$ is a family $(g_i : G_i \to F)$ of open subfunctors of F such that the fiber product projections $G_i \times_b X@ \to X@$ for morphisms $b : X@ \to F$ from the representable functor of a scheme X define an open covering of X. For example, the open subfunctors of an affine scheme $Spec(R)$ are the functors $F_I : \mathbf{ComRings} \to \mathbf{Ens}$ of form $F_I(S) = \{f : R \to S | f(I)S = S\}$, where I is an ideal in R.

The Grassmann scheme $Grass_{r,n}$ is defined for any couple $0 \leqslant r \leqslant n$ of natural numbers by the functor

$$Grass_{r,n}(R) = \{V \subseteq R^n | R^n/V \text{ locally free of rank } r\},$$

which for a ring homomorphism $R \to S$ maps the exact sequence

$$0 \to V \to R^n \to R^n/V \to 0$$

to the exact sequence

$$0 \to Im(S \otimes_R V) \to S^n \to S \otimes_R (R^n/V) \to 0,$$

where the image of the tensorized space $S \otimes_R V$ is the image of V under this map. The locally free quotient remains locally free since the localization on R carries over to a localization over S: For $f \in R$, and its image $f' \in S$, we have $(S \otimes_R R^n/V)_{f'} \xrightarrow{\sim} S_{f'} \otimes_{R_f} (R^n/V)_f$.

The functor $Grass_{r,n}$ is covered by the following open subfunctors. Let $i. = i_1, i_2, \ldots i_r$ be an increasing subsequence $1 \leqslant i_1 < i_2 < \ldots i_r \leqslant n$. We have the affine open subfunctors $Grass_{n,r,i.}(R)$ of those submodules $V \subset R^n$ such that the factor $R^{i.} = \bigoplus_{j=1,\ldots r} R.e_{i_j}$ of R^n projects isomorphically onto the quotient R^n/V. If i' denotes the complementary increasing sequence, $Grass_{n,r,i.}(R)$ identifies to the set of graphs Γf of linear maps in $Lin_R(R^{i'.}, R^{i.})$, i.e., to $n \times (n-r)$-matrices with columns $(e_{i'_k}, f(e_{i'_k}))^t$. In fact, the isomorphism $R^{i.} \xrightarrow{\sim} R^n/V$ corresponds to an isomorphism $V \xrightarrow{\sim} R^{i'.}$, and this makes V a graph of a linear map in $Lin_R(R^{i'.}, R^{i.})$. The fact that these open subfunctors (represented by affine schemes $\mathbb{A}^{r \times (n-r)}$) cover the Grassmannian results from the situation over a field, where the covering is evident.

Proposition 108 *Let n be a positive natural number. Then the subfunctor*

$$B_n : R \mapsto \{x \in R^n | \ x \text{ is part of a basis of } R^n\} \tag{F.2}$$

of the affine n-space $\mathbb{A}^n$ over $\mathbb{Z}$ is an open subscheme.

Proof. Consider the open subscheme GL_n of $\mathbb{A}^{n^2}$. Then B_n is the image of GL_n under the projection onto the first column (n_{1j}) which by [396, IV/2, 2.4.7] is (universally) open. QED.

Lemma 97 *If X is an S-scheme, E a quasi-coherent O_X-module, then any section $s : S \to Grass_r(E)$ is a closed immersion.*

Proof. By [395, Proposition (9.7.7)], $Grass_r(E)$ separated over S, and by [395, Corollaire (5.2.4)], a section of such a structural morphism is a closed immersion. QED.

Lemma 98 *If R is a product of local rings of finite length, then for two elements x, y of an R-module M, $R^\times x = R^\times y$ iff these elements generate the same space, i.e., $R.x = R.y$.*

Without loss of generality, we may suppose that R is local with maximal nilpotent ideal m. Clearly, the condition is sufficient. Suppose now that $R^\times x \neq R^\times y$, and therefore $R^\times x \cap R^\times y = \varnothing$. Then, $R.x = R.x$ implies $R^\times x \subset m.y$, since $R = R^\times \cup m$. But then we have $x \in m.y$, and symmetrically $y \in m.x$, which gives $x \in m^k.y$ for all powers k, and m being nilpotent yields $x = y = 0$, a contradiction. QED.

F.6 Quotients

If G is a finite group, and if $(X, \mathcal{O}_X)$ is a scheme, a group action of G on X can be given by a group homomorphism $\alpha : G \to Aut(X)$. This can also be seen as a morphism of schemes $\alpha' : G_{\mathbb{Z}} \times_{Spec(\mathbb{Z})} X \to X$ with the functorially described axioms of group actions associated with the functors of the schemes $G_{\mathbb{Z}}, X$, the scheme $G_{\mathbb{Z}} = Spec(\mathbb{Z}^G)$ is a group scheme whose multiplication is associated with the group multiplication $\mu : G \times G \to G$ via the ring homomorphism $\mu' : \mathbb{Z}^G \to \mathbb{Z}^{G \times G} \xrightarrow{\sim} \mathbb{Z}^G \otimes_{\mathbb{Z}} \mathbb{Z}^G$. The scheme $G_{\mathbb{Z}}$ is finite and locally free over $\mathbb{Z}$. The *set-theoretic orbits* of the action α are the equivalence classes defined by the relation on the product set $X \times X$, image of the set map $G \times X \to X \times X : (g, x) \mapsto (\alpha(g).x, x)$. If we use the schema-theoretic map $\alpha' : G_{\mathbb{Z}} \times_{Spec(\mathbb{Z})} X \to X$, the cokernel functor of the pair $pr_2, \alpha' : G_{\mathbb{Z}} \times_{Spec(\mathbb{Z})} X \rightrightarrows X$ of functor morphisms, if it exists, is called the *scheme functor of orbits of X under the action of G*. We have this particular case of [259, III,2.6.1]:

Theorem 67 *With the above notation, if G is a finite group and $\alpha' : G_{\mathbb{Z}} \times_{Spec(\mathbb{Z})} X \to X$ the group action associated with an 'abstract' action $\alpha : G \to Aut(X)$ on the scheme X, such that every set-theoretic orbit is contained in an affine open subscheme of X, then there is a scheme-functor of orbits $Y = coker(pr_2, \alpha')$ and the associated diagram of schemes (qua locally ringed spaces)*

$$G \times_{Spec(\mathbb{Z})} X \rightrightarrows X \to Y$$

is exact.

G

Categories, Topoi, and Logic

For a comprehensive introduction to category theory, see [637]. For topos theory and sheaves see [639], for topos theory and logic, see [376].

G.1 Categories Instead of Sets

One may rebuild mathematics from categories rather than from sets. In this framework, the most radical approach is the arrow-only definition of a category[1]:

Definition 163 *A category* **C** *is a collection of objects* $f, g, h, \ldots$ *which are called* morphisms, *together with a partial composition* $f \circ g$ *which yields morphisms of* **C**. *An identity is a morphism* e *such that, whenever defined, we have* $e \circ f = f$ *and* $g \circ e = g$. *We have these axioms:*

1. *Whenever one of the two compositions* $(f \circ g) \circ h, f \circ (g \circ h)$ *is defined, both are defined and they are equal; we denote the resulting morphism by* $f \circ g \circ h$.
2. *If* $f \circ g, g \circ h$ *are both defined,* $(f \circ g) \circ h$ *is defined.*
3. *For every morphism* f *there are two identities, a 'left' identity* e_L *and a 'right' identity* e_R, *such that* $e_L \circ f, f \circ e_R$ *are defined (and necessarily equal to* f).

It is easily seen that two right (left) identities of a morphism f are necessarily equal; they are called the *domain of f (codomain of f)* and are denoted by $dom(f)$ $(codom(f))$. To make domain and codomain evident, one also writes $f : a \to b$ with $a = dom(f), b = codom(f)$ instead of f. For two morphisms a, b, the collection of those f with $dom(f) = a, codom(f) = b$ is denoted by $Hom(a, b), Hom_{\mathbf{C}}(a, b), \mathbf{C}(a, b), \ldots$ according to the specific situation. Evidently, no morphism can be a member of $Hom(a, b)$ and of $Hom(a', b')$ if either $a \neq a'$ or $b \neq b'$, i.e., the Hom collections form a partition of **C** (in the non-set-theoretic common sense).

Exercise 88 *Two identities* e, e' *of* **C** *can be composed iff they are equal, and then* $e \circ e = e$ *(identities are idempotent).*

In a more conservative understanding of categories, the identities are associated with the "objects" of a category, which are a second type of concepts, but do not enrich the category except in the way it is constructed. The identification of objects and identities is carried out as often as possible in our present text.

---

[1] Mac Lane calls this type of set-less categories "metacategories", and reserves the proper term "category" for metacategories which are built upon sets. We do however preconize the foundational character of metacategories and therefore omit the "meta" prefix. However, we then should provide a germ for existing categories, in order to get off ground as with axiomatic set theory. See [639, VI.10] for a discussion of the foundation of mathematics via topoi.

© Springer International Publishing AG, part of Springer Nature 2017
G. Mazzola, *The Topos of Music IV: Roots*, Computational Music Science,
https://doi.org/10.1007/978-3-319-64495-0_7

In either case, the collection of identities (qua objects) is denoted by $_0\mathbf{C}$ or $Ob(\mathbf{C})$, whereas the morphisms are denoted by $_1\mathbf{C}$ or $Mor(\mathbf{C})$. To stress the morphic character of an identity e (in contrast to the underlying object in the conservative understanding), one also writes Id_e instead of e.

In a category, a morphism f is *mono, a monomorphism*, iff for any two compositions $f \circ g, f \circ g'$, the equality $f \circ g = f \circ g'$ implies $g = g'$. The morphism f is *epi, an epimorphism*, iff for any two compositions $g \circ f, g' \circ f$, the equality $g \circ f = g' \circ f$ implies $g = g'$. The morphism f is called a *section* if there is a left inverse g, i.e., $g \circ f = dom(f)$; f is called a *retraction* if it has a right inverse h, i.e., $f \circ h = dom(h)$. A morphism f that is a section and a retraction is *iso, an isomorphism*. If $dom(f) = codom(f)$, the morphism is called *endo, an endomorphism*. An endomorphism which is an isomorphism is called *auto, an automorphism*. The collection of endomorphisms for a domain c is denoted by $End(c)$, whereas the collection of automorphisms for c is denoted by $Aut(c)$. If these collections are sets, they define monoids $End(c)$ and groups $Aut(c)$ with the identity Id_c as unit.

Exercise 89 *The composition of two monomorphisms, epimorphisms, isomorphisms, endomorphisms, and automorphisms, if defined, shares, each of these properties.*

G.1.1 Examples

Example 114 The category **Ens** of all sets. The morphisms are the set maps between existing sets, and the composition is the usual composition of set maps.

Remark 33 *Usually, the delicate comprehension axiom which can cause contradictory constructions of sets, is avoided by a strong restriction of the available sets. One takes a very large set U, which has the properties of a "universe", i.e., it is stable in the following sense:*

- *If $x \in U$, then $x \subset U$;*
- *If $x, y \in U$, then $\{x, y\} \in U$;*
- *If $x \in U$, then $2^x \in U$ (the set of all subsets, the powerset);*
- *A set of all natural[2] numbers $\mathbb{N}$ is element of U;*
- *If $f : x \to y$ is a surjective function with $x \in U, y \subset U$, then $y \in U$*

*One then restricts the **Ens** objects to the elements of the universe U and says that these are* small *sets. We denote such a category of small sets by $\mathbf{Ens}_U$.*

Example 115 Given a quiver $Q = (\text{head}, \text{tail} : A \rightrightarrows V)$ (see Section C.2.2), the *path category* $P(Q)$ has the paths as morphisms, the identities are the lazy paths, and the composition is the path composition. Here, the vertexes are separate concepts which can be identified (and in fact *are* identified in our construction) with the lazy paths. All paths are mono and epi, but only the identities are isomorphisms. The terminology "quiver" stems from algebra, in category theory, a quiver is more known as a "diagram scheme".

Relations among paths give rise to *quotient categories* as follows: Suppose that we are given any binary relation $\sim$ between some paths of equal domain and codomain. Consider the smallest equivalence relation $\sim'$ among paths which contains $\sim$ and is a 'two-sided ideal' in the sense that for $f \sim' g$ with $dom(f) = dom(g) = d, codom(f) = codom(g) = c$ and h, k with $dom(h) = c, codom(k) = d$, we have $c \circ f \sim' c \circ g$ and $f \circ d \sim' g \circ d$. Then we obtain a new category, the quotient category $P(Q)/ \sim$, and its morphisms are the equivalence classes of paths, while the composition is the composition of representatives of these classes. In the language of category theory, the relation $\sim$ is called a *commutativity relation* of the given diagram scheme.

Example 116 Fix a ring R, the *matrix category over R* is the collection $\mathbb{M}_R$ of all $m \times n$-matrices $M = (m_{i,j})$ with coefficients in R and for any row and column numbers m, n, together with the usual matrix multiplication $M \cdot N$ as composition. The identities are all the identity matrices $E_n, n = 1, 2, \ldots$ (over R). We evidently

---

[2] For example the set of finite ordinals $0 = \varnothing, 1 = \{0\}, 2 = \{0, 1\}, \ldots n, n^+ = n \cup \{n\}, \ldots$

have $Hom_{\mathbb{M}_R}(E_n, E_m) = \mathbb{M}_{m,n}(R)$. In particular, the vectors in R^n are identified with the morphisms in $Hom_{\mathbb{M}_R}(E_1, E_m) = \mathbb{M}_{1,n}(R)$.

Example 117 Given a category $\mathbf{C}$, the isomorphism classes of $\mathbf{C}$-objects define a *skeleton* category $\mathbf{C}/iso$: For each isomorphism class, select a representative and then consider the full subcategory[3] of $\mathbf{C}$ on these representative objects. Clearly $\mathbf{C}/iso$ is defined up to isomorphism of categories (see below G.1.2) and no two skeleton objects are isomorphic.

Example 118 Common examples of categories are the categories $\mathbf{Mon}$ of monoids, $\mathbf{Grp}$ groups, $\mathbf{Rings}$ of rings with ring homomorphisms, $\mathbf{LinMod}_R$ R-modules with linear homomorphisms, $\mathbf{LinMod}$ modules with dilinear homomorphisms, $\mathbf{Mod}_R$ R-modules with affine homomorphisms, $\mathbf{Mod}$ of modules with diaffine homomorphisms, or $\mathbf{Top}$ of topological spaces with continuous maps.

Example 119 For every category $\mathbf{C}$ we have the *opposite category* $\mathbf{C}^{opp}$. Its morphisms are the same, but composition works via $f \circ^{opp} g = g \circ f$, i.e., it is defined iff the composition with opposite factors is defined in $\mathbf{C}$. This opposite construction exchanges the domains and codomains of morphisms. Intuitively, an arrow $f : x \to y$ in $\mathbf{C}$ becomes a arrow $f : y \to x$ in $\mathbf{C}^{opp}$.

G.1.2 Functors

Functors are the morphisms between categories:

Definition 164 *If* $\mathbf{C}, \mathbf{D}$ *are categories, a functor* $F : \mathbf{C} \to \mathbf{D}$ *is a function which assigns to every morphism* c *in* $\mathbf{C}$ *a morphism* $F(c)$ *in* $\mathbf{D}$ *such that*

(i) $F(c)$ *is an identity if* c *is so,*
(ii) *if* $c \circ c'$ *is defined in* $\mathbf{C}$, *then* $F(c) \circ F(c')$ *is defined and* $F(c \circ c') = F(c) \circ F(c')$.

In particular, functors carry isomorphisms to isomorphisms. Moreover, the composition $F \circ G : \mathbf{C} \to \mathbf{E}$ of two functors $F : \mathbf{C} \to \mathbf{D}, G : \mathbf{D} \to \mathbf{E}$ is a functor. Two categories $\mathbf{C}, \mathbf{D}$ are called *isomorphic* if there exists a functor isomorphism, i.e., two functors $F : \mathbf{C} \xrightarrow{\sim} \mathbf{D}, F^{-1} : \mathbf{D} \xrightarrow{\sim} \mathbf{C}$ such that $F^{-1} \circ F = Id_{\mathbf{C}}, F \circ F^{-1} = Id_{\mathbf{C}}$. A functor is called *full* iff the $F(Hom_{\mathbf{C}}(x, y)) = Hom_{\mathbf{D}}(F(x), F(y))$ for all object pairs c, d. It is called *faithful* iff $F : Hom_{\mathbf{C}}(x, y) \to Hom_{\mathbf{D}}(F(x), F(y))$ is injective for all pairs c, d. It is called fully faithful iff it is full and faithful, i.e., the map $F : Hom_{\mathbf{C}}(x, y) \to Hom_{D}(F(x), F(y))$ is a bijection. Functors are also called "covariant" since they are opposed to functors $F : \mathbf{C}^{opp} \to \mathbf{D}$, which are called "contravariant" but then also denoted by $F : \mathbf{C} \to \mathbf{D}$.

One often considers systems of morphisms in a category $\mathbf{C}$, which are defined by a graphical approach: diagrams. Here is the precise definition.

Definition 165 *A* diagram *in a category* $\mathbf{C}$ *is a functor* $\Delta : P(Q) \to \mathbf{C}$, *where* Q *is a quiver. The diagram* Δ *is said to* commute *with respect to a relation* $\sim$ *among* Q-paths, *iff* Δ *factorizes through* $P(Q)/\sim$. *If the relation is* maximal *(it identifies all paths having common domain and codomain), then the diagram is said to be* commutative *without specification of* $\sim$.

By the very definition of a path category, diagrams are given by systems of morphisms in $\mathbf{C}$ which cope with the domain-codomain configuration in the underlying quiver (i.e., diagram scheme).

Example 120 If $\mathbf{C}$ is a category, a subcategory is a sub-collection $\mathbf{C}'$ of $\mathbf{C}$ such that for each morphism f in $\mathbf{C}'$, its domain and codomain are also in $\mathbf{C}'$, and such that for any two f, g in $\mathbf{C}'$ such that $f \circ g$ is defined in $\mathbf{C}$, the composition is also a morphism in $\mathbf{C}'$. A category can be defined by an arbitrary selection of objects (identities) out of $\mathbf{C}$ and the full collections of morphisms having these identities as domains or codomains.

---

[3] See Example 120 below.

Such a subcategory is called a *full* subcategory of **C**. A subcategory obviously induces an *embedding* functor $\mathbf{C}' \to \mathbf{C}$ by the identity on the morphisms in $\mathbf{C}'$. For any collection S of morphisms in **C**, the smallest subcategory of **C** containing S is denoted by $\langle S \rangle$ and called the subcategory *generated by S*.

Example 121 If **C**, **D** are two categories, the product category $\mathbf{C} \times \mathbf{D}$ consists of all ordered pairs (c, d) of morphisms c in **C** and d in **D**. The composition $(c, d) \circ (c', d')$ is possible iff it is possible in each component and then evaluates to $(c, d) \circ (c', d') = (c \circ c', d \circ d')$. One has the canonical projection functors $p_1 : \mathbf{C} \times \mathbf{D} \to \mathbf{C}, p_2 : \mathbf{C} \times \mathbf{D} \to \mathbf{D}$ with $p_1(c, d) = c, p_2(c, d) = d$. The same procedure allows the definition of any finite product of categories.

G.1.3 Natural Transformations

Natural transformations are the morphisms between functors.

Definition 166 *If $F, G : \mathbf{C} \to \mathbf{D}$ are two functors, a natural transformation $t : F \to G$ is a system of morphisms $t(c) : F(c) \to G(c)$ in **D**, for each object c in **C**, such that for every morphism $f : x \to y$ in **C**, we have $G(f) \circ t(x) = t(y) \circ F(f)$. One can also rephrase this property by requiring the following commutative diagram in **D**:*

$$
\begin{array}{ccc}
F(x) & \xrightarrow{\;t(x)\;} & G(x) \\
{\scriptstyle F(f)}\downarrow & & \downarrow{\scriptstyle G(f)} \\
F(y) & \xrightarrow{\;t(y)\;} & G(y)
\end{array}
\tag{G.1}
$$

Natural transformations can be composed in an evident way, and the composition is associative. For every functor F we have the natural identity Id_F. We therefore have the category $Func(\mathbf{C}, \mathbf{D})$ of functors $F : \mathbf{C} \to \mathbf{D}$ and natural transformations $Nat(F, G)$ between two functors $F, G : \mathbf{C} \to \mathbf{D}$. Properties between such functors are said to be *natural* if they relate to the category $Func(\mathbf{C}, \mathbf{D})$, for example, $F \xrightarrow{\sim} G$ is a natural isomorphism iff it is an isomorphism among the natural transformations from F to G.

If two categories **C**, **D** satisfy the following properties, they are called *equivalent*, equivalence is an equivalence relation which is weaker than isomorphism.

Lemma 99 *For categories **C**, **D** the following properties are equivalent:*

(i) *There are two functors $F : \mathbf{C} \to \mathbf{D}, G : \mathbf{D} \to \mathbf{C}$ such that $G \circ F \xrightarrow{\sim} Id_{\mathbf{C}}$ and $F \circ G \xrightarrow{\sim} Id_{\mathbf{D}}$, where these isomorphisms are natural.*

(ii) *There is a functor $F : \mathbf{C} \to \mathbf{D}$ which is fully faithful and essentially surjective, i.e., every object (identity) in $_0\mathbf{D}$ is isomorphic to an image $F(c)$ of an object of **C**.*

Example 122 If **C** is a category with sets as hom collections $Hom(x, y)$, we have two types of hom functors as follows: For fixed object x, we have the functor $Hom(x, ?) : \mathbf{C} \to \mathbf{Ens} : y \mapsto Hom(x, y)$, which sends a morphism $f : y \to z$ to $Hom(x, f) : Hom(x, y) \to Hom(x, z) : u \mapsto f \circ u$. We further have the contravariant functor $Hom(?, y) : \mathbf{C}^{opp} \to \mathbf{Ens} : x \mapsto Hom(x, y)$, which sends a morphism $f : x \to z$ to $Hom(f, y) : Hom(z, y) \to Hom(x, y) : u \mapsto u \circ f$. The category $Func(\mathbf{C}^{opp}, \mathbf{Ens})$ of contravariant set-valued functors on **C** is denoted by $\mathbf{C}^{@}$; its elements are called *(set-valued) presheaves over C*. In the theory of denotators, one works with $\mathbf{Mod}^{@}$ and for a module M, we have the notation $Hom_{\mathbf{Mod}}(M, ?) = M@$, whereas the contravariant hom functor is $Hom_{\mathbf{Mod}}(?, M) = @M$.

Example 123 For two categories **C** and **D** and an object S of **D**, we have the constant functor $[S] : \mathbf{C} \to \mathbf{D}$ with $[S](X) = S$ and $[S](f) = Id_S$ for all $X \in Ob(\mathbf{C})$ and all $f \in Mor(\mathbf{C})$. In particular, if S is a set, then we write $[S]$ for the constant functor in $\mathbf{Mod}^{@}$ if the contrary is not stressed.

Given a quiver G, if we fix an object c in a category **C**, we have the constant diagram $\Delta_c = [c]$. It associates every vertex of G with c and every arrow with Id_c. For a diagram Δ in **C**, a natural transformation

$[c] \to \Delta$ is called a *cone* on Δ, whereas a natural transformation $\Delta \to [c]$ is called a *cocone on Δ*. In a cone, all arrows starting from c must commute with the arrows of the diagram, whereas in a cocone all arrows arriving at c must commute with the arrows of the diagram.

G.2 The Yoneda Lemma

For a given category $\mathbf{C}$ with sets as hom collections $Hom(x, y)$, the *Yoneda embedding* Y is the functor

$$Y : \mathbf{C} \to \mathbf{C}^@ : x \mapsto Y(x) = Hom(?, x) \tag{G.2}$$

with the natural transformations $Y(f : x \to y) : Y(x) \to Y(y)$ being defined by $u \mapsto f \circ u$ for $u : z \to x \in Y(x)(z) = Hom(z, x)$. For $\mathbf{C} = \mathbf{Mod}$, we also write $Y = ?@$, i.e., $Y(M) = @M$ for a module M.

A functor F in $\mathbf{C}^@$ is called *representable* iff there is an object c in $_0\mathbf{C}$ such that $F \overset{\sim}{\to} Y(c)$. Yoneda's Lemma states that the full subcategory of representable functors in $\mathbf{C}^@$ is equivalent to $\mathbf{C}$, and that such an equivalence is given by the Yoneda embedding. More precisely:

Lemma 100 *For every functor F in $\mathbf{C}^@$ and object c in $_0\mathbf{C}$, the map*

$$\epsilon : Nat(Y(c), F) \to F(c) : h \mapsto h(c)(Id_c) \tag{G.3}$$

is a bijection.

The proof is an easy exercise, but see also [395, 637]. In particular, if $F = Y(d)$, we have a bijection $\epsilon : Nat(Y(c), Y(d)) \overset{\sim}{\to} Hom(c, d)$. More precisely, this means that the Yoneda functor Y is fully faithful, so we obtain an equivalence of categories as announced. For $\mathbf{C} = \mathbf{Mod}$, we also write $F(A) = A@F$, even if F is not representable. We then have the bijection $Nat(@A, F) \overset{\sim}{\to} A@F$. This means that the evaluation of F at "address" A is the same as the calculation of the morphisms from $@A$ to F. This is a justification of the name "address" for the argument A: Evaluating F at A means "observing F under all morphisms" when being "positioned on (the functor $@A$ of) A". And the Yoneda philosophy means that F is known, when it is known while observed from all addresses.

G.2.1 Universal Constructions: Adjoints, Limits, and Colimits

Definition 167 *We suppose that for two categories $\mathbf{C}, \mathbf{D}$, the hom collections are sets. Given two functors $F : \mathbf{C} \to \mathbf{D}, G : \mathbf{D} \to \mathbf{C}$, we say that F is* left adjoint *to G or (equivalently) that G is* right adjoint *to F, in signs $F \dashv G$ iff the functors $Hom_D(F(?), ?) : \mathbf{C}^{opp} \times \mathbf{D} \to \mathbf{Ens}$ and $Hom_\mathbf{C}(?, G(?)) : \mathbf{C}^{opp} \times \mathbf{D} \to \mathbf{Ens}$ are isomorphic. One also writes this fact in these symbols:*

$$\frac{c \to G(d)}{F(c) \to d}$$

meaning that morphisms in the numerator correspond one-to-one to morphisms in the denominator.

In particular, if we are given an adjoint pair of functors $F \dashv G$, when fixing the variable d in $\mathbf{D}$, the adjointness isomorphism means that the contravariant functor $c \mapsto Hom_\mathbf{D}(F(c), d)$ is representable by the object $G(d)$.

Example 124 For $\mathbf{C} = \mathbf{D} = \mathbf{Ens}$, fix a set A. We have the functors $A \times ? : \mathbf{Ens} \to \mathbf{Ens} : X \mapsto A \times X$ and $?^A : \mathbf{Ens} \to \mathbf{Ens} : X \mapsto X^A$, which are an adjoint pair $A \times ? \dashv ?^A$ via the isomorphism that sends $f : A \times X \to B$ to $f \dashv : X \to B^A : f \dashv (x)(a) = f(a, x)$. This adjointness property is crucial in the definition of exponential objects in topoi. The "exponential set" B^A represents the functor $X \mapsto Hom(A \times X, B)$. See Section G.3.2 for this subject.

A *terminal* object 1 in a category $\mathbf{C}$ is one that admits exactly one morphism, denoted by $! : x \to 1$ from each object x of $\mathbf{C}$. An *initial* object 0 is a terminal object in the opposite category. For example, in **Ens**, every singleton, such as $1 = \{0\}$, is a terminal object, while the empty set 0 is initial.

Example 125 A terminal object in $\mathbf{C}^@$ is defined by the constant $1_{\mathbf{C}^@} = [1]$ (of the set 1). For a presheaf $P \in \mathbf{C}^@$, a *global section* γ is a natural transformation $\gamma : 1_{\mathbf{C}^@} \to P$. In other words, the global sections Γ correspond to the hom functor $\Gamma(P) = Nat(1_{\mathbf{C}^@}, P)$. The global section functor $\Gamma : \mathbf{C}^@ \to \mathbf{Ens}$ is right adjoint to the constant functor $[\ ] : \mathbf{Ens} \to \mathbf{C}^@$.

Universal objects in categories, such as limits and colimits, are related to terminal or initial objects as follows. Given a "basis" object b of $\mathbf{C}$, the *comma category* $\mathbf{C}/b$ has all morphisms $f : x \to b$ as objects, and for two objects $f : x \to b, g : y \to b$, we have

$$Hom_{\mathbf{C}/b}(f, g) = \{u | g \circ u = f\},$$

the set of "commutative triangles above b" with the evident composition. The *cocomma category* $\mathbf{C}/^{opp}b$ is the comma category $(\mathbf{C}^{opp}/b)^{opp}$, in other words, for two objects $f : b \to x, g : b \to y$, we have

$$Hom_{\mathbf{C}/^{opp}b}(f, g) = \{u | u \circ f = g\}.$$

Given a quiver Q and a category $\mathbf{C}$, we have in the category $Func(P(Q), \mathbf{C})$ of diagrams in $\mathbf{C}$, and given such a diagram Δ, the comma category $Func(P(Q), \mathbf{C})/\Delta$. In this category, take the full subcategory $cones(\Delta)$ of cones $[c] \to \Delta$. Then a *limit of Δ* is a terminal object $lim(\Delta)$ in $cones(\Delta)$. Since terminal objects are evidently unique up to isomorphisms, a limit is also unique up to isomorphism. A *colimit $colim(\Delta)$ of a diagram Δ* is an initial object in the subcategory $cocones(\Delta)$ of cocones on Δ in the cocomma category $Func(P(Q), \mathbf{C})/^{opp}\Delta$.

If the diagram is a pair $f : a \to c, g : b \to c$, the limit is called the *fiber product* or *pullback* of f, g, or (more sloppily) of a and b if f, g are clear; it is denoted by $a \times_c b$. If the diagram is a pair $f : c \to a, g : c \to b$, the colimit is called the *fiber sum* or *pushout* of f, g, or (more sloppily) of a and b if f, g are clear; it is denoted by $a \sqcup_c b$. The limit of two isolated objects a, b (discrete diagram with two points) is called the *(cartesian) product of a, b* and denoted by $a \times b$. The colimit of two isolated objects a, b is called the *(disjoint) sum of a, b* and denoted by $a \sqcup b$.

Theorem 68 *The category of sets* **Ens** *has arbitrary limits and colimits. For a category* $\mathbf{C}$ *with sets as hom collections, the category* $\mathbf{C}^@$ *of presheaves over* $\mathbf{C}$ *has arbitrary limits and colimits.*

Proof. If Δ is a diagram of set morphisms $f_{i,j,k} : X_i \to X_j$, the limit is the subset in $\prod_i X_i$ consisting of all families (x_i) such that for any pair $(x_i, x_j) \in X_i \times X_j$ and any $f_{i,j,k}$, we have $f_{i,j,k}(x_i) = x_j$. The projections $lim(\Delta) \to X_i$ are the restrictions of the canonical projections from the product to X_i. The colimit is the set $colim(\Delta)$ of equivalence classes $\coprod_i X_i/ \sim$ defined by the equivalence relation generated by the relation $x_i \sim x_j$ iff there is $f_{i,j,k}(x_i) = x_j$. The morphisms $X_i \to colim(\Delta)$ are the injections $X_i \to \coprod_i X_i$, followed by the quotient map $\coprod_i X_i \to colim(\Delta)$. The universal properties are immediate and left as an exercise.

For a diagram Δ of presheaves F_i, we take for each argument c in C the set-theoretic limit or colimit, respectively, of the set diagram $\Delta(c)$ of the sets $F_i(c)$ and the corresponding maps to define the limit or colimit of Δ, respectively, QED.

The following proposition makes sure that the category of presheaves over $\mathbf{C}$ is not too large with respect to its Yoneda embedding of $\mathbf{C}$:

Proposition 109 *Every presheaf F in $\mathbf{C}^@$ is a colimit of representable presheaves.*

See [639, pp.41/42] for a proof. The idea of this proof uses the so-called *category of elements* $\int_{\mathbf{C}} F$ of a functor F. Its objects are all pairs (C, p) where C is an object of $\mathbf{C}$, and $p \in C@F$. The morphisms $(C, p) \to (C', p')$ are the morphisms $u : C \to C'$ in $\mathbf{C}$ such that $p'.u = p$.

Definition 168 *A category is called* ⟨finitely⟩ (co)complete *iff it has* (co)limits *for all* ⟨finite⟩ *diagrams* ⟨diagrams with finitely many objects and arrows⟩.

Proposition 109 turns out to make the Yoneda embedding into a universal device for making a category **C** cocomplete:

Proposition 110 *For each functor* $f : \mathbf{C} \to \mathbf{E}$ *to a cocomplete category* $\mathbf{E}$, *there exists an essentially unique colimit preserving functor* $L : \mathbf{C}^{@} \to \mathbf{E}$ *such that* $f = L \circ Y$.

See [639, p.43] for a proof.

G.2.2 Limit and Colimit Characterizations

Proposition 111 *For any category* **C**, *the following statements are equivalent:*

(*i*) **C** *is finitely complete.*
(*ii*) **C** *has finite products and equalizers*[4].
(*iii*) **C** *has a terminal object and fiber products.*

For a proof, see [953, I, 7.8.8].

Proposition 112 *For any category* **C**, *the following statements are equivalent:*

(*i*) **C** *is finitely cocomplete.*
(*ii*) **C** *has finite sums and coequalizers*[5].
(*iii*) **C** *has an initial object and fiber sums.*

This is just the dual statement of Proposition 111.

Proposition 113 *Let* **C** *be a finitely complete category. A morphism* $f : A \to B$ *in* **C** *is mono iff the canonical projections* p_1 *and* p_2 *in the pullback*

$$
\begin{array}{ccc}
X & \xrightarrow{\ p_1\ } & A \\
{\scriptstyle p_2}\downarrow & & \downarrow{\scriptstyle f} \\
A & \xrightarrow{\ f\ } & B
\end{array}
\tag{G.4}
$$

coincide and are isomorphisms.

Proof. Clearly, if f is mono, then $X = A$ and $p_1 = p_2 = 1_A$ define a fiber product. Conversely, if $p_1 = p_2$ is an isomorphism, then any couple $u, v : Z \to A$ with $f \cdot u = f \cdot v$ creates factorizations $u = p_1 \cdot t$ and $v = p_2 \cdot t$ through $t : Z \to X$ which therefore also coincide. QED.

Therefore we have the dual result:

Corollary 36 *Let* **C** *be a finitely cocomplete category. A morphism* $f : A \to B$ *in* **C** *is epi iff the canonical morphisms* i_1 *and* i_2 *in the pushout*

$$
\begin{array}{ccc}
A & \xrightarrow{\ f\ } & B \\
{\scriptstyle f}\downarrow & & \downarrow{\scriptstyle i_1} \\
B & \xrightarrow{\ i_2\ } & X
\end{array}
\tag{G.5}
$$

coincide and are isomorphisms.

---

[4] An equalizer is a limit of a pair $f, g : x \rightrightarrows y$ of arrows.
[5] A coequalizer is a colimit of a pair $f, g : x \rightrightarrows y$ of arrows.

Proposition 114 *For any category* **C**, *let* $f : H \to G$ *be a morphism in* $\mathbf{C}^@$. *Then:*

(*i*)   *The morphism* f *is mono iff* $A@f : A@H \to A@G$ *is injective for all objects* A *of* **C**.
(*ii*) *The morphism* f *is epi iff* $A@f : A@H \to A@G$ *is surjective for all objects* A *of* **C**.
(*iii*) *The morphism* f *is iso iff it is mono and epi iff* $A@f : A@H \to A@G$ *is bijective for all objects* A *of* **C**.

Proof. Observe that $\mathbf{C}^@$ is finitely complete and cocomplete and that limits and colimits are calculated pointwise. Let us first look at point (iii). Clearly, f is iso iff its evaluations $A@f : A@H \to A@G$ are all bijective. Further, we know from Proposition 113 that f is mono iff the fiber product projections p_1 and p_2 coincide and are iso. But with (iii) this is true iff this is true for all evaluations at objects A of **C**, i.e., iff this is true set-theoretically, and this means having an injection for every object A of **C**, and (i) is done; the dual argument shows (ii). Finally, iso always implies mono and epi; conversely, mono and epi means being in- and surjective, i.e., bijective at every object A of **C**, whence f is iso. QED.

G.2.2.1 Special Results for Mod$^@$

Lemma 101 *Let* H *be a functor in* $\mathbf{Mod}^@$, *M, N be two addresses, and* $f : N \to M$ *a morphism of addresses. Then the map*

$$M \mapsto M@Fin(H) := \{F \subset M@H, card(F) < \infty\}, \tag{G.6}$$

together with the maps

$$f@Fin(H) : M@Fin(H) \to N@Fin(H) : X \mapsto f@H(X), \tag{G.7}$$

defines a functor $Fin(H)$ *in* $\mathbf{Mod}^@$.

Lemma 102 *For any functor H in* $\mathbf{Mod}^@$ *and address M, the maps*

$$sing_H(M) : M@H \to M@Fin(H) : x \mapsto \{x\} \tag{G.8}$$

defines a monomorphism $sing_H : H \rightarrowtail Fin(H)$ *of functors.*

Lemma 103 *The map*
$$Fin : \mathbf{Mod}^@ \to \mathbf{Mod}^@ : H \mapsto Fin(H) \tag{G.9}$$

defines an endofunctor on $\mathbf{Mod}^@$, *and the monomorphism sing defines a natural transformation*

$$sing : Id_{\mathbf{Mod}^@} \rightarrowtail Fin. \tag{G.10}$$

Lemma 104 *Let* $\mathbf{D} = H_0 \underset{f_0}{\to} H_1 \underset{f_1}{\to} H_2...$ *be a natural sequence diagram in* $\mathbf{Mod}^@$. *Then we have*

$$colim(Fin(\mathbf{D})) \overset{\sim}{\to} Fin(colim(\mathbf{D}). \tag{G.11}$$

This yields an important proposition for the construction of circular forms and denotators.

Proposition 115 *Let H be a functor in* $\mathbf{Mod}^@$. *Then there are functors X and Y in* $\mathbf{Mod}^@$ *such that*

$$X \overset{\sim}{\to} Fin(H \times X) \text{ and} \tag{G.12}$$
$$Y \overset{\sim}{\to} H \times Fin(Y). \tag{G.13}$$

Proof. For the first isomorphism, let $(X_n)_{0 \leqslant n}$ be the following sequence of functors in $\mathbf{Mod}^@$. We recursively define

$$X_n = \begin{cases} \varnothing & \text{for } n = 0, \\ Fin(H \times X_{n-1}) & \text{for } n > 0. \end{cases} \tag{G.14}$$

Then we have a diagram of subfunctors

$$f_n : X_n \hookrightarrow X_{n+1} \tag{G.15}$$

for all $0 \leqslant n$. In fact, clearly $X_0 \hookrightarrow X_1$. Now, let $0 < n$ and take an address M. We have $M@X_n = M@Fin(H \times X_{n-1}) = Fin(M@H \times M@X_{n-1})$. Since by induction $X_{n-1} \hookrightarrow X_n$, we have $M@X_{n-1} \subset M@X_n$ and hence $M@X_n \subset M@X_{n+1}$. Now, we know from [953] that the product commutes with the colimit over a sequence diagram. Taking the diagram $\mathbf{D} = X_0 \underset{f_0}{\to} X_1 \underset{f_1}{\to} X_2...$ and setting $X = colim(\mathbf{D})$, Lemma 104 yields

$$Fin(H \times X) = \tag{G.16}$$
$$Fin(H \times colim(\mathbf{D})) \overset{\sim}{\to}$$
$$Fin(colim(H \times \mathbf{D})) \overset{\sim}{\to}$$
$$colim(Fin(H \times \mathbf{D})) \overset{\sim}{\to}$$
$$colim(\mathbf{D}) =$$
$$X \tag{G.17}$$

and we are done for the first isomorphism. For the second, take

$$Y_n = \begin{cases} \varnothing & \text{for } n = 0, \\ H \times Fin(Y_{n-1}) & \text{for } n > 0. \end{cases} \tag{G.18}$$

We again have a diagram $\mathbf{E} = Y_0 \underset{g_0}{\to} Y_1 \underset{g_1}{\to} Y_2...$ of subfunctors

$$g_n : Y_n \hookrightarrow Y_{n+1} \tag{G.19}$$

for all $0 \leqslant n$. Setting $Y = colim(\mathbf{E})$, our second isomorphism results:

$$H \times Fin(Y) = \tag{G.20}$$
$$H \times Fin(colim(\mathbf{E})) \overset{\sim}{\to}$$
$$H \times colim(Fin(\mathbf{E})) \overset{\sim}{\to}$$
$$colim(H \times Fin(\mathbf{E})) \overset{\sim}{\to}$$
$$colim(\mathbf{E}) =$$
$$Y \tag{G.21}$$

and we are done.

G.3 Topoi

Topoi are special categories which imitate the crucial constructions of set theory, such as cartesian products, disjoint unions, power sets, and characteristic maps. In our context, topoi play two roles: (1) the role of basic mathematical realities which are instantiated to get off ground in denotator theory, i.e., to build compound concept spaces and their points; (2) the more technical role of topoi of sheaves associated with presheaves for Grothendieck topologies.

G.3.1 Subobject Classifiers

Definition 169 *Given a category* **C** *which is finitely complete, with the terminal object* 1*, a monomorphism* $true : 1 \rightarrowtail \Omega$ *in* **C** *is called a* subobject classifier *iff given any monomorphism* $\sigma : S \rightarrowtail X$ *in* **C***, there is a unique morphism* $\chi_\sigma : X \to \Omega$ *such that the diagram*

$$
\begin{array}{ccc}
S & \xrightarrow{\ \sigma\ } & X \\
{\scriptstyle !}\big\downarrow & & \big\downarrow{\scriptstyle \chi_\sigma} \\
1 & \xrightarrow{\ true\ } & \Omega
\end{array}
\tag{G.22}
$$

is a pullback.

Subobject classifiers are unique up to isomorphism. If a subobject classifier exists, the morphism χ must be the same if we replace σ by $\sigma \circ q$ for any isomorphism $q : S' \xrightarrow{\sim} S$ since an isomorphic object to a pullback is also a pullback. A *subobject of* X is an equivalence class of monomorphisms $\sigma : S \rightarrowtail X$ under the relation $\sigma \sim \sigma'$ iff there is an isomorphism q such that $\sigma' = \sigma \circ q$. Suppose that the collection of subobjects of X is a set $Sub_{\mathbf{C}}(X)$ for each object X in **C**. Then this is a presheaf in $\mathbf{C}^@$ by this map: take a morphism $f : Y \to X$. Then we define

$$Sub_{\mathbf{C}}(f) : Sub_{\mathbf{C}}(X) \to Sub_{\mathbf{C}}(Y) : \sigma \mapsto \sigma_f$$

where $\sigma_f : S \times_X Y \to Y$ is the canonical projection of the pullback under f, σ. It is straightforward that this is a monomorphism. Then we have:

Proposition 116 *A category* **C** *which is finitely complete and such that the subobject presheaf* $Sub_{\mathbf{C}}$ *is defined, has a subobject classifier iff the* $Sub_{\mathbf{C}}$ *is representable,* $Sub_{\mathbf{C}}(X) \xrightarrow{\sim} Hom(X, \Omega)$ *for all* X*. If so, the subobject classifier can be set to the inverse image* $\epsilon \rightarrowtail \Omega$ *of* Id_Ω *in* $Sub_{\mathbf{C}}(\Omega)$*.*

See [639, p.33] for a proof.

Example 126 In the category **Ens**, the ordinal number inclusion $true : 1 \rightarrowtail 2 = \{0, 1\} : 0 \mapsto 0$ is a subobject classifier. Since the subobjects of a set X in **Ens** identify to the subsets $S \subseteq X$, we have the classical result that subsets S of X are characterized by their characteristic maps $\chi_S : X \to 2$, a fact that is also traced in the notation 2^X for the set of subsets, the powerset of X.

Example 127 The equivalence classes of monomorphisms of presheaves $S \rightarrowtail X$ in $\mathbf{C}^@$ are defined by their images $Im(S) \subseteq X$ (take everything pointwise). So $sub_{\mathbf{C}^@}(X) \xrightarrow{\sim} \{S \subseteq X\}$, the set of subfunctors of X (supposing that it exists as a set). By the Yoneda Lemma, if a subobject classifier in $\mathbf{C}^@$ exists, we must have[6] $Sub_{\mathbf{C}^@}(@Y) \xrightarrow{\sim} Hom(@Y, \Omega) \xrightarrow{\sim} \Omega(Y)$. So the functor $Y \mapsto Sub_{\mathbf{C}^@}(@Y)$ is a canonical candidate for Ω, and it in fact does the job, see [639, pp.37/38]. The final presheaf being the constant presheaf $1_{\mathbf{C}^@} : X \mapsto 1$, we get the *true* morphism (natural transformation) $true(0) = @Y$. A subfunctor of $@Y$ is called a *sieve in* Y, so a candidate for the *subobject classifier is the functor of sieves* (verify that it is a functor!).

Exercise 90 *The categories* **Ab** *of abelian groups or* $\mathbf{Mod}_R$ *of* R*-modules have no subobject classifiers.*

In denotator theory, sieves and more general subfunctors replace local compositions (which are essentially subsets of ambient modules) in the functorial setup. This is also necessitated since module categories are no topoi (since they have no subobject classifiers, see definition 170 in Appendix G.3.3), so the passage to the presheaves over modules, i.e., the category $\mathbf{Mod}^@$ is mandatory in order to recover the subobject classifier structure.

---

[6] Writing the shorter $@Y$ instead of $Hom(?, Y)$.

G.3.2 Exponentiation

Recall Example 124 in Appendix G.2.1 of exponential sets. More generally, a category $\mathbf{C}$ is called *cartesian closed* iff it has finite products[7] and each element A is *exponentiable*, which means that the functor $A \times ?$ has a right adjoint $?^A$, i.e., we have an adjoint pair of functors $A \times ? \dashv ?^A$.

Example 128 The category of sets **Ens** is cartesian closed. And any product of cartesian closed categories is cartesian closed.

Example 129 A category of presheaves $\mathbf{C}^{@}$ is cartesian closed by the following discussion. Again, we use the Yoneda Lemma to find a canonical candidate of the exponentiation X^Y of two presheaves X, Y. If the exponential X^Y exists, we must have $U@X^Y \overset{\sim}{\to} Nat(@U, X^Y) \overset{\sim}{\to} Nat(@U \times Y, X)$. So one canonical definition must be

$$U@X^Y = Nat(@U \times Y, X) \tag{G.23}$$

for any object U in $\mathbf{C}$, which is evidently a presheaf. The proof that this formula does the job is found in [639, p.47].

In every cartesian closed category $\mathbf{C}$ one has these standard formulas

$$1^X \overset{\sim}{\to} 1, X^1 \overset{\sim}{\to} X, (Y \times Z)^X \overset{\sim}{\to} Y^X \times Z^X, X^{Y \times Z} \overset{\sim}{\to} (X^Y)^Z,$$

which follow from the universal adjointness property of exponentiation.

G.3.3 Definition of Topoi

There are several equivalent definitions of a topos, which we first summarize in the following proposition:

Proposition 117 *For a category $\mathbf{C}$, the following group properties are equivalent:*

1. a) $\mathbf{C}$ *is cartesian closed,*
    b) $\mathbf{C}$ *has a subobject classifier $1 \rightarrowtail \Omega$.*
2. a) $\mathbf{C}$ *is cartesian closed,*
    b) $\mathbf{C}$ *is finitely cocomplete,*
    c) $\mathbf{C}$ *has a subobject classifier $1 \rightarrowtail \Omega$.*
3. a) $\mathbf{C}$ *has a terminal object and pullbacks,*
    b) $\mathbf{C}$ *has exponentials,*
    c) $\mathbf{C}$ *has a subobject classifier $1 \rightarrowtail \Omega$.*
4. a) $\mathbf{C}$ *has a terminal object and pullbacks,*
    b) $\mathbf{C}$ *has an initial object and pushouts,*
    c) $\mathbf{C}$ *has exponentials,*
    d) $\mathbf{C}$ *has a subobject classifier $1 \rightarrowtail \Omega$.*
5. a) $\mathbf{C}$ *is finitely complete,*
    b) $\mathbf{C}$ *has power objects[8].*

Definition 170 *A category $\mathbf{C}$ which has the equivalent groups of properties in Proposition 117 is called a* (elementary) *topos.*

Here are some general properties and examples of topoi:

Proposition 118 *For a topos $\mathbf{C}$ a comma category $\mathbf{C}/b$ is also a topos.*

---

[7] Equivalently: binary products and a terminal object.
[8] See [376, p.106] for this group of properties.

Proposition 119 *For a category* **C** *the presheaf category* $\mathbf{C}^@$ *is a topos.*

This is immediate from the previous discussion of the presheaf category.

Proposition 120 *Let* **C** *be a topos. Then we have these properties:*

(*i*)   *Every morphism f has an image, i.e., factors as $f = i \circ e$ with i mono and e epi. For any two such factorizations $f = i \circ e, f = i' \circ e'$, there is an isomorphism t such that $e' = t \circ e, i' = i \circ t$.*

(*ii*)   *A morphism is iso iff it is mono and epi.*

(*iii*) *The pullback of an epi is an epi.*

(*iv*) *Every arrow $X \to 0$ is iso.*

(*v*)   *Every arrow $0 \to X$ is mono.*

Definition 171 Logical *morphisms between topoi are functors which preserve (up to isomorphism) finite limits, exponentials, and subobject classifiers.*

For example, the canonical base change functor $\mathbf{C}/b \to \mathbf{C}/c$ of comma topoi for a base change morphism $c \to b$ is logical, see [639, p.193].

G.4 Grothendieck Topologies

Grothendieck topologies and associated topoi of sheaves are a classical example for the geometric aspects of topoi. Here is the context.

Given a finitely complete category (a small one for those who like universes) **C** with the subobject classifier functor of sieves $X@\Omega = Sub_{\mathbf{C}^@}(@X)$ (see Example 127). Recall that given a morphism $f : Y \to X$ the functor maps a sieve $S \subseteq X$ to the pullback sieve $f^*(S) = S \times_X Y$.

Definition 172 *A* Grothendieck topology *on a category* **C** *is a function J which for each X is a subset $X@J \subseteq X@\Omega$ of sieves in X with these properties:*

(*i*) $@X \in X@J$,

(*ii*) *(Stability) If $S \in X@J$, then for $f : Y \to X$, $f^*S \in Y@J$,*

(*iii*) *(Transitivity) If $S \in X@J$ and $R \in X@\Omega$ with $f^*R \in Y@J$ for all $f : Y \to X$ in S, then $R \in X@J$.*

A site *is a pair $(\mathbf{C}, J)$ of a Grothendieck topology J on a category* **C**. *A sieve in $X@J$ is called a* covering sieve, *one also says that "it covers X".*

The first two requirements mean that J is a subfunctor of Ω through which the *true* arrow factorizes. Very often, Grothendieck topologies are not given directly, but via a so-called basis:

Definition 173 *For a finitely complete category* **C**, *a* basis *(for a Grothendieck topology) is a function K which assigns to each object X a collection $K(X)$ of families of morphisms with codomain X such that:*

(*i*)   *For every isomorphism $f : X' \xrightarrow{\sim} X$, the singleton $\{f\}$ is in $K(X)$;*

(*ii*) *(Stability) If $(f_i : X_i \to X) \in K(X)$, and $h : Y \to X$, then $(h^*f_i : X_i \times_X Y \to Y) \in K(Y)$;*

(*iii*) *(Transitivity) If $(f_i : X_i \to X) \in K(X)$ and, for each index i, $(f_{ij} : X_{ij} \to X_i) \in K(X_i)$, then $(f_i \circ f_{ij} : X_{ij} \to X) \in K(X)$.*

A pair $(\mathbf{C}, K)$ is again called a site *(see below for a justification!); whereas the families in the sets $K(X)$ are called* covering families.

Here is the relation to Grothendieck topologies: Given a basis K as above, one defines

$$J_K(X) = \{S|\ \text{there is } R \in K(X) \text{ with } R \subseteq S\}, \tag{G.24}$$

where $R \subseteq S$ means that R is in the union of the evaluations $\bigcup_{\mathbf{C}} Z@S$ of S.

And the converse: Given a family R of morphisms with codomain X, we denote by (R) the sieve generated by R, i.e., the smallest sieve in X containing all arrows of R. Then a Grothendieck topology J can be defined by the following basis K which is this set at X:

$$K(X) = \{R \subseteq @X|(R) \in X@J\}. \tag{G.25}$$

G.4.1 Sheaves

Definition 174 *Given a site* $(\mathbf{C}, J)$, *a presheaf* P *in* $\mathbf{C}$ *is a sheaf for* J *iff for every covering sieve* $S \subseteq @X$, *the inclusion induces a bijection* $Nat(@X, P) \xrightarrow{\sim} Nat(S, P)$.

This condition can be rephrased for a basis K of J in a more effective and classical way. To this end recall that a (co)equalizer of a pair $f, g : x \rightrightarrows y$ of parallel arrows is the (co)limit of this diagram.

Proposition 121 *A presheaf* P *on* $\mathbf{C}$ *is a sheaf for the topology* J, *iff for any covering family* $(f_i : X_i \to X) \in K(X)$, *the canonical diagram*

$$P(X) \to \prod_i P(X_i) \rightrightarrows \prod_{i,j} P(X_i \times_X X_j) \tag{G.26}$$

is an equalizer.

Here the two arrows to the right stem from the two projections from $X_i \times_X X_j$ to X_i and to X_j, whereas the arrow to the left stems from the covering morphisms f_i.

Definition 175 *A* Grothendieck topos *is a category which is equivalent to the full subcategory* $Sh(\mathbf{C}, J)$ *of sheaves in* $\mathbf{C}^@$

The justification of this terminology lies in the following theorem:

Theorem 69 *A Grothendieck topos is an elementary topos.*

The proof evidently splits in the verification of finite completeness, existence of exponentials and of a subobject classifier. For details, see [639, pp.128-144]. Finite completeness is easy since:

Lemma 105 *Limits of sheaves are sheaves.*

Lemma 106 *If* P *is a presheaf and* F *is a* J-*sheaf in* $\mathbf{C}^@$, *then the presheaf exponential* F^P *is a sheaf and therefore an exponential in* $Sh(\mathbf{C}, J)$.

Definition 176 *A sieve* $L \subseteq @X$ *is* closed *with respect to* $(\mathbf{C}, J)$ *iff* $f : Y \to X$ *with* $f^*(L) \in Y@J$ *implies* $f \in Y@L$.

Proposition 122 *The function* $X \mapsto X@\Omega_{Sh} = \{closed\ sieves\ in\ X\} \subset X@\Omega$ *contains* $@X$ *defines a subpresheaf of the subobject classifier of* $\mathbf{C}^@$ *and is a sheaf. Together with the morphism* $true : 1_{\mathbf{C}^@} \to \Omega_{Sh}$ *it defines a subobject classifier of* $Sh(\mathbf{C}, J)$.

The topos $Sh(\mathbf{C}, J)$ is a subtopos of $\mathbf{C}^@$ with the natural inclusion $i : Sh(\mathbf{C}, J) \to \mathbf{C}^@$. This natural transformation has a left adjoint of sheafification which we shall discuss now. If P is a presheaf over $\mathbf{C}$, the sheafification operator $P \mapsto P^+$ evaluates as follows. For an object $X \in \mathbf{C}$ and a sieve $S \in X@J$, consider the limit $Match_P(S) = \prod_{f:Y \to X \in S} Y@P$. Consider the diagram $(Match_P(S))_{S \in X@J}$ with canonical restriction maps $Match_P(S) \to Match_P(T)$ for $T \subseteq S$, and define $X@P^+ = lim_{S \in X@J} Match_P(S)$. For a morphism $g : X_1 \to X_2$, we have a map $P^+(g) : X_2@P^+ \to X_1@P^+$. It takes a "matching family" $(x_f)_{f \in S}$ to the matching family $(x_{g.h})_{h \in g^\star S}$. This evidently defines a presheaf, and we have a canonical morphism $\eta : P \to P^+$. Then:

Theorem 70 *With the above notation, we have*

(i)   The presheaf P^+ is separated[9].
(ii) The presheaf P is separated iff η is mono.
(iii) The presheaf P is a sheaf iff η is iso.

For a proof, see [639, III.5]. In particular, the double application $\mathbf{a}P = ((P)^+)^+$ yields a sheaf and a natural presheaf morphism $P \to \mathbf{a}P$. We have

Theorem 71 *The map $P \mapsto \mathbf{a}P$ defines a left adjoint of the inclusion i; $\mathbf{a} \dashv i$. The composition $\mathbf{a} \circ i$ is isomorphic to the identity on the sheaf category $Sh(\mathbf{C}, J)$.*

For a proof, see [639, III.5, Theorem 1] and [639, III.5, Corollary 6].

Corollary 37 *If $f : F \to G$ is a morphism of sheaves, f is mono iff it is an injection for each argument.*

The proof follows from the fact that this is true for presheaves, and that by the adjunction Theorem 71, i preserves and reflects[10] monomorphisms, QED.

G.5 Formal Logic

Formal logic does not replace absolute logic which is built upon the non-formalizable theorem of identity (A is identical to A), of contradiction (A and non-A exclude each other), and of the excluded third (there is no third choice except A or non-A). It does however model the way a specific domain of knowledge can handle its formal truth mechanisms.

G.5.1 Propositional Calculus

Sentences in propositional calculus are defined from a set $\Phi = \{\pi_0, \pi_1, \ldots\}$ of symbols, called *propositional variables*; a set $\Xi = \{!, \&, |, ->\}$ of logical connective symbols[11] (! for negation, & for conjunction, | for disjunction, and $->$ for implication); a set $\Delta = \{(,)\}$ of two brackets; mutually disjoint from each other. Let $EX = FM(\Phi \cup \Xi \cup \Delta)$ be the free monoid of word *expressions* above these symbols. Let $S(EX)$, the set of *sentences*, be the smallest subset of EX with these properties:

Property 4 *Given the symbol sets Φ, Ξ, Δ, we require:*

(i) $\Phi \subset S(EX)$;
(ii) if $\alpha \in S(EX)$, then $(!\alpha) \in S(EX)$;
(iii) if $\alpha, \beta \in S(EX)$, then $(\alpha\&\beta) \in S(EX)$;

---

[9] The left arrow in equation (G.26) for P^+ is injective.

[10] A functor f reflects a property of morphisms if the property for the image morphism $f(x)$ implies the property for x.

[11] Our symbols are near to programming symbols, where the formalism is really needed.

(iv) if $\alpha, \beta \in S(EX)$, then $(\alpha | \beta) \in S(EX)$;
(v) if $\alpha, \beta \in S(EX)$, then $(\alpha \,\text{->}\, \beta) \in S(EX)$.

Clearly, in $S(EX)$, the building blocks of a sentence are uniquely determined, so it makes sense to define set-valued functions $\epsilon : S(EX) \to A$ on such sentences by recursion of the building blocks. Suppose that A is a lattice, i.e., a partially ordered set $(A, \leqslant)$ with a *join* operation $\vee : A \times A \to A$, a *meet* operation $\wedge : A \times A \to A$, minimum (False) $\bot$, a maximum (True) $\top$, further a unary *negation* operation $\neg : A \to A$, and a binary *implication* operation $\Rightarrow : A \times A \to A$. Call such an A a *logical algebra*. Then any set function $\epsilon_0 : \Phi \to A$ extends in a unique way to the *evaluation* $\epsilon = \epsilon(\epsilon_0) : S(EX) \to A$ by these rules:

Property 5 *For all sentences α, β, we set*

(i) $\epsilon(!\alpha) = \neg \epsilon(\alpha)$;
(ii) $\epsilon((\alpha \& \beta)) = \epsilon(\alpha) \wedge \epsilon(\beta)$;
(iii) $\epsilon((\alpha | \beta)) = \epsilon(\alpha) \vee \epsilon(\beta)$;
(iv) $\epsilon((\alpha \,\text{->}\, \beta)) = \epsilon(\alpha) \Rightarrow \epsilon(\beta)$.

Propositional calculus deals with the evaluation map on special logical algebras. A sentence α *is called A-valid*, in symbols: $A \models \alpha$, iff $\epsilon(\alpha) = \top$ for all evaluations $\epsilon_0 : \Phi \to A$ on the propositional variables. It is called *classically valid* or a *tautology* iff it is 2-valid for the well-known Boolean algebra $2 = \{0, 1\}$ of classical truth values, where we set $\top = 1, \bot = 0$. The symbol for classical validity is $\models \alpha$. Here are typical classes of logical algebras:

Boolean Algebras. A Boolean algebra is a distributive logical algebra such that $x \vee \neg x = \top$ and $x \wedge \neg x = \bot$. Distributivity means that $x \wedge (y \vee z) = x \wedge y \vee x \wedge z$ and $x \vee (y \wedge z) = x \wedge y \vee x \wedge z$. Further, implication is defined by $x \Rightarrow y = \neg x \vee y$. In a Boolean algebra (BA), one has these properties: $\neg \neg x = x$, $x \wedge y = \bot$ iff $y \leqslant \neg x$, $x \leqslant y$ iff $\neg y \leqslant \neg x$, $\neg (x \wedge y) = \neg x \vee \neg y$, $\neg (x \vee y) = \neg x \wedge \neg y$.

Heyting Algebras. A Heyting algebra A is a partially ordered set which, as a category whose morphisms $x \to y$ are the pairs $x \leqslant y$, has all finite products and coproducts, and which has exponentials, so it is cartesian closed. In other words, a Heyting algebra is a lattice with minimum $\bot$ and maximum $\top$ which has exponentials x^y. One writes the product as meet ($\wedge$) and the coproduct as join ($\vee$). The exponential x^y is written as $y \Rightarrow x$, and the adjunction property of exponentiation reads

$$z \leqslant y \Rightarrow x \text{ iff } z \wedge y \leqslant x. \tag{G.27}$$

For a Heyting algebra, we define a negation $\neg x = x \Rightarrow \bot$, which is equivalent to $y \leqslant \neg x$ iff $y \wedge x = \bot$.

Proposition 123 *For a Heyting algebra, we have these identities:* $x \leqslant \neg \neg x$, $x \leqslant y$ *implies* $\neg y \leqslant \neg x$, $\neg x = \neg \neg \neg x$, $\neg \neg (x \wedge y) = \neg \neg x \wedge \neg \neg y$, $(x \Rightarrow x) = \top$, $x \wedge (x \Rightarrow y) = x \wedge y$, $y \wedge (x \Rightarrow y) = y$, $x \Rightarrow (y \wedge z) = (x \Rightarrow y) \wedge (x \Rightarrow z)$.

Proposition 124 *A Heyting algebra is distributive, and it is Boolean iff* $x = \neg \neg x$ *for all x, or iff* $x \vee \neg x = \top$ *for all x.*

Proposition 125 *For a presheaf category $\mathbf{C}^@$, the partially ordered set $Sub_{\mathbf{C}^@}(P)$ of an object P is a Heyting algebra. The connectives are defined as follows. If S, T are two subfunctors of P, then:*
(i) $X @ (S \vee T) = X @ S \cup X @ T$;
(ii) $X @ (S \wedge T) = X @ S \cap X @ T$;
(iii) $X @ (S \Rightarrow T) = \{x \in X @ P | \text{ for every morphism } f : Y \to X, \text{ if } x \cdot f \in Y @ S, \text{ then } x \cdot f \in Y @ T\}$;
(iv) $X @ (\neg S) = \{x \in X @ P | \text{ for every morphism } f : Y \to X, x \cdot f \notin Y @ S\}$;

More generally (see [639, IV.8] for details):

Theorem 72 *For every topos $\mathbf{C}$, the partially ordered set $Sub(X) \xrightarrow{\sim} Hom(X, \Omega)$ of subobjects of X is a Heyting algebra.*

To the left, this structure stems from the canonical Heyting algebra structure on the subobjects of X. To the right, this structure is induced by the following operations on Ω:

1. Negation $\neg : \Omega \to \Omega$ is the characteristic map of the false arrow $false : 1 \rightarrowtail \Omega$, which is the characteristic map of the zero arrow $0 \rightarrowtail 1$.
2. Disjunction $\wedge : \Omega \times \Omega \to \Omega$ is the character of the diagonal morphism $\Delta(true, true) : 1 \to \Omega$.
3. Conjunction $\vee : \Omega \times \Omega \to \Omega$ is the character of the image of the universal morphism $a \sqcup b : \Omega \sqcup \Omega \to \Omega \times \Omega$, where $a = \Delta(Id_\Omega, true) : \Omega \to \Omega \times \Omega$ and $b = \Delta(true, Id_\Omega) : \Omega \to \Omega \times \Omega$.
4. Implication $\Rightarrow : \Omega \times \Omega \to \Omega$ is the character of the equalizer of $p_1, \wedge : \Omega \times \Omega \rightrightarrows \Omega$.

If a sentence α is valid for all Boolean algebras, we write $BA \models \alpha$. If it is valid for all Heyting algebras, we write $HA \models \alpha$. If it is valid in the Heyting algebra $Hom(1, \Omega)$ of a topos $\mathbf{C}$, we write $\mathbf{C} \models \alpha$.

Validity is also described by a recursive construction process of valid sentences. One gives a set AX of sentences, called *axioms*, and defines *theorems* as those sentences s which are at the end of *proof chains*, i.e., finite sequences of sentences $(s_0, s_1, \ldots s_n, s)$ such that each member of this sequence is either an axiom or can be inferred from earlier members by a set $RULES$ of rules.

The classical setup is this. AX consists of 12 types of sentences:

Axiom 7 *Axioms of classical logic (CL):*

(i) $\alpha \to (\alpha \& \alpha)$
(ii) $(\alpha \& \beta) \to (\beta \& \alpha)$
(iii) $(\alpha \to \beta) \to ((\alpha \& \gamma) \to (\beta \& \gamma))$
(iv) $((\alpha \to \beta) \& (\beta \to \gamma)) \to (\alpha \to \gamma)$
(v) $\beta \to (\alpha \to \beta)$
(vi) $(\alpha \& (\alpha \to \beta)) \to \beta$
(vii) $\alpha \to (\alpha | \beta)$
(viii) $(\alpha | \beta) \to (\beta | \alpha)$
(ix) $((\alpha \to \beta) \& (\beta \to \gamma)) \to ((\alpha | \beta) \to \gamma)$
(x) $(!\alpha) \to (\alpha \to \beta)$
(xi) $((\alpha \to \beta) \& (\alpha \to (!\beta))) \to (!\alpha)$
(xii) $\alpha | (!\alpha)$

The system CL has one single rule of inference:

Principle 29 (Modus ponens) *From α and $\alpha \to \beta$, β may be derived.*

The property of a sentence α of being a CL-theorem is denoted by $\vdash_{CL} \alpha$. Following Heyting, the intuitionistic logic (IL) is the (CL) with the axiom (xii) omitted, and the same inference rule.

Theorem 73 *The following validity statements are equivalent:*

(i) $\vdash_{CL} \alpha$.
(ii) $\models \alpha$.
(iii) *There exists a Boolean algebra B such that $B \models \alpha$.*
(iv) $BA \models \alpha$.

See [376] for details. We have this weaker relation:

Theorem 74 *The topos validity $\mathbf{C} \models \alpha$ implies classical validity $\vdash_{CL} \alpha$.*

Definition 177 *A topos is Boolean iff the Heyting algebra $Sub(X)$ of each object X is Boolean.*

Theorem 75 *The following statements for a topos $\mathbf{C}$ are equivalent:*

(*i*) **C** *is Boolean.*
(*ii*) $Sub(\Omega)$ *is a Boolean algebra.*
(*iii*) $true : 1 \to \Omega$ *has a complement in* $Sub(\Omega)$.
(*iv*) $false : 1 \to \Omega$ *is the complement of* $true$ *in* $Sub(\Omega)$.
(*v*) $true \cup false = Id_\Omega$ *in* $Sub(\Omega)$.
(*vi*) **C** *is classic, i.e.,* $true \sqcup false : 1 \sqcup 1 \to \Omega$ *is iso.*
(*vii*) *The first inclusion* $\iota_1 : 1 \to 1 \sqcup 1$ *is a subobject classifier.*

For a proof, see [376, p.156 ff.].

Theorem 76 *If the topos* **C** *is Boolean, then* $\mathbf{C} \models \alpha | !\alpha$ *for all sentences* α.

Theorem 77 *For a topos* **C**, *the following are equivalent:*

(*i*) $\mathbf{C} \models \alpha$ *iff* $|\underset{CL}{\overline{}} \alpha$ *for all* α.
(*ii*) $\mathbf{C} \models \alpha | !\alpha$ *for all* α.
(*iii*) $Sub(1)$ *is a Boolean algebra.*

Theorem 78 *We have* $HA \models \alpha$ *iff* $|\underset{IL}{\overline{}} \alpha$.

Theorem 79 *For all topoi* **C**, *the validity* $|\underset{IL}{\overline{}} \alpha$ *implies* $\mathbf{C} \models \alpha$.

G.5.2 Predicate Logic

Predicate calculus generates a richer set of sentences whose validity is a function of the interpretation of predicate variables and individual variables and not only of abstract propositional variables. We are given a set $\Upsilon = \{\iota_0, \iota_1, \ldots\}$ of *individual variables*, a set $\Xi = \{!, \&, |, \; ->, \exists, \forall\}$ of *predicate connectives*, a set $\Pi = \coprod_{i=0,1,2,\ldots} \Pi_i, \Pi_i = \{A^i, B^i, \ldots\}$ of *i-ary predicate variables*, and a set $\Delta = \{(,)\}$ of brackets as above. Within the free monoid $PEX = FM(\Upsilon \sqcup \Xi \sqcup \Pi \sqcup \Delta)$ of *predicate expressions*, one exhibits the subset $FO(PEX)$ of *formulae* as follows:

Property 6 *Given the symbols* $\Upsilon, \Xi, \Pi, \Delta$, *we require*

(*i*) (Atomic formulae) $A^i \iota_{i_1} \iota_{i_2} \ldots \iota_{i_i} \in FO(PEX)$ *for any* $A^i \in \Pi_i$ *and* $\iota_{i_k} \in \Upsilon$, *and* 0*-ary predicate variables are constants.*
(*ii*) (Propositional formulae) *If* $\alpha, \beta \in FO(PEX)$, *then*

$$(!\alpha) \in FO(PEX),$$
$$(\alpha \& \beta) \in FO(PEX),$$
$$(\alpha | \beta) \in FO(PEX),$$
$$(\alpha \; -> \beta) \in FO(PEX).$$

(*iii*) (Quantifier formulae) *If* $\alpha \in FO(PEX), x \in \Upsilon$, *then*

$$(\forall x)\alpha \in FO(PEX),$$
$$(\exists x)\alpha \in FO(PEX).$$

An individual variable x which appears in the formula after an expression of type $(\forall x)$ or $(\exists x)$ is *bound*, otherwise it is *free*.

A model $\mathfrak{M}$ of a predicate logic $FO(PEX)$ is a set M, together with i-ary relations $a^i, b^i \ldots \subseteq M^i$ (elements $a^0 \in M$ for constants). In such a model, the formal predicate expressions are interpreted via the interpretation of atomic formulae $A^i \iota_{i_1} \iota_{i_2} \ldots \iota_{i_i}$ by the truth value of $(x_{i_1}, x_{i_2}, \ldots x_{i_i}) \in a^i$. Whereas the interpretation of a quantifier formula $(\forall x)\alpha$ means "true" if the truth value of the interpretation of α is "true" for all valuations in M of all occurrences of the variable x, and the interpretation of a quantifier formula $(\exists x)\alpha$ means "true" if the truth value of the interpretation of α is "true" for at least one value in M of the occurrences of the variable x—of course, this is only decided if no free variables are left, in which case one calls the formula a *sentence*, otherwise, no truth value is defined and the formula is just a truth-value-function of the left free variables. We write $\mathfrak{M} \models \alpha[x]$ for truth value "true" for the evaluation $[x]$ of the free variables of α.

The recursive calculation of truth values of compound formulas relates to the Boolean algebra $Sub(M^m)$ as follows: To begin with, if a logical combination of two formulae is considered, one may suppose that both variable sets of these formulas coincide by just taking their union if they do not coincide. If we fix such a variable set of cardinality m, say, the truth evaluation of a formula α with (at most) these m free variables can be described by the inverse image $supp(\alpha) \subseteq M^m$ of *true*. Then evidently, $supp(!\alpha) = M^m - supp(\alpha)$, $supp((\alpha \& \beta)) = supp(\alpha) \cap supp(\beta)$, $supp((\alpha|\beta)) = supp(\alpha) \cup supp(\beta)$, $supp((\alpha \to \beta)) = supp(!\alpha) \cup supp(\beta)$. For the quantifiers, we have this situation: If a variable x is bound by a quantifier, we have the support $supp(\alpha) \subseteq M^m$ of the given formula α and the support $supp((\forall x)\alpha) \subseteq M^{m-1}$ or $supp((\exists x)\alpha) \subseteq M^{m-1}$, respectively. Suppose that x is the ith variable, then we have the projection $p_i : M^m \to M^{m-1}$, which omits the ith coordinate, and the inverse image map $p^* : 2^{M^{m-1}} \to 2^{M^m}$. If S is a support of a formula α in M^m, then the support $(\forall x)(S)$ of $(\forall x)\alpha$ is the set $\{(y_1, \ldots, y_{m-1})|(y_1, \ldots y_{i-1}, x, y_{i+1}, \ldots y_{m-1}) \in S$ for all $x \in M\}$, while the support $(\exists x)(S)$ of $(\exists x)\alpha$ is the set $\{(y_1, \ldots y_{m-1})|(y_1, \ldots y_{i-1}, x, y_{i+1}, \ldots y_{m-1}) \in S$ for at least one $x \in M\}$.

Proposition 126 *The functor of partially ordered sets* $\forall x : 2^{M^m} \to 2^{M^{m-1}}$ *is a right adjoint of* $p^* : 2^{M^{m-1}} \to 2^{M^m}$, *while* $\exists x : 2^{M^m} \to 2^{M^{m-1}}$ *is its left adjoint, in other words,* $p^*(X) \subset Y$ *iff* $X \subset \forall(x)(Y)$ *and* $Y \subset p^*(X)$ *iff* $\exists(x)(Y) \subset X$.

The topos-theoretic generalization of this result is the following theorem; for a proof, see [639, p.209,p.206].

Theorem 80 *If* $f : A \to B$ *is a morphism in the topos* $\mathbf{C}$, *then the functor* $Sub(B) \to Sub(A)$ *of Heyting algebras (which are viewed as categories via their partial orders as morphisms) associated with the natural morphism* $\Omega^f : \Omega^B \to \Omega^A$ *has a right adjoint functor* $\forall_x$ *and a left adjoint functor* $\exists_x$.

In order to rewrite the predicate calculus in general topoi, one uses the characteristic maps associated with supports of predicates as follows: If M is a non-zero object of the topos $\mathbf{C}$, we consider the characters $\chi_{a^m} : M^m \to \Omega$ of the "supports" $a^m \subset M^m$ of m-ary predicates. Their recombination via logical constructions runs as follows: Using the morphisms of negation, conjunction, disjunction, and implication defined above in Section G.5.1 for Ω, one has the evident combination of supports of formulas via their characters. The new thing here is the definition of quantifier supports. Given a character $\chi_\alpha : M^m \to \Omega$, we have the adjoint morphism $ad_i(\chi_\alpha) : M^{m-1} \to \Omega^M$ with respect to the ith coordinate. Then we have two arrows $\forall_M, \exists_M : \Omega^M \to \Omega$. The first $\forall_M$ is the character of the adjoint of the composite $true \circ ! \circ pr_M : 1 \times M \to M \to \Omega$. The second $\exists_M$ is the character of the image of the composed map $pr_{\Omega^M} \circ \varepsilon_M : \varepsilon \rightarrowtail \Omega^M \times M \to \Omega^M$, where $\varepsilon_M : \varepsilon \rightarrowtail \Omega^M$ is the subobject whose character is the evaluation map $ev_M : \Omega^M \times M \to \Omega$ adjoint to the identity on Ω^M. We then have these formulas:

$$\forall(x)\alpha \text{ has the character } \forall_M \circ ad_i(\chi_\alpha),$$
$$\exists(x)\alpha \text{ has the character } \exists_M \circ ad_i(\chi_\alpha).$$

G.5.3 A Formal Setup for Consistent Domains of Forms

Since forms do not automatically exist if we allow circularity, it is important to set up a formal mathematical context in order to describe what a logically consistent domain of forms should be. This mathematical formalism turns out to be valid in an interesting general context. We have been working in the topos $\mathbf{Mod}^{@}$ of presheaves over the category $\mathbf{Mod}$, where we have the Yoneda embedding $Y : \mathbf{Mod} \rightarrow \mathbf{Mod}^{@}$. Without loss of generality, we may identify $\mathbf{Mod}$ with the full subcategory of represented presheaves $@M$, M a module.

More generally, we may consider *Yoneda pairs* $\mathcal{R} \subset \mathcal{E}$, where $\mathcal{R}$ is a full subcategory of a topos $\mathcal{E}$, $\mathcal{R}$ playing the role of represented modules (we also say that $\mathcal{R}$ is a *Yoneda subcategory*). This means that we require that the canonical Yoneda functor $\mathcal{E} \rightarrow \mathcal{E}^{@} \rightarrow \mathcal{R}^{@}$ be fully faithful. By Yoneda's Lemma, we may identify the evaluation $M@F$ of a "presheaf" $F \in \mathcal{E}$ at a "module" $M \in \mathcal{R}$ by the morphism set from M to F: $M@F = Hom_{\mathcal{E}}(M, F)$. This setup in particular includes the classical case of $\mathcal{E} = \mathbf{Ens}$ and $\mathcal{R}$ the one-element category consisting of a singleton 1 (the terminal object in $\mathbf{Ens}$), say $1 = \{\varnothing\}$, and its unique identity morphism. In this case, we may identify $1@F$ and the set F.

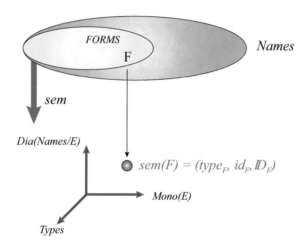

Fig. G.1. The formal setup of a semiotic of $\mathcal{E}$-forms.

To achieve the intended formalism, we consider the set $Mono(\mathcal{E})$ of monomorphisms in $\mathcal{E}$. We further consider the set

$$Types = \{\mathbf{Simple}, \mathbf{Syn}, \mathbf{Limit}, \mathbf{Colimit}, \mathbf{Power}\}$$

of form types. And we need the free monoid $Names = FM(UNICODE)$ over the $UNICODE$ alphabet (which is an extension of the $ASCII$ alphabet to non-European letters). We next need the set $Dia(Names)$ of all diagram schemes with vertexes in $Names$. More precisely, a diagram scheme over $Names$ is a finite directed multigraph whose vertexes are the elements of a subset of $Names$, and whose arrows $i : A \rightarrow B$ are triples (i, A, B), with $i = 1, \ldots$ natural numbers to identify arrows for given vertexes.

Next, consider the set $Dia(Names/\mathcal{E})$ of diagrams on $Dia(Names)$ with values in $\mathcal{E}$. Such a diagram is a map

$$dia : D \rightarrow \mathcal{E}$$

which to every vertex of the diagram scheme D associates an object of $\mathcal{E}$ and to every arrow associates a morphism in $\mathcal{E}$ between corresponding vertex objects. So $i : A \rightarrow B$ is mapped to the morphism $dia(i) : dia(A) \rightarrow dia(B)$. We also will identify two such diagrams iff their arrows for given names A, B are permutations of each other, i.e., we only consider the orbits of diagrams modulo the permutation group of arrows on given names. Why? Because any construction of limits or colimits is invariant under this group since the limit condition is a logical conjunction which does not depend on the numbering of the arrows.

So this identification will always be valid unless explicitly suspended. Observe further that a multiple appearance of a vertex in a diagram scheme is not allowed, so when constructing diagram schemes upon form names, one must add synonymous forms when multiple appearance of one and the same form in a diagram is desired. This is the advantage of form names: the annoying indexing of mathematical names can be absorbed by intrinsic renaming on the level of form names.

With these notation, we can define a *semiotic of $\mathcal{E}$-forms* as follows (see also Figure G.1):

Definition 178 *A semiotic of $\mathcal{E}$-forms is a set map*

$$sem : FORMS \rightarrow Types \times Mono(\mathcal{E}) \times Dia(Names/\mathcal{E})$$

defined on a subset $FORMS \subset Names$ with the following properties (i) *to* (iii). *To ease language, we use the following notation and terminology:*

- *An element $F \in FORMS$ is called a* form name, *and the pair (F, sem) a* form,
- $pr_1 \cdot sem(F) = t(F)$ *(=type of F),*
- $pr_2 \cdot sem(F) = id(F)$ *(= identifier of F),*
- $domain(id(F)) = fun(F)$ *(= topor[12] or "space" of F),*
- $codomain(id(F)) = frame(F)$ *(= frame or "frame space" of F),*
- $pr_3 \cdot sem(F) = coord(F)$ *(= coordinator of F).*

Then these properties are required:

(*i*) *The empty word $\emptyset$ is not a member of $FORMS$*

(*ii*) *For any vertex X of the coordinator diagram $coord(F)$, if $X \in FORMS$, then we have*

$$coord(F)(X) = fun(X).$$

(*iii*) *If the type $t(F)$ is given, we have the following for the corresponding frames:*

- *For* **Syn** *and* **Power**, *the coordinator has one vertex $G \in FORMS$ and no arrows, i.e., $coord(F) : G \rightarrow fun(G)$, which means that in these cases, the coordinator is determined by a form name G. Further, for* **Syn**, *we have $frame(F) = fun(G)$, and for* **Power**, *we have $frame(F) = \Omega^{fun(G)}$, if $coord(F) : G \rightarrow fun(G)$, as above.*
- *For* **Limit** *and* **Colimit**, *the coordinator is any diagram $coord(F)$ whose names are all in $FORMS$. Further, for* **Limit**, *we have $frame(F) = lim(coord(F))$, and for* **Colimit**, *we have $frame(F) = colim(coord(F))$.*
- *For type* **Simple**, *the coordinator has the unique vertex $\emptyset$, and a value $coord(F) : \emptyset \rightarrow M$ for a 'module' $M \in \mathcal{R}$ (i.e., a represented presheaf $M = @X$ in the case of presheaves over* **Mod**), *or, in a more sloppy notation: $coord(F) = M$.*

Here, circular forms are evidently included via form names which refer to themselves in their diagrams or in deeper recursion structures. With this definition we may discuss the existence and size of form semiotics, i.e., the extent of the $FORMS$ set, maximal such sets, gluing such sets together along compatible intersections, etc. However, we shall not pursue this interesting and logically essential branch for simple reasons of space and time.

G.5.3.1 Morphisms Between Semiotics of Forms

Although the theory of form semiotics is in its very beginnings, it is clear that two form semiotics with intersecting domains $FORM_1$ and $FORM_2$ nee not be contradictory even if the semiotic maps do not coincide on the intersection $FORM_1 \cap FORM_2$. In fact, it could happen that on this intersection, the maps are just "equivalent" semiotics. More generally, it could happen that two form semiotics have subsemiotics

---

[12] The functor in the special case $\mathcal{E} = \mathbf{Mod}^@$.

which are in complete correspondence and therefore we may glue them to a global semiotic structure. In other words: It is reasonable and feasible to consider morphisms and then categories of form semiotics and therefore isomorphisms of semiotics, which enables us to construct global semiotics just by gluing together local "charts" as usual.

Let us abbreviate $Sema(\mathcal{E}) = Types \times Mono(\mathcal{E}) \times Dia(Names/\mathcal{E})$, $Sema$ being an abbreviation for semantic target space. Suppose that we are given two form semiotics $sem_1 : FORMS_1 \rightarrow Sema(\mathcal{E}_1), sem_2 : FORMS_2 \rightarrow (\mathcal{E}_2)$. We correspondingly denote by

$$fun_1, fun_2, t_1, t_2, id_1, id_2, frame_1, frame_2, \text{ and } coord_1, coord_2$$

the respective maps. Consider pairs (u, v) where $u : FORMS_1 \rightarrow FORMS_2$ is a set map, and where $v : \mathcal{E}_1 \rightarrow \mathcal{E}_2$ is a logical functor (see Appendix G.3) sending $\mathcal{R}_1$ to $\mathcal{R}_2$. We say that the pair (u, v) *is morphic* (for $FORMS_1, FORMS_2$) iff

1. We have $u(\varnothing) = \varnothing$.
2. The functors commute with u, v, i.e., we have $v \cdot fun_1 = fun_2 \cdot u$.
3. The type is invariant under u, i.e., $t_2 \cdot u = t_1$.

In particular, mono- and epimorphisms on $\mathcal{E}_1$ are preserved (see Appendix G.2.2). Suppose that we are given a diagram scheme

$$C = coord_1(F) : D \rightarrow \mathcal{E}_1$$

(modulo permutations on the numberings of the arrows between fixed names, as announced!) associated with the form name $F \in FORM_1$. Let $|D|$ be the vertex names of D. We define a diagram E as follows. Its vertexes are the image $|E| = u(|D|)$. For every vertex pair (X, Y) of $|E|$ we take all arrows $i : A \rightarrow B$ with $X = u(A), Y = u(B)$. By lexicographic order on the triples (A, B, i), we can order all these arrows and index them with positive natural numbers $j = 1, \ldots n(X, Y)$. This defines a unique new diagram scheme. Secondly, we define a new diagram $C' : E \rightarrow \mathcal{E}_2$ as follows: If the arrow $i : A \rightarrow B$ gives arrow $j(i) : X \rightarrow Y$, the new diagram C' maps this arrow to the morphism

$$v(C(i)) : fun_2(X) = v(fun_1(A)) \rightarrow v(fun_1(B)) = fun_2(Y).$$

Denote this diagram by $(u, v)(C)$. Clearly, since we only retain orbits of diagram schemes, we have functoriality, i.e., if $(u_1, v_1), (u_2, v_2)$ are two such morphic pairs for $FORMS_1, FORMS_2$, and $FORMS_2, FORMS_3$, respectively we have

$$(u_2, v_2)((u_1, v_1)(C)) = (u_2 \cdot u_1, v_2 \cdot v_1)(C). \tag{G.28}$$

Definition 179 *A morphic pair (u, v) for the pair $FORMS_1, FORMS_2$ is said to be a* morphism of form semiotics *$(u, v) : sem_1 \rightarrow sem_2$ if the following semiotic data with each given form name $F \in FORMS_1$ are verified:*

1. *Let F be simple, i.e., $F \xrightarrow{Id} \mathbf{Simple}(M)$. Then $u(F) \xrightarrow{Id'} \mathbf{Simple}(N)$[13], and we require that*
   - *$N = v(M)$,*
   - *$Id'=v(Id)$, i.e., the monomorphism of Id' is the v-image of the monomorphism of Id—the domains and codomains are already the right ones, only the morphism (a monomorphism by the conservation of limits) has to fit—so*

   $$Id' = v(Id) : fun_2(u(F)) = v(fun_1(F)) \rightarrowtail N = v(M).$$

2. *Let F be synonymous, i.e., $F \xrightarrow{Id} \mathbf{Syn}(G)$. Then $u(F) \xrightarrow{Id'} \mathbf{Syn}(G')$, and we require that $G' = u(G)$ and $Id' = v(Id) = fun_2(u(F)) \rightarrowtail fun_2(u(G))$.*

---

[13] Observe that in the case $\mathcal{R} = \mathbf{Mod}, \mathcal{E} = \mathbf{Mod}^{@}$ of presheaves, we usually write the module as a coordinator, but we mean its represented functor $@M$.

3. *Let F be of power type, i.e., $F \xrightarrow[Id]{} \textbf{Power}(G)$. Then $u(F) \xrightarrow[Id']{} \textbf{Power}(G')$, and we require that $G' = u(G)$ and $Id' = v(Id) = fun_2(u(F)) \rightarrowtail v(\Omega^{fun_1(G)})$*
   $\xrightarrow{\sim} \Omega^{v(fun_1(G))} = \Omega^{fun_2(u(G))}$.

4. *Let F be of limit (resp. colimit) type, i.e., $F \xrightarrow[Id]{} \textbf{Limit}(C)$ (resp. $F \xrightarrow[Id]{} \textbf{Colimit}(C)$) with $C = coord(F)$. Then we have $u(F) \xrightarrow[Id']{} \textbf{Limit}(C')$ (resp. $u(F) \xrightarrow[Id']{} \textbf{Colimit}(C')$). We then require that $C' = (u, v)(C)$ and that $Id' = v(Id) = fun_2(u(F)) = v(fun_1(F))$*
   $\rightarrowtail v(lim(C)) \xrightarrow{\sim} lim(v \cdot C) \xrightarrow{\sim} lim((u, v)(C))$ (resp. the analogous expression with colimits).

In any case, the associated form $u(F)$ is related to its ingredients through the given functor v and the recursive constructions on the coordinators via u.

Clearly, the evident composition of two morphisms from formula (G.28) is again a morphism, and we obtain the category *ForSem* of form semiotics.

G.5.3.2 Local and Global Form Semiotics

It is clear what one should understand by a *global form semiotic*: This is a set G, together with a covering I and an atlas $f_i : I_i \xrightarrow{\sim} FORMS_i$ of bijections onto domains of form semiotics $semi_i : FORMS_i \to Sema$ such that all the induced bijections $u_{i,j} : FORMS_i|_j \xrightarrow{\sim} FORMS_j|_i$ extend to isomorphisms $(u_{i,j}, v_{i,j})$ of form semiotics. This means in particular that all intersections $FORMS_i|_j$ are form domains of form sub-semiotics in sem_i, and that the underlying functors on $\mathcal{E}_i$ are compatible. We leave the details to the interested reader.

G.5.3.3 Connotator Form Semiotics

Denotator and form names were very simple word objects in the previous setup. But name spaces may also be required to encompass more articulated structures, in other words: we want names to be denotators as well, thereby turning the denotator concept into a 'connotator' concept. Here is the formal setup.

We again suppose given a Yoneda pair $\mathcal{R}, \mathcal{E}$. We also retain the set $Mono(\mathcal{E})$. We are given two sets $\mathcal{D}$ of denotators and $\mathcal{F}$ of forms, they are supposed to parametrize denotators and forms according to the following system of maps. We have three maps on $\mathcal{D}$:

$$coordinate : \mathcal{D} \to \mathcal{E},$$
$$form : \mathcal{D} \to \mathcal{F},$$
$$denotatorName : \mathcal{D} \to \mathcal{D}.$$

The coordinate C of a denotator is supposed to be any morphism with domain A within $\mathcal{R}$, which is called the denotator's address. We require that a denotator be uniquely determined by its coordinate C, form F, and denotatorName N. This is why denotators are also written as quadruples $N : A@F(C)$, where the address is denoted for comfort since it is important information.

The denotator's form mimics the space where the denotator lives. To this end, we need two more sets. The set of types is $\mathcal{T} = \{\textbf{Limit}, \textbf{Colimit}, \textbf{Power}, \textbf{Simple}\}$, it contains the basic constructors of objects in a topos. But we omit synonymy in this generic setup because it can be mimicked by a limit with just one vertex. We also need the set $Diagrams(\mathcal{D}/\mathcal{E})$ of finite diagrams whose vertexes are denotators, and whose arrows are numbered by $1, 2, 3, \ldots$ as above. This means that the diagram schemes are these symbols, and that the evaluation of the diagram scheme yields objects and morphisms in $\mathcal{E}$. Forms have, by hypothesis, uniquely determined values under these four maps:

$$formName : \mathcal{F} \to \mathcal{D},$$
$$identifier : \mathcal{F} \to Mono(\mathcal{E}),$$
$$diagram : \mathcal{F} \to Diagrams(\mathcal{D}, \mathcal{E}),$$
$$type : \mathcal{F} \to \mathcal{T}.$$

This means that a form can be written as $FN : Id.T(Dg)$, where FN is the form's name denotator, Id its identifier, T its type, and Dg its diagram.

We impose a small number of axioms for these structures. To this end, we call the domain $dom(Id)$ of a form $FN : Id.T(Dg)$ the form's *space*, whereas the codomain $cod(Id)$ is called its *frame space*. Accordingly, for a denotator $N : A@F(C)$, $codom(C) = dom(Id)$, and the composition $Id \circ C$ with its form's identifier Id is called the *frame coordinate*, it uniquely determines the denotator's coordinate.

Axiom 8 *Here are the conditions for this setup:*

(*i*) *The map formName is injective, i.e., the form's name is a key.*

(*ii*) *For all form diagrams, except for simple type, the vertex denotators of the diagram schemes are form names, and their values are the spaces of the respective forms.*

(*iii*) *If the form's type is* **Limit** *or* **Colimit***, its frame space is the limit or colimit of the diagram.*

(*iv*) *If the form's type is* **Power***, the diagram has just one vertex and no arrows, and the frame space is Ω^S, where S is the space of the vertex form.*

(*v*) *There is a denotator $\varnothing$ which is not a form name, and for a simple type form, the diagram has exactly the vertex $\varnothing$, no arrow, and the value is a 'representable' object X in $\mathcal{R}$. In other words, the simple type frame space is just a representable object in disguise. Such a diagram is represented by $\varnothing X$.*

The language of forms and denotators has been encoded in an ASCII-based textual form, like TeX, which is therefore called *Denotex* and is available in BNF[14]. In RUBATO®, a Denotex parser is available for communication with Denotex files. Our present notation in this section, such as $FN : Id.T(Dg)$ for forms and $N : A@F(C)$ for denotators, is an illustration of the Denotex notation.

Example 130 An elementary form for names can be set up as follows: The form NF represented by $fn : Id.\mathbf{Simple}(Dg)$ is simple with the diagram $\varnothing \mathbb{Z}\langle UNICODE \rangle$. The identifier Id is the identity on the representable presheaf $@\mathbb{Z}\langle UNICODE \rangle$, and the name fn is a denotator $fn : 0@FN(C)$, whose coordinate C is the zero-addressed homomorphism $C : 0 \to \mathbb{Z}\langle UNICODE \rangle$ with value $C(0) = $ "NameForm" with $denotatorName(fn : 0@NF(C)) = fn$, i.e., it is its proper name denotator. So its identification resides on its coordinate value "NameForm" and the form named fn. This identifies the entire NF form. Then, general $UNICODE$ names n may be defined by $n : 0@NF(Cn)$, where the value $Cn(0) = $ "anyName" is any $UNICODE$ string combination, such as "3.Violin+4.Piano", and which are their proper name denotators, i.e., $denotatorName(n : 0@NF(C)) = n$.

---

[14] Denotex was developed in collaboration with Thomas Noll, Jörg Garbers, Stefan Göller, and Stefan Müller.

H

Complements on General and Algebraic Topology

H.1 Topology

Refer to [527] for general topology, and to [993] for algebraic topology.

H.1.1 General

A topological space is a pair $(X, Open_X)$ of a set X and a set $Open_X$ of *open* subsets of X such that X is open, $U \cap V$ is open if U, V are so, and $\bigcup_i U_i$ is open for any family (U_i) of open sets, in particular the union of the empty family, the empty set, is open. The complement $X - U$ of an open set U is called *closed*. Therefore the collection $Closed_X$ of closed sets fits with the corresponding axioms: the union of any two closed sets is closed, the intersection of any family of closed sets is closed[1], and the empty set is closed. If we define the *closure* $\overline{Y}$ of any subset of X as the intersection of all closed sets containing Y, then the topology is again defined by the axioms for the Kuratowski closure operator $: 2^X \to 2^X$, i.e., $\overline{\varnothing} = \varnothing$, is idempotent, $Y \subseteq \overline{Y}$, and $\overline{Y \cup Z} = \overline{Y} \cup \overline{Z}$.

Given two topologies $Open_X, Open'_X$, one says that $Open_X$ *is coarser than* $Open'_X$ or that $Open'_X$ *is finer than* $Open_X$ iff $Open_X \subseteq Open'_X$. On any set X, the coarsest topology consists just of X and of the empty set, it is called the *indiscrete topology*, whereas the finest topology is the powerset of X, it is called the *discrete topology*. The intersection of any family of topologies on X is the finest topology which is coarser than each member of the family. Every set of subsets $\mathcal{S}$ of X is contained in the intersection of all topologies containing this subset, a family containing at least the discrete topology. It consists of all unions of finite intersections (the empty intersection gives X) of members of $\mathcal{S}$ and is denoted by $Open(\mathcal{S})$.

A *neighborhood* W of $x \in X$ is a subset containing an open set U which contains x. Finite intersections of neighborhoods of x are neighborhoods, supersets of neighborhoods are neighborhoods. An *accumulation point* of a subset Y of X is a point not in Y which intersects Y in each of its neighborhoods. The closure of a subset Y of X is the union of Y and of its accumulation points. The *interior U^o* of a subset is the union of all open subsets of U. It is also the complement of the closure of its complement. The interior operator has an evident set of axioms corresponding to the closure axioms which also characterize the topology. The *boundary ∂U* of a subset is the difference $\overline{U} - U^o$.

A subset $\mathcal{B}$ of open sets is called a *base for the topology* iff any open set is the union of a family of $\mathcal{B}$ members, or, equivalently, every neighborhood of a point contains a neighborhood from $\mathcal{B}$. The axioms for a set of subsets $\mathcal{B}$ of X to be a base for a topology is that $X = \bigcup \mathcal{B}$, and that for any two $U, B \in \mathcal{B}$, $U \cap V$ is the union of members of $\mathcal{B}$. A *subbase for a topology* $Open_X$ on X is a set $\mathcal{S}$ of subsets of X such that $Open_X = Open(\mathcal{S})$.

---

[1] The intersection of the empty family being defined as the total space X.

© Springer International Publishing AG, part of Springer Nature 2017
G. Mazzola, *The Topos of Music IV: Roots*, Computational Music Science,
https://doi.org/10.1007/978-3-319-64495-0_8

H.1.2 The Category of Topological Spaces

Suppose that $(X, Open_X), (Y, Open_Y)$ are topological spaces. A set map $X \to Y$ is *continuous* iff the inverse map $2^Y \to 2^X$ induces a map $Open_Y \to Open_X$. The set-theoretic composition of continuous maps is continuous, the identity map is so, and therefore, we have the *category* **Top** *of topological spaces and continuous maps*. An isomorphism of topological spaces is called a *homeomorphism*. Any subset W of a topological space becomes a topological space by the coarsest topology $Open_W = Open_X | W$ such that the inclusion $W \subset X$ is continuous; its open sets are just the intersections of open sets of X with W, this topology is called the *relative topology on W*. More generally, given any set map $f : X \to Y$ into a topological space $(Y, Open_Y)$ the coarsest topology $Open_Y | f$ (smallest set of open sets) on X such that f becomes continuous is given by the set of inverse images of open sets of Y, we also call it the *relative topology with respect to f*. Conversely, for a set map $f : X \to Y$, where $(X, Open_X)$ is a topological space, we have a finest topology such that f becomes continuous, it is given by the set of all subsets of Y such that their inverse image is open in X. This is the *quotient topology* $Open_X / f$. If (X_i) is a family of topological spaces, the cartesian product $\prod_i X_i$ has the coarsest topology such that the projections to all factors become continuous. A base of this *product topology* is given by the products $\prod_i U_i$ of open sets $U_i \subseteq X_i$ with $U_i = X_i$ except for a finite number of indices. This is a limit in the category **Top**. The coarsest topology on the set-theoretical limit $lim()$ of a diagram of continuous maps is the limit in **Top**, a similar construction (this time with the finest topology) yields the colimit of a diagram of continuous maps.

If we are given a family $f_i : X_i \to X$ of set maps whose domains are topological spaces, there is a finest topology which makes these maps continuous. Its universal property is that with this topology on X, a map $g : X \to Y$ into a topological space Y is continuous iff all compositions $g \circ f_i$ are so. This is a particular case of a quotient topology for the situation $\coprod_i X_i \to X$. This topology is called the *coinduced topology*. If the maps f_i are inclusions of subspaces X_i of a topological space X, the topology of X is called *coherent* or *weak* if it is coinduced from the relative topologies on the spaces X_i.

If we are given a set X, together with a collection of subsets C_i of X which are topological spaces such that for all indexes i, j, the intersections $C_i \cap C_j$ have the same relative topology as inherited from C_i or from C_j, and that these intersections are closed in both, C_i and C_j. Then the coinduced topology is coherent with this family, in other words, the coinduced topology relativizes to the given topologies on all C_i.

H.1.3 Uniform Spaces

Topologies are often defined by relations that stem from metrical distance functions. The axiomatics is as follows:

Definition 180 *A* uniformity *on a set X is a set $\mathcal{U}$ of* uniform sets $U \subseteq X^2$ *such that:*

(i) *Each uniform set contains the diagonal Δ.*
(ii) *If U is uniform, so is U^{-1}.*
(iii) *If U is uniform, then there is a uniform V such that $V \circ V \subset U$.*
(iv) *If U, V are uniform, then so is $U \cap V$.*
(v) *If U is uniform, then so is every superset in X^2.*

The prototype of a uniformity is given by a distance function, i.e., a pseudo-metric $d : X \times X \to \mathbb{R}$ as defined in definition 185 in Appendix I.1.1. The uniformity contains all $U \subseteq X^2$ which contain a set of type $U_\epsilon = \{(x, y) | d(x, y) < \epsilon\}, \epsilon > 0$.

Each uniformity $\mathcal{U}$ gives rise to a *uniform topology* $Open(\mathcal{U})$ whose open sets are those V such that for each $x \in V$, there is a uniform set U with $U[x] \subset V$, where $U[x] = \{y | (x, y) \in U\}$. So the uniform topology imitates metrical neighborhoods.

H.1.4 Special Issues

Definition 181 *A topological space X is said to be:*

(*i*) T_0 *iff for any two different points $x, y \in X$, at least one of them is not the specialization of the other;*
(*ii*) T_1 *iff every point is closed, i.e., no other point dominates it;*
(*iii*) T_2 *(Hausdorff) iff every two different points have disjoint neighborhoods.*

Definition 182 *A subset $L \subset X$ of a topological space X is said to be* locally closed *iff one of the equivalent properties holds:*

(*i*) $L = O \cap C$, O open, C closed.
(*ii*) *Every point $l \in L$ has an open neighborhood U_l such that $U_l \cap L$ is closed in U_l.*
(*iii*) L *is open in its closure in X.*

See [147, I,§3.3] for a proof.

Definition 183 *A topological space X is called* quasi-compact *iff every covering of X by open sets admits a finite subcovering. A Hausdorff quasi-compact space is called* compact.

Typically, prime spectra of commutative rings are quasi-compact but not compact.

H.2 Algebraic Topology

Refer to [993] for this section.

H.2.1 Simplicial Complexes

A *simplicial complex K* is a set V of *vertexes*, together with a subset K of 2^V whose elements are called *simplexes* such that (1) each singleton $\{v\}, v \in V$ is a simplex, (2) each non-empty subset of a simplex is a simplex. If for a simplex s of K, $card(s) = q + 1$, one says that s is a *q-simplex* or a *q-dimensional simplex*. A subsimplex $s' \subseteq s$ of a simplex s is called a *face of s*; it is called a *q-face* if it is a q-dimensional simplex, we also write $s' \leqslant s$ instead of $s' \subseteq s$. Evidently, a simplicial complex is completely determined by its simplex set K and may be identified with it.

Example 131 Let U be a covering of a set X by non-empty subsets. The *nerve $n(U)$* of U is the simplicial complex with $V = U$, and the simplexes s being those finite sets $s = \{u_0, u_1, \ldots u_p\}$ in U which have non-empty intersection $\cap s = \bigcap_i u_i$.

The *dimension $dim(K)$* of a simplicial complex K is the maximal dimension of its simplexes, including the special cases $dim(\varnothing) = -1$ $dim(K) = \infty$ if no maximal dimension exists. A simplicial map $f : K_1 \to K_2$ is a set map $f : V_1 \to V_2$ on the underlying vertex sets such that the induced map $2^f : 2^{V_1} \to 2^{V_2}$ carries simplexes to simplexes, i.e., restricts to a map $f : K_1 \to K_2$, meaning that if $s \in K_1$, then $f(s) \in K_2$. One may also say that it is a set map $F : K_1 \to K_2$ which is induced by a map f on the underlying vertex sets. The simplicial complexes and their simplicial maps define the *category* **Simpl** *of simplicial complexes*.

A *subcomplex L* of a simplicial complex K is a subset of simplexes which is also a simplicial complex. L is *full* iff a simplex of K whose vertexes belong to L is also in L. For example, given a simplicial complex K and a natural number k, the *k-dimensional skeleton $K|k$* is the subcomplex of all simplexes of dimension $\leqslant k$. For a covering U, the k-dimensional skeleton of its nerve is denoted by $n_k(U)$.

Example 132 Let *Covens* be the category of set coverings, whose objects are pairs (X, I) of sets X and coverings I of X by non-empty subsets. The morphisms are pairs $(f, \phi) : (X, I) \to (Y, J)$ with $f : X \to Y, \phi : I \to J$ two maps such that for all $i \in I$, $f(i) \subset \phi(i)$. We then have the nerve functor $n : Covens \to$ **Simpl** : $(X, I) \mapsto n(I)$.

H.2.2 Geometric Realization of a Simplicial Complex

We have a functorially defined geometric representation of simplicial complexes K by topological spaces $|K|$ as follows. The set $|K|$ is the subset of those functions $\alpha : V(K) \to I = [0,1]$ into the real unit interval I such that

1. the support $supp(\alpha) = \{v \in V(K)|\alpha(v) \neq 0\}$ of α is a simplex,
2. $\sum_{v \in V(K)} \alpha(v) = 1$.

The value $\alpha(v)$ is called the vth *barycentric coordinate* of α. On the set $I^{(V(K))}$ of functions with finite support, one has the Euclidean metric $d(\alpha, \beta) = \|\alpha - \beta\|_2$. We induce this metric and its associated topology (see Section H.1.3) on $|K|$ and denote it by $|K|_d$.

For a simplex $s \in K$, the closed simplex $|s|$ is defined by

$$|s| = \{\alpha \in K | supp(\alpha) \subset s\}.$$

Evidently, if $dim(s) = q$, there is a homeomorphism $|s|_d \xrightarrow{\sim} \Delta^q = \{x \in I^{q+1}| \sum x_i = 1\}$ onto the "standard closed q-simplex". If $s, t \in K$, either $s \cap t = \varnothing$ or a common face, and then $|s \cap t| = |s| \cap |t|$, so $|s|_d \cap |t|_d$ is closed in both, $|s|_d, |t|_d$, and the relative topologies from $|s|_d, |t|_d$ coincide on the intersection. By the remarks on coinduced topologies in Section H.1.2, we have the *coherent topology on* $|K|$ which is coinduced from the topologies on the closed simplexes. This means that

Fact 23 *A subset $E \subseteq |K|$ is closed/open iff each intersection $E \cap |s|_d$ is closed/open.*

Therefore, a function $f : |K| \to X$ into a topological space X is continuous iff its restrictions $f||s|$ are so for all simplexes s of K. In particular, the identity $|K| \to |K|_d$ is continuous, therefore, $|K|$ is Hausdorff, it is also normal, see [993, 3.1, Th.17]. Also, $|K|$ is compact iff K is finite. Call K *locally finite*, iff every vertex belongs to a finite number of simplexes. Then

Theorem 81 *For a simplicial complex K the following statements are equivalent:*

(i) *K is locally finite.*
(ii) *The identity $|K| \to |K|_d$ is a homeomorphism.*
(iii) *$|K|$ is metrizable, i.e., there is a metric whose topology is the coherent topology.*

See [993, 3.2, Th.8] for a proof.

If $f : K_1 \to K_2$ is a simplicial map, we have the continuous map

$$|f|(\alpha)(v) = \sum_{f(w)=v} \alpha(w)$$

which is continuous for both topologies on $|K|$. We are therefore given two functors $|?|, |?|_d : \mathbf{Simpl} \to \mathbf{Top}$ and a natural transformation $Id : |?| \to |?|_d$.

A continuous map $f : |K| \to X \subset \mathbb{R}^n$ is said to be *linear* iff $f(\alpha) = \sum_{v \in V(K)} \alpha(v) f(v)$ for all $\alpha \in |K|$. Any function on the vertexes may uniquely be extended to a continuous linear map, this is the universal property of affine pointsets in general position. In particular, the map $|f|$ associated with a simplicial map f is linear.

Definition 184 *A* geometric realization *of a simplicial complex K in $\mathbb{R}^n$ is a linear embedding (injection) of $|K|$ in $\mathbb{R}^n$.*

Theorem 82 *If a simplicial complex K has a geometric realization in $\mathbb{R}^n$, then it is countable, locally finite and has dimension $\leqslant n$. Conversely, if it is countable, locally finite, and has dimension $\leqslant n$, then it has a geometric realization as a closed subset of $\mathbb{R}^{2n+1}$.*

Example 133 For the nerve $n(U)$ of a finite covering U, we write $N(U)$ for the geometric realization $|n(U)|$, we also write $N_k(U)$ for $|n_k(U)|$.

H.2.3 Contiguity

A *simplicial pair* is a couple (K, L), where K is a subcomplex of L. A simplicial map of pairs $f : (K_1, L_1) \rightarrow (K_2, L_2)$ is a simplicial map $f : L_1 \rightarrow L_2$ which induces a simplicial map on the respective subcomplexes. Two simplicial maps $f, f' : (K_1, L_1) \rightarrow (K_2, L_2)$ are called *contiguous* if for every simplex s in K_1 or L_1, the union $f(s) \cup f'(s)$ is a simplex in K_2 or L_s. Contiguity is an equivalence relation and defines *contiguity classes* of simplicial maps.

Two continuous maps $f, g : X \rightrightarrows Y$ of topological spaces are called *homotopic* iff there is a continuous map (a homotopy) $F : X \times I \rightarrow Y$ such that $f = F(?, 0), g = F(?, 1)$; the homotopy relation is an equivalence relation. If $X' \subseteq X$ is a subspace, and if $f|X' = g|X'$, a homotopy is called *relative to this subspace*, iff $F|X' \times t = f|X' = g|X'$, all $t \in I$.

Lemma 107 ([993, Lemma 2, p.130]) *Contiguous simplicial maps which agree on a subcomplex define contiguous maps which are homotopic relative to the space of the subcomplex.*

H.3 Simplicial Coefficient Systems

A simplicial complex K can be viewed as a category whose objects are the simplexes s of K, and whose morphisms are the inclusions $s \subseteq t$ of simplexes. For a commutative ring R, a *coefficient system* of R-modules is a covariant functor $M : K \rightarrow {}_R\mathbf{Mod}$ with values in the category ${}_R\mathbf{Mod}$ of R-modules and affine homomorphisms. Let $\Delta_q = \{0, 1, 2, \ldots q\}$ be the standard simplex of dimension q. A *singular simplex of dimension q* is a simplicial map $s : \Delta_q \rightarrow K$, i.e., a sequence $s_0, s_1, \ldots s_q$ of points in K which define a simplex. If we have any set map $f : \Delta_p \rightarrow \Delta_q$, we have the singular p-simplex $\overline{f}(s) = s \circ f : \Delta_p \rightarrow K$. For a singular simplex s, we denote $M(s) = M(Im(s))$. Clearly $Im(\overline{f}(s)) \subseteq Im(s)$. Therefore we have an affine homomorphism $f_s : M(\overline{f}(s)) \rightarrow M(s)$.

Denote by $S_n(K)$ the set of singular simplexes of dimension n in K. Then we have a module $C^n(K; M) = \prod_{s \in S_n(K)} M(s)$, whose elements are called the *singular cochains of dimension n*. For a map $f : \Delta_p \rightarrow \Delta_q$, we have an affine map

$$\overline{f} : C^p(K; M) \rightarrow C^q(K; M) \tag{H.1}$$

which has $\overline{f}((a_s)_{s \in S_p(K)}) = (b_t)_{t \in S_q(K)}$ and $b_t = f_t(a_{\overline{f}(t)})$. In other words, $C^*(K; M) = (C^n(K; M))_n$ is a *simplicial cochain complex*.

H.3.1 Cohomology

Suppose now that the simplicial cochain complex stems from a system of coefficients with linear maps. Then all the transition maps of equation (H.1) are linear. Consider now the strictly increasing i^{th}-face maps $F_n^i : \Delta_{n-1} \rightarrow \Delta_n$ leaving aside index i in Δ_n, i.e., mapping Δ_{n-1} onto the subset $\{0, 1, 2, \ldots \hat{i} \ldots n\}$. Then we have the coboundary map

$$d_n : C^n(K; M) \rightarrow C^{n+1}(K; M), \tag{H.2}$$

$$d_n(a) = \sum_{j=0}^{n+1} (-1)^j \overline{F_{n+1}^j}(a),$$

and $d_{n+1} \circ d_n = 0$. This means that $Im(d_n) \subseteq Ker(d_{n+1})$, and we may consider the *cohomology groups*

$$H^n(K; M) = Ker(d_n)/Im(d_{n-1}) \tag{H.3}$$

for $n \geqslant 0$, with the trivial extension to $C^{-1}(K; M) = 0$.

I

Complements on Calculus

I.1 Abstract on Calculus

I.1.1 Norms and Metrics

Definition 185 *A* pseudo-metric *on a set V is a* (pseudo-distance) *function $d : V \times V \to \mathbb{R}$ such that:*

1. *(Positivity) $0 \leqslant d(x, y)$, and $d(x, x) = 0$ for all $(x, y) \in V \times V$;*
2. *(Symmetry) $d(x, y) = d(y, x)$ for all $(x, y) \in V \times V$;*
3. *(Triangle inequality) $d(x, z) \leqslant d(x, y) + d(y, z)$ for all $(x, y, z) \in V \times V \times V$.*

If conversely $d(x, y) = 0$ implies $x = y$, the pseudo-metric (pseudo-distance function) *is called a* metric (distance function).

Definition 186 *For a pseudo-metric space (X, d), if $0 < r, x \in X$, the* open ball *of radius r around x is $B_r(x) = \{y | d(y, x) < r\}$. The system of open balls*

$$\{B_r(x) | 0 < r, x \in X\}$$

is a base of a topology[1], the (uniform) topology *associated with the pseudo-metric d. Evidently, this topology is Hausdorff iff the pseudo-metric is a metric.*

A map $f : V \to V$ of a pseudo-metric space V is called an isometry *iff $d(f(x), f(y)) = d(x, y)$, for all $(x, y) \in V \times V$.*

Lemma 108 ([163, Lemma 4]) *Given an action $\mu : G \times V \to V$ of a group G on a pseudo-metric space (V, d) by isometries, then*

$$\inf_{g \in G} d(g.x, y) = \inf_{g \in G} d(g.x', y) \quad \text{whenever } G.x = G.x', \tag{I.1}$$

$$\inf_{g \in G} d(g.x, y) = \inf_{g \in G} d(g.x, y') \quad \text{whenever } G.y = G.y'. \tag{I.2}$$

With the above notation, we may define

$$d^*(G.x, G.y) = \inf_{g \in G} d(g.x, y), \tag{I.3}$$

and Lemma 108 guarantees that this is a well-defined function $d^* : G\backslash V \times G\backslash V \to \mathbb{R}$.

---

[1] In fact, the system $\{B_r = \{(x, y) \in V^2 | d(x, y) < r\} | 0 < r\}$ is a base of a uniformity.

© Springer International Publishing AG, part of Springer Nature 2017
G. Mazzola, *The Topos of Music IV: Roots*, Computational Music Science,
https://doi.org/10.1007/978-3-319-64495-0_9

Definition 187 *If $d(x, y)$ is a pseudo-metric on a set V, and if we have a group action $G \times V \to V$, we say that g acts by isometries, iff each map $g. : V \to V$ is an isometry, i.e., iff $d(g.x, g.y) = d(x, y)$, for all $x, y \in V, g \in G$.*

Lemma 109 ([163, Lemma 5]) *Let d be a pseudo-metric on V, and $\mu : G \times V \to V$ a group action by isometries. Then the function d^* defined in (I.3) is a pseudo-distance on the orbit space $G \backslash V$.*

I.1.2 Completeness

A *Cauchy sequence* in a uniform space $(X, \mathcal{U})$ is a sequence $(x_i)_{i=0,1,2,\ldots}$ of elements in X such that for every uniform set $U \in \mathcal{U}$, there is an index t such that $(x_i, x_j) \in U$ for all $i, j > t$. A uniform space is (sequentially) *complete* iff every Cauchy sequence converges.

Lemma 110 *A closed subspace of a complete uniform (in particular: a metric space) space is complete.*

Definition 188 *A norm on real vector space X is a function $\| \ \| : X \to \mathbb{R}$ such that for all $(x, y) \in V \times V$:*

1. *(Positivity)* $0 \leqslant \|x\|$, and $\|x\| = 0$ iff $x = 0$;
2. *(Homogeneity)* $\|\lambda.x\| = |\lambda|.\|x\|$;
3. *(Triangle inequality)* $\|x + y\| \leqslant \|x\| + \|y\|$.

Every norm gives rise to an associated metric $d(x, y) = \|x - y\|$, *and therefore to an* associated topology. *A normed vector space with a complete associated (uniform) topology is called a* Banach space.

Example 134 On $\mathbb{R}^n$, we have three well-known norms. If $x = (x_1, \ldots x_n) \in \mathbb{R}^n$, then

1. the absolute or 1-norm is $\|x\|_1 = \sum_i |x_i|$,
2. the Euclidean norm is $\|x\|_2 = \sqrt{\sum_i x_i^2}$,
3. the uniform norm is $\|x\|_\infty = max\{|x_i| \| \ i = 1, \ldots n\}$.

For real numbers $a < b$, we have the vector space $C^0[a, b]$ of continuous real-valued functions on the interval $[a, b]$. On $C^0[a, b]$, we have three well-known norms (corresponding to the above three norms). For $f \in C^0[a, b]$, we have:

1. the absolute or 1-norm is $\|f\|_1 = \int_a^b |f|$,
2. the Euclidean norm is $\|f\|_2 = (\int_a^b f^2)^{1/2}$,
3. the uniform norm is $\|f\|_\infty = Max_{[a,b]}|f|$.

Two norms $\| \ \|_1, \| \ \|_2$ on a real vector space X are called equivalent iff there are two positive constants a, b such that $\| \ \|_1 \leqslant a.\| \ \|_2, \| \ \|_2 \leqslant b.\| \ \|_1$. Equivalent norms give rise to the same associated uniformities and topologies, so they have the same Cauchy sequences.

Theorem 83 *Any two norms on a finite-dimensional real vector space are equivalent.*

See [617, Th.3.4.1] for a proof. The theorem implies that every finite-dimensional normed real vector space is Banach, since the standard $\mathbb{R}^n$ is so under the Euclidean norm. We shall therefore mainly work in $\mathbb{R}^n$.

I.1.3 Differentiation

We say that two functions $f, g : U \to \mathbb{R}^m$ that are defined in a neighborhood U of $0 \in \mathbb{R}^n$ define the same *germ* iff they coincide on a common neighborhood of 0. (We are in fact considering the colimit of function spaces on the neighborhood system of 0.) The set of germs in 0 of functions f with $f(0) = 0$ is denoted by F_0. Within this vector space, we have the vector subspace DF_0 of those f with $f(0) = 0$ and $\|f(z)\|/\|z\| \to 0$ if $z \to 0$. We evidently have $Lin_\mathbb{R}(\mathbb{R}^n, \mathbb{R}^m) \cap DF_0 = \{0\}$.

Definition 189 *A function $f : U \to \mathbb{R}^m$ which is defined in a neighborhood $U \subseteq \mathbb{R}^n$ of a point x is differentiable in x iff there is a linear map $D \in Lin_\mathbb{R}(\mathbb{R}^n, \mathbb{R}^m)$ such that $\Delta_x f - D \in DF_0$, where $\Delta_x f(z) = f(x + z) - f(x)$. By the above, D is uniquely determined and is denoted by Df_x. The coefficient of row i and column j of the matrix of Df_x in the canonical basis is denoted by $\partial f_i / \partial x_j$, whereas the matrix is called the Jacobian of f in x. A function $f : O \to V$ on an open set $O \subseteq \mathbb{R}^n$ with values in an open set $V \subseteq \mathbb{R}^m$ is differentiable if it is differentiable in each point of its domain O.*

A differentiable function on O defines its *derivative* $Df : O \to Lin_\mathbb{R}(\mathbb{R}^n, \mathbb{R}^m) \overset{\sim}{\to} \mathbb{R}^{nm}$, which may again be differentiated according to the norm on the space of linear maps. Inductively we define $D^{t+1} f = D(D^t f)$, if it exists. The function f is $\mathcal{C}^r$ iff all derivatives $Df, D^2 f, \dots D^r f$ exist and are continuous, $\mathcal{C}^0$ denotes just the set of continuous functions. This definition is however not in the right shape for functorial behavior. One therefore adds the linear behavior to the function as follows: Let $TO = O \times \mathbb{R}^n$ be the *tangent bundle* of the open set O. Then we define $Tf : TO \to T\mathbb{R}^m$ by $Tf(x, u) = (f(x), Df_x(u))$. This implies that if $g : U \to \mathbb{R}^l$ is a second differentiable function on an open set $U \subseteq \mathbb{R}^m$ with $f(O) \subset U$, then $g \circ f$ is differentiable and

$$T(g \circ f) = Tg \circ Tf.$$

So we have a functor $T : f \mapsto Tf$ and the natural transformation $pr_1 : T \to Id$ of first projection. More generally, defining $T^{r+1} f = T(T^r f)$, we also have

$$T^r(g \circ f) = T^r g \circ T^r f.$$

Moreover, if we identify $Lin_\mathbb{R}(\mathbb{R}^n, Lin_\mathbb{R}(\mathbb{R}^n, \mathbb{R}^m))$ with $Bil_\mathbb{R}(\mathbb{R}^n, \mathbb{R}^m)$), etc. for higher multilinear maps, the higher derivatives $D^r f_x$ identify to r-linear maps $(\mathbb{R}^n)^r \to \mathbb{R}^m$.

Proposition 127 *If f is $\mathcal{C}^r$, then $D^r f_x$ is a symmetric r-linear matrix.*

The category of r times differentiable or $\mathcal{C}^r$ functions has the property that the linear parts of the tangent maps compose as normal linear maps do, and this means that the Jacobians of isomorphisms are invertible quadratic matrices.

A curve in $\mathbb{R}^n$ is a $\mathcal{C}^1$-map $y : U \to \mathbb{R}^n$, $U \subset \mathbb{R}$. Its derivative Dy_t in a point $t \in U$ is a linear map $\mathbb{R} \to \mathbb{R}^n$ which identifies to the image of 1 in $Dy_t(1) \in \mathbb{R}^n$, meaning that the derivative can be identified with a continuous map $y' : U \to \mathbb{R}^n : t \mapsto y'(t) = Dy_t(1)$.

I.2 Ordinary Differential Equations (ODEs)

Throughout this section, D denotes an open set in $\mathbb{R}^n$, and $f : D \to \mathbb{R}^n$ denotes a continuous vector field (a function) with components $f_i, i = 1, \dots n$.

Definition 190 *Let $\zeta \in \mathbb{R}, \eta \in D$, $J(\zeta)$ an open interval containing ζ, and $U(\eta) \subset D$ an open neighborhood of η in D. Denote by $A(f, \zeta, \eta, J(\zeta), U(\eta))$ the set of all C^1-functions $y : J(\zeta) \to U(\eta)$ such that*

$$y' = f \circ y \text{ and } y(\zeta) = \eta. \tag{I.4}$$

Denote by $B(f, \zeta, \eta, J(\zeta), U(\eta))$ the set of all C^1-functions $y : J(\zeta) \to U(\eta)$ such that

$$y = \eta + \int_\zeta^? f \circ y. \tag{I.5}$$

Lemma 111 *With the above definitions, we have*

$$A(f, \zeta, \eta, J(\zeta), U(\eta)) = B(f, \zeta, \eta, , J(\zeta), U(\eta)).$$

The easy proof is left to the reader.

I.2.1 The Fundamental Theorem: Local Case

The following theorem is called the local case of the fundamental theorem of ordinary differential equations.

Theorem 84 *With the preceding notation and definitions, suppose that f is locally Lipschitz, i.e., for every $x \in D$, there is a neighborhood $U(x) \subset D$ and a positive number L such that $x_1, x_2 \in U(x)$ implies $|f(x_1) - f(x_2)| \leq L.|x_1 - x_2|$. Then for any "initial condition" $\zeta \in \mathbb{R}, \eta \in D$, there is an open interval $J(\zeta)$ containing ζ, and an open neighborhood $U(\eta)$ of η such that $A(f, \zeta, \eta, J(\zeta), U(\eta))$ is a singleton. The element of A is called the* local solution *of the differential equation $y' = f \circ y$ at $J(\zeta), U(\eta)$.*

The proof uses Lemma 111 and refers to the set B. In fact, it is shown that the operator

$$T_{\zeta, \eta, f}(y) = \eta + \int_{\zeta}^{?} f \circ y$$

is a contraction, and contractions have a unique fixpoint.

Proposition 128 *Let $T : X \to X$ be a contraction on a complete metric space[2] (X, d), i.e., there is a constant $0 < c < 1$ such that $d(T(x), T(y)) \leq c.d(x, y)$ for all $x, y \in X$. Then, T has a unique fixpoint $z = T(z)$.*

Proof. It suffices to show that the sequence $(x_n = T^n(x))$ is Cauchy. In fact, setting $k = |n - m|$, we have $d(x_n, x_m) = d(T^n(x), T^m(x)) = c^{Min(n,m)}.d(x, T^k)$. But

$$d(x, T^k) \leq d(x, T(x)) + d(T(x), T^2(x)) + \dots d(x^{k-1}, T^k(x))$$

$$\leq (1 + c + \dots c^{k-1})d(x, T(x)) \leq \frac{1}{1-c}d(x, T(x)).$$

So this term is limited, while $c^{Min(n,m)}$ tends to zero as n, m tend to infinity, QED.

Corollary 38 *Let X be a complete metric space, and $B = \overline{B_r(x)}$ the closed ball of radius $r > 0$ around x. Let $T : B \to X$ a contraction with $d(T(x), x) \leq (1 - c)r, 0 < c < 1$. Then T has a unique fixpoint in B.*

Proof. We know from Lemma 110 that B is complete. Further, for $y \in B$, we have $d(T(y), x) \leq d(T(y), T(x)) + d(T(x), x) \leq c.d(y, x) + (1 - c)r \leq r$. Therefore, T leaves B invariant and the claim follows from Proposition 128, QED.

Corollary 39 *With the notation of Corollary 38, suppose that $T : B_r(x) \to X$ is a contraction with $d(T(x), x) < (1 - c)r, 0 < c < 1$. Then there is a unique fixpoint of T in $B_r(x)$.*

Next, we need some auxiliary results concerning uniform convergence of continuous functions. Let W be a Banach space (in our case $W = \mathbb{R}^n$), A a set, then we set

$$B(A, W) = \{f : A \to W | \|f\|_\infty < \infty\}.$$

Proposition 129 *The set $B(A, W)$ with the usual scalar multiplication and addition of functions is a Banach space.*

---

[2] See this Appendix, Section I.1.1.

It is clear that $B(A, W)$ is a vector space. Let $(f_n)_n$ be a Cauchy sequence in $B(A, W)$. Since for any $x \in A$, $\|f_n(x) - f_m(x)\| \leqslant \|f_n - f_m\|$, and the right term converges to zero, the left term is also a Cauchy sequence in W and converges to $\lim_{n \to \infty} f_n(x) = f(x)$. We first show that $\lim_{n \to \infty} f_n = f$. For $0 < \epsilon$, let N be such that $n, m > N$ implies $\|f_m - f_n\| < \epsilon$. Then by definition, for all $x \in A$, $\|f_n(x) - f(x)\| = \|f_n(x) - \lim_{m > N} f_m(x)\| = \lim_{m > N} \|f_n(x) - f_m(x)\| \leqslant \epsilon$. Therefore $\|f_n - f\| \leqslant \epsilon$, and $f = (f - f_n) + f_n$ is a sum of two elements of $B(A, W)$ and therefore lives in $B(A, W)$, whereas $\lim_{n \to \infty} f_n = f$, QED.

Theorem 85 *Let A be a metric space, W a Banach space, and let*

$$BC(A, W) = B(A, W) \cap C^0(A, W)$$

be the set of continuous functions with limited norm. Then $BC(A, W) \subset B(A, W)$ is a closed sub-vector space, and therefore also Banach.

Proof. It is clearly a sub-vector space. Let (f_n) be a Cauchy sequence in $BC(A, W)$. It converges to f in $B(A, W)$. We have to show that it is also continuous. In fact, given $0 < \epsilon$ select n such that $\|f - f_n\| < \epsilon/3$. Let $a \in A$. Take $0 < \delta$ such that $d(x, a) < \delta$ implies $\|f_n(x) - f_n(a)\| < \epsilon/3$. Then $\|f(x) - f(a)\| \leqslant \|f(x) - f_n(x)\| + \|f_n(x) - f_n(a)\| + \|f_n(a) - f(a)\| < \epsilon/3 + \epsilon/3 + \epsilon/3$, QED.

We are now ready for the proof of the local theorem. Recall that we are given a locally Lipschitz vector field function $f : D \to \mathbb{R}^n$. Consider the Banach space $BC = BC(J(\zeta), \mathbb{R}^n)$ for an interval $J(\zeta)$ whose length δ will be determined in the course of the proof. Select $0 < r$ such that (1) the closed ball $B_r(\eta)^- \subset D$, and (2) $f|B_r(\eta)^-$ is Lipschitz with a constant L. Then f is evidently limited on $B_r(\eta)^-$, let m be an upper bound. Let $\bar{\eta} : J(\zeta) \to B_r(\eta)^- : t \mapsto \eta$ be the constant map. Consider the closed ball $B_r(\bar{\eta})^- \subset BC$ around $\bar{\eta}$. For every $g \in B_r(\bar{\eta})^-$, $f \circ g : J(\zeta) \to \mathbb{R}^n$ lives in BC.

We now show that the operator $T(g) = \eta + \int_\zeta^? f \circ g$ defines a contraction

$$T : B_r(\bar{\eta})^- \to BC$$

with contraction constant c such that $d(T(\bar{\eta}), \bar{\eta}) < (1 - c)r$. According to Corollary 38, this will imply that T has a unique fixpoint in $B_r(\bar{\eta})^-$ and we are done.

Evidently, $T(g)$ is continuous. Further, for any $x \in J(\zeta)$, we have $|T(g)(x)| \leqslant |\eta| + |\int_\zeta^x f \circ g| \leqslant |\eta| + |x - \zeta| . \|f \circ g\|_\infty$ which evidently is finite.

We are left with the contraction claims. We have

$$\|T(\bar{\eta}) - \bar{\eta}\|_\infty = lub_{J(\zeta)} \|T(\bar{\eta})(t) - \bar{\eta}(t)\| = lub_{J(\zeta)} \left\| \int_\zeta^t f(\eta) \right\|$$

$$= lub_{J(\zeta)} |t - \zeta| . |f(\eta)| \leqslant \delta . |f(\eta)| \leqslant \delta . m.$$

For two functions $g_1, g_2 \in B_r(\bar{\eta})^-$, we have

$$\|T(g_1) - T(g_2)\| = lub_{J(\zeta)} \left| \int_\zeta^t f(g_1) - f(g_2) \right| \leqslant \delta . \|f \circ g_1 - f \circ g_2\|$$

$$= \delta . lub_{J(\zeta)} |f(g_1(s)) - f(g_2(s))| \leqslant \delta . L . lub_{J(\zeta)} |g_1(s) - g_2(s)|$$

$$= \delta . L . \|g_1 - g_2\|.$$

This means that T is a contraction with $c = \delta . L$ if δ is such that $\delta . L < 1$. Further, we need $\delta . m < (1 - c)r = (1 - \delta . L)r$, i.e., $\delta < \frac{r}{m + Lr}$ solves the problem, QED.

I.2.2 The Fundamental Theorem: Global Case

The global fundamental theorem deals with maximal integral curves $y : J \to D$ for the differential equation $y' = f \circ y$.

Definition 191 *We say that $u \sim v$ for $u, v \in D$ iff there is a curve $y : J \to D$, defined on an open interval J for the differential equation $y' = f \circ y$, and such that $\{u, v\} \subset y(J)$.*

Lemma 112 *The relation $\sim$ is an equivalence relation. The equivalence class of an element $x \in D$ is denoted by $[x]$.*

It is clearly reflexive and symmetric. It is transitive for the following reason. Let $y_i : J_i \to D, i = 1, 2$ be two integral curves such that $y_1(t_1) = x, y_1(t_2) = y, y_2(t_3) = y, y_2(t_4) = z$. By an evident parameter shift, we may suppose $t_2 = t_3$. We claim that

$$y_1 | J_1 \cap J_2 = y_2 | J_1 \cap J_2.$$

Suppose that $y_1(t) \neq y_2(t)$ for a $t > t_2$. Let $t_2 \leqslant t_0$ be the infimum of these t. Since our curves are continuous, we have $y_1(t_0) = y_2(t_0)$. But then, according to the local Theorem 84, there is an ϵ-ball $U_\epsilon(t_0)$ around t_0 and a neighborhood $U(y_1(t_0) = y_2(t_0))$ such that there is a unique integral curve $y : U_\epsilon(t_0) \to U(y_1(t_0))$. But we may suppose WLOG that ϵ is so small that both $y_1 | U_\epsilon(t_0), y_2 | U_\epsilon(t_0)$ have their codomains in $U(y_1(t_0))$. Evidently, these solutions must then coincide with the unique solution on the open interval $U_\epsilon(t_0)$, but this contradicts the choice of t_0. A symmetric argument holds for the supremum $s_0 \leqslant t_2$ of those arguments with $y_1(t) \neq y_2(t)$. Therefore, $y_1 | J_1 \cap J_2 = y_2 | J_1 \cap J_2$, and we may extend the integral curves y_1, y_2 to the domain $J_1 \cup J_2$, whence the transitivity of the $\sim$-relation, QED.

Theorem 86 *Let $x \in D$. Then there is a unique integral curve $y : J \to D$ with $y(0) = x, y' = f \circ y$, and such that J contains all domains of any integral curve $z, z(0) = x, z' = f \circ z$. We have $y(J) = [x]$ and write $\int_x f$ for this curve; it is called the* global solution through x.

Proof. Let $\Gamma = \{\Gamma_{y_i} \subset \mathbb{R} \times D | \Gamma_{y_i} = $ graph of solution y_i of $y_i' = y_i \circ f, x = y_i(0)\}$. Since two solutions coincide on the intersection of their domains, the union $\bigcup \Gamma$ is functional, and the union of the domains is an open interval J. Further, the function y of this graph is a solution of the differential equation $y' = f \circ y$ which reaches all elements equivalent to x, QED.

Corollary 40 *Let $x_1 \sim x_2$ and $\int_{x_1} f(t_2) = x_2$. Then $\int_{x_2} f = \int_{x_1} f \circ e^{t_2}$ and $J_2 = e^{-t_2} J_1$.*

Definition 192 *The quotient $D/ \sim = \{Im(\int_x f) | x \in D\}$ is called the* phase portrait of the vector field f *and denoted by D/f. An integral curve which is not an injective function of its parameter is called a* cycle of the field f.

Proposition 130 *Let $\int_w f$ be a cycle with $\int_w f(t_1) = \int_w f(t_1 + T)$. Then the cycle's domain is $\mathbb{R}$ and $\int_w f$ is T-periodic.*

Proof. Let $y : J \to D$ be the cycle $\int_w f$ with $y(t_1) = z$. Consider the function $\hat{y} = y \circ e^T : J - T \to D$. Evidently, $\hat{y}(t_1) = y(t_1)$, and $\hat{y}$ also solves the differential equation since $\hat{y}'(t) = y'(t + T) = f \circ y(t + T) = f \circ \hat{y}(t)$. So, since $\hat{y}$ has a common value with y at t_1, by maximality of y, we have $\hat{y} = y | J - T$, and $J - T \subset J$, whence $J =] - \infty, b[$. Symmetrically, exchanging t_1 with $t_1 + T$, and T with $-T$, we obtain $J =]a, \infty[$, i.e., $J = \mathbb{R}$. Now, for any $t \in \mathbb{R}$, with $\hat{y} = y \circ e^T$, uniqueness guarantees $\hat{y} = y$, whence the periodicity of y, QED.

Proposition 131 *Suppose that D^- is compact (e.g.: D is bounded), and that f is locally Lipschitz on D^-. If the domain $J =]a, b[$ of a maximal curve $\int_x f$ has finite upper bound b, then $t \to b$ implies $\int_x f(t) \to \partial D$*

Sketch of proof: Write $y = \int_x f$, and suppose that the closure of $y(J)$ were in D. Then, since $y(J)^-$ is compact, there is a convergent sequence $t_n \to b$ with a convergent image sequence $y(t_n) \to q, q \in D$. It can be shown that q is uniquely determined, i.e., another such sequence yields the same limit. We then set $y(b) = lim_{t \to b} y(t) = q$, and y may be extended to a local solution containing b, a contradiction, QED.

I.2.3 Flows and Differential Equations

On an open set $O \subseteq \mathbb{R}^n$, a vector field (a $\mathcal{C}^1$-map) $f : O \to \mathbb{R}^n$ can also be viewed by its graph as a section $O \to TO : x \mapsto F(x) = (x, f(x))$. If $x \in O$, an *integral curve of F in x* is a curve $y : U(0) \to O$ defined on an open neighborhood $U(0)$ of 0 such that $y(0) = x$ and $y' = f \circ y$. By the main Theorem 86 of ODEs, there is a unique maximal integral curve $\int_x f$ for every point $x \in O$.

For a vector field F on $O \subseteq \mathbb{R}^n$, a *flow box* is a triple (U, a, W) where $U \subseteq O$ is open, a is a positive real number of ∞, and $W : U \times\,] - a, a [\to O$ is $\mathcal{C}^1$ such that for all $x \in U$, $W_x :\,] - a, a [\to O : t \mapsto W(x, t)$ is an integral curve of F at x. Two flow boxes $(U, a, W), (U', a', W')$ always coincide in their maps W, W' on the intersection $(U \cap U') \times\, (] - a, a [\cap\,] - a', a' [)$ of their domains. For each point $x \in O$, there is a flow box (U, a, W) with $x \in U$. Let $\mathcal{D}_F = \{(x, t) \in \mathbb{R} \times O |$ there is an integral curve $\int_x$ whose domain contains $t\}$. Then (1) $\mathcal{D}_F$ is open in $\mathbb{R} \times \mathbb{R}^n$; (2) there is a unique map $W_F : \mathcal{D}_F \to O$ such that $t \mapsto W_F(t, x)$ is an integral curve at x for all $x \in O$.

I.2.4 Vector Fields and Derivations

For a $\mathcal{C}^1$-function $f : O \to \mathbb{R}$, we have the derivative $Tf : TO \to T\mathbb{R}$, whose second component evaluates to linear forms on $\mathbb{R}^n$. This map $df = pr_2 \circ Tf = Df$ is called the *differential of f*. If $F : O \to TO$ is a vector field, the composition $L_F f = df \circ F : O \to \mathbb{R}$ is called the *Lie derivative of f with respect to F*. If we denote by $grad(f)$ the differential of f as a tangent vector $(\partial_{x_1} f, \ldots \partial_{x_n} f)$ (the old-fashioned gradient of f), the Lie derivative is just the scalar product of $grad(f)$ with the vector field. If $\mathcal{F}(O)$ denotes the real algebra[3] of $\mathcal{C}^1$-function on O, the map $L_F : \mathcal{F}(O) \to \mathcal{F}(O)$ is a *derivation* in the sense that:

(i) L_F is linear;
(ii) for $f, g \in \mathcal{F}(O)$, we have $L_F(f.g) = f.L_F(g) + L_F(f).g$;
(iii) If $c \in \mathcal{F}(O)$ is constant, then $L_F c = 0$.

Therefore, we also have $d(f.g) = df.g + f.dg$ and $dc = 0$ for a constant c. Denote by $VF(O)$ the vector space of all $\mathcal{C}^\infty$-vector fields on O. Then:

Theorem 87 *The Lie map*
$$L_? : VF(O) \to Der(\mathcal{F}(O)) : F \mapsto L_F$$

is an isomorphism of vector spaces.

See [2, Th.8.10] for a proof.

In particular, the Lie bracket $[L_F, L_G] = L_F \circ L_G - L_G \circ L_F$ which is a derivation, must be the Lie derivative of a unique vector field which is denoted by $[F, G]$, the *Lie bracket of the vector fields F and G*. The Lie bracket makes the vector space $VF(O)$ into a real Lie algebra, see Section E.4.4.

I.3 Partial Differential Equations

For this section, refer to [509].

We only need a short review of quasi-linear first order partial differential equations (PDE). Recall that a PDE is an equation of type $E(x_1, x_2, \ldots u, u_{x_1}, u_{x_2}, \ldots u_{x_1 x_1}, u_{x_1 x_2}, \ldots) = 0$ where u is a function of the n real variables $x_1, x_2, \ldots x_n$, with its partial derivatives $u_{x_1}, \ldots$, the higher partial derivatives $u_{x_1 x_1} \ldots$ etc. A solution is meant to be such a function u which is defined in an open set O of $\mathbb{R}^n$. Its *order m* is the highest number of iterated partial derivatives, whereas E is called *quasi-linear* iff it is an affine function of the derivatives of u of highest order m, with coefficients that are functions of the variables $x_1, x_2, \ldots u, u_{x_1} \ldots$ until derivatives of order $m - 1$.

A first-order quasi-linear PDE has the shape

---

[3] Multiplication goes pointwise.

$$\sum_i a_i(x_1, x_2, \ldots u)u_{x_i} = c(x_1, x_2, \ldots u)$$

and can be solved by a system of ODEs, this is the *method of characteristics*.

We illustrate the method for two variables, i.e., for the equation

$$a(x, y, u)u_x + b(x, y, u)u_y = c(x, y, u). \tag{I.6}$$

The solution $u(x, y)$ is represented as a surface $z = u(x, y)$ in $\mathbb{R}^3$. Such a surface is called an *integral surface* of the equation (I.6). We have a vector field

$$F(x, y, z) = (a(x, y, z), b(x, y, z), c(x, y, z))$$

on the common domain U of the three functions a, b, c. The tangent space of an integral curve at $x, y, u(x, y)$ is spanned by the vectors $X = (1, 0, u_x)$ and $Y = (0, 1, u_y)$. Their vector product $Y \wedge X = (u_x, u_y, -1)$ is the normal vector to the integral surface. Therefore equation (I.6) just means that the scalar product $(F(x, y, z), Y \wedge X)$ vanishes identically, i.e., the vector field F is tangent to the integral surface. Clearly, only the direction of the vectors of the vector field F, the *characteristic directions* matter for the equation (I.6). It is easily seen that an integral curve of F, if it crosses a point of an integral surface, is entirely contained in this surface. Therefore an *integral surface is the union of integral curves of the directional vector field F*. An integral surface can be constructed by finding a curve Γ which lies in an integral surface, and which is never parallel to an integral curve of F. This is the *Cauchy problem for the equation* (I.6). Then, the parameter of Γ and the curve parameter of a flow box (see Section I.2.3) around Γ describe the integral surface. Technically, the existence condition for $\Gamma(t) = (\Gamma_x(t), \Gamma_y(t), \Gamma_z(t))$ to generate a surface is that the projection $\Gamma_{xy}(t) = (\Gamma_x(t), \Gamma_y(t))$ is never parallel to the projection F_{xy} of the directional field on the xy plane. The existence of a curve Γ is again guaranteed by the main theorem of ODEs, and we are done.

J

More Complements on Mathematics

Summary. This appendix is not self-contained, it completes the mathematical Appendix Part XXI and is built upon those topics.

$$- \Sigma -$$

J.1 Directed Graphs

Directed graphs, also called *digraphs* or, in Gabriel's language *quivers*, define a basic category in mathematics and computer science. Their category is also one of the most accessible topoi (see Appendix Section G.3 for the concept of a topos). The sketchy Appendix Section C.2.2 about graphs and quivers can be skipped when reading the following section.

J.1.1 The Category of Directed Graphs (Digraphs)

The category **Digraph** of directed graphs, short: digraphs. This is a basic category for algebra as well as for topology. In the naive setup, its objects are functions $\Gamma : A \to V^2$ from a set $A = A_\Gamma$ of *arrows* to the Cartesian square $V^2 = V \times V$ of the set $V = V_\Gamma$ of *vertices*. The first projection $t = pr_1 \circ \Gamma$ is called the *tail* function, the second $t = pr_2 \circ \Gamma$ is called the *head* function of the digraph. For an arrow a, the vertices $t(a), h(a)$ are called its head and tail, respectively, and denoted by $t(a) \xrightarrow{a} h(a)$. A morphism $f : \Gamma \to \Delta$ of digraphs is a couple $f = (u, v)$ of functions $u : A_\Gamma \to A_\Delta, v : V_\Gamma \to V_\Delta$ such that $v^2 \circ \Gamma = \Delta \circ u$.

Example 135 The *initial object* in the category **Digraph** is the empty digraph $\varnothing$, whose arrow and vertex sets are both empty. The *terminal object* in **Digraph** is the digraph $1 = t \circlearrowleft T$ with one vertex t and a single loop T on t. A third important elementary digraph is the *arrow digraph* $\uparrow = \bullet \to \bullet$, having two vertices and one connecting arrow. A generalization of $\uparrow$ is the n-fold concatenation $\uparrow^n$ of $\uparrow$, consisting of $n + 1$ vertices, and having one arrow from vertex i to vertex $i + 1$ for all $i = 0, 1, 2, \ldots n - 1$. One also extends this power to $n = 0$, meaning that $\uparrow^0$ is the digraph with one single vertex and no arrow.

Example 136 Denote the switch bijection on V^2 which maps (x, y) to (y, x) by $?^*$. Then for every digraph $\Gamma : A \to V^2$, we have its *dual digraph* Γ^* that is defined by $\Gamma^* = ?^* \circ \Gamma$. Evidently, $\Gamma^{**} = \Gamma$. The dual digraph construction defines an functor automorphism on **Digraph**.

Example 137 In this book, we often need a special subcategory of digraphs, the *spatial digraphs*. Such a digraph is associated with a topological space X (see Appendix Section H.1) and is denoted by $\vec{X}$. By definition, the arrow set is $A_{\vec{X}} = I@_{\mathbf{Top}}X$, the set of continuous curves $c : I = [0, 1] \to X$ in X, while the vertex set is $V_{\vec{X}} = X$, with $h(c) = c(1)$, and $t(c) = c(0)$. A spatial morphism is a digraph morphism

© Springer International Publishing AG, part of Springer Nature 2017
G. Mazzola, *The Topos of Music IV: Roots*, Computational Music Science,
https://doi.org/10.1007/978-3-319-64495-0_10

$\overrightarrow{f} : \overrightarrow{X} \to \overrightarrow{Y}$ canonically induced by a continuous map $f : X \to Y$. The subcategory of spatial digraphs and morphisms is denoted by *SpaceDigraph*.

A spatial digraph is more than a digraph. It is also a topological digraph in the following sense. The set $A_{\overrightarrow{X}} = I@X$ of arrows of $\overrightarrow{X}$ is a topological space by the compact-open topology (see Appendix Section J.4.1.2), and the head and tail maps $h, t : I@X \to X$ are continuous. Moreover, for a continuous map $f : X \to Y$, the arrow map $I@f : I@X \to I@Y$ is continuous.

Every digraph Γ gives rise to a (small) category, the *path category Path(Γ)*. It is defined as follows.

Definition 193 *A path of length $n \geq 0$ in a digraph Γ is a morphism $p :\uparrow^n \to \Gamma$, we the write $l(p) = n$ for the length of p. A path of length zero is also called a* lazy path. *It is given by the single vertex $p(0)$ in Γ. The first vertex of a path p is denoted by $d(p)$ and called the path's* domain, *the last vertex of p is denoted by $c(p)$ and called the path's* codomain. *A path p with $d(p) = c(p)$ is called a* cycle, *and if $l(p) = 1$, the cycle is called a* loop. *A vertex in Γ that is not the tail of an arrow is called a* leaf *of Γ. A vertex v of Γ that has a path p for every vertex w such that $d(p) = v, c(p) = w$ is called a* root *or* source *of Γ, and if v is a root of the dual Γ^*, it is called a* co-root *or* sink *of Γ. A digraph with a (necessarily unique) root and without directed cycles is called a* directed tree.

The category *Path(Γ)* has the paths $p :\uparrow^n \to \Gamma$ as their morphisms. If $d(q) = c(p)$, then we may concatenate q after p in the evident way, yielding a path $q \circ p$ of length $l(q \circ p) = l(q) + l(p)$, the *composition of p with q*. This defines the category *Path(Γ)*, the lazy paths being its objects.

J.1.1.1 Unordered Graphs

Unordered graphs will play a minor role in this book, so we keep the information minimal here. Whereas a digraph needs the cartesian product $V \times V = V^2$ of its vertex set V, an undirected graph, or simple *graph*, needs the *edge set* $^2V = \{a \subset V | 1 \leq card(a) \leq 2\}$. These sets a represent intuitively edges between vertices, and a loop in case $card(a) = 1$. There is an evident surjection $|V| : V^2 \to {}^2V : (x, y) \mapsto \{x, y\}$, which has a number of sections that we denote somewhat ambigusouly by $\overrightarrow{V} : {}^2V \to V^2$, i.e., $|V| \circ \overrightarrow{V} = Id_{{}^2V}$.

Definition 194 *A graph (or an* unordered graph*) is map $\Gamma : A \to {}^2V$. The elements of A are called* edges, *the elements of V are called* vertices *of Γ. For $a \in A$ and $\Gamma(a) = \{x, y\}$, we also write $x \overset{a}{\rule{2em}{0.4pt}} y$, which is the same as $y \overset{a}{\rule{2em}{0.4pt}} x$.*

Much as digraphs, graphs also have morphisms as follows.

Definition 195 *If $\Gamma : A \to {}^2V$ and $\Delta : B \to {}^2W$ are two graphs, a morphism $f : \Gamma \to \Delta$ is a couple $f = (u, v)$ of maps, $u : A \to B, v : V \to W$ such that $^2v \circ \Gamma = \Delta \circ u$.*

Graphs and their morphisms the clearly define the category *Graph* of graphs. Moreover, the surjections $|V|$ define a functor $|?| : \mathbf{Digraph} \to Graph$, and the graph $|?|(\Gamma)$ is called the *associated graph of Γ* and denoted by $|\Gamma|$. Conversely, given a graph Δ, a digraph Γ such that its associated graph $|\Gamma|$ is Δ is called *associated digraph of Δ*, even though this one is not uniquely determined.

Definition 196 *Similar to the chain digraphs $\uparrow^n$, we have chain graphs, which are just the associated graphs $|\uparrow^n|$, we denote them by $|^n$. Corresponding to directed paths we have undirected paths in a graph Δ, called* walks, *namely the morphisms $w : |^n \to \Delta$. A (undirected) cycle in graph Δ is a walk that starts where it ends, and a lazy walk is a cycle of length 0. An* undirected tree *is a graph that has no undirected cycles of positive length and a (automatically uniquely determined) root. Undirected trees are associated with directed trees, bot not vice versa. A* spanning tree *in a graph Δ is a subgraph that is a tree and contains all vertices of Δ.*

Here is a useful lemma:

Lemma 113 *Every connected graph (any two vertices admit a walk that connects them) admits a spanning tree.*

J.1.2 Two Standard Constructions in Graph Theory

The following constructions of digraphs are crucial in graph theory, and will be used in section J.1.3.

Proposition 132 *For any three digraphs Γ, Δ, Σ and a couple of morphisms $\Gamma \to \Delta, \Delta \to \Sigma$, there is a fiber product $\Gamma \times_\Sigma \Delta$. For any three digraphs Γ, Δ, Σ and a couple of morphisms $\Sigma \to \Gamma, \Sigma \to \Delta$, there is a fiber sum $\Gamma \sqcup_\Sigma \Delta$.*

These constructions are meant in the sense of category theory, sharing the universal properties. Here are the two constructions for the simpler case where $\Sigma = \emptyset$, the general case works similerly: If $\Gamma : A \to V^2, \Delta : B \to W^2$, then we have the cartesian product map $A \times B \to V^2 \times W^2$, but we have a canonical bijection $V^2 \times W^2 \xrightarrow{\sim} (V \times W)^2$, which yields a digraph $\Gamma \times \Delta : A \times B \to (V \times W)^2$, this is the cartesian product digraph. On the other hand, we have the coproduct map $A \sqcup B \to V^2 \sqcup W^2$, and there is an evident injection $V^2 \sqcup W^2 \to (V \sqcup W)^2$, whence the coproduct digraph $\Gamma \sqcup \Delta : A \sqcup B \to (V \sqcup W)^2$.

As is well known from category theory (see Appendix Section G.2.2), if we have fiber products and fiber sums, and also initial and final objects, then all finite limits and colimits also exist, whence this theorem:

Theorem 88 *The category **Digraph** is finitely complete and co-complete.*

Remark 34 A special case is this fact: Any digraph Δ is the colimit of the following diagram $\mathcal{D}$ of digraphs: We take one arrow digraph $\uparrow_a = \uparrow$ for each arrow $a \in A_\Delta$ and one bullet digraph $\bullet_x = \bullet$ for each vertex $x \in V_\Delta$. We take as morphisms the tail or head injections $\bullet_x \to \uparrow_a$ whenever $x = t(a)$ or $x = h(a)$. Then evidently, $\Delta \xrightarrow{\sim} \mathrm{colim}\mathcal{D}$.

J.1.3 The Topos of Digraphs

In this section we want to prove that **Digraph** is a topos. Recall from Definition 170 the concept of a(n elementary) topos. Let us recapitulate a short definition (Appendix Section G.3):

Definition 197 *A topos is a category $\mathcal{C}$ that*

(i) $\mathbf{C}$ is cartesian closed,
(ii) $\mathbf{C}$ has a subobject classifier $1 \rightarrowtail \Omega$.

Let us recall these two properties from Appendix Section G.3.

Definition 198 *A category $\mathcal{C}$ is cartesian closed if it has finite cartesian products (which is equivalent to having binary products and a terminal object), and if every functor $A \times ? : X \mapsto A \times X$ has a right adjoint, denoted by $?^A$, i.e., $(A \times X)@_{\mathcal{C}}Y \xrightarrow{\sim} X@_{\mathcal{C}}Y^A$ as a bifunctor in X and Y.*

Definition 199 *Given a category $\mathcal{C}$ which is finitely complete, with the terminal object 1, a monomorphism $true : 1 \rightarrowtail \Omega$ in $\mathcal{C}$ is called a subobject classifier iff given any monomorphism $\sigma : S \rightarrowtail X$ in $\mathcal{C}$, there is a unique morphism $\chi_\sigma : X \to \Omega$ such that the diagram*

$$
\begin{array}{ccc}
S & \xrightarrow{\;\sigma\;} & X \\
{\scriptstyle !}\downarrow & & \downarrow{\scriptstyle \chi_\sigma} \\
1 & \xrightarrow{\;true\;} & \Omega
\end{array}
\tag{J.1}
$$

is a fiber product.

There is an easy proof of the fact that **Digraph** is a topos. It resides on a more creative redefinition of **Digraph** as follows. We may view it as the category of presheaves[1] $ht^@$ over the small category $ht =$

$$\bullet \underset{t}{\overset{h}{\rightrightarrows}} \bullet$$ with two objects and just two parallel morphisms h, t between these points. Then our claim follows immediately[2] from a general proposition:

Proposition 133 *For a category $\mathcal{C}$ the presheaf category $\mathcal{C}^@$ is a topos.*

The description of the subobject classifier in **Digraph** is as follows. The final object $1 = t\,\circlearrowright$ is embedded by the true morphism $T : 1 \to \Omega$ into the subobject classifier

$$\Omega = f \underset{N}{\overset{P}{\rightleftarrows}} t$$

Every evaluation $\Gamma @ \Omega$ describes the set of subdigraphs of Γ, together with its canonical Heyting logic (see Appendix Section G.5.1, Theorem 72).

J.2 Galois Theory

This section is not an introduction to classical Galois theory, but rather focuses on the relationship of Galois theory to the classical problems of cube doubling, angle trisection and π construction using compass and straightedge. For a reference to Galois theory, see [1079].

The main theorem of Galois theory deals with extensions of commutative fields an their relations to corresponding groups of field automorphisms. We only deal with commutative fields here. A field extension is symbolized by L/K, this means that we have an inclusion of two fields $K \subset L$. An extension is called algebraic if every element $x \in L$ is algebraic over K, i.e., if it is a zero, $f(x) = 0$, of a polynomial $f(X) \in K[X]$. A non-algebraic element of an extension is called *transcendental* (over K). A finite extension is one where the dimension $[L : K] = dim_K(L) < \infty$. Such an extension is automatically algebraic. A separable extension is an extension L/K such that every element $x \in L$ is a zero of an irreducible polynomial that has no multiple zeros.

Definition 200 *An separable finite field extension L/K is called* Galois *or* normal *iff it verifies one of the following equivalent conditions:*

1. *Every irreducible polynomial $f \in K[X]$ that has a zero in L factorizes in linear factors.*
2. *The field L is generated from K by adjunction of all zeros of a polynomial $f \in K[X]$.*

A Galois field extension L/K always has a *primitive element* ξ, i.e., it is the algebraic extension generated by adjunction of a single element ξ to K. $L = K(\xi)$. The *Galois group* of a Galois extension L/K is the group $Gal(L/K)$ of all field automorphisms $f : L \overset{\sim}{\to} L$ that leave K pointwise fixed, also called *relative automorphisms*. We then have the fact that if $L = K(\xi)$ with ξ being the zero of an irreducible polynomial $g(X)$ of degree n, then we have

$$|Gal(L/K)| = [L : K] = n.$$

This is the main theorem of Galois theory:

---

[1] For any category $\mathcal{C}$, $\mathcal{C}^@$ denotes the category of presheaves, i.e., contravariant set-valued functors, on $\mathcal{C}$. Our notation stems from the Yoneda embedding $y : \mathcal{C} \to \mathcal{C}^@$, which associates with an object X in $\mathcal{C}$ the presheaf $@X$ which yields the set $@X(Y) = Y@X$ of morphisms $f : Y \to X$ in $\mathcal{C}$.

[2] Observe that $ht^{op} \overset{\sim}{\to} ht$, so it is also legitimate to view $ht^@$ as being built from the covariant functors on ht.

Theorem 89 *Let $Ex(L/K)$ be the set of intermediate field extensions M, $K \subset M \subset L$. Let $SubGal(L/K)$ be the set of all subgroups H of $Gal(L/K)$. Then we have a bijection*

$$Ex(L/K) \xrightarrow{\sim} SubGal(L/K)$$

defined by $K \subset M \subset L$ being sent to the group of relative automorphisms of the extension L/M, while the inverse bijection sends a subgroup $H \in SubGal(L/K)$ to the field of those elements of L that remain fixed under H. We have $|H| = [M : K]$.

The problem of constructing the double of a cube's volume or the trisection of an angle can be reduced to a problem of Galois theory as follows: Suppose the geometric situation of the plane can be described in a coordinate system by a set of numbers that represent points, straight lines (defined by two points) or circles (defined by their center and the diameter), call the field extension $K/\mathbb{Q}$ defined by these numbers K. Then a new geometric object x can be constructed using compass and straightedge if its coordinate numbers can be calculated using rational operations (sum, difference, product, quotient) and square roots since all these operations can be done using compass and straightedge. These operations are not necessarily real ones, but working in $\mathbb{C}$ does not make the problem more difficult since rational operations and square roots on complex numbers can be managed using compass and straightedge, this is an easy exercise. The converse is also true: If we can construct a number using compass and straightedge, we only perform rational operations or calculate square roots. Think of intersecting two curves, a curve and a circle, or two circles.

Theorem 90 *Given the basic data in the field extension $K/\mathbb{Q}$, in order to be constructible using compass and straightedge, a number x must be contained in a Galois field extension L/K with $[L : K] = 2^m$. Conversely, if x is contained in such an extension, it can be constructed using compass and straightedge.*

The second statement follows from the fact that if $[L : K] = 2^m$, then $Gal(L : K)$ is a solvable group which means that it has a composition series $1 \subset H_1 \subset H_2 \subset \ldots Gal(L : K)$ of subgroups H_i that are normal pairs $H_i \triangleleft H_{i+1}$ with H_{i+1}/H_i being abelian of order a power of 2, which (using the main theorem on finitely generate abelian groups [714, Appendix C.3.4.2]) implies $H_{i+1}/H_i \xrightarrow{\sim} \mathbb{Z}_2$, i.e., using the main theorem of Galois theory, the corresponding extensions are quadratic and can be constructed using rational operations and square roots.

But our problems of doubling the volume or angle trisection lead to cubic irreducible polynomials. Their solutions must generate field extensions of K that have degree 3, and such extensions cannot be contained in extensions of degree 2^m, therefore these problems cannot be solved using compass and straightedge. The construction of π is even simpler to deal with if one accepts that π is not even algebraic over K, it is a transcendental number (the proof of this fact is quite hard and not a topic of pure algebra), and therefore cannot be contained in any algebraic extension of K.

J.3 Splines

Splines are functions that are built from a special type of functions (polynomials, exponential functions, etc.) that are defined on simple types of subspaces of their domains, whence the name "spline": these special functions are "splined" together.

J.3.1 Some Simplex Constructions for Splines

Recall from Appendix Section H.2 that the standard (closed) simplex Δ_d of dimension d (also called d-simplex) is the set of points $(\xi_0, \xi_1, \ldots \xi_d) \in \mathbb{R}^{d+1}$ such that $\sum_i \xi_i = 1$ and $0 \leqslant \xi_i \leqslant 1$ for all $i = 0, 1, \ldots d$.

Definition 201 *Given natural numbers n, d and a sequence $(P_i)_{i=0,\ldots d}$ of points $P_i \in \mathbb{R}^n$, the affine simplex $\Delta(P_0, \ldots P_d)$ is the unique map*

$$\Delta(P_0, \ldots P_d) : \Delta_d \to \mathbb{R}^n$$

with $\Delta(P_0, \ldots P_d)(e_i) = P_i$ for all standard basis vectors $e_i = (0, \ldots 0, 1, 0, \ldots 0)$, which is induced by an affine map $\mathbb{R}^{d+1} \to \mathbb{R}^n$. We denote the image of this affine simplex by $\Delta[P_0, \ldots P_d]$.

The image is also described by this formula:

$$\Delta[P_0, \ldots P_d] = \{\sum_{i=0}^{d} \lambda_i P_i | (\lambda.) \in \Delta_d\}.$$

If $f : \mathbb{R}^n \to \mathbb{R}^m$ is an affine map, then we have (with the above notations)

$$f \circ \Delta(P_0, \ldots P_d) = \Delta(f(P_0), \ldots f(P_d)).$$

Definition 202 *A subset $X \subset \mathbb{R}^n$ is called* convex *iff for all $x, y \in X$ the closed line $\Delta[x, y]$ is a subset of X. For any subset D of $\mathbb{R}^n$, its* convex hull $Conv(D)$ *is the intersection of all convex sets in $\mathbb{R}^n$ containing D. $Conv(D)$ is in fact the minimal convex set containing D.*

In particular,

Proposition 134 *The convex hull of a finite sequence $(P_i)_{i=0,\ldots d}$ of points $P_i \in \mathbb{R}^n$ is the image set $\Delta[P_0, \ldots P_d]$. More generally, the convex hull of any set $X \subset \mathbb{R}^n$ is the union of all affine images $\Delta[P_0, \ldots P_d]$ of finite sequences $(P_i)_{i=0,\ldots d}$ of points in D.*

Proposition 135 *If $f : \mathbb{R}^n \to \mathbb{R}^m$ is an affine map and $D \subset \mathbb{R}^n$, then*

$$f(Conv(D) = Conv(f(D)).$$

J.3.2 Definition of General Splines

Given an n-dimensional cube $K(a., b.) = K(a_1, \ldots a_n; b_1, \ldots b_n)$ for n intervals $a_i \leq x_i \leq b_i$, denote by $Part(a., b.)$ the set of *partitions of $K(a., b.)$.* A partition is defined as follows: First, a partition of a one-dimensional cube, i.e., an interval $[a, b]$, is a finite subset $P \subset [a, b]$ that contains a and b. This means that we have an increasing set of points $a < x_1 < \ldots x_k < b$, the elements of P. Call $vol(x_i, x_{i+1}) = x_{i+1} - x_i$ the *volume* of this interval. An n-dimensional partition is a cartesian product $P = P_1 \times \ldots P_n$ of partitions P_i of $[a_i, b_i]$. This means that the elements P_j of P define cartesian products of one-dimensional intervals, closed cubes c, and the volume $vol(c)$ of such a cube is the product of its one-dimensional interval volumes. Denote by $I)(P)$ the set of these cubes of P. Figure J.1 shows a three-dimensional partition.

Let partition P, Q be two partitions in $Part(a., b.)$. We say that *Q is a refinement of P* and write $P \to Q$ if every defining factor P_i of P is a subset of the defining factor Q_i of Q.

The definition of types of functions for splines is less precise, but here is what they mean: If $K \subset \mathbb{R}^n$ is an n-dimensional cube, we consider a set $\mathcal{T}_n(K)$ of functions $f : K \to \mathbb{R}^m$, usually a real vector space (with the point-wise addition and scalar multiplication). Typical examples are polynomial functions, differentiable functions, polynomial functions of limited degree, etc. One prominent example are the functions that are polynomial functions in every coordinate of $\mathbb{R}^m$.

Definition 203 *Given natural numbers n, m, a partition $P \in Part(a., b.)$, and a type $\mathcal{T}$ of functions, a m-dimensional* spline *of type $\mathcal{T}$ is a function $f : K(a., b.) \to \mathbb{R}^m$ such that everyone of its restrictions to a cube $K \in I(P)$ is of type $\mathcal{T}$. If we require an additional condition C of f about its behavior on the partition's cubes, we say that f is a spline of type $\mathcal{T}$ with condition C. In particular, if $n = 1$ we speak of* spline curves, *for $n = 2$ we speak of* spline surfaces.

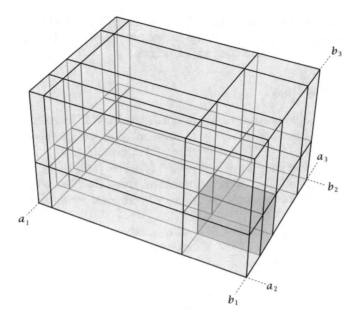

Fig. J.1. A three-dimensional partition.

Let us discuss the popular role of cubic splines, i.e., splines that relate to the function type $\mathcal{T} = Pol^3_m$, the polynomial functions of at most degree 3 in every coordinate of the target space $\mathbb{R}^m$. Let us first consider such a function $f : [0,1] \to \mathbb{R}$ with these conditions:

$$f(0) = y_0, f(1) = y_1, f'(0) = t_0, f'(1) = t_1.$$

If $f(X) = a + bX + cX^2 + dX^3$, then these conditions mean that

$$a = y_0, b = t_0, c = 3(y_1 - y_0) - 2t_0 - t_1, d = 2(y_0 - y_1) + t_0 + t_1.$$

This means that such polynomials allow for a unique solution for values of the functions and their first derivatives on the boundary points of intervals. This implies the general result:

Theorem 91 *For every one-dimensional partition* $P = (a = x_0 < x_1 < \ldots x_k = b)$ *and two sequences* $(y_0, y_1, \ldots y_k), (t_0, t_1, \ldots t_k)$ *of vectors in* $\mathbb{R}^m$, *with the intervals* $K_i = [x_i, x_{i+1}], i = 0, \ldots k - 1$, *there is a unique spline function* $f : [a, b] \to \mathbb{R}^m$ *of type* Pol^3_m *with the condition* $f(x_i) = y_i, i = 0, \ldots k$ *and such that* $Df_{i-1}(x_i) = Df_i(x_i) = t_i, i = 1, \ldots k - 1$, *and* $Df_0(x_0) = t_0, Df_{k-1}(x_k) = t_k$, *where* $f_j = f|_{K_j}$.

J.4 Topology and Topological Categories

J.4.1 Topology

J.4.1.1 Generators for Topologies

(Recalling some basic concepts from Appendix Secction H.1) For any non-empty family $(T_i)_i$ of topologies $T_i \subset 2^X$ of a set X, the intersection $\bigcap_i T_i$ is again a topology. If $T \subset 2^X$ is any set of subsets of X, the family of topologies T_i such that $T \subset T_i$ is not empty (the discrete topology is a member), and this family's intersection is called the *topology generated by* T and denoted by $\langle T \rangle$. The elements of T are also called a *subbase* of $\langle T \rangle$. The open sets of $\langle T \rangle$ are the unions of finite intersections of members of T. A set B of open sets of a topology is called a *base* of this topology iff every open set is the union of a family of open sets from B. For example, the set of finite intersections of members of T in the above example are a base of $\langle T \rangle$. For a point $x \in X$ of a topological space X, a *neighborhood base* is a set of neighborhoods such that every neighborhood of x contains at least one of such neighborhoods.

J.4.1.2 Compact-Open Topology

The *compact-open topology* is an important topological construction for function spaces. It is defined on any space $X@_{\mathbf{Top}}Y$ of continuous functions $f : X \to Y$ between two topologic spaces X, Y. It is the topology generated by the sets $\langle K; U \rangle := \{f \in X@_{\mathbf{Top}}Y | f(K) \subset U\}$, where $K \subset X$ is quasi-compact and $U \subset Y$ is open. Denote by E the evaluation map $X@Y \times X \to Y : (f, x) \mapsto f(x)$.

A *locally compact* topological space X is a topological space that has a base of quasi-compact neighborhoods for every point $x \in X$. For example, metric spaces are locally compact.

Theorem 92 (Exponential Correspondence) *If X is a locally compact Hausdorff space, and Y and Z are topological spaces, then a map $g : Z \to X@Y$ is continuous (for the compact-open topology on $X@Y$) iff $E \circ (g \times Id_X) : Z \times X \to Y$ is continuous.*

Theorem 93 (Exponential Law) *If X is locally compact Hausdorff, Z is Hausdorff, and Y is a topological space, then the function $\epsilon : Z@(X@Y) \to (Z \times X)@Y$ defined by $\epsilon(g) = E \circ (g \times Id_X)$ is a homeomorphism.*

Remark 35 In particular, this theorem proves that the functor $Z \mapsto Z \times X$ is left adjoint to the functor $Y \mapsto X@Y$ (see Appendix Section G.2.1).

Theorem 94 *If X is a compact Hausdorff space and Y is metrized by a metric d, then $X@Y$ is metrized by the metric d^* defined by*

$$d^*(f, g) = \sup\{d(f(x), g(x)) | x \in X\}$$

J.4.2 Topological Categories

Intuitively, a topological[3] category is a category that is also a topological space. This is a concept which appears in the theory of internal categories, i.e., categories that are "internal" to a given category. Our case is a category that is internal to the category **Top** of topological spaces. Internal categories are simple constructions as follows:

Definition 204 *Given a category $\mathcal{E}$ with fiber products (also called pullbacks), an internal category in $\mathcal{E}$ is a pair $(\mathbf{C_0}, \mathbf{C_1})$ of objects (the first being called "object of objects", the second "object of morphisms", together with four arrows in $\mathcal{E}$:*

1. *two morphisms for category domain (d) an codomain (c)*

$$d : \mathbf{C_1} \to \mathbf{C_0}, c : \mathbf{C_1} \to \mathbf{C_0}$$

2. *one identities morphism*

$$e : \mathbf{C_0} \to \mathbf{C_1}$$

simulating the identification of objects and identity morphisms,

3. *and an arrow*

$$m : \mathbf{C_1} \times_{\mathbf{C_0}} \mathbf{C_1} \to \mathbf{C_1}$$

for composition of "composable" morphisms (having connecting domains and codomains). This fiber product relates to the domain and codomain morphisms d, c.

---

[3] There is also a second understanding of this concept which we shall not discuss here. It has nothing to do with the one we deal with.

These morphisms are subject to a number of rules that are the diagrammatic restatement of the definition of a category:

1. Associativity, the following diagram commutes:

$$
\begin{array}{ccc}
\mathbf{C_1} \times_{\mathbf{C_0}} \mathbf{C_1} \times_{\mathbf{C_0}} \mathbf{C_1} & \xrightarrow{Id \times m} & \mathbf{C_1} \times_{\mathbf{C_0}} \mathbf{C_1} \\
{\scriptstyle m \times Id} \downarrow & & \downarrow {\scriptstyle m} \\
\mathbf{C_1} \times_{\mathbf{C_0}} \mathbf{C_1} & \xrightarrow{m} & \mathbf{C_1}
\end{array}
$$

2. Identity domains/codomains:

$$
d \circ e = c \circ e = Id_{\mathbf{C_0}}
$$

3. Domains and codomains of compositions:

$$
d \circ m = d \circ pr_2, c \circ m = c \circ pr_1
$$

4. Identity neutrality

$$
m \circ (Id_{\mathbf{C_1}} \times e) = pr_1, m \circ (e \times Id_{\mathbf{C_1}}) = pr_2.
$$

If two internal categories $(\mathbf{C_0}, \mathbf{C_1})$, $(\mathbf{D_0}, \mathbf{D_1})$ are given, an *internal functor* $f : (\mathbf{C_0}, \mathbf{C_1}) \to (\mathbf{D_0}, \mathbf{D_1})$ is a pair $f = (f_0 : \mathbf{C_0} \to \mathbf{D_0}, f_1 : \mathbf{C_1} \to \mathbf{D_1})$ of morphisms such that they define functors of the internal categories, i.e., commute with the defining morphisms e, d, c, m. This defines the *category $Int(\mathcal{E})$ of categories internal to $\mathcal{E}$*.

To apply such a situation to the category of topological spaces, i.e., $\mathcal{E} = \mathbf{Top}$, we need to know that $\mathbf{Top}$ has fiber products, but such objects exist, they are the set-theoretical fiber products together with the product topologies and their evident restrictions.

J.5 Complex Analysis

For an reference to complex analysis we refer to Lars V. Alfohrs' excellent book [22]. This theory is the complex number analogue to calculus on the field of real numbers. The core construction is the complex derivative of a function $f : U \to \mathbb{C}$ defined on an open set $U \subset \mathbb{C}$. If $z \in U$, the derivative of f in z is defined by

$$
f'(z) = \lim_{h \to 0} \frac{f(z+h) - f(z)}{h}
$$

if this limit exist for complex h. This in particular implies that the limit is independent of the way h approaches 0 in $\mathbb{C}$, and this implies that if $f'(z)$ exists, then we have the famous *Cauchy-Riemann equation* with the real and imaginary arguments x, y in $z = x + iy$:

$$
\partial_x f = -i \partial_y f.
$$

And we then have $f'(z) = \partial_x f$, in which case we say that f is *analytic* or *holomorphic* in z. This differential equation is also sufficient for the existence of the complex derivative of f in z. The function f is analytic (holomorphic) if it is so on all of U.

Being analytical is a very strong property of f, as is shown by the following theorem:

Theorem 95 *Let f be analytic in an open set $U \subset \mathbb{C}$, let $z \in U$, and let $C \subset U$ be a circle curve centered in z, the disk it defines is entirely in U, then for any natural number n we have*

$$
f^{(n)}(z) = \frac{n!}{2\pi i} \int_C \frac{f(\zeta)}{(\zeta - z)^{n+1}}
$$

where $f^{(n)}$ is the nth derivative of f, especially $f^{(0)} = f$. In particular, f has all higher derivatives.

It is important to know that analytical functions are *conformal*, i.e., they locally conserve angles and therefore forms. Here is the precise fact. Suppose $\gamma :]a, b[\to U \subset \mathbb{C}$ is a differentiable curve, defined on an open neighborhood $]a, b[\subset \mathbb{R}$ containing 0, with $\gamma(0) = z$ and $T\gamma(0) = t$, and let $f : U \to \mathbb{C}$ be analytic with $f'(z) \neq 0$, then the tangent w of $f \circ \gamma$ at 0 is $w = f'(z)t$ (product of complex numbers), i.e., the argument of w is increased by the argument of $f'(z)$, independently of the t's direction. In particular, the f-image of any orthogonal basis of the tangent space Tz is again an orthogonal basis for $f'(z) \neq 0$.

Although con formality is a seemingly strong property of a function, the Riemann mapping theorem allows us to map any two simply connected open sets $U, V \subsetneq \mathbb{C}$ bijectively onto each other, more precisely:

Theorem 96 (Riemann Mapping Theorem) *Let $U \subsetneq \mathbb{C}$ be a proper simply connected open set and $z \in U$, then there exists a unique analytic function f on U that maps U bijectively onto the unit disk around 0, and such that $f(z) = 0$ and $f'(z) \in \mathbb{R}_+$ are fixed.*

Observe that the size of $f'(z)$ is not specified. For our applications to world-sheets of complex time (see Section 78.2.1), we shall impose non-automatic conditions on the derivatives of associated analytic functions. The Riemann Mapping Theorem is very strong for it connects any two simply connected proper regions in $\mathbb{C}$ by conformal mappings, but the size of the derivatives is not further controllable and is of course a function of the sizes and shapes of involved regions.

J.6 Differentiable Manifolds

Differentiable Manifolds are differentiable space structures that locally look like an open set of $\mathbb{R}^n$. Here is the definition.

Definition 205 *Let X be a topological space and n, r natural numbers. A n-dimensional C^r-atlas for X is given by*

1. *A family A of $(f_\kappa : U_\kappa \to X)_{\kappa \in K}$ of homeomorphisms on open sets $U_\kappa \subset \mathbb{R}^n$ onto their images $Im(f_\kappa) =: X_\kappa \subset X$ that are open in X (called charts of the atlas) such that*
2. $\bigcup_\kappa X_\kappa = X$ *(i.e., the family $(X_\kappa)_K$ is a covering of X), and*
3. *for any pair $\kappa, \lambda \in K$ with the (automatically) open $X_\kappa \cap X_\lambda \neq \varnothing$, the bijection*

$$X_{\kappa, \lambda} \to X_{\lambda, \kappa}$$

induced by the charts on the inverse images $X_{\kappa, \lambda} = f_\kappa^{-1}(X_\kappa \cap X_\lambda), X_{\lambda, \kappa} = f_\lambda^{-1}(X_\kappa \cap X_\lambda)$ of the intersection $X_\kappa \cap X_\lambda$ is a diffeomorphism in C^r.

Definition 206 *Two atlases A, B for X are called* equivalent *is their direct sum $A \sqcup B$ is also an atlas.*

Equivalence of atlases for X is evidently an equivalence relation. Therefore we may define

Definition 207 *Given a topological space X, an n-dimensional C^r-differentiable manifold structure on X is an equivalence class of n-dimensional C^r-atlases for X.*

In short, a differentiable manifold is a topological space X together with a representative of an atlas class. Observe that the characteristic difference between differentiable manifolds and global compositions (see [714, 13]) is that for global compositions the nerve of its covering by an atlas is fixed, it cannot be refined ad libitum like for differentiable manifolds, but see our corresponding remarks in Sections 13.1 & 19.1. C^r-differentiable manifolds define a category Man^r whose morphisms are defined as follows[4].

---

[4] The category Man^r can be generalized to Banach spaces instead of $\mathbb{R}^n$, i.e., complete topological vector spaces that are defined by a norm. But we don't need this generality except for some calculus of variations.

Definition 208 *If X and Y are C^r-differentiable manifolds, a morphism $f : X \to Y$ is a set map $f : X \to Y$ such that for every $x \in X$, and for every chart $g_\mu : V_\mu \to Y$ of Y with $f(x) \in Y_\mu$, there is a chart $f_\kappa : U_\kappa \to X$ with $f(X_\kappa) \subset Y_\mu$ and such that the corresponding map $U_\kappa \to V_\mu$ (between open sets in real spaces) is C^r-differentiable.*

J.6.1 Manifolds with Boundary

For the Stokes Theorem we need also define compact manifolds that have a "boundary". Here is the definition:

Definition 209 *A compact manifold with boundary is a triple (X, X_0, BdX_0) where*

1. *X is a n-dimensional manifold,*
2. *X_0 is an open submanifold of X,*
3. *the closure of X_0 is compact,*
4. *BdX_0 is the topological boundary of X_0 (closure of X_0 minus (interior of) X_0) and a $n-1$-dimensional submanifold of X.*

The well-known Möbius strip, including its boundary line, is a famous two-dimensional manifold with boundary. The n-dimensional standard simplex Δ_n or the cube I^n are other examples, the faces being their boundary.

J.6.2 The Tangent Manifold

Locally, an n-dimensional manifold X looks like an open set in $\mathbb{R}^n$. In particular, if $m \in X$ is a point and $f_\kappa : U_\kappa \to X$ a chart, the corresponding point $x = f_\kappa^{-1}(m) \in \mathbb{R}^n$ allows to consider the "tangent"vectors at x, which we denote as pairs $t_x = (x, t), t \in \mathbb{R}^n$. This set $^f T_x$ is a function of the selected chart. But for manifolds in Man^r, $r \geq 1$, we can create a chart-independent tangent concept as follows: Take the disjoint union $\coprod_f \, ^f T_x$ and consider the relation $(x, t) \sim (y, s)$ for charts f, g around m iff the Jacobi matrix at x of the isomorphism $g^{-1} \circ f$ that is defined on the intersection of the two charts maps t to s. This is an equivalence relation and the *tangent space of X at m* is defined to be the set of equivalence classes $T_m X = \coprod_f \, ^f T_x / \sim$; it is obviously also a real vector space of dimension n. The tangent manifold TX is the disjoint union $TX = \coprod_{m \in X} T_m X$ with the $2n$-dimensional manifold structure that is given by the evident charts $Tf : U \times \mathbb{R}^n \to TX$, the transition morphisms being the original ones in the first component, while the second component is transformed by the Jacobian. We have the base point projection morphism $p_X : TX \to X$ sending tangent (m, t) to m.

The tangent construction is functorial, i.e., if $f : X \to Y$ is a morphism of manifolds, then we have a corresponding tangent morphism $Tf : TX \to TY$ which commutes with the projections, i.e., the diagram

$$
\begin{array}{ccc}
TX & \xrightarrow{\;Tf\;} & TY \\
{\scriptstyle p_X}\big\downarrow & & \big\downarrow{\scriptstyle p_Y} \\
X & \xrightarrow{\;f\;} & Y
\end{array}
$$

is commutative. On the tangent spaces it is defined by the Jacobian of f on its charts. The functor $T : X \mapsto TX$ is called the *tangent functor*, and the projection $p_X : TX \to X$ defines a natural transformation $p : T \to Id_{Man^r}$. Tangent morphisms are vector bundle morphisms: they are linear on every tangent space $T_m X$.

Sorite 17 *All manifolds in this sorite are in Man^r, $r \geq 1$.*

1. *If X, Y are two manifolds, then $T(X \times Y) \xrightarrow{\sim} TX \times TY$.*
2. *In particular, for any point $(m, n) \in X \times Y$, we have $T_{(m,n)} X \times Y \xrightarrow{\sim} T_m X \oplus T_n Y$.*

J.7 Tensor Fields

On n-dimensional vector space T and for every pair r, s of natural numbers, we consider the vector space of multilinear forms $T_s^r(T)$, this by definition the vector space of multilinear forms $t : T^* \times T^* \times \ldots T^* \times T \times T \times \ldots T \to \mathbb{R}$, the first r factors being the linear dual space T^* and the last s factors being T. By the universal property of tensor products, this is also the linear dual of $(T^*)^{r\otimes} \otimes T^{s\otimes}$. Denote this n^{r+s}-dimensional vector space by T_s^r, and therefore call its elements tensors, r *times contravriant and s times covariant.*

J.7.1 Alternating Tensors

Within T_s^0, we have the subspace $\Lambda^s T$ of alternating tensors, i.e., by definition those tensors $t : T^{s\otimes} \to \mathbb{R}$ such that for any permutation $\pi \in \mathfrak{S}_s$, we have $t(x_{\pi(1)}, x_{\pi(2)}, \ldots x_{\pi(s)}) = sgn(\pi)t(x_1, x_2, \ldots x_s)$. The tensor product of two alternating tensors is not alternating in general, but there is corresponding construction, namely the wedge product. To this end, one needs an auxiliary construction that is defined for any covariant tensor $t \in T_s^0$:

$$Alt(s)(x_1, x_2, \ldots x_s) = \frac{1}{s!} \sum_{\pi \in \mathfrak{S}_s} sgn(\pi)s(x_{\pi(1)}, x_{\pi(2)}, \ldots x_{\pi(s)}).$$

$Alt(s)$ is an alternating tensor. The *wedge product* of two alternating tensors v, w or degree r, s, respectively, is defined by

$$v \wedge w = \frac{(r+s)!}{r!s!} Alt(v \otimes w).$$

This new alternating tensor construction is associative, distributive, bilinear, and $w \wedge v = (-1)^{rs} v \wedge w$. This construction entails the calculation of the dimension of $\Lambda^s T$:

Lemma 114 *If $v_1, v_2, \ldots v_k$ is a basis of T, and if $v_1^*, v_2^*, \ldots v_k^*$ is the dual basis, then a basis of $\Lambda^s T$ is given by the sequence of all wedge products*

$$v_{i_1}^* \wedge v_{i_1}^* \wedge \ldots v_{i_s}^*$$

with increasing indices $i_1 < i_2 < \ldots i_s$. And this implies $dim(\Lambda^s T) = \binom{k}{s}$, in particular $dim(\Lambda^k T) = 1$.

Proposition 136 *If $dim(T) = k$, then for a basis $v_1^*, v_2^*, \ldots v_k^*$ the base-change on $\Lambda^k T$ is covered by the following formula. If $w_i = \sum_i a_{ij}v_j, i = 1, 2 \ldots k$, are any k vectors in T, then*

$$w_1 \wedge w_2 \wedge \ldots w_k = det(a_{ij})v_1 \wedge v_2 \wedge \ldots v_k.$$

This result splits the bases of $\Lambda^k T$ into two classes: those which have a positive transition determinant, and those having a negative determinant. The choice of a determinant class is called *orientation* on T. On $\mathbb{R}^k$, the *usual orientation* is defined by the canonical basis $e_1, e_2, \ldots e_k$.

J.7.2 Tangent Tensors

For the tangent space $T = T_m X$ of a n-dimensional manifold X, the disjoint union of the tensors $T_s^r{}_m X := (T_m X)_s^r$ of the tangent spaces, $T_s^r X = \coprod_{m \in X} T_s^r{}_m X$, can be given a vector bundle structure whose charts are the evident injections $T_s^r f : U \times (\mathbb{R}^n)_s^r \to T_s^r X$. The gluing isomorphisms on these charts are the canonical multilinear extensions of the isomorphisms (!) on the tangent charts.

Definition 210 *Given a differentiable manifold $X \in Man^t$, a tensor field of type $\binom{r}{s}$ is a C^t-section $t : X \rightarrowtail T_s^r X$. A differential s-form is a tensor field of type $\binom{0}{s}$. A vector field is a tensor field of type $\binom{1}{0}$, where one uses the canonical isomorphism $T \xrightarrow{\sim} T^{**}$ for finite dimensional vector spaces to identify a vector field with a section $X \rightarrowtail TX$. Denote by $\mathcal{T}_s^r(X)$ the set of tensor fields of type $\binom{r}{s}$. The set of vector fields over X is denoted by $\mathcal{V}(X)$.*

Tensor fields of a given type can be added pointwise, and also be multiplied pointwise by scalar functions, i.e., manifold morphisms $f : X \to \mathbb{R}$. Denote the ring of these functions by $\mathcal{F}X$ (and also denote it by $\mathcal{T}_0^0(X)$). If $f \in \mathcal{T}_s^r(X), g \in \mathcal{T}_{s'}^{r'}(X)$, then their pointwise product is in $\mathcal{T}_{s+s'}^{r+r'}(X)$, we denote it by $f \otimes g$ in view of the tensor product formalism we recalled above. With these operations, the direct sum $\mathcal{T}X = \bigoplus_{r,s \geqslant 0} \mathcal{T}_s^r(X)$ is a bigraded algebra over the ring $\mathcal{F}X$ of functions on X, the *tensor algebra of X*.

If $f \in \mathcal{F}X$, then we have $Tf : TX \to T\mathbb{R} = \mathbb{R}^2$. If we look at the second component of Tf, we get a function $df : TX \to T\mathbb{R} = \mathbb{R}$, i.e., a 1-form, the *differential of f*. Given a vector field V on X, we may then define a new function $L_V f(m) = df_m(V(m))$ in $\mathcal{F}X$. This function is called the *Lie derivative of f with respect to V*. Here are the characteristic properties of this assignment:

Proposition 137 *The map $L_V : \mathcal{F}X \to \mathcal{F}X$ is a derivation on the function ring, i.e., it is $\mathbb{R}$-linear and $L_V(fg) = (L_V f)g + f(L_V g)$. Moreover, $L_V c = 0$ for a constant function c.*

Denote by $Der(X)$ the real vector space of derivations on X. This result defines a function $L : \mathcal{V}(X) \to Der(X) : V \mapsto L_V$.

Theorem 97 *The Lie derivative function*

$$L : \mathcal{V}(X) \to Der(X)$$

is a linear isomorphism of real vector spaces.

This means that derivations and vector fields are essentially the same thing. This fact is very important in musical performance theory where for important performance operators, the Lie derivative is taken as a function of a performance field that acts upon performance weights, generating a new performance field, but see Section 39.7.

On a n-dimensional differentiable manifold X, an orientation is defined by a section $D : X \rightarrowtail \Lambda^n X$ that has non-vanishing values D_m in all the one-dimensional spaces $\Lambda^n T_m X$. A manifold need not have any orientation, it can be *non-orientable*; an example is the Möbius strip (without its boundary line).

J.8 Stokes' Theorem

This theorem connects integrals over differential forms on a compact manifold with boundary with integrals of the derivative of such forms on the manifold's boundary.

The first concept is the integral of a differential form. We cannot recapitulate the extensive setup here and refer to [617] for a precise reference. The essential steps are these: First consider the integral $\int_{I^n} f$ of a sufficiently tame function f on a cube I^n, then take the integral $\int_{I^n} w$ on a differential form $w = f.d$ on the manifold I^n with respect to a basis $d \in \Lambda^n I^n$, which is by definition $\int_{I^n} f$ for that function f of points in I^n. Next, one takes a singular cube[5], i.e., a morphism $c : I^n \to X$ into an oriented n-dimensional manifold X with orientation form d. One integrates a form $w = f.d$ by taking the inverse image $c^* w$ of w, and then integrates this one, i.e., $\int_c w = \int_{I^n} c^* w$. Next, one takes a general oriented manifold and calculates integrals using partitions of unity, which enable reduction to the singular cube situation.

We next need to make precise the manifolds with boundary and orientation needed for Stokes.

Definition 211 *An oriented compact manifold with boundary is a compact manifold with boundary (see Section J.6.1) (X, X_0, BdX_0) in which X is oriented, and for each $b \in BdX_0$, there is a positively oriented chart $f : U \to X$ with the following properties (see Figure J.2):*

1. *There is a submanifold chart $g : U' \to BdX_0$ with $U \subset U' \times I \subset \mathbb{R}^{n-1} \times \mathbb{R}$, b being represented by 0,*
2. *$f^{-1}(f(U) \cap BdX_0) = U' \times \{0\}$,*
3. *$f^{-1}(f(U) \cap X_0) \subset U' \times]0, 1[$.*

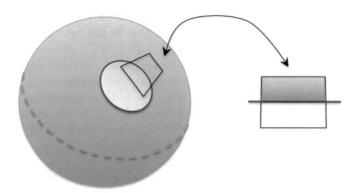

Fig. J.2. The opened spherical surface manifold with a boundary chart construction.

The last ingredient is about the differential of a differential form: We already know the differential df of a function, which is a 1-form. The general definition is s follows: The idea is to define the differential $dw \in \Lambda^{k+1} X$ of a k-form $w \in \Lambda^k X$, $dim(X) = n$, on charts, to show that the definition is independent of the chart, and then to glue the local definitions to a global one. If we have local coordinates $x_1, x_2, \ldots x_n$ on a chart, then we have

$$w = \sum_{i_1 < i_2 < \ldots i_k} w_{i_1, i_2, \ldots i_k} dx_{i_1} \wedge dx_{i_2} \wedge \ldots dx_{i_k},$$

and we define the $k + 1$ form, the *differential of w* by

$$dw = \sum_{i_1, i_2, \ldots i_k} dw_{i_1, i_2, \ldots i_k} dx_{i_1} \wedge dx_{i_2} \wedge \ldots dx_{i_k} +$$

$$\sum_{i_1, i_2, \ldots i_k} \sum_{j=1}^{n} \frac{\partial}{\partial x_j} w_{i_1 < i_2 < \ldots i_k} dx_j \wedge dx_{i_1} \wedge dx_{i_2} \wedge \ldots dx_{i_k}.$$

And here is the basic theorem about the differential operator:

Theorem 98 *The differential operator $d : \Lambda^k X \to \Lambda^{k+1} X$ is a $\mathbb{R}$-linear map with*

1. *$d^2 = 0$,*
2. *for a k-form w and a l-form y we have*

$$d(w \wedge y) = dw \wedge y + (-1)^{kl} w \wedge dy,$$

if $f : X \to Y$ is a morphism of manifolds and $w \in \Lambda^k Y$, then

$$f^* dw = df^* w.$$

The first statement implies that we have a cochain complex (see Example 139 in Section J.13) of differential forms which defines the associated cohomology modules (in fact real vector spaces).

Theorem 99 (Stokes' Theorem) *If $w \in \Lambda^{k-1} X$ and if (X, X_0, BdX_0) is an oriented compact k-dimensional manifold with boundary, and denote by $i : BdX_0 \to X$ the inclusion, then we have*

$$\int_{BdX_0} i^* w = \int_{\overline{X_0}} dw$$

where $\overline{X_0}$ is the topological closure of X_0.

---
[5] See this Appendix, Section J.12.1 for this concept.

J.9 Calculus of Variations

The calculus of variations deals with the calculus of local extrema of functionals, which are defined as being numerical functions J (usually real-valued) on spaces of continuous functions $f : [a, b] \to \mathbb{R}^n$ on a closed interval $[a, b] \subset \mathbb{R}$. The space $\mathcal{C}_n^0[a, b]$ of $\mathbb{R}^n$-valued continuous functions is a Banach space with norm $\| \, \|$ (see Appendix Section I.1.2 for norms and Banach spaces), defined by

$$_0\|f\| = max\{\|f(x)\| \mid x \in [a, b]\}.$$

and if we restrict to the subspace $\mathcal{C}_n^1[a, b] \subset \mathcal{C}_n^0[a, b]$ of $\mathcal{C}^1$-functions, we have the norm[6]

$$_1\|f\| = max(_0\|f\|, \, _0\|f'\|).$$

With these notations, we define extremal values:

Definition 212 *A function $f \in \mathcal{C}_n^0[a, b]$ is a* strong *maximum(minimum) of a functional J on $\mathcal{C}_n^0[a, b]$ if the values $J(g) - J(f) \leqslant 0(J(g) - J(f) \geqslant 0)$ for all $g \in U(f)$ for some neighborhood $U(f) \subset \mathcal{C}_n^0[a, b]$ of f. A function $f \in \mathcal{C}_n^1[a, b]$ is a* weak *maximum(minimum) of a functional J on $\mathcal{C}^1[a, b]$ if the values $J(g) - J(f) \leqslant 0(J(g) - J(f) \geqslant 0)$ for all $g \in U(f)$ for some neighborhood $U(f) \subset \mathcal{C}_n^1[a, b]$ of f.*

The following *Fundamental Lemma of Calculus of Variations* is useful for finding extremal functions for functionals that involve integrals.

Lemma 115 *Let $f \in \mathcal{C}^k[a, b]$, and suppose that we have*

$$\int_a^b f(x)h(x) = 0$$

for every $h \in \mathcal{C}^k[a, b]$ with $h(a) = h(b) = 0$. Then $f = 0$.

Its proof is an easy exercise (choose a special function $h(x) = f(x)(x - a)(b - x)$). The lemma is useful for the calculation of Lagrangians in theoretical physics (see Section K.1.1).

J.10 Partial Differential Equations

Quasi-linear first order partial differential equations have been discussed in Appendix Section I.6. Here, we need to look at a special type of second order partial differential equations: $\Delta w + U = 0$, where w and U are functions of two real variables x, y. This one is known as Poisson's equation with the Laplace operator $\Delta = \partial_x^2 + \partial_y^2$. This equation appears in the performance world-sheet discussion in Section 78.

We shall discuss here the solution of such an equation for $0 \leqslant x, y \leqslant 1$, the more general frame $0 \leqslant x \leqslant a, 0 \leqslant y \leqslant b$ being a straightforward generalization. The solution is given in several steps of special cases. First, we solve the equation for zero potential (a Laplace equation). Then, we solve the equation with $w|\partial I^2 = 0$, and then we add these two special solutions and get the general one.

To begin with, in [768, §7.1], an explicit solution of the Poisson equation $\Delta w + U = 0$ in a special case is given[7]. The potential is supposed to be zero. The function w is supposed to vanish on the boundary ∂I^2 except for the upper line $y = 1$ where the function $w_1(x)$ is given. The solution is

$$w(x, y) = \int_{\xi=0}^1 w_1(\xi)G^{y=1}(x, y|\xi), \tag{J.2}$$

---

[6] Attention: the usual index notation conflicts with the absolute or 1-norm described in Section I.1.2, we therefore write the index to the left of the norm symbol.

[7] We are grateful to theoretical physicist Emilio Fiordilino for his help to understand the Green function in our situation.

where ([768, formula 7.1.8]) the Green function is defined by

$$G^{y=1}(x, y|\xi) = 2 \sum_{n=0}^{\infty} \frac{\sinh(\pi n y)}{\sinh(\pi n)} \sin(\pi n x) \sin(\pi n \xi).$$

The fact that $w(x, 1) = w_1(x)$ relies on the standard integral expression of Fourier series coefficients and on the classical orthogonality relations.

Next, we find a solution of $\Delta w + U = 0$ with $w|\partial I^2 = 0$. In [768, §7.1, equation (7.1.12)], we are given this solution:

$$w(x, y) = \int_{I^2} G(x, y|\xi, \eta) U(\xi, \eta)$$

with the Green function

$$G(x, y|\xi, \eta) = 8 \sum_{n=0}^{\infty} \frac{\sin(\pi n x) \sin(\pi n \xi)}{n \sinh(\pi n)} H_n(y, \eta)$$

where

$$H_n(y, \eta) = \begin{cases} \sinh(p_n \eta) \sinh(p_n(1 - y)) & \text{for } 1 \geqslant y > \eta \geqslant 0, \\ \sinh(p_n y) \sinh(p_n(1 - \eta)) & \text{for } 1 \geqslant \eta \geqslant y \geqslant 0. \end{cases}$$

The general case can be deduced in the following way. To begin with, the given solution can be found also for w that vanishes on all but one side of I^2, simply switch x and y or exchange 1 and 0. Then, if the four solutions $w^i, i = 1, 2, 3, 4$, are found for the four situations $y = 1, y = 0, x = 1, x = 0$ and the potential $U/4$, their sum $w = w^1 + w^2 + w^3 + w^4$ solves the original equation $\Delta w + U = 0$. But this w is still special in that its values $w(x, y)$ vanish for all four corner points $(0, 0), (0, 1), (1, 0), (1, 1)$.

To solve the general case, we need a function $c(x, y)$ that solves the Laplace equation $\Delta c = 0$ and has arbitrary values on the four corner points. Then the general w can be corrected to $w - c$ and we may apply the given solution for this function that vanishes on the four corner points.

To begin with, we find an affine function $a(x, y) = rx + sy + t$ such that $a(0, 0) = w(0, 0), a(1, 0) = w(1, 0), a(0, 1) = w(0, 1)$. With the subtraction $w - a$, and observing that $\Delta a = 0$, we reduce w to a function that has all corner values equal to zero except possibly $w(1, 1)$. Take the function $p(x, y) = w(1, 1)xy$. Evidently $\Delta p = 0$, and p vanishes for the three corners of I^2 except for corner $(1, 1)$ where its value is $w(1, 1)$. Therefore $w - p$ has vanishing values on all four corners.

J.10.1 Explicit Calculation

We calculate explicitly the solution with the previous method. Recapitulate the idea of solving the Poisson equation $\Delta w + U = 0$:

1. Build $w_1 = w + l$, $l(x, y) = ax + by + c$ such that $w_1(0, 0) = w_1(1, 0) = w_1(0, 1) = 0$ and $\Delta l = 0$,
2. Build $w_2 = w_1 - q$, $q(x, y) = dxy$ such that w_2 vanishes on all four angles of I^2, and $\Delta q = 0$.
3. Then solve two partial problems: (1) $w_{2,1} + U = 0$ with $w_{2,1}|\partial I^2 = 0$, and (2) $\Delta w_{2,2}^i = 0$ with $w_{2,2}^i|\partial I_i^2 = w_2|\partial I_i^2$ and $w_{2,2}^i|\partial I_j^2 = 0$ for $j \neq i$.
4. Then $w_2 = w_{2,1} + \sum_{i=1}^{4} w_{2,2}^i$ solves the equation $\Delta w_2 + U = 0$, and the original function is $w = w_2 - Q$, $Q = l - q$.

The first two steps yield this representation:

$$a = w(0,0) - w(1,0)$$
$$b = w(0,0) - w(0,1)$$
$$c = -w(0,0)$$
$$d = w(1,1) + w(0,0) - w(1,0) - w(0,1)$$
$$w_2(x,y) = w(x,y) + Q(x,y)$$
$$= w(x,y) + ax + by + c - dxy$$

Next, we have to define the Green functions for the four sides of the second part of the third step. For side $y = 1$, the Green function has been defined above by the expression $G^{y=1}(x,y|\xi)$. For $x = 1$ we have to exchange the roles of x and y, whence $G^{x=1}(x,y|\eta) = G^{y=1}(y,x|\eta)$. For $y = 0$ we switch from y to $1 - y$: $G^{y=0}(x,y|\xi) = G^{y=1}(x, 1 - y|\xi)$, and for $x = 0$ we get $G^{x=0}(x,y|\eta) = G^{x=1}(1 - x, y|\eta)$.

And here are the functions on the four sides of ∂I^2:

$$w_2|_{y=0} = w|_{y=0} + ax + c$$
$$w_2|_{x=0} = w|_{x=0} + by + c$$
$$w_2|_{y=1} = w|_{y=1} + ex - w(0,1)$$
$$w_2|_{x=1} = w|_{x=1} + fy - w(1,0)$$

with $e = w(0,1) - w(1,1)$, $f = w(1,0) - w(1,1)$.

With this data, we can now calculate the solutions in the second part of step three:

$$w_{2,1}(x,y) = \int_{I^2} G(x,y|\xi,\eta) U(\xi,\eta)$$

$$w_{2,2}^{y=1}(x,y) = \int_{\xi=0}^{1} w_2|_{y=1} G^{y=1}(x,y|\xi)$$

$$= \int_\xi w|_{y=1} G^{y=1}(x,y|\xi) + e \int_\xi \xi G^{y=1}(x,y|\xi) - w(0,1) \int_\xi G^{y=1}(x,y|\xi)$$

$$w_{2,2}^{y=0}(x,y) = \int_{\xi=0}^{1} w_2|_{y=0} G^{y=0}(x,y|\xi)$$

$$= \int_\xi w|_{y=0} G^{y=0}(x,y|\xi) + a \int_\xi \xi G^{y=0}(x,y|\xi) + c \int_\xi G^{y=0}(x,y|\xi)$$

$$w_{2,2}^{x=1}(x,y) = \int_{\eta=0}^{1} w_2|_{x=1} G^{x=1}(x,y|\eta)$$

$$= \int_\eta w|_{x=1} G^{x=1}(x,y|\eta) + f \int_\eta \eta G^{x=1}(x,y|\eta) - w(1,0) \int_\eta G^{x=1}(x,y|\eta)$$

$$w_{2,2}^{x=0}(x,y) = \int_{\eta=0}^{1} w_2|_{x=0} G^{x=0}(x,y|\eta)$$

$$= \int_\eta w|_{x=0} G^{x=0}(x,y|\eta) + b \int_\eta \eta G^{x=0}(x,y|\eta) + c \int_\eta G^{x=0}(x,y|\eta)$$

This means that $w_{2,1}$ only depends linearly on the potential, $w_{2,1} = P(U)$, while the four functions $w_{2,2}^{y=1}, w_{2,2}^{y=0}, w_{2,2}^{x=1}, w_{2,2}^{x=0}$ depend in a linear way on the restriction of w to the four sides $w|y = 1, w|y = 0, w|x = 1, w|x = 0$, respectively, i.e., to the components of the boundary function ∂w. The values of the constants a, b, c, d are only a linear function of the values of w on the four angles of I^2, i.e., we have

$$w = P(U) + S(\partial w)$$

for two linear functions P, S.

J.11 Algebraic Topology

Refer also to Appendix Section H.2.

J.11.1 Homotopy Theory

Homotopy theory is the basic conceptual setup of algebraic topology. It deals with continuous curves $I \to X$, defined on the unit interval $I = [0,1] \subset \mathbb{R}$ in topological spaces X and so doing it is hoped to understand X in the best way possible. This is a special case of the Yoneda Lemma (Appendix Section G.2), where the sheaf @X is used to classify X. The continuous curves $I \to X$ are not sufficient to completely understand X, but they provide us with a lot of information, and one could say that Algebraic Topology is the effort to approximate Yoneda's Lemma by a special type of morphisms (their domains being special objects that one understands well).

Definition 213 *A* topological pair *is a pair (X, A) of a topological space X, together with a subspace $A \subset X$. If $card(A) = 1$, we call the pair a* pointed topological space (X), *and the point in question is the singleton $a \in A$. If (Y, B) is a second topological pair, a morphism $f : (X, A) \to (Y, B)$ is a continuous function $f : X \to Y$ such that $f(A) \subset B$. The* category of topological pairs *is denoted by* **Top2**.

Homotopy is about comparing different morphisms in **Top2**:

Definition 214 *If $(X, A), (Y, B)$ are two topological pairs and if $X' \subset X$ is a subspace of X, a* homotopy *from a morphism $f_0 : (X, A) \to (Y, B)$ to a morphism $f_1 : (X, A) \to (Y, B)$ relative to X' is a continuous map $F : X \times I \to Y$ such that $F(?, t) : (X, A) \to (Y, B)$ is a morphism for all $t \in I$, with $F(?, t)|_{X'} = f_0|_{X'}$, all $t \in I$, $f_0 = F(?, 0)$ and $f_1 = F(?, 1)$. In symbols: $F : f_0 \simeq_{X'} f_1$. In particular, f_0 is called* null homotopic *if it is homotopic to a constant map f_1. The attribute "relative to X'" is omitted for $X' = \varnothing$.*

The symbol choice $\simeq$ is justified be this easy lemma:

Lemma 116 *The relation of homotopy among the morphisms $f : (X, A) \to (Y, B)$ relative to a fixed $X' \subset X$ is an equivalence relation.*

Definition 215 *The set of homotopy equivalence classes of morphisms $f : (X, A) \to (Y, B)$ relative to a fixed $X' \subset X$ is denoted by $[X, A; Y, B]_{X'}$. We denote by $[f]_{X'}$ the class of morphism f. If $X' = \varnothing$, we omit the index X'.*

Homotopic maps can be composed:

Theorem 100 *If $f_0 \simeq_{X'} f_1$ in $[X, A; Y, B]_{X'}$ and $g_0 \simeq_{Y'} g_1$ in $[Y, B; Z, C]_{Y'}$ with $f_0(X') \subset Y'$, then $g_0 \circ f_0 \simeq g_1 \circ f_1$ in $[X, A; Z, C]_{X'}$.*

This theorem shows that we have categories of homotopy classes of topological pair morphisms, so-called *homotopy categories*. In particular, we have the homotopy subcategory of pointed topological spaces or the homotopy category of topological spaces (pairs with empty subsets). Clearly the map $f \mapsto [f]$ is a functor from topological pairs to homotopy categories. A diagram of topological pairs that becomes commutative after passing to the homotopy classes is called *homotopy commutative*. A pair morphism that becomes an isomorphism after passage to homotopy is called a *homotopy equivalence*, and a pair morphism that becomes the inverse to another morphism after passage to homotopy is called a *homotopy inverse*. Pairs that become isomorphic after passage to homotopy are said to have the *same homotopy type*.

Here is the construction of a terminal object in the homotopy category: A topological space X is called *contractible* if its identity morphism is homotopic to a constant endomorphism of X. A homotopy from the identity Id_X to a constant map at $x_0 \in X$ is called a *contraction of X to x_0*.

Lemma 117 *Any two maps of an arbitrary topological space X to a contractible space are homotopic, i.e., there is (always) one and only one morphism in the homotopy category to a contractible space. In particular, any two endomorphisms of a contractible space are homotopic.*

This means that contractible spaces define homotopy terminal objects. And here its an easy consequence:

Theorem 101 *A space is contractible iff its homotopy type is that of a one-point space. In particular, any two contractible spaces have the same homotopy type, and any morphism between contractible spaces is a homotopy equivalence (isomorphism).*

J.11.2 The Fundamental Group(oid)

The fundamental groupoid $\mathcal{P}(X)$ of a topological space X is a groupoid, i.e., by definition, a small category whose morphisms are all isomorphisms. Here is its construction: It has as objects the elements of X, and for any two such elements x, y, a morphism $f : x \to y$ is the homotopy class of a map $f : I \to X$ with $f(0) = x, f(1) = y$, the homotopy being relative to the boundary $\partial I = \{0, 1\}$. The identity on a point x is represented by the class of the constant map $\epsilon_x : I \to X$ in x. The composition $[g] \circ [f]$ of two morphisms $[f : I \to X], [g : I \to X]$ with $f(1) = g(0)$ is $[g * f]$, where

$$g * f(t) = f(2t) \ for \ 0 \leqslant t \leqslant 1/2, \ and \ g * f(t) = g(2t - 1) \ for \ 1/2 \leqslant t \leqslant 1.$$

A proof of the category-theoretical properties of this construction is found in [993, Chap. 1, Sec. 7].

This specializes to the fundamental group construction:

Definition 216 *Let (X, x_0)be a pointed topological space, then the* fundamental group of X at x_0 *is the group $\pi(X, x_0) = x_0 @_{\mathcal{P}(X)} x_0$ of automorphisms of the fundamental groupoid of X at x_0.*

This defines the *fundamental group functor* $\pi :$ **Top.** $\to$ **Grp** from the category **Top.** of pointed topological spaces to the category **Grp** of groups.

J.12 Homology

Homology is a vast field that can be understood from different points of view. An evident historical one is that homology generalizes the concept of a hole in a topological space. From a more category-theoretical point of view, it is the theory which quantifies the Yoneda philosophy, namely the idea of understanding/classifying an object X (here: a topological space) by considering classes of morphisms $D \to X$ from special simple objects D, hoping to extract characteristic information about X via these special arrows. Let us recall that in set theory, classifying a set X can be done by the maps $D \to X$, where D is a singleton, since the cardinality of X is given by the cardinality of the set $D @_{\mathbf{Ens}} X$.

The formalism of homology starts from a *chain complex* over a commutative ring R. This is by definition a sequence

$$C = \ldots C_{q+1} \xrightarrow{\partial_{q+1}} C_q \xrightarrow{\partial_q} C_{q-1} \ldots$$

of r-modules $C_q, q \in \mathbb{Z}$ and module "differential" homomorphisms ∂_q such that for all q, $\partial_q \circ \partial_{q+1} = 0$. The elements of C_q are called *q-chains*, and C is called nonnegative if $C_q = 0$ for $q < 0$. If all chain modules are free over R, C is called a *free chain complex*.

Observe that because of $\partial_q \circ \partial_{q+1} = 0$, we have the inclusion $Im(\partial_{q+1}) \subset Ker(\partial_q)$. The quotient module $H_q(C) = Ker(\partial_q)/Im(\partial_{q+1})$ is called the *qth homology module of C*. The *complex $Z(C)$ of cycles of C* is the subcomplex $Z(C) = (Ker(\partial_q))_q$ of C with the zero homomorphisms everywhere. The *complex $B(C)$ of boundaries of C* is the subcomplex $(Im(\partial_{q+1}))_q$ of C with the zero homomorphisms everywhere. We then have $B(C) \subset Z(C)$, and the quotient complex $H(C) = Z(C)/B(C)$, i.e., $H_q(C) = Z_q(C)/B_q(C)$ for all q.

This construction is functorial in the following sense. Consider the category of chain complexes $Chain_R$ over R. Its objects are the chain complexes over R while the morphisms $f : C \to C'$ are the sequences $f = (f_q)_q$ of R-module homomorphisms $f_q : C_q \to C'_q$ with commutative diagrams

$$
\begin{array}{ccc}
C_q & \xrightarrow{f_q} & C'_q \\
\partial_q \downarrow & & \downarrow \partial'_q \\
C_{q-1} & \xrightarrow{f_{q-1}} & C'_{q-1}
\end{array}
$$

for all q. Then we have a *homology functor* $H : Chain_R \to \mathbf{Mod}_R : C \mapsto H(C)$ to the category of R-modules.

There are several homology theories (defining chain complexes in different contexts). We shall only need a small selection here, see [635, 993] for a more complete discussion.

J.12.1 Singular Homology

This is a classical concrete theory. We denote by $|\Delta_q|$ the standard (topological) simplex space of dimension q associated with the standard simplex $\Delta_q = \{0, 1, 2, \ldots q\}$, see Appendix Section H.3. For each index $i \in \Delta_q$, there is an affine embedding $e_q^i : |\Delta_{q-1}| \to |\Delta_q|$ sending the vertex e_j to e_j if $j < i$, and to e_{j+1} if $j \geq i$. The image of e_q^i is called the *ith face of* $|\Delta_q|$ and denoted by $|\Delta_q|^i$. Standard simplex spaces are one type of "simple" special topological spaces D mentioned above.

A second type of "simple" spaces are the q-dimensional cubes I^q, the q powers of the real unit interval I. Here we also have the face morphism: $e_q^{i\alpha} : I^{q-1} \to I^q, \alpha = +, -$, with $e_q^{i\alpha}(x_0, \ldots x_{q-1})_j = x_j$ for $j < i$ and $e_q^{i-}(x_0, \ldots x_{q-1})_i = 0$, $e_q^{i+}(x_0, \ldots x_{q-1})_i = 1$, and $e_q^{i\alpha}(x_0, \ldots x_{q-1})_j = x_{j+1}$ for $j > i$. e_q^{i+} id called the front face while e_q^{i-} id called the back face.

Given a topological space X, the q-chains for the *simplex* approach are defined starting by the basic continuous maps $c : |\Delta_q| \to X$. The simplex chain complex R-module $\Delta_R(X)_q$ for $0 \leq q$ is the free R-module generated by the basis of the maps c. In other words, the q-chains in $\Delta(X)_q$ are the formal linear combinations $\sum_k r_k c_k$ with $r_k \in R$ and $c_k : |\Delta_q| \to X$. The chain modules for negative q vanish by definition.

The differential homomorphism is defined as follows: Suppose that $c : |\Delta_q| \to X$ is a base element in $\Delta_R(X)_q$ for $q > 0$. Then

$$\partial_q(c) = \sum_i (-1)^i c \circ e_q^i,$$

the alternate sum of c's (restriction to its) faces. This definition is extend linearly to general q-chains.

For the cube approach, the q-chains are defined starting from the basic continuous maps $c : I^q \to X$. The cube chain complex R-module $\square_R(X)_q$ is the free R-module generated by the basis of the maps c, similarly to the above simplex approach.

The differential homomorphism is defined as follows: Suppose that $c : I^q \to X$ is a base element in $\square_R(X)_q$ for $q > 0$. Then

$$\partial_q(c) = \sum_i (-1)^i (c \circ e_q^{i+} - c \circ e_q^{i-}),$$

the alternate sum of c's (restriction to its) +- and --faces. This definition is extend linearly to general q-chains. The important fact about the homology modules of these two approaches is this:

Theorem 102 *For a topological space X and the two singular homology approaches with simplexes and cubes we have isomorphic homology modules:*

$$H(\Delta_R(X)) \xrightarrow{\sim} H(\square_R(X)).$$

A proof can be found in [470]. While the simplex approach is standard in algebraic topology, although Jean-Pierre Serre used the cubic approach in his thesis, the cube approach is more useful for the generalized homology theory of gestures, see Chapter 63. See also [661] for the cube approach.

J.13 Cohomology

Formally speaking, cohomology is derived from a *cochain complex* over a commutative ring R. This is essentially about giving all arrows a reversed direction. But this reversal can also be induced by a chain complex, an example will be given after the definitions. A cochain complex is by definition a sequence

$$C^* = \ldots C^q \xrightarrow{\partial^q} C^q \xrightarrow{\partial^{q+1}} C^{q+1} \ldots$$

of r-modules $C^q, q \in \mathbb{Z}$ and module "differential" homomorphisms ∂^q such that for all q, $\partial^{q+1} \circ \partial^q = 0$. The elements of C^q are called *q-cochains*, and C^* is called nonnegative if $C^q = 0$ for $q < 0$. If all chain modules are free over R, C^* is called a *free cochain complex*.

Observe that because of $\partial^{q+1} \circ \partial^q = 0$, we have the inclusion $Im(\partial^q) \subset Ker(\partial^{q+1})$. The quotient module $H^q(C) = Ker(\partial^q)/Im(\partial^{q-1})$ is called the *qth cohomology module of C*. The *complex $Z^*(C^*)$ of* cocycles of C^* is the subcomplex $Z^*(C) = (Ker(\partial^q))_q$ of C^* with the zero homomorphisms everywhere. The *complex $B^*(C)$ of* coboundaries of C^* is the subcomplex $(Im(\partial^{q-1}))_q$ of C^* with the zero homomorphisms everywhere. We then have $B^*(C^*) \subset Z^*(C^*)$, and the quotient complex $H^*(C^*) = Z^*(C^*)/B^*(C^*)$, i.e., $H^q(C^*) = Z^q(C^*)/B^q(C^*)$ for all q.

This construction is functorial in the following sense. Consider the category of cochain complexes $Cochain_R$ over R. Its objects are the cochain complexes over R while the morphisms $f : C^* \to C'^*$ are the sequences $f = (f^q)_q$ of R-module homomorphisms $f^q : C^q \to C'^q$ with commutative diagrams

$$
\begin{array}{ccc}
C^q & \xrightarrow{f^q} & C'^q \\
\partial^q \downarrow & & \downarrow \partial'^q \\
C^{q+1} & \xrightarrow{f^{q+1}} & C'^{q+1}
\end{array}
$$

for all q. Then we have a *cohomology functor* $H^* : Cochain_R \to \mathbf{Mod}_R : C^* \mapsto H^*(C^*)$ to the category of R-modules.

Example 138 A generic example of a cohomology complex is to start with a chain complex $C \in Chain_R$ and to take the Hom_R functor, mapping every chain module C_q to its dual $C^q = C_q^* = Hom_R(C_q, R)$, and same for the differentials, reverting the original directions, and also keeping the vanishing of the composition of successive differential homomorphisms.

Example 139 A second example is the *de Rham cohomology* defined by the differential operator on differential forms on manifolds, as introduced in Section J.8.

There are several cohomology theories (defining chain complexes in different contexts). We shall only need a small selection here, see [635, 993] for a more complete discussion. Let us refer to the examples we have already discussed previously: Section 16.1.2, Section 19.1.1, and Appendix Section H.3.1.

Part XXII

Appendix: Complements in Physics

K

Complements on Physics

Summary. The appendix recalls some special topics in theoretical physics that are referred to in the main text. They are neither self-contained nor complete.

$$- \Sigma -$$

K.1 Hamilton's Variational Principle

Often physical laws are not deduced from mathematical results as theorems or corollaries, they are found without logical stringency from thought or real experiments in the framework already existing mathematical models of physical phenomena. This procedure is one of two possible approaches, as they have been described in [2, Section 19]. The second approach is more axiomatic, and it enables physicists to deduce physical laws as corollaries of metaphysical principles, i.e., of principles that are not directly present in physical reality, but which enable a deduction of physical laws. *Hamilton's variational principle* is such a metaphysical principle. It is applied in classical mechanics and in electrodynamics, but also in string theory (see Appendix K.2).

Hamilton's classical principle can be stated using the formalism of Lagrangian functions $L(t, x, \dot{x})$ of time t, space point x, and velocity $\dot{x} = dx/dt$. The typical Lagrangian for a classical point particle of mass m is $L = T - U$, where $T(\dot{x}) = \frac{m}{2}\dot{x}^2$ is the particle's kinetic energy[1] and $U(x)$ is a potential whose gradient defines forces that act upon the point particle.

The Lagrangian action is a functional of the $\mathcal{C}^1$ path function $x : t \mapsto x(t)$ of time t:

$$S(x) = \int_{t_0}^{t_1} L,$$

an integral over a time interval $[t_0, t_1]$, $t_0 < t_1$. Hamilton's principle uses the calculus of variations (see Appendix J.9) for the functional $J = S$ acting on paths $x : t \mapsto x(t)$. We consider all $\mathcal{C}^1$ paths $y : t \mapsto y(t)$ on $[t_0, t_1]$ which coincide with x on the endpoints t_0, t_1.

Principle 30 *Hamilton's action principle claims that a physical solution for such a path x must be a weak minimum of the functional S.*

In particular, the weak minimum condition can be applied to small variations $x + \varepsilon\eta$ of x defined by $\varepsilon\eta$, where η is a $\mathcal{C}^1$ path on $[t_0, t_1]$ and vanishes at the endpoints, and where ε is close to zero. The variational component $\varepsilon\eta$ is usually denoted by δx, whence the condition $S(s + \delta x) \geqslant S(x)$ in terms of variational calculus. But observe that this a local condition on x, it does not automatically imply that S is globally minimal. In particular, this means that for a fixed η, the function $\Sigma_\eta(\varepsilon) = S(x + \varepsilon\eta)$ has a local minimum

---

[1] The expression $\dot{x}^2$ means the scalar product $\dot{x} \cdot \dot{x}$, and we also have evident scalar products in the following products of vectorial functions.

© Springer International Publishing AG, part of Springer Nature 2017
G. Mazzola, *The Topos of Music IV: Roots*, Computational Music Science,
https://doi.org/10.1007/978-3-319-64495-0_11

for $\varepsilon = 0$. That implies $d\Sigma_\eta/d\varepsilon = 0$ for $\varepsilon = 0$. This is the basis for the Euler-Lagrange equation for Lagrange functions.

K.1.1 Euler-Lagrange Equations for a Non-relativistic Particle

We have

$$d\Sigma_\eta/d\varepsilon \Big|_{\varepsilon=0} = \int_{t_0}^{t_1} \frac{dL}{d\varepsilon} \Big|_{\varepsilon=0} = 0$$

and, with $y = x + \varepsilon\eta$ instead of x,

$$\frac{dL}{d\varepsilon} = \frac{\partial L}{\partial y}\frac{dy}{d\varepsilon} + \frac{\partial L}{\partial \dot{y}}\frac{d\dot{y}}{d\varepsilon}$$

as t does not vary with ε. But $\frac{dy}{d\varepsilon} = \eta$ and $\frac{d\dot{y}}{d\varepsilon} = \dot{\eta}$, whence

$$\frac{dL}{d\varepsilon} = \frac{\partial L}{\partial y}\eta + \frac{\partial L}{\partial \dot{y}}\dot{\eta}$$

Using integration by parts, this implies

$$0 = \int_{t_0}^{t_1} \frac{dL}{d\varepsilon} \Big|_{\varepsilon=0}$$
$$= \int_{t_0}^{t_1} \left(\frac{\partial L}{\partial x}\eta + \frac{\partial L}{\partial \dot{x}}\dot{\eta} \right)$$
$$= \int_{t_0}^{t_1} \eta \left(\frac{\partial L}{\partial x} - \frac{d}{dt}\frac{\partial L}{\partial \dot{x}} \right) + \frac{\partial L}{\partial \dot{x}}\eta \Big|_{t_1}^{t_0}$$

But the second summand vanishes because $\eta(t_0) = \eta(t_1) = 0$. The fundamental Lemma 115 of calculus of variations (applied to every component of the vectorial function η) implies the *Euler-Lagrange equation*

$$\frac{\partial L}{\partial x} - \frac{d}{dt}\frac{\partial L}{\partial \dot{x}} = 0.$$

Taking the classical Lagrange function $L(x, \dot{x}) = \frac{m}{2}\dot{x}^2 - U(x)$, we get

$$-\nabla U = m\ddot{x},$$

the classical second Newton law of mechanics, using the *nabla operator* $\nabla = (\frac{\partial}{\partial x_1}, \ldots \frac{\partial}{\partial x_n})$.

K.2 String Theory

In string theory, which is still (when this book is being written) a hypothetical theory despite its elegance and explanatory power, the elementary objects are not point particles but string particles. We have a corresponding Lagrange function and a Euler-Lagrange equation. As pointed out by Barton Zwiebach in [1159, p.74], already the non-relativistic Euler-Lagrange equation for a string shows the phenomenon of *D-branes*, i.e., the necessity of extended physical objects that are connected to a string and interact with its momentum, see also [440] for a short introduction to D-branes.

Because a string is determined by two spatial positions, horizontal x between $x = 0$ and $x = a$, and vertical y, a function of x and time t, its kinetic and potential energies are integrals that are derived from local data, the *Lagrange density* $\mathcal{L}$. Denote $y' = \frac{\partial y}{\partial x}$ and $\dot{y} = \frac{\partial y}{\partial t}$. Then

$$\mathcal{L}(y', \dot{y}) = \frac{1}{2}(\mu_0 y'^2 - T_0 \dot{y}^2),$$

where μ_0 is the string's mass density and T_0 the horizontal tension, both constant. With this structure, the Lagrangian $L = T - U$ at time t is

$$L(t) = \int_{x=0}^{x=a} \mathcal{L}$$

while the action is

$$S = \int_{t=t_0}^{t=t_1} L(t) = \int_{t=t_0}^{t=t_1} \int_{x=0}^{x=a} \mathcal{L}.$$

The variational calculus (similar to the calculation for point particles) yields a wave equation

$$0 = \mu_0 \frac{\partial^2 y}{\partial t^2} - T_0 \frac{\partial^2 y}{\partial x^2}$$

and two possible solutions for the boundary values of the string:

1. the *Neumann condition* $y'(t, 0) = y'(t, a) = 0$,
2. the *Dirichlet condition* $\dot{y}(t, 0) = \dot{y}(t, a) = 0$.

The first condition is what one gets when the boundary positions are free to move. The second condition fixes the y boundary positions over time, and this one is remarkable: It can be shown that the momentum of the string is conserved for the Neumann condition, but not necessarily for the Dirichlet condition. In this situation momentum can be exchanged with the points where the string is attached. This leads to the idea that such a string is connected to extended physical objects (one at $x = 0$, one at $x = a$) that may absorb some of the string's momentum. Accordingly, such objects are called *Dirichlet-branes* or *D-branes*.

While point particles, as they move in space-time, generate a *world-line*, strings create a *world-sheet*, see Figure K.1, that can be an open sheet (left in that figure) or a closed sheet, i.e., the trace of a closed string.

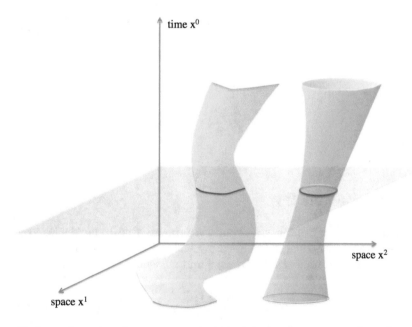

Fig. K.1. The world-sheet of a string, for open strings to the left, for closed strings (loops) to the right. The strings at a given time are the intersections of the world-sheet with the horizontal constant time hyperplane.

For the general relativistic theory of strings, the Lagrange action is defined using the surface of the world-sheet in terms of the Minkowski metric. The important fact here is to ensure that this surface is defined intrinsically, i.e., independent of the parametrization of the world-sheet. More precisely, the world-sheet is represented as a mapping $x : [a, b] \times [c, d] \to \mathbb{R}^4$ from a parameter rectangle $[a, b] \times [c, d]$ with arguments (ξ_1, ξ_2) (that could be normalized to I^2) to the physical space-time with the Minkowski metric. Using the classical scalar product[2] of two vectors, and the dot product in the Minkowski metric, the surface of the image of x is

$$A = \int \sqrt{\left(\frac{\partial x}{\partial \xi_1} \cdot \frac{\partial x}{\partial \xi_2} \right)^2 - \left(\frac{\partial x}{\partial \xi_1} \cdot \frac{\partial x}{\partial \xi_1} \right) \left(\frac{\partial x}{\partial \xi_2} \cdot \frac{\partial x}{\partial \xi_2} \right)}$$

This surface is independent of the parametrization. It can be expressed in a more compact way that also makes parametrization invariance more evident, using the matrix

$$g = \begin{pmatrix} \frac{\partial x}{\partial \xi_1} \cdot \frac{\partial x}{\partial \xi_1} & \frac{\partial x}{\partial \xi_1} \cdot \frac{\partial x}{\partial \xi_2} \\ \frac{\partial x}{\partial \xi_2} \cdot \frac{\partial x}{\partial \xi_1} & \frac{\partial x}{\partial \xi_2} \cdot \frac{\partial x}{\partial \xi_2} \end{pmatrix}$$

and then, we have

$$A = \int \sqrt{-det(g)}$$

which is transformed according to classical determinant rules and therefore is invariant under a change of parameters. Then the Lagrangian density is defined by

$$\mathcal{L}(\dot{x}, x') = -\frac{T_0}{c} \sqrt{-det(g)}$$

where $\dot{x}$ is the partial derivative with respect to the first parameter and x' is the partial derivative with respect to the second parameter. Reparametrizing the parameters to time τ and a length parameter σ, we get the *Nambu-Goto action* as

$$S = \int_{\tau_0}^{\tau_1} L = \int_{\tau_0}^{\tau_1} \int_{\sigma_0}^{\sigma_1} \mathcal{L},$$

where the Lagrangian is $L = \int_{\sigma_0}^{\sigma_1} \mathcal{L}$. The equations of motion, again deduced from Hamilton's principle, are as follows. Let

$$\mathcal{P}_\mu^\tau = \frac{\partial \mathcal{L}}{\dot{x}_\mu}, \mathcal{P}_\mu^\sigma = \frac{\partial \mathcal{L}}{x'_\mu}.$$

Then we have

$$\frac{\mathcal{P}_\mu^\tau}{\partial \tau} + \frac{\mathcal{P}_\mu^\sigma}{\partial \sigma} = 0.$$

for all μ.

K.3 Duality and Supersymmetry

In string theory, *strong-weak duality*, S-duality for short, is a duality that in particular exchanges strings and D-branes. Duality in theoretical physics is exemplified for Maxwell's equations for electric field E, magnetic field B, light speed constant c, and in absence of charge and current:

$$\nabla \cdot E = 0, \ \nabla \times B = \frac{1}{c} \frac{\partial E}{\partial t}$$

$$\nabla \cdot B = 0, \ \nabla \times E = \frac{1}{c} \frac{\partial B}{\partial t}.$$

---

[2] The negative sign is due to the Minkowski metric's specificity.

The exchange of variables $(E, B) \mapsto (-B, E)$ leaves Maxwell's equations invariant, i.e., electromagnetism is *self-dual*. It can be shown that type IIB string theory with coupling constant g is dual to the same theory with coupling constant $\frac{1}{g}$, same for type I string theory with coupling constant g, which is dual to SO(32) theory with coupling constant $\frac{1}{g}$. The inversion of the coupling constant, turning a strong coupling into a weak one, and vice versa, is the reason for the name "S-duality". This type of relationship among different string theories is one reason for Edward Witten's 1995 suggestion that the five known string theories might be special cases of a single limit string theory, coined M-theory, "M" for Matrix, Mystery, Magic [1136].

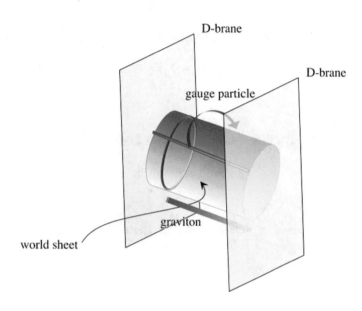

Fig. K.2. S-duality exchanging particles and D-branes.

Such a duality can exchange strings and D-branes, as shown in Figure K.2. Here a first interpretation is that an open gauge particle string, connecting two D-branes, by S-duality is interpreted as the exchange of a closed graviton string between two D-branes, this duality sending the fermonic gauge particle into a bosonic graviton, and reinterpreting D-branes as particles that exchange gravitational force. *Fermions* are those particles we usually associate with matter, e.g. electrons, protons, neutrons, quarks, whereas *bosons* are particles that transfer specific types of force, such as photons for the electromagnetic force, two W-bosons and one Z-boson for the weak force, gluons for the strong force, and (the still hypothetical) gravitons for gravitation; the Higgs boson (discovered in 2012) is involved in the unified electroweak force that unites electromagnetic and weak force types.

In theoretical physics, a theory that describes a symmetry between bosonic and fermionic particles is called supersymmetric. In string theory this approach is called a *superstring theory*. For this reason the above S-duality relates to superstring theory (but is of course not the same).

K.4 Quantum Mechanics

This part of the physical appendix refers to two books on mathematical methods in Quantum Mechanics (QM): Gerhard Teschl's well-written book [1034], and (much less) Leon A. Takhtajan's rather sketchy but also inspiring book [1033].

K.4.1 Banach and Hilbert Spaces

The mathematics of QM resides on Banach spaces and in particular Hilbert spaces. Banach spaces, i.e., complete topological $\mathbb{C}$-vector spaces with norms $\|\ \|$, were defined in Appendix Section I.1.2.

Example 140 A standard example is the space $\mathcal{C}^0[a,b]$ of continuous functions $f : [a,b] \to \mathbb{C}$ defined on the real closed interval $[a,b]$, the norm being

$$\|f\|_\infty = max_{x \in [a,b]} f(x).$$

Example 141 This Banach space $l^1(\mathbb{N})$ is the subspace of $\mathbb{C}^{\mathbb{N}}$ of all sequences $a. = (a_j)_j$ such that the norm

$$\|a.\|_1 = \sum_j |a_j|$$

is finite.

Example 142 The space $l^\infty(\mathbb{N})$ of all sequences in $\mathbb{C}^{\mathbb{N}}$ with finite norm

$$\|a.\|_\infty = sup_j |a_j|$$

is a Banach space.

Definition 217 *A Schauder basis of a normed $\mathbb{C}$-vector space X is a sequence $u. = (u_n)_{n \in \mathbb{N}}$ of linearly independent vectors such that every x is a limit $x = \lim_{n \to \infty} \sum_{m \leq n} c_n.u_n$. The space X is separable iff it contains a countable dense subset.*

Lemma 118 *A normed vector space is separable iff it has a Schauder basis.*

Theorem 103 (Weierstrass) *The subset of polynomial functions of $\mathcal{C}^0[a,b]$ is dense.*

Theorem 104 *$\mathcal{C}^0[a,b]$ with the maximum norm is a separable Banach space.*

While $\mathcal{C}^0[a,b]$ is separable, $l^\infty(\mathbb{N})$ is not.

A Hilbert space is a complex vector space H together with a *sesquilinear* form (usually called *scalar product*) $\langle , \rangle : H \times H \to \mathbb{C}$ which by definition means:

1. $\langle x, y \rangle$ it is linear in the right argument and conjugate linear in the left argument, i.e., $\langle x, \lambda y \rangle = \langle x, y \rangle \lambda$ while $\langle \lambda x, y \rangle = \overline{\lambda} \langle x, y \rangle$,
2. it is *positive definite*, i.e., $\langle x, x \rangle \geq 0$ and $\langle x, x \rangle = 0$ iff $x = 0$,
3. we have *symmetry*: $\langle x, y \rangle = \overline{\langle y, x \rangle}$.

A *Hilbert space* is a Banach space whose norm is defined by the sesquilinear form via $\|x\| = \sqrt{\langle x, x \rangle}$. A vector space with an inner product is called an *inner product space*.

Example 143 The finite dimensional space $\mathbb{C}^n$ is a Hilbert space with the usual scalar product

$$\langle x, y \rangle = \sum_j^n \overline{x_j} y_j.$$

Example 144 The space $l^2(\mathbb{N})$ of sequences $a. \in \mathbb{C}^{\mathbb{N}}$ with $\sum_j |a_j|^2 < \infty$ is a separable Hilbert space with scalar product

$$\langle x, y \rangle = \sum_j^\infty \overline{x_j} y_j.$$

In an inner product space H, a vector with $\|x\| = 1$ is called *normalized* or *unit* vector. If $\langle x, y \rangle = 0$ one writes $x \perp y$ and calls them *orthogonal*, and if x is a multiple of y or vice versa, they are called *parallel*.

Proposition 138 (Cauchy-Schwarz-Bunjakowski) *For two vectors x, y in an inner product space we have*

$$|\langle x, y \rangle| \leq \|x\| \|y\|$$

with equality iff x and y are parallel.

The relationship between norms and scalar products is described by

Proposition 139 (Jordan-von Neumann) *A norm is associate with a scalar product iff the* parallelogram law

$$\|x + y\|^2 + \|x - y\|^2 = 2\|x\|^2 + \|y\|^2$$

holds. In that case the scalar product is given by

$$\langle x, y \rangle = \frac{1}{4}(\|x + y\|^2 - \|x - y\|^2 + i\|x - iy\|^2 - i\|x + iy\|^2).$$

Example 145 On $\mathcal{C}^0[a, b]$, there is a scalar product

$$\langle x, y \rangle = \int_a^b \overline{x(t)} y(t),$$

defining the inner product space $\mathcal{L}^2_{cont}([a, b])$.

If we are given two norms $\| \|_1, \| \|_2$ on one space, $\| \|_2$ is called *stronger than* $\| \|_1$ iff there is a positive real constant s such that $\|x\|_1 \leq s\|x\|_2$. Then since $\|x\| \leq \sqrt{|b - a|}\|x\|_\infty$, the maximum norm is stronger than the $\mathcal{L}^2_{cont}$ norm.

Lemma 119 *If, for a normed space X, $\| \|_2$ is stronger than $\| \|_1$, than a Cauchy sequence for $\| \|_2$ is also a Cauchy sequence for $\| \|_1$, and if a set is dense in $(X, \| \|_2)$ it is also dense in $(X, \| \|_1)$.*

Therefore $\mathcal{L}^2_{cont}$ is separable, but not complete.

Proposition 140 *For a finite dimensional space, all norms are equivalent, i.e., mutually stronger. In particular, they define the same complete topology.*

Remark 36 *Normed spaces that are not complete can be turned into complete vector spaces by the standard process from calculus: Take the space of all Cauchy sequences and divide it by the subspace of zero convergent sequences.*

K.4.1.1 Bounded Operators

Definition 218 *A* linear operator A *between two normed spaces X, Y is a linear map $A : D(A) \to Y$, where $D(A)$, the domain of A isa dense linear subspace of X. We set $Ker(A)$ and $Rank(A)$ to be the kernel and image of the map $A : D(A) \to Y$. The operator A is* bounded *iff the operator norm*

$$\|A\| = sup_{\|x\|_X = = 1} \|A(x)\|_Y$$

is finite. The set of bounded linear operators from X to Y is denoted by $\mathcal{L}(X, Y)$; we write $\mathcal{L}(X, X) = \mathcal{L}(X)$.

Theorem 105 *The space $\mathcal{L}(X, Y)$ together with the operator norm is a normed space that is Banach if Y is so. An operator A is bounded iff it is continuous.*

Here is the extension theorem:

Theorem 106 (Bounded Linear Transformation Theorem) *For $A \in \mathcal{L}(X, Y)$, if Y is Banach space, then there is a unique continuous extension of the domain $D(A)$ to all of X which has the same norm as A.*

Therefore it is no restriction to assume that an operator is defined on all of X.

The special space $X^* = \mathcal{L}(X, \mathbb{C})$ is called the *dual space of X*, and an operator in X^* is called a *bounded linear functional*.

The space $\mathcal{L}(X)$ has also the structure of a (unitary) $\mathbb{C}$-algebra defined by composition of operators. Moreover, one has $\|AB\| \leqslant \|A\|\|B\|$. A Banach space with such an algebra structure is called a *Banach algebra*.

K.4.1.2 Lebesque Integration

Definition 219 *A set $\Sigma \subset 2^X$ (X any set) is a σ-algebra of X if*

1. *$X \in \Sigma$,*
2. *Σ is closed under countable unions,*
3. *Σ is closed under complements.*

Therefore $\varnothing \in \Sigma$ and Σ is closed under countable intersections. Since the intersection of any non-empty family of σ-algebras of X is a σ-algebra, one can define the σ-algebra $\Sigma(S)$ generated by a subset $S \in 2^X$, which is the intersection of all σ-algebras containing S. Elements of Σ are called *measurable*.

Example 146 If X is a topological space, the *Borel σ-algebra* is the σ-algebra generated by the open sets. Its elements are called *Borel sets*. For $X = \mathbb{R}^n$, the Borel σ-algebra is denoted by $\mathcal{B}^n$, for $n = 1$ simply by $\mathcal{B}$.

Definition 220 *A measure μ is a map $\mu : \Sigma \to [0, \infty]$ on a σ-algebra Σ such that*

1. *$\mu(\varnothing) = 0$,*
2. *σ-addtitivity: $\mu(\bigcup_{j=1}^{\infty} A_j) = \sum_{j=1}^{\infty} \mu(A_j)$ if The A_j are disjoint from each other.*
3. *μ is called σ-finite if there is a countable cover of X by measurable sets of Σ whose elements have finite measure. It is called finite if $\mu(X)$ is finite.*
4. *The pair (Σ, μ) is called a measure space (observe that X is determined by Σ).*
5. *A measure on the Borel σ-algebra is called a Borel measure if the measure of any quasi-compact set is finite.*

Given a σ-algebra Σ on X, a function $f : X \to \mathbb{R}^n$ is called *measurable* if $f^{-1}(B) \in \Sigma$ for every Borel set $B \in \mathcal{B}^n$. Measurable functions defined on the Borel σ-algebra of a topological space X are called *Borel functions*.

Lemma 120 *A function $f : X \to \mathbb{R}^n$ is measurable iff each projection $f_i : X \to \mathbb{R}$ is so.*

Lemma 121 *Borel functions are continuous. If $f : X \to Y$ with $Y \subset \mathbb{R}^n$ and $g : Y \to \mathbb{R}^m$ are Borel, then so is $g \circ f$. Let X be a topologic space and Σ its Borel σ-algebra. If $f, g : X \to \mathbb{R}$ are measurable functions, then so is $f + g$ and $f \cdot g$.*

Definition 221 *A measurable function $s : X \to \mathbb{R}$ is called* simple *if its image is finite. If its values are $\alpha_1, \ldots a_p$, using characteristic functions, we have $s = \sum_{i=1}^{p} \alpha_i \chi(A_i)$, $A_i = s^{-1}(\alpha_i) \in \Sigma$.*

Definition 222 *With the above notations, for a non-negative simple function s and a measurable set A, we define its integral by*

$$\int_A s \, d\mu = \sum_{i=1}^{p} \alpha_i \mu(A_i \cap A)$$

with the convention $0 \cdot \infty = 0$.

Sorite 18 *With the above notations, the integral has these basic properties:*

(i) $\int_A s d\mu = \int_X \chi(A) s d\mu$.
(ii) $\int_{\bigcup_{i=1}^\infty A_j} s d\mu = \sum_j \int_{A_j} s d\mu$ *for pairwise disjoint* A_j.
(iii) $\int_A \alpha s d\mu = \alpha \int_A s d\mu$ *for* $\alpha \geqslant 0$.
(iv) $\int_A (s + t) d\mu = \int_A s d\mu + \int_A t d\mu$.
(v) $A \subset B$ *implies* $\int_A s d\mu \leqslant \int_B s d\mu$.
(vi) $s \leqslant t$ *implies* $\int_A s d\mu \leqslant \int_A t d\mu$.

Definition 223 *IF f is any positive function on X, one defines*

$$\int_A f d\mu = \sup_{s \leqslant f} \int_A s d\mu,$$

the supremum being taken over all simple functions $s \leqslant f$. If both, the positive part f^+ and the negative part f^- of a measurable function f (defined by $f = f^+ - f^-$) are such that their integrals are finite, we define

$$\int_A f d\mu = \int_A f^+ d\mu - \int_A f^- d\mu$$

and call f integrable. The set of integrable functions is denoted by $\mathcal{L}^1(X, d\mu)$.

Then Sorite 18 holds for integrable functions. Many standard theorems for Riemann integrals are also valid for Lebegue integrals, e.g. change of variables, Fubini, but see [1034, A.4] for examples.

K.4.1.3 Lebesgue L^p Spaces

Given a σ-finite measure space (Σ, μ), we denote for $p \geqslant 1$ by $\mathcal{L}^p(X, d\mu)$ the $\mathbb{C}$-linear space of $\mathbb{C}$-valued measurable functions with finite norm-like expression

$$\|f\|_p = \left(\int_X |f|^p d\mu \right)^{1/p}$$

In order to construct a Banach space from this data, one needs a result concerning vanishing of the norm:

Lemma 122 *Let f her measurable, then*

$$\int_X |f|^p d\mu = 0$$

iff $f(x) = 0$ almost everywhere with respect to μ, i.e., if the null set of f has measure zero.

Denote by $\mathcal{N}(X, d\mu)$ the set of measurable functions which are zero almost everywhere on X. This is a $\mathbb{C}$-linear subspace of $\mathcal{L}^p(X, d\mu)$, and we consider the quotient

$$L^p(X, d\mu) = \mathcal{L}^p(X, d\mu) / \mathcal{N}(X, d\mu)$$

which we also write as $L^P(X)$ if $X \subset \mathbb{R}^n$. Obviously, we now have a well-defined norm $\|f\|_p$ on equivalence classes of functions.

Theorem 107 *$L^p(X, d\mu)$ is a Banach space. If X has a countable basis (second countable) and μ is a regular Borel measure, then $L^p(X, d\mu)$ is separable.*

Regular Borel measures are defined in [1034, p.261].

K.4.2 Geometry on Hilbert Spaces

The phase spaces in QM are always Hilbert spaces, and their geometry is essential. We therefore discuss geometry on Hilbert spaces.

Example 147 The typical Hilbert space is $\mathcal{H} = L^2(X, d\mu)$ with the scalar product

$$\langle f, g \rangle = \int_X \overline{f(x)} g(x) d\mu$$

that induces the norm $\|\|_2$ discussed in Section K.4.1.3.

We already introduced unit or normalized vectors $\psi \in \mathcal{H}$, parallel vectors, and orthogonality in Section K.4.1. Given a unit vector ϕ, one defines two associated vectors to ψ, namely (1) the *ϕ-projection of ψ*: $\psi_\| = \langle \phi, \psi \rangle \phi$, and (2) the *associated orthogonal projection of ψ*: $\psi_\perp = \psi - \psi_\|$, whence $\psi = \psi_\| + \psi_\perp$.

Definition 224 *An orthonormal family of vectors is a family $(\phi_i)_i$ such that for all j, $\langle \phi_i, \phi_j \rangle = \delta_{ij}$, in particular, all are units.*

Lemma 123 *Given an orthonormal family $(\phi_1 \ldots \phi_n)$, for a vector $\psi \in \mathcal{H}$, we set $\psi_\| = \sum_i \psi_{\|,\phi_i}$ and $\psi_\perp = \psi - \psi_\|$. Then $\psi_\| \perp \psi_\perp$ and $\phi_i \perp \psi_\perp$. We have* Bessel's inequality

$$\|\psi\|^2 \geq \sum_{i=1}^n |\langle \phi_i, \psi \rangle|^2,$$

equality holding iff $\psi \in \mathbb{C}\langle \phi_1 \ldots \phi_n \rangle$. We have this formula for norms:

$$\|\psi\|^2 = \sum_i |\langle \phi_i, \psi \rangle|^2 + \|\psi_\perp\|^2.$$

The orthogonal projection is a well-known extremal solution as follows: If $\psi^* \in \mathbb{C}\langle \phi_1 \ldots \phi_n \rangle$, then $\|\psi - \psi^*\| \geq \|\psi_\perp\|$, equality holding iff $\psi^* = \psi_\|$.

Bessel's inequality helps generalize these results to general orthonormal families:

Lemma 124 *Given an orthonormal family $(\phi_i)_i$, for a vector $\psi \in \mathcal{H}$, we set $\psi = \psi_\| + \psi_\perp$ with $\psi_\| = \sum_i \langle \phi_i, \psi \rangle \phi_i$, and we have $\psi_\| \perp \psi_\perp$. Also $\phi_i \perp \psi_\perp$. This implies (generalizing the same formula in Lemma 123)*

$$\|\psi\|^2 = \sum_i |\langle \phi_i, \psi \rangle|^2 + \|\psi_\perp\|^2.$$

Further, if $\psi^* \in \mathbb{C}\langle (\phi_i)_i \rangle$, then $\|\psi - \psi^*\| \geq \|\psi_\perp\|$, equality holding iff $\psi^* = \psi_\|$.

Definition 225 *Let $(\phi_i)_i$ be an orthonormal family of vectors in $\mathcal{H}$. Then they are an* orthonormal basis *of $\mathcal{H}$ if anyone of the following equivalent conditions holds.*

(i) *$(\phi_i)_i$ is a maximal orthonormal family.*
(ii) *For all $\psi \in \mathcal{H}$, we have $\psi = \sum_i \langle \phi_i, \psi \rangle \phi_i$.*
(iii) *For all $\psi \in \mathcal{H}$, we have $\|\psi\|^2 = \sum_i |\langle \phi_i, \psi \rangle|^2$.*
(iv) *If $\langle \phi_i, \psi \rangle = 0$ for all i, then $\psi = 0$.*

Example 148 A classical example is the orthornormal basis $\phi_n(x) = \frac{1}{\sqrt{2\pi}} e^{ins}$, $n \in \mathbb{Z}$ of $\mathcal{H} = L^2(0, 2\pi)$. The orthonormal series is the ordinary Fourier series.

Theorem 108 *A Hilbert space $\mathcal{H}$ is separable iff it has a countable orthonormal basis (then all orthonormal bases are countable[3]).*

Definition 226 *A linear isomorphism $f : \mathcal{H}_1 \to \mathcal{H}_2$ is called* unitary *iff it conserves the scalar products.*

Theorem 109 *Every infinite-dimensional separable Hilbert Space $\mathcal{H}$ is unitary isomorphic to $l^2(\mathbb{N})$. An isomorphism can be defined as follows. Take an orthonormal basis $(\phi_i)_i$ of $\mathcal{H}$ and define $f : \mathcal{H} \overset{\sim}{\to} l^2(\mathbb{N})$ by $f(\psi) = (\langle \phi_i, \psi \rangle)_i$*

If $M \subset \mathcal{H}$ is a subset, its *orthogonal complement* is $M^\perp = \{x | x \perp M\}$. It is a closed linear subspace. In particular, $0^\perp = \mathcal{H}$ by Definition 225.

Theorem 110 *Let M be a closed linear subspace of Hilbert space $\mathcal{H}$. Then we have a direct decomposition*

$$\mathcal{H} = M \oplus M^\perp$$

which means that every vector ψ has the unique representation $\psi = \psi_\| + \psi_\perp, \psi_\| \in M, \psi_\perp \in M^\perp$. Also $M^{\perp\perp} = M$.

Evidently, we then have the *orthogonal projection* $p_M : \mathcal{H} \to \mathcal{H} : \psi \mapsto \psi_\|$ which is idempotent and verifies $\langle p_M \psi, \phi \rangle = \langle \psi_\|, \phi_\| \rangle = \langle \psi, p_M \phi \rangle$. The other projection $p_{M^\perp} : \mathcal{H} \to \mathcal{H} : \psi \mapsto \psi_\perp = \psi - p_M \psi$ is also idempotent and verifies $\langle p_{M^\perp} \psi, \phi \rangle = \langle \psi_\perp, \phi_\perp \rangle = \langle \psi, p_{M^\perp} \phi \rangle$. If $P_M \neq 0$ then $\|P_M\| = 1$.

Lemma 125 (Riesz Lemma) *For every bounded linear functional $l : \mathcal{H} \to \mathbb{C}$ there is a unique vector $\lambda \in \mathcal{H}$ with*

$$l(\psi) = \langle \psi, \lambda \rangle$$

This means that the map $\mathcal{H} \to \mathcal{H}^* : \lambda \mapsto \langle ?, \lambda \rangle$ is a linear isomorphism.

Corollary 41 *Suppose that s is a bounded sesquilinear form, i.e., $|s(\psi, \phi)| \leqslant C\|\psi\|\|\phi\|$, then there is a unique bounded operator B such that $s(\psi, \phi) = \langle B\psi, \phi \rangle$, moreover $\|B\| \leqslant C$. In particular, if $A \in \mathcal{L}(\mathcal{H})$, then there is the* adjoint operator $A^* \in \mathcal{L}(\mathcal{H})$ *such that $\langle \phi, A^* \psi \rangle = \langle A\phi, \psi \rangle$.*

Lemma 126 *Let $A, B \in \mathcal{L}(\mathcal{H})$. Then*

(i) (conjugate linearity) $(A + B)^* = A^* + B^*$ and $(\lambda A)^* = \bar{\lambda} A^*$,
(ii) (involution) $A^{**} = A$,
(iii) (antimultiplicative) $(AB)^* = B^* A^*$,
(iv) $\|A\| = \|A^*\|$ and $\|A\|^2 = \|A^* A\| = \|AA^*\|$.

Definition 227 *If $\mathcal{H}_1, \langle, \rangle_1$ and $\mathcal{H}_2, \langle, \rangle_2$ are two Hilbert spaces, their* orthogonal direct sum $\mathcal{H}_1 \oplus^\perp \mathcal{H}_2$ *is the algebraic direct sum with the orthogonal sum of their scalar products*

$$\langle (\phi_1, \phi_2), (\psi_1, \psi_2) \rangle_{1 \oplus 2} = \langle \phi_1, \psi_1 \rangle_1 + \langle \phi_2, \psi_2 \rangle_2.$$

The direct sum $\bigoplus_i^\perp \mathcal{H}_i$ of a countable family $(\mathcal{H}_i)_i$ of Hilbert spaces is the subspace of the algebraic direct sum consisting of the vectors (ψ_i) with $\sum_i \|\psi_i\|^2 < \infty$ and the orthogonal sum of the scalar products of the factors.

Example 149 *We have $\bigoplus_i \mathbb{C} = l^2(\mathbb{N})$.*

---

[3] This is automatic for linear bases, but here we deal with Schauder bases.

Definition 228 *If* $\mathcal{H}_1, \langle, \rangle_1$ *and* $\mathcal{H}_2, \langle, \rangle_2$ *are two Hilbert spaces, their* (orthogonal) *tensor product* $\mathcal{H}_1 \otimes^\perp \mathcal{H}_2$ *is the completion of the algebraic tensor product for the norm that is induced by the scalar product*

$$\langle \phi_1 \otimes \phi_2, \psi_1 \otimes \psi_2 \rangle_{1 \otimes 2} = \langle \phi_1, \psi_1 \rangle_1 \cdot \langle \phi_2, \psi_2 \rangle_2.$$

If $(\phi_i^1)_i), (\phi_i^2)_i)$ *are orthonormal bases for* $\mathcal{H}_1, \mathcal{H}_2$, *respectively, then the tensor family* $(\phi_i^1 \otimes \phi_j^2)_{i,j}$ *is an orthonormal basis of* $\mathcal{H}_1 \otimes^\perp \mathcal{H}_2$.

Example 150 We have $\mathcal{H} \otimes \mathbb{C}^n = \mathcal{H}^n$.

K.4.2.1 The C^*-Algebra of Bounded Linear Operators

The properties (i) to (iv) in Lemma 126 turn $\mathcal{L}(\mathcal{H})$ into a C^*-*algebra*. A subalgebra that is closed under the * involution is called a *-*subalgebra*. An *ideal* in such an algebra is an ordinary two-sided ideal. It is a 8-ideal if it is stable under the adjunction map. An element a in a C^*-algebra is called *normal* if $aa^* = a^*a$ (it commutes with its adjoint), *self-adjoint* if $a = a^*$, *unitary* if it is normal and $aa^* = Id$, an *(orthogonal) projection* if it is self-adjoint and idempotent, and *positive* if $a = bb^*$ for some b.

Lemma 127 *If* $A \in \mathcal{L}(\mathcal{H})$, *then* A *is normal iff* $\|A\psi\| = \|A^*\psi\|$ *for all* ψ.

Lemma 128 *A transformation* $T : \mathcal{H} \to \mathcal{H}$ *is unitary iff* $T^{-1} = T^*$.

K.4.3 Axioms for Quantum Mechanics

Here is the axiomatic setup of QM. To define a quantum system we are given a Hilbert space $\mathcal{H}$ which is called *phase space* or *configuration space*. The *states* of the quantum system are the unit vectors $\psi \in \mathcal{H}$ (the elements of the unit sphere $S^1 \subset \mathcal{H}$). An *observable* is a self-adjoint linear operator A on $\mathcal{H}$.

The expectation of an observable A in state ψ is defined by the quadratic form value, i.e.,

$$E_\psi(A) = \langle \psi, A\psi \rangle.$$

Example 151 Here is an example that illustrates this definition. It is the single particle in $\mathbb{R}^4 = \mathbb{R}^3 \times \mathbb{R}$, i.e., three-space and time. It is described by a $\mathbb{C}$-valued *wave function* $\psi : \mathbb{R}^4 \to \mathbb{C}$. The square absolute value $|\psi(x,t)|^2 = \overline{\psi(x,t)}\psi(x,t)$ defines the *probability density* of the particle at (x,t). This quantity must be defined such that the space integral of density at any time t is one (the particle must for sure be somewhere at any time):

$$\int_{\mathbb{R}^3} |\psi(x,t)|^2 = 1.$$

Here, the observable corresponds to the location x of the particle. This location is however not a point but an extended spatial area $\Omega \subset \mathbb{R}^3$, such as the interior of a detector. We may then want to calculate the probability for the particle to live in Ω at time t. This probability is the integral of the probability density over Ω:

$$E_\psi(\chi(\Omega)) = \int_\Omega |\psi(x,t)|^2.$$

But this expression can be rephrased by

$$\int_\Omega |\psi(x,t)|^2 = \int_{\mathbb{R}^3} \overline{\psi(x,t)} \chi(\Omega) \psi(x,t)$$

which is the above quadratic form expression for the scalar product $\langle \phi, \psi \rangle = \int_{\mathbb{R}^3} \overline{\phi}\psi$ and the operator $A(\psi) = \chi(\Omega)\psi$.

The time evolution of a system is given by a *one parameter unitary group*, which means the following: It is a one-parameter family $(U(t))_{t \geqslant 0}$ of linear unitary operators $U(t) : \mathcal{H} \to \mathcal{H}$ such that $U(t) \circ U(s) = U(t+s)$, in particular $U(0) = Id$, and such that the group is *strongly continuous*, i.e., $\lim_{t \to t_0} U(t)\psi = U(t_0)\psi$.

The group is given a *Hamilton operator*

$$H\psi = \lim_{t \to 0} \frac{i}{t}(U(t)\psi - \psi),$$

and the domain $\mathcal{D}(H)$ is the subset of $\mathcal{H}$ where $H\psi$ exists. If $\psi \in \mathcal{D}(H)$, then we have the *Schrödinger equation*

$$i \frac{d}{dt}\psi(t) = H\psi(t).$$

These structures define the axioms of QM: (1) The configuration space is a separable Hilbert space $\mathcal{H}$, the system states are the unit vectors on the unit sphere in $\mathcal{H}$. (2) An observable corresponds to a linear selfadjoint operator. (3) The expectation value for a measurement of an observable A in state ψ is the real valued quadratic form $\langle A\psi, \psi \rangle$. (4) the time evolution of the system, starting at state ψ, is given by a strongly continuous one-parameter unitary group $(U(t))$. The Hamilton operator $H\psi$ corresponds to the *energy of the system at state ψ*.

K.4.3.1 Resolvents and Spectra

Eigenvectors and -values play an important role in QM. For infinite-dimensional spaces, these concepts needs to be defined in a generalized way compared to the finite-dimensional situation.

Definition 229 *The* resolvent set *of a selfadjoint operator A on $\mathcal{H}$ is defined as $\rho(A) = \{z \in \mathbb{C} | (A - z)^{-1} \in \mathcal{L}(\mathcal{H})\}$. The function $R_A : \rho(A) \to \mathcal{L}(\mathcal{H}) : z \mapsto (A - z)^{-1}$ is called the* resolvent *of A.*

The complement $\sigma(A) = \mathbb{C} - \rho(A)$ is called the *spectrum of A*. If $z \in \sigma(A)$, this means that there are non-zero vectors ψ with $A\psi = z\psi$, z is called an *eigenvalue* and ψ an *eigenvector for z*.

Lemma 129 *For a selfadjoint operator all eigenvalues are real, and eigenvectors to different eigenvalues are orthogonal.*

Theorem 111 *Let U be unitary, then $\sigma(U)$ consists of unit vectors (elements of the unit circles $S^1 \subset \mathbb{C}$). Eigenvectors for different eigenvalues are orthogonal.*

K.4.4 The Spectral Theorem

The spectral theorem relates to the fact that a measurement of an observable A results in an eigenvalue of A (which is a real number according to Lemma 129).

K.4.4.1 Projection-valued Measures

Definition 230 *Let $\mathcal{B}$ be the Borel σ-algebra of $\mathbb{R}$. A* projection-valued measure *is a function $P : \mathcal{B} \to \mathcal{L}(\mathcal{H})$ to the set of orthogonal projections on $\mathcal{H}$ (recall from Section K.4.2.1: such a projection p is idempotent and self-adjoint) with the following conditions:*

(i) $P(\mathbb{R}) = Id$.
(ii) If $\Omega = \bigcup_n \Omega_n$ with $\Omega_n \cap \Omega_m = \emptyset$ for $m \neq n$, then $\sum_n P(\Omega_n)\psi = P(\Omega)\psi$ for all ψ.

Example 152 —em *Let* $\mathcal{H} = \mathbb{C}^n$ *and let* $A \in GL(n)$ *be a symmetric matrix. If* $\lambda_1, \dots \lambda_m$ *are its different eigenvalues, denote by* $P_1, \dots P_m$ *the orthogonal projections onto the corresponding eigenspaces. Then*

$$P_A(\Omega) = \sum_{j|\lambda_j \in \Omega} P_j$$

is a projection valued measure.

Sorite 19 *If P is a projection-valued measure, then*

(i) $P(\varnothing) = 0$,
(ii) $P(\mathbb{R} - \Omega) = Id - P(\Omega)$,
(iii) $P(\Omega_1 \cup \Omega_2) + P(\Omega_1 \cap \Omega_2) = P(\Omega_1) + P(\Omega_2)$,
(iv) $P(\Omega_1)P(\Omega_2) = P(\Omega_1 \cap \Omega_2)$,
(v) *if* $\Omega_1 \subset \Omega_2$, *then* $P(\Omega_1) \leqslant P(\Omega_2)$.

In particular, one can define a *resolution of the identity* for every projection-valued measure P, namely $P(\lambda) = P(]-\infty, \lambda])$, $\lambda \in \mathbb{R}$. This enables us to extend the concept of Lebegue integration to projection-valued measures. For a simple function $f = \sum_{i=1}^{n} \alpha_i \chi(\Omega_j)$ with the usual $\Omega_j = f^{-1}(\alpha_j)$, one sets

$$P(f) = \int_{\mathbb{R}} f(\lambda) dP(\lambda) = \sum_{i=1}^{n} \alpha_i P(\Omega_j),$$

which defines an operator instead of a complex number. For two state vectors ϕ, ψ, one then has a complex Borel measure $\mu_{\phi, \psi}(\Omega) = \langle \phi, P(\Omega) \rangle$, and this relates our operator-valued integration to the complex-valued one by the formula

$$\langle \phi, P(f)\psi \rangle = \int_{\mathbb{R}} f(\lambda) d\mu_{\phi, \psi}(\lambda).$$

The definition of this operator-valued integral can be extended to the Banach space of bounded Borel functions $B(\mathbb{R})$ and yields a function

$$P : B(\mathbb{R}) \to \mathcal{L}(\mathcal{H}) : f \mapsto \int_{\mathbb{R}} f(\lambda) dP(\lambda)$$

and the above equation relating operator-valued integration to the complex-valued one remains valid.

Theorem 112 *Given a projection-valued measure P on* $\mathcal{H}$, *the transformation*

$$P : B(\mathbb{R}) \to \mathcal{L}(\mathcal{H}) : f \mapsto \int_{\mathbb{R}} f(\lambda) dP(\lambda)$$

is a C^* *algebra homomorphism with norm one such that*

$$\langle P(g)\phi, P(f)\psi \rangle = \int_{\mathbb{R}} g^*(\lambda) f(\lambda) d\mu_{\phi, \psi}(\lambda).$$

The following *Spectral Theorem* (John von Neumann) decomposes any self-adjoint operator A into its "projection spectrum".

Theorem 113 (Spectral Theorem) *For every self-adjoint operator A there is a unique projection-valued measure* P_A *such that*

$$A = \int_{\mathbb{R}} \lambda P_A(\lambda).$$

This enables in particular the calculation of A's spectrum:

Proposition 141 *The spectrum of operator A is described by*

$$\sigma(A) = \{\lambda \in \mathbb{R} | P_A(\ ]\lambda - \varepsilon, \lambda + \varepsilon[\ ) \neq 0 \text{ for all } \varepsilon > 0\}.$$

Part XXIII

Appendix: Tables

L

Euler's Gradus Function

This table lists the rational numbers x/y with Euler's gradus suavitatis $\Gamma(x/y) \leqslant 10$, see also [161].

Γ	Intervals
2	1/2
3	1/3, 1/4
4	1/6,2/3,1/8
5	1/5,1/9,1/12,3/4,1/16
6	1/10,2/5,1/18,2/9,1/24,3/8,1/32
7	1/7,1/15,3/5,1/20,4/5,1/27,1/36,4/9,1/48,3/16,1/64
8	1/14,2/7,1/30,2/15,3/10,5/6,1/40,5/8,1/54,2/27,1/72,8/9,1/96,3/32,1/128
9	1/21,3/7,1/25,1/28,4/7,1/45,5/9,1/60,3/20,4/15,5/12,1/80,5/16,1/81,1/108, 4/27,1/144,9/16,1/192,3/64,1/256
10	1/42,2/21,3/14,6/7,1/50,2/25,1/56,7/8,1/90,2/45,5/18,9/10,1/120,3/40,5/24, 8/15,1/160,5/32,1/162,2/81,1/216,8/27,1/288,9/32,1/384,3/128,1/512

© Springer International Publishing AG, part of Springer Nature 2017
G. Mazzola, *The Topos of Music IV: Roots*, Computational Music Science,
https://doi.org/10.1007/978-3-319-64495-0_12

M

Just and Well-Tempered Tuning

This table lists the just coordinates of the just tuning intervals (with respect to c, second tone in first column) according to Vogel [1089], see Section 7.2.1.4, together with the value in Cents, and the deviation in % from the tempered tuning with 100, 200, 300, etc. Cents.

Tone name	Frequency ratio	Octave coord.	Fifth coord.	Third coord.	Pitch (Ct)	% deviation
c	1	0	0	0	0	0
$d_\flat$	16/15	4	-1	-1	111.73	+11.73
d	9/8	-3	2	0	203.91	+1.96
$e_\flat$	6/5	1	1	-1	315.65	+5.22
e	5/4	-2	0	1	386.31	-3.42
f	4/3	2	-1	0	498.05	-0.39
$f_\sharp$	45/32	-5	2	1	590.22	-1.63
g	3/2	-1	1	0	701.96	+0.28
$a_\flat$	8/5	3	0	-1	813.69	+1.71
a	5/3	0	-1	1	884.36	-1.74
$b_\flat$	16/9	4	-2	0	996.09	-0.39
b	15/8	-3	1	1	1088.27	-1.07

© Springer International Publishing AG, part of Springer Nature 2017
G. Mazzola, *The Topos of Music IV: Roots*, Computational Music Science,
https://doi.org/10.1007/978-3-319-64495-0_13

N

Chord and Third Chain Classes

N.1 Chord Classes

This section contains the list of all isomorphism classes of zero-addressed chords in $PiMod_{12}$. The meanings of the column items are explained in Section 11.3.7; here we give a short definition.

- *Class Nr.* is the number of the isomorphism class, numbers with extension ".1" indicate the class number for classification under symmetries from $\mathbb{Z}$ (no fifth or fourth transformations). Autocomplementary classes have a star after the number.
- *Representative* of Nr. without hat is the number's representative in full circles, the one with hat is the complementary chord.
- *Group of symmetries* is $Sym(Nr.)$. To keep notation readable, we use the notation with linear factor to the left.
- *Conj. Class* denotes the conjugacy class symbol of $Sym(Nr.)$ and refers to the numbering $1, 2, \ldots 19$ from [799].
- *Card. End. Cl. $Nr.|\widehat{Nr.}$* is the pair of numbers of conjugacy classes of endomorphisms in $Nr.$ and in its complement $\widehat{Nr.}$, respectively.

© Springer International Publishing AG, part of Springer Nature 2017
G. Mazzola, *The Topos of Music IV: Roots*, Computational Music Science,
https://doi.org/10.1007/978-3-319-64495-0_14

Chord Classes

Class Nr.	Representative Nr. = ●, $\widehat{Nr.}$ = ○	Group of Symmetries	Conj. Class	♯ End. Nr.|$\widehat{Nr.}$
1	●●●●●●●●●●●●	$\overrightarrow{GL}(\mathbb{Z}_{12})$	19	28|28
One/Eleven Element				
2	●○○○○○○○○○○○	$\mathbb{Z}_{12}^{\times}$	8	1|31
Two/Ten Elements				
3	●●○○○○○○○○○○	$\langle -1e^{-1}\rangle$	3	3|23
3.1	●○○○○●○○○○○○			
4	●○●○○○○○○○○○	$\{1,7,-1e^{-2},5e^{-2}\}$	8	3|25
5	●○○●○○○○○○○○	$\{1,5,7e^{-3},-1e^{-3}\}$	8	3|19
6	●○○○●○○○○○○○	$\{1,7,5e^8,-1e^8\}$	8	3|31
7	●○○○○○●○○○○○	$\mathbb{Z}_{12}^{\times}\ltimes e^{6\mathbb{Z}_{12}}$	13	3|28
Three/Nine Elements				
8	●●●○○○○○○○○○	$\langle -1e^{-2}\rangle$	2	4|14
8.1	●○●○○○○○●○○○			
9	●●○●○○○○○○○○	$\{1\}$	1	4|30
9.1	●○○○○○○○○○○○			
10	●●○○●○○○○○○○	$\{1\}$	1	8|36
10.1	●○○●○○○●○○○○			
11	●●○○○●○○○○○○	$\langle 5\rangle$	4	4|20
12	●●○○○○○●○○○○	$\langle 7e^6\rangle$	6	5|29
13	●○●○●○○○○○○○	$\{1,7,-1e^8,5e^8\}$	8	4|18
14	●○●○○○○●○○○○	$\langle 7\rangle$	6	8|31
15	●○○●○○●○○○○○	$\{1,5,-1e^6,7e^6\}$	8	5|32
16	●○○○●○○○●○○○	$\mathbb{Z}_{12}^{\times}\ltimes e^{4\mathbb{Z}_{12}}$	15	4|20
Four/Eight Elements				
17	●●●●○○○○○○○○	$\langle -1e^{-3}\rangle$	3	4|8
17.1	●○●○○●○○●○○○			
18	●●●○●○○○○○○○	$\{1\}$	1	5|19
18.1	●○●○●○○●○○○○			
19	●●●○○●○○○○○○	$\{1\}$	1	5|19
19.1	●●○●○○○○○●○○			
20	●●●○○○●○○○○○	$\{1\}$	1	7|23
20.1	●●○○○●○●○○○○			
21	●●●○○○○●○○○○	$\{1,7,-1e^{-2},5e^{-2}\}$	9	7|9
22	●●○●●○○○○○○○	$\langle -1e^{-4}\rangle$	2	6|20
22.1	●○●○○●○○○●○○			
23	●●○●○●○○○○○○	$\langle 5\rangle$	4	5|13
24	●●○●○○●○○○○○	$\langle 7e^6\rangle$	6	6|17

Chord Classes—Continued				
Class Nr.	Representative Nr. = •, $\widehat{Nr.}$ = ○	Group of Symmetries	Conj. Class	♯ End. $Nr.\|\widehat{Nr.}$
25	● ● ○ ● ○ ○ ○ ● ○ ○ ○ ○	$\{1\}$	1	0|3
25.1	● ● ○ ● ○ ● ○ ○ ○ ○ ○○			
26	● ● ○ ● ○ ○ ○ ○ ○ ● ○○	$\{1\}$	1	12|31
26.1	● ○ ● ○ ○ ● ○ ○ ● ○○			
27	● ● ○ ● ○ ○ ○ ○ ○ ○ ●○	$\{1, 7e^{-3}, 5e^2, -1e^{-1}\}$	11	5|13
28	● ● ○ ○ ● ● ○ ○ ○ ○ ○○	$\langle -1e^7 \rangle$	3	6|14
28.1	● ● ○ ○ ○ ● ○ ○ ● ○○			
29	● ● ○ ○ ● ○ ○ ● ○ ○ ○○	$\langle 7 \rangle$	6	10|23
30	● ● ○ ○ ● ● ○ ○ ○ ● ○○	$\langle 5e^4 \rangle$	4	11|23
31	● ● ○ ○ ● ○ ○ ○ ○ ● ○○	$\{1, -1e^{-1}, 5e^{-4}, 7e^3\}$	10	9|19
32	● ● ○ ○ ○ ● ● ○ ○ ○ ○○	$\{1, 5, -1e^6, 7e^6\}$	8	7|15
33	● ● ○ ○ ○ ○ ● ● ○ ○ ○○	$\{1, 7, -1e^{-1}, 5e^{-1},$ $e^6, 7e^6, 5e^5, -1e^5\}$	14	7|14
34	● ○ ● ○ ● ○ ● ○ ○ ○ ○○	$\{1, 7, -1e^6, 5e^6\}$	9	6|17
35	● ○ ● ○ ● ○ ○ ○ ● ○ ○○	$\{1, 7, -1e^4, 5e^4\}$	8	11|19
36	● ○ ● ○ ○ ○ ● ○ ● ○ ○○	$\{1, 7, -1e^{-2}, 5e^{-2},$ $e^6, 7e^6, 5e^4, -1e^4\}$	13	9|28
37	● ○ ○ ● ○ ○ ● ○ ○ ● ○○	$\mathbb{Z}_{12}^{\times} \ltimes e^{3\mathbb{Z}_{12}}$	17	7|21
Five/Seven Elements				
38	● ● ● ● ● ○ ○ ○ ○ ○ ○○	$\langle -1e^{-4} \rangle$	2	5|7
38.1	● ○ ● ○ ● ○ ○ ● ○ ●○○			
39	● ● ● ● ○ ● ○ ○ ○ ○ ○○	$\{1\}$	1	6|10
39.1	● ● ○ ● ○ ● ○ ○ ○ ○ ●○			
40	● ● ● ● ○ ○ ● ○ ○ ○ ○○	$\{1\}$	1	8|12
40.1	● ● ○ ● ○ ○ ○ ● ○ ● ○○			
41	● ● ● ● ○ ○ ○ ● ○ ○ ○○	$\{1\}$	1	8|12
41.1	● ● ● ○ ○ ● ○ ● ○ ○○			
42	● ● ● ○ ● ● ○ ○ ○ ○ ○○	$\{1\}$	1	6|16
42.1	● ● ○ ● ○ ● ○ ○ ● ○○			
43	● ● ● ○ ● ○ ● ○ ○ ○ ○○	$\{1\}$	1	8|20
43.1	● ● ○ ● ○ ● ○ ● ○ ○○			
44	● ● ● ○ ● ○ ○ ● ○ ○ ○○	$\langle 7 \rangle$	6	7|9
45	● ● ● ○ ● ○ ○ ○ ● ○ ○○	$\{1\}$	1	16|22
45.1	● ● ○ ○ ● ○ ● ○ ● ○○			
46	● ● ● ○ ● ○ ○ ○ ○ ● ○○	$\langle 5e^{-4} \rangle$	4	5|12
47	● ● ● ○ ● ○ ○ ○ ○ ○ ●○	$\langle -1e^{-2} \rangle$	2	8|14
47.1	● ○ ● ○ ● ○ ● ○ ○ ● ○○			

Chord Classes—Continued				
Class Nr.	Representative $Nr. = \bullet$, $\widehat{Nr.} = \circ$	Group of Symmetries	Conj. Class	♯ End. $Nr.|\widehat{Nr.}$
48	●●●○○●●○○○○○○	$\{1\}$	1	8|18
48.1	●●●○●○○○●●○○○			
49	●●●○○●○○●○○○○	$\{1\}$	1	10|18
49.1	●●●○●○○○○○●●○○			
50	●●●○○●○○○●○○○	$\langle -1e^{-2}\rangle$	2	9|13
50.1	●●●○●●○○○●○○○			
51	●●●○○○●●○●○○○○	$\langle 7\rangle$	6	9|11
52	●●●○○○●○●○○○○	$\{1, -1e^{-2}, 7e^{6}, 5e^{4}\}$	8	7|17
53	●●○●●●○●○○○○	$\{1\}$	1	10|20
53.1	●●○●○○●○○○●○			
54	●●○●●○●○○●○○○	$\{1\}$	1	14|26
54.1	●●○○●○●○○●○○			
55	●●○●○●●○○○○○	$\{1, 5, -1e^{6}, 7e^{6}\}$	8	8|8
56	●●○●○●○○○●○○○	$\langle 5\rangle$	4	16|16
57	●●○●○○●●○○○○	$\langle 7e^{6}\rangle$	6	12|16
58	●●○●○○●○○●○○○	$\langle 7e^{6}\rangle$	6	18|23
59	●●○●○○○○●○●○○	$\langle 7\rangle$	6	13|29
60	●●○○○●●○○●○○○○	$\langle 5\rangle$	4	11|19
61	●●○○●○○●●○○○○	$\{1, 7, -1e^{4}, 5e^{4}\}$	8	14|14
62	●○●○●○●○●○○○	$\{1, 7, -1e^{4}, 7e^{4}\}$	8	11|19
	Six/Six Elements			
63*	●●●●●●○○○○○○○○	$\langle -1e^{-5}\rangle$	3	5|5
63.1*	●●○●○●○○●○●○			
64*	●●●●●○●○○○○○○	$\{1\}$	1	9|9
64.1*	●●○●○●○●○○●○			
65	●●●●●○○●○○○○○	$\{1\}$	1	9|9
65.1	●●●○●●○○●○●○			
66	●●●●●○○○●○○○○	$\langle -1e^{-4}\rangle$	2	12|6
66.1	●●●○○●○●○●○○			
67*	●●●●○●○●○○○○	$\langle 5e^{-2}\rangle$	5	6|6
68*	●●●●○●○○●○○○○	$\{1\}$	1	9|9
69	●●●●○●○○○○●○○	$\{1\}$	1	15|11
69.1	●●○●●○●○●○○○			
70*	●●●●○○○○○○○●○	$\{1, 5, -1e^{-3}, 7e^{-3}\}$	10	6|6
71*	●●●●○●●○●○○○○	$\{1\}$	1	11|11
71.1*	●●●○○●○●●○○○			
72	●●●●○○●○●○○○○	$\langle 7e^{6}\rangle$	6	8|10

Chord Classes—Continued				
Class Nr.	Representative Nr. = ●, $\widehat{Nr.}$ = ○	Group of Symmetries	Conj. Class	♯ End. Nr.|$\widehat{Nr.}$
73	●●●●○○●○○●○○	$\langle -1e^{-3} \rangle$	3	13|9
73.1	●●○●○○●○●●○○			
74*	●●●●○○○○●●○○	$\langle -1e^{-3} \rangle$	3	7|7
75*	●●●○●●○○●○○○	$\{1\}$	1	17|17
75.1*	●●○○○●●○○●○●			
76*	●●●○●●○○○●○○	$\langle 5e^{-5} \rangle$	4	10|10
77*	●●●○●○●○●○○○	$\langle 5e^{4} \rangle$	4	14|14
78*	●●●○●○●○○○○●	$\{1\}$	1	23|23
78.1*	●●○●○●○●○●○○			
79	●●●○●○○●●○○○	$\langle 7 \rangle$	6	18|10
80	●●●○●○○●○○●○	$\{1, 7, 5e^{-2}, -1e^{-2}\}$	9	15|11
81*	●●●○●○○○○●●○○	$\langle 5e^{-4} \rangle$	4	11|11
82*	●●●○○●●○○●○○	$\{1\}$	1	17|17
82.1*	●●○●●○○●●○○○			
83*	●●●○○○○●●●○○○	$\{1, -1e^{-2}, 5e^{-2}, 7\} \ltimes e^{6\mathbb{Z}_{12}}$	13	12|12
84*	●●○●●○●○○●○○	$\langle 7e^{3} \rangle$	7	14|14
85	●●○●●○○○●○●○○	$\{1, -1e^{-4}, 5e^{-4}, 7\}$	8	15|23
86*	●●○●○○●●○●○○	$\langle 7 \rangle \ltimes e^{6\mathbb{Z}_{12}}$	12	20|20
87*	●●○○○●●○○●●○○	$\{1, e^{4}, e^{8}, 5, 5e^{4}, 5e^{8}, -1e^{-1}, -1e^{3}, -1e^{7}, 7e^{-1}, 7e^{3}, 7e^{7}\}$	16	12|12
88*	●○●○●○●●○●○○●	$\mathbb{Z}_{12}^{\times} \ltimes e^{\mathbb{Z}_{12}}$	18	12|12

N.2 Third Chain Classes

The following list of third chain translation classes shows the class number in the first column, where equivalence ($\sim$) means that the same *pc* set is generated. The second column shows the pitch classes in the order of appearance along the third chain. The third column shows the third chain, the fourth column shows the chord class of the *pc* set, and the fifth column shows lead-sheet symbols as systematically derived in Section 25.2.1.

		Third Chains		
Chain Nr.	*Pitch Classes*	*Third*	*Chord*	*Lead-Sheet*
$\sim$ *equiv.*	*from 0*	*Chain*	*Class*	*Symbols*
Two Pitch Classes				
1	0,3	3	5	*trd*
2	0,4	4	6	*Trd*
Three Pitch Classes				
3	0,3,6	33	15	$C0, Cm5-$
4	0,3,7	34	10.1	Cm
5	0,4,7	43	10.1	C
6	0,4,8	44	16	$C+, C5+$
Four Pitch Classes				
7	0,3,6,9	333	37	$C07-$
8	0,3,6,10	334	26.1	$C07$
9	0,3,7,10	343	22.1	$Cm7$
10	0,3,7,11	344	30	$Cm7+$
11	0,4,7,10	433	26.1	$C7$
12	0,4,7,11	434	28.1	$C7+$
13	0,4,8,11	443	30	$C+7+$
Five Pitch Classes				
14	0,3,6,9,1	3334	58	$C07-/9-$
15	0,3,6,10,1	3343	53.1	$C09-$
16	0,3,6,10,2	3344	56	$C09$
17	0,3,7,10,1	3433	53.1	$Cm9-$
18	0,3,7,10,2	3434	42.1	$Cm9$
19	0,3,7,11,2	3443	59	$Cm7+/9, Cmmaj7/9$
20	0,4,7,10,1	4333	58	$C9-$
21	0,4,7,10,2	4334	47.1	$C9$
22	0,4,7,11,2	4343	42.1	$C7+/9, Cmaj7/9$
23	0,4,7,11,3	4344	60	$C7+/9+$
24	0,4,8,11,2	4433	56	$C+7+/9$
25	0,4,8,11,3	4434	60	$C+7+/9+$
Six Pitch Classes				
26	0,3,6,9,1,4	33343	84*	$C07-/9-/11-$
27	0,3,6,9,1,5	33344	85^	$C07-/9-/11$
28	0,3,6,10,1,4	33433	79	$C09-/11-$
29	0,3,6,10,1,5	33434	65.1^	$C09-/11$

Chain Nr. ~ equiv.	Pitch Classes from 0	Third Chain	Chord Class	Lead-Sheet Symbols
Third Chains—Continued				
30	0,3,6,10,2,5	33443	69.1	C011
31	0,3,7,10,1,4	34333	84*	Cm9-/11-
32	0,3,7,10,1,5	34334	64.1*	Cm9-/11
33	0,3,7,10,2,5	34343	63.1*	Cm11
34	0,3,7,10,2,6	34344	75.1*	Cm11+
35	0,3,7,11,2,5	34433	69.1	Cm7+/11
36	0,3,7,11,2,6	34434	82*	Cm7+/11+
37	0,4,7,10,1,5	43334	73.1	C9-/11
38	0,4,7,10,2,5	43343	64.1*	C11
39	0,4,7,10,2,6	43344	78.1*	C11+
40	0,4,7,11,2,5	43433	65.1^	C7+/11
41	0,4,7,11,2,6	43434	66.1^	C7+/11+
42	0,4,7,11,3,6	43443	82*	C7+/9+/11+
43	0,4,8,11,2,5	44333	85^	C+7+/11
44	0,4,8,11,2,6	44334	78.1*	C+7+/11+
45	0,4,8,11,3,6	44343	75.1*	C+7+/9+/11+
46	0,4,8,11,3,7	44344	87*	C+7+/9+/(11)/13-
Seven Pitch Classes				
47	0,3,6,9,1,4,7	333433	58^	C07-/9-/11-/13-
48	0,3,6,9,1,4,8	333434	54.1^	C07-/9-/11-/13
49	0,3,6,9,1,5,8	333443	54.1^	C07-/9-/13
50	0,3,6,10,1,4,7	334333	58^	C09-/11-/13-
51	0,3,6,10,1,4,8	334334	47.1	C09-/11-/13
52	0,3,6,10,1,5,8	334343	38.1	C09-/13
53	0,3,6,10,1,5,9	334344	54.1^	C09-/13+
54	0,3,6,10,2,5,8	334433	47.1^	C013
55	0,3,6,10,2,5,9	334434	54.1^	C013+
56	0,3,7,10,1,4,8	343334	54.1^	Cm9-/11-/13
57	0,3,7,10,1,5,8	343343	38.1^	Cm9-/13
58	0,3,7,10,1,5,9	343344	47.1^	Cm9-/13+
59	0,3,7,10,2,5,8	343433	38.1^	Cm13
60	0,3,7,10,2,5,9	343434	38.1^	Cm13+
61	0,3,7,10,2,6,9	343443	54.1^	Cm11+/13+
62	0,3,7,11,2,5,8	344333	54.1^	Cm7+/13
63	0,3,7,11,2,5,9	344334	47.1^	Cm7+/13+
64	0,3,7,11,2,6,9	344343	54.1^	Cm7+/11+/13+
65	0,3,7,11,2,6,10	344344	60^	Cm7+/11+/(13)/15-
66	0,4,7,10,1,5,8	433343	54.1^	C9-/13
67	0,4,7,10,1,5,9	433344	54.1^	C9-/13+
68	0,4,7,10,2,5,8	433433	47.1^	C13
69	0,4,7,10,2,5,9	433434	38.1^	C13+

	Third Chains—Continued			
Chain Nr. ~ equiv.	Pitch Classes from 0	Third Chain	Chord Class	Lead-Sheet Symbols
70	0,4,7,10,2,6,9	433443	47.1^	C11+/13+
71	0,4,7,11,2,5,8	434333	54.1^	C7+/13
72	0,4,7,11,2,5,9	434334	38.1^	C7+/13+
73	0,4,7,11,2,6,9	434343	38.1^	C7+/11+/13+
74	0,4,7,11,2,6,10	434344	45.1^	C7+/11+/(13)/15-
75	0,4,7,11,3,6,9	434433	54.1^	C7+/9+/11+/13+
76	0,4,7,11,3,6,10	434434	55^	C7+/9+/11+/(13)/15-
77	0,4,8,11,2,5,9	443334	54.1^	C+7+/13+
78	0,4,8,11,2,6,9	443343	47.1^	C+7+/11+/13+
79	0,4,8,11,2,6,10	443344	62^	C+7+/11+/(13)/15-
80	0,4,8,11,3,6,9	443433	54.1^	C+7+/9+/11+/13+
81	0,4,8,11,3,6,10	443434	45.1^	C+7+/9+/11+/(13)/15-
82	0,4,8,11,3,7,10	443443	60^	C+7+/9+/(11)/13-/15-
	Eight Pitch Classes			
83	0,3,6,9,1,4,7,10	3334333	37^	C07-/9-/11-/13-...
84	0,3,6,9,1,4,7,11	3334334	26.1^	C07-/9-/11-/13-...
85	0,3,6,9,1,4,8,11	3334343	22.1^	C07-/9-/11-/13...
86	0,3,6,9,1,5,8,11	3334433	26.1^	C07-/9-/13...
87	0,3,6,10,1,4,7,11	3343334	29^	C09-/11-/13-...
88	0,3,6,10,1,4,8,11	3343343	18.1^	C09-/11-/13...
89	0,3,6,10,1,5,8,11	3343433	17.1^	C09-/13...
90	0,3,6,10,2,5,8,11	3344333	26.1^	C013...
91	0,3,6,10,2,5,9,1	3344344	31^	C013+...
92	0,3,7,10,1,4,8,11	3433343	31^	Cm9-/11-/13...
93	0,3,7,10,1,5,8,11	3433433	18.1^	Cm9-/13...
94	0,3,7,10,2,5,8,11	3434333	22.1^	Cm13...
95	0,3,7,10,2,5,9,1	3434344	18.1^	Cm13+...
96	0,3,7,10,2,6,9,1	3434434	29^	Cm11+/13+...
97	0,3,7,11,2,5,9,1	3443344	34^	Cm7+/13+...
98	0,3,7,11,2,6,9,1	3443434	29^	Cm7+/11+/13+...
99	0,3,7,11,2,6,10,1	3443443	28^	Cm7+/11+/(13)/15-...
100	0,4,7,10,1,5,8,11	4333433	29^	C9-/13...
101	0,4,7,10,2,5,8,11	4334333	18.1^	C15
102	0,4,7,10,2,5,9,1	4334344	22.1^	C13+...
103	0,4,7,10,2,6,9,1	4334434	26.1^	C11+/13+...
104	0,4,7,11,2,5,9,1	4343344	18.1^	C7+/13+...
105	0,4,7,11,2,6,9,1	4343434	17.1^	C7+/11+/13+...
106	0,4,7,11,2,6,10,1	4343443	25.1^	C7+/11+/(13)/15-...
107 ~ 84	0,4,7,11,3,6,9,1	4344334	26.1^	C7+/9+/11+/13+...
108 ~ 87	0,4,7,11,3,6,10,1	4344343	29^	C7+/9+/11+/(13)/15-...
109	0,4,7,11,3,6,10,2	4344344	30^	C7+/9+/11+/(13)/15-...

Chain Nr. ~ equiv.	Pitch Classes from 0	Third Chain	Chord Class	Lead-Sheet Symbols
Third Chains—Continued				
110	0,4,8,11,2,5,9,1	4433344	31^	$C+7+/13+\dots$
111	0,4,8,11,2,6,9,1	4433434	18.1^	$C+7+/11+/13+\dots$
112	0,4,8,11,2,6,10,1	4433443	34^	$C+7+/11+/(13)/15-\dots$
113 ~ 85	0,4,8,11,3,6,9,1	4434334	22.1^	$C+7+/9+/11+/13+\dots$
114 ~ 88	0,4,8,11,3,6,10,1	4434343	18.1^	$C+7+/9+/11+/(13)/15-\dots$
115	0,4,8,11,3,6,10,2	4434344	35^	$C+7+/9+/11+/(13)/15-\dots$
116 ~ 92	0,4,8,11,3,7,10,1	4434433	31^	$C+7+/9+/(11)/13-/15-\dots$
117	0,4,8,11,3,7,10,2	4434434	30^	$C+7+/9+/(11)/13-/15-\dots$
Nine Pitch Classes				
118	0,3,6,9,1,4,7,10,2	33343334	15^	$C07-/9-/11-/13-\dots$
119	0,3,6,9,1,4,7,11,2	33343343	9.1^	$C07-/9-/11-/13-\dots$
120	0,3,6,9,1,4,8,11,2	33343433	9.1^	$C07-/9-/11-/13\dots$
121	0,3,6,9,1,5,8,11,2	33344333	15^	$C07-/9-/13\dots$
122	0,3,6,10,1,4,7,11,2	33433343	10^	$C09-/11-/13-\dots$
123	0,3,6,10,1,4,8,11,2	33433433	13^	$C09-/11-/13\dots$
124	0,3,6,10,1,5,8,11,2	33434333	9.1^	$C09-/13\dots$
125	0,3,6,10,2,5,9,1,4	33443443	10^	$C013+\dots$
126	0,3,7,10,1,4,8,11,2	34333433	10^	$Cm9-/11-/13\dots$
127	0,3,7,10,1,5,8,11,2	34334333	9.1^	$Cm9-/13\dots$
128	0,3,7,10,2,5,9,1,4	34343443	9.1^	$Cm13+\dots$
129	0,3,7,10,2,6,9,1,4	34344343	15^	$Cm11+/13+\dots$
130	0,3,7,11,2,5,9,1,4	34433443	13^	$Cm7+/13+\dots$
131 ~ 119	0,3,7,11,2,6,9,1,4	34434343	9.1^	$Cm7+/11+/13+\dots$
132	0,3,7,11,2,6,9,1,5	34434344	14^	$Cm7+/11+/13+\dots$
133 ~ 122	0,3,7,11,2,6,10,1,4	34434433	10^	$Cm7+/11+/(13)/15-\dots$
134	0,3,7,11,2,6,10,1,5	34434434	11^	$Cm7+/11+/(13)/15-\dots$
135	0,4,7,10,1,5,8,11,2	43334333	15^	$C9-/13\dots$
136	0,4,7,10,1,5,8,11,3	43334334	10.1^	$C9-/13\dots$
137	0,4,7,10,2,5,8,11,3	43343334	10.1^	$C17$
138	0,4,7,10,2,6,9,1,5	43344344	10.1^	$C11+/13+\dots$
139	0,4,7,11,2,6,9,1,5	43434344	8.1^	$C7+/11+/13+\dots$
140	0,4,7,11,2,6,10,1,5	43434434	12^	$C7+/11+/(13)/15-\dots$
141	0,4,7,11,3,6,9,1,5	43443344	14^	$C7+/9+/11+/13+\dots$
142	0,4,7,11,3,6,10,1,5	43443434	12^	$C7+/9+/11+/(13)/15-\dots$
143	0,4,7,11,3,6,10,2,5	43443443	11^	$C7+/9+/11+/(13)/15-\dots$
144	0,4,8,11,2,6,9,1,5	44334344	10.1^	$C+7+/11+/13+\dots$
145	0,4,8,11,2,6,10,1,5	44334434	14^	$C+7+/11+/(13)/15-\dots$
146	0,4,8,11,3,6,9,1,5	44343344	10.1^	$C+7+/9+/11+/13+\dots$
147	0,4,8,11,3,6,10,1,5	44343434	8.1^	$C+7+/9+/11+/(13)/15-\dots$
148	0,4,8,11,3,6,10,2,5	44343443	14^	$C+7+/9+/11+/(13)/15-\dots$
149 ~ 136	0,4,8,11,3,7,10,1,5	44344334	10.1^	$C+7+/9+/(11)/13-/15-\dots$

		Third Chains—Continued		
Chain Nr.	*Pitch Classes*	*Third*	*Chord*	*Lead-Sheet*
~ equiv.	*from 0*	*Chain*	*Class*	*Symbols*
150 ~ 137	0,4,8,11,3,7,10,2,5	44344343	10.1^	$C+7+/9+/(11)/13-/15-\ldots$
151	0,4,8,11,3,7,10,2,6	44344344	16^	$C+7+/9+/(11)/13-/15-\ldots$
		Ten Pitch Classes		
152	0,3,6,9,1,4,7,10,2,5	333433343	5^	$C07-/9-/11-/13-\ldots$
153	0,3,6,9,1,4,7,11,2,5	333433433	4^	$C07-/9-/11-/13-\ldots$
154	0,3,6,9,1,4,8,11,2,5	333434333	5^	$C07-/9-/11-/13\ldots$
155	0,3,6,10,1,4,7,11,2,5	334333433	3^	$C09-/11-/13-\ldots$
156	0,3,6,10,1,4,8,11,2,5	334334333	4^	$C09-/11-/13\ldots$
157 ~ 152	0,3,6,10,2,5,9,1,4,7	334434433	5^	$C013+\ldots$
158	0,3,6,10,2,5,9,1,4,8	334434434	6^	$C013+\ldots$
159	0,3,7,10,1,4,8,11,2,5	343334333	5^	$Cm9-/11-/13\ldots$
160	0,3,7,10,1,4,8,11,2,6	343334334	6^	$Cm9-/11-/13\ldots$
161	0,3,7,10,1,5,8,11,2,6	343343334	3.1^	$Cm9-/13\ldots$
162	0,3,7,10,2,5,9,1,4,8	343434434	3.1^	$Cm13+\ldots$
163	0,3,7,10,2,6,9,1,4,8	343443434	7^	$Cm11+/13+\ldots$
164	0,3,7,11,2,5,9,1,4,8	344334434	6^	$Cm7+/13+\ldots$
165	0,3,7,11,2,6,9,1,4,8	344343434	3.1^	$Cm7+/11+/13+\ldots$
166	0,3,7,11,2,6,9,1,5,8	344343443	7^	$Cm7+/11+/13+\ldots$
167 ~ 160	0,3,7,11,2,6,10,1,4,8	344344334	6^	$Cm7+/11+/(13)/15-\ldots$
168 ~ 161	0,3,7,11,2,6,10,1,5,8	344344343	3.1^	$Cm7+/11+/(13)/15-\ldots$
169	0,3,7,11,2,6,10,1,5,9	344344344	6^	$Cm7+/11+/(13)/15-\ldots$
170	0,4,7,10,1,5,8,11,2,6	433343334	7^	$C9-/13\ldots$
171	0,4,7,10,1,5,8,11,3,6	433343343	3.1^	$C9-/13\ldots$
172	0,4,7,10,2,5,8,11,3,6	433433343	6^	$C19$
173	0,4,7,10,2,6,9,1,5,8	433443443	6^	$C11+/13+\ldots$
174	0,4,7,11,2,6,9,1,5,8	434343443	3.1^	$C7+/11+/13+\ldots$
175 ~ 170	0,4,7,11,2,6,10,1,5,8	434344343	7^	$C7+/11+/(13)/15-\ldots$
176	0,4,7,11,2,6,10,1,5,9	434344344	3.1^	$C7+/11+/(13)/15-\ldots$
177 ~ 171	0,4,7,11,3,6,9,1,5,8	434433443	3.1^	$C7+/9+/11+/13+\ldots$
178	0,4,7,11,3,6,10,1,5,8	434434343	6^	$C7+/9+/11+/(13)/15-\ldots$
179	0,4,7,11,3,6,10,1,5,9	434434344	7^	$C7+/9+/11+/(13)/15-\ldots$
180 ~ 172	0,4,7,11,3,6,10,2,5,8	434434433	6^	$C7+/9+/11+/(13)/15-\ldots$
181	0,4,7,11,3,6,10,2,5,9	434434434	3.1^	$C7+/9+/11+/(13)/15-\ldots$
182	0,4,8,11,2,6,10,1,5,9	443344344	6^	$C+7+/11+/(13)/15-\ldots$
183	0,4,8,11,3,6,10,1,5,9	443434344	3.1^	$C+7+/9+/11+/(13)/15-\ldots$
184	0,4,8,11,3,6,10,2,5,9	443434434	7^	$C+7+/9+/(11)/13-/15-\ldots$
185	0,4,8,11,3,7,10,1,5,9	443443344	6^	$C+7+/9+/(11)/13-/15-\ldots$
186	0,4,8,11,3,7,10,2,5,9	443443434	3.1^	$C+7+/9+/(11)/13-/15-\ldots$
187	0,4,8,11,3,7,10,2,6,9	443443443	6^	$C+7+/9+/(11)/13-/15-\ldots$
		Eleven Pitch Classes		
188	0,3,6,9,1,4,7,10,2,5,8	3334333433	2^	$C07-/9-/11-/13-\ldots$

Chain Nr. ~ equiv.	Pitch Classes from 0	Third Chain	Chord Class	Lead-Sheet Symbols
Third Chains—Continued				
189	0,3,6,9,1,4,7,11,2,5,8	3334334333	2^	$C07-/9-/11-/13-\ldots$
190	0,3,6,10,1,4,7,11,2,5,8	3343334333	2^	$C09-/11-/13-\ldots$
191	0,3,6,10,1,4,7,11,2,5,9	3343334334	2^	$C09-/11-/13-\ldots$
192	0,3,6,10,1,4,8,11,2,5,9	3343343334	2^	$C09-/11-/13\ldots$
193 ~ 191	0,3,6,10,2,5,9,1,4,7,11	3344344334	2^	$C013+\ldots$
194 ~ 192	0,3,6,10,2,5,9,1,4,8,11	3344344343	2^	$C013+\ldots$
195	0,3,7,10,1,4,8,11,2,5,9	3433343334	2^	$Cm9-/11-/13\ldots$
196	0,3,7,10,1,4,8,11,2,6,9	3433343343	2^	$Cm9-/11-/13\ldots$
197	0,3,7,10,1,5,8,11,2,6,9	3433433343	2^	$Cm9-/13\ldots$
198 ~ 195	0,3,7,10,2,5,9,1,4,8,11	3434344343	2^	$Cm13+\ldots$
199	0,4,7,10,1,5,8,11,2,6,9	4333433343	2^	$C9-/13\ldots$
200	0,4,7,10,1,5,8,11,3,6,9	4333433433	2^	$C9-/13\ldots$
201	0,4,7,10,2,5,8,11,3,6,9	4334333433	2^	$C21$
202 ~ 199	0,4,7,10,2,6,9,1,5,8,11	4334434433	2^	$C11+/13+\ldots$
203 ~ 191	0,4,7,11,3,6,10,2,5,9,1	4344344344	2^	$C7+/9+/11+/(13)/15-\ldots$
204 ~ 192	0,4,8,11,3,6,10,2,5,9,1	4434344344	2^	$C+7+/9+/(11)/13-/15-\ldots$
205 ~ 195	0,4,8,11,3,7,10,2,5,9,1	4434434344	2^	$C+7+/9+/(11)/13-/15-\ldots$
206 ~ 196	0,4,8,11,3,7,10,2,6,9,1	4434434434	2^	$C+7+/9+/(11)/13-/15-\ldots$
Twelve Pitch Classes				
207	0,3,6,9,1,4,7,10,2,5,8,11	33343334333	1^	$C07-/9-/11-/13-\ldots$
208 ~ 207	0,4,7,10,2,5,8,11,3,6,9,1	43343334334	1^	$C23$
209 ~ 207	0,4,7,10,2,6,9,1,5,8,11,3	43344344334	1^	$C11+/13+\ldots$
210 ~ 207	0,4,8,11,3,7,10,2,6,9,1,5	44344344344	1^	$C+7+/9+/(11)/13-/15-\ldots$

O

Two, Three, and Four Tone Motif Classes

O.1 Two Tone Motifs in $OnPiMod_{12,12}$

ClassNr.	Representative
1	$(0,0),(0,1)$
2	$(0,0),(0,2)$
3	$(0,0),(0,3)$
4	$(0,0),(0,4)$
5	$(0,0),(0,6)$

O.2 Two Tone Motifs in $OnPiMod_{5,12}$

ClassNr.	Representative
1	$(0,0),(0,1)$
2	$(0,0),(0,2)$
3	$(0,0),(0,3)$
4	$(0,0),(0,4)$
5	$(0,0),(0,6)$
6	$(0,0),(1,0)$
7	$(0,0),(1,1)$
8	$(0,0),(1,2)$
9	$(0,0),(1,2)$
10	$(0,0),(1,4)$
11	$(0,0),(1,6)$

© Springer International Publishing AG, part of Springer Nature 2017
G. Mazzola, *The Topos of Music IV: Roots*, Computational Music Science,
https://doi.org/10.1007/978-3-319-64495-0_15

O.3 Three Tone Motifs in $OnPiMod_{12,12}$

Refer to the discussion in Section 11.3.8 for the entries of this table. The order of these representatives is a historical one. After this table, the representatives are also visualized on a 12×12 square in list O.1.

	Three-Element Motif Classes in $OnPiMod_{12,12}$			
Class Nr.	*Representative*	*Kernel*	*Class Weight*	*Volume*
1	$(0,0),(1,0),(2,0)$	$\mathbb{Z}.(1,2) \times \mathbb{Z}.(1,1)$	$(1,1,2)$	0
2	$(0,0),(1,0),(3,0)$	$\mathbb{Z}.(1,2) \times \mathbb{Z}.(0,1)$	$(1,2,3)$	0
3	$(0,0),(1,0),(4,0)$	$\mathbb{Z}.(1,0) \times \mathbb{Z}.(0,1)$	$(1,3,4)$	0
4	$(0,0),(1,0),(5,0)$	$\mathbb{Z}.(1,2) \times \mathbb{Z}.(1,1)$	$(1,1,4)$	0
5	$(0,0),(1,0),(6,0)$	$\mathbb{Z}.(1,2) \times \mathbb{Z}.(1,0)$	$(1,1,6)$	0
6	$(0,0),(2,0),(4,0)$	$(\mathbb{Z}_4 \times 2\mathbb{Z}_4) \times \mathbb{Z}.(1,1)$	$(2,2,4)$	0
7	$(0,0),(2,0),(6,0)$	$(\mathbb{Z}_4 \times 2\mathbb{Z}_4) \times \mathbb{Z}.(0,1)$	$(2,4,6)$	0
8	$(0,0),(3,0),(6,0)$	$\mathbb{Z}.(1,2) \times \mathbb{Z}_3^2$	$(3,3,6)$	0
9	$(0,0),(4,0),(8,0)$	$(\mathbb{Z}_4^2) \times \mathbb{Z}.(1,1)$	$(4,4,4)$	0
10	$(0,0),(1,0),(0,1)$	0×0	$(1,1,1)$	1
11	$(0,0),(2,0),(0,1)$	$\mathbb{Z}.(2,0) \times 0$	$(1,1,2)$	2
12	$(0,0),(3,0),(0,1)$	$0 \times \mathbb{Z}.(1,0)$	$(1,1,3)$	3
13	$(0,1),(0,2),(3,0)$	$0 \times \mathbb{Z}.(1,1)$	$(1,1,1)$	3
14	$(0,0),(0,1),(4,0)$	$\mathbb{Z}.(1,0) \times 0$	$(1,1,4)$	4
15	$(0,0),(1,2),(2,0)$	$\mathbb{Z}.(1,2) \times 0$	$(1,1,2)$	4
16	$(0,0),(2,0),(0,2)$	$2\mathbb{Z}_4^2 \times 0$	$(2,2,2)$	4
17	$(0,0),(6,0),(0,1)$	$\mathbb{Z}.(2,0) \times \mathbb{Z}.(1,0)$	$(1,1,6)$	6
18	$(0,0),(3,0),(0,2)$	$\mathbb{Z}.(2,0) \times \mathbb{Z}.(0,1)$	$(1,2,3)$	6
19	$(0,0),(0,2),(3,1)$	$\mathbb{Z}.(2,0) \times \mathbb{Z}.(1,1)$	$(1,1,2)$	6
20	$(0,0),(4,0),(0,2)$	$(\mathbb{Z}_4 \times 2\mathbb{Z}_4) \times 0$	$(2,2,4)$	4
21	$(0,0),(4,0),(0,4)$	$0 \times \mathbb{Z}_3^2$	$(3,3,3)$	3
22	$(0,0),(6,0),(0,2)$	$2\mathbb{Z}_4^2 \times \mathbb{Z}.(1,0)$	$(2,2,6)$	0
23	$(0,2),(0,4),(6,0)$	$2\mathbb{Z}_4^2 \times \mathbb{Z}.(1,1)$	$(2,2,2)$	0
24	$(0,0),(4,0),(0,4)$	$\mathbb{Z}_4^2 \times 0$	$(4,4,4)$	4
25	$(0,0),(6,0),(0,3)$	$\mathbb{Z}.(2,0) \times \mathbb{Z}_3^2$	$(3,3,6)$	6
26	$(0,0),(6,0),(0,6)$	$2\mathbb{Z}_4^2 \times \mathbb{Z}_3^2$	$(6,6,6)$	0

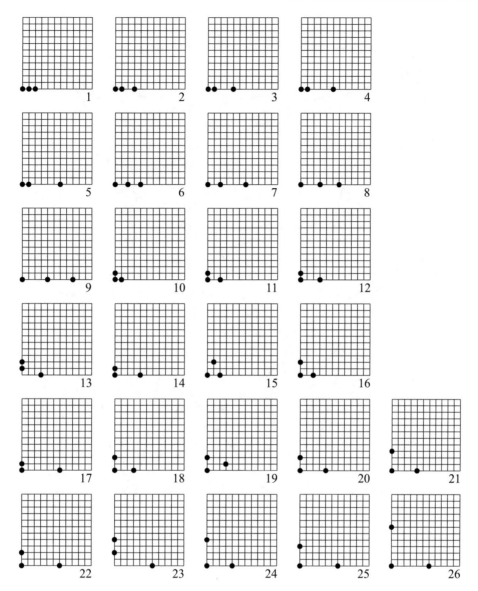

Fig. O.1. Representatives of the 26 isomorphism classes of three-element motives in $OnPiMod_{12,12}$.

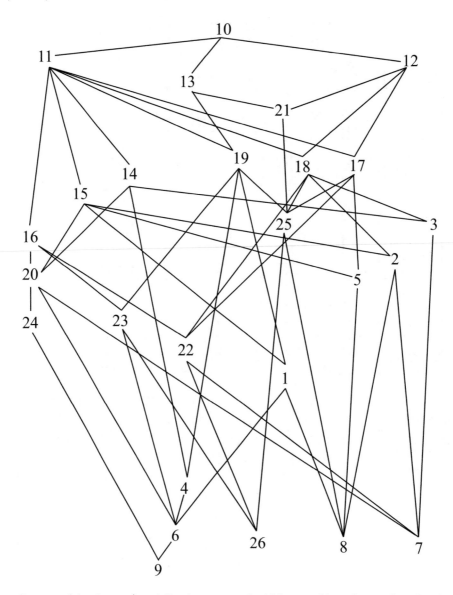

Fig. O.2. Hasse diagram of dominance/specialization among the 26 isomorphism classes of motives in $OnPiMod_{12,12}$.

O.4 Four Tone Motifs in $OnPiMod_{12,12}$

This list was calculated by Straub in [1017], refer to Section 11.3.8 for details. The list's numbering follows Straub's algorithm; * denotes classes which are not determined by volume and class weight.

	Four-Element Motif Classes		
Class Nr.	Representative	Class Weight	Volume
0	(0,0),(0,1),(0,2),(0,7)	(1,1,5,5)	0
1	(0,0),(0,1),(0,2),(0,3)	(1,1,2,2)	0
2	(0,0),(0,1),(0,2),(0,6)	(1,4,5,7)	0
3	(0,0),(0,1),(0,2),(0,5)	(1,4,2,3)	0
4	(0,0),(0,1),(0,2),(0,4)	(1,2,3,6)	0
5	(0,0),(1,0),(0,5),(0,6)	(4,4,5,5)	0
6	(0,0),(0,1),(0,4),(0,5)	(4,4,3,3)	0
7	(0,0),(0,1),(0,3),(0,5)	(4,2,2,6)	0
8	(0,0),(0,1),(0,4),(0,8)	(4,3,3,9)	0
9	(0,0),(0,1),(0,6),(0,7)	(5,5,5,5)	0
10	(0,0),(0,1),(0,3),(0,6)	(5,2,2,8)	0
11	(0,0),(0,1),(0,3),(0,7)	(5,2,3,7)	0
12	(0,0),(0,1),(0,4),(0,7)	(5,3,3,8)	0
13	(0,0),(0,1),(0,3),(0,10)	(2,2,2,2)	0
14	(0,0),(0,1),(0,3),(0,4)	(2,2,3,3)	0
15	(0,0),(0,1),(0,3),(0,9)	(2,3,7,8)	0
16	(0,0),(0,1),(0,4),(0,9)	(3,3,3,3)	0
17	(0,0),(0,2),(6,0),(6,10)	(23,23,22,22)	0
18	(0,0),(0,2),(0,4),(6,0)	(23,6,22,7)	0
19	(0,0),(0,2),(0,4),(6,2)	(6,6,22,22)	0
20	(0,0),(0,2),(0,4),(0,6)	(6,6,7,7)	0
21	(0,0),(0,2),0,4),(0,8)	(6,7,7,9)	0
22	(0,0),(0,2),(6,0),(6,2)	(22,22,22,22)	0
23	(0,0),(0,2),(6,0),(6,6)	(22,22,22,26)	0
24	(0,0),(0,2),(0,6),(6,2)	(22,22,7,7)	0
25	(0,0),(0,2),(0,6),(6,0)	(22,7,7,26)	0
26	(0,0),(0,2),(0,6),(0,8)	(7,7,7,7)	0
27	(0,0),(0,3),(0,6),(0,9)	(8,8,8,8)	0
28	(0,0),(0,6),(6,0),(6,6)	(26,26,26,26)	0
29	(0,0),(0,1),(0,2),(1,0)	(1,10,10,11)	1
30	(0,0),(0,1),(0,5),(1,0)	(4,10,10,14)	1
31	(0,0),(0,1),(0,6),(1,0)	(5,10,10,17)	1
32	(0,0),(0,1),(0,3),(1,0)	(2,10,11,12)	1
33	(0,0),(0,1),(0,4),(1,0)	(3,10,12,14)	1
34*	(0,0),(0,1),(1,0),(1,5)	(10,10,10,10)	1

Four-Element Motif Classes—Continued			
Class Nr.	Representative	Class Weight	Volume
35*	(0,0),(0,1),(1,0),(7,7)	(10,10,10,10)	1
36*	(0,0),(0,1),(1,0),(1,1)	(10,10,10,10)	1
37*	(0,0),(0,1),(1,0),(3,5)	(10,10,10,13)	1
38*	(0,0),(0,1),(1,0),(3,11)	(10,10,10,13)	1
39*	(0,0),(0,1),(1,0),(1,2)	(10,10,11,11)	1
40*	(0,0),(0,1),(1,0),(5,10)	(10,10,11,11)	1
41*	(0,0),(0,1),(1,0),(4,10)	(10,10,11,15)	1
42*	(0,0),(0,1),(1,0),(2,4)	(10,10,11,15)	1
43	(0,0),(0,1),(1,0),(2,5)	(10,10,11,19)	1
44*	(0,0),(0,1),(1,0),(1,3)	(10,10,12,12)	1
45*	(0,0),(0,1),(1,0),(7,9)	(10,10,12,12)	1
46*	(0,0),(0,1),(1,0),(3,3)	(10,10,12,12)	1
47	(0,0),(0,1),(1,0),(6,8)	(10,10,15,19)	1
48*	(0,0),(0,1),(1,0),(1,4)	(10,10,14,14)	1
49*	(0,0),(0,1),(1,0),(4,4)	(10,10,14,14)	1
50	(0,0),(0,1),(1,0),(1,6)	(10,10,17,17)	1
51	(0,0),(0,1),(1,0),(2,2)	(10,11,11,13)	1
52	(0,0),(0,1),(1,0),(2,3)	(10,11,13,15)	1
53	(0,0),(0,1),(1,0),(6,9)	(10,11,12,18)	1
54	(0,0),(0,1),(1,0),(8,8)	(10,13,14,14)	1
55	(0,0),(0,1),(1,0),(3,4)	(10,12,15,18)	1
56	(0,0),(0,1),(2,0),(3,1)	(11,11,12,12)	1
57	(0,0),(0,1),(2,0),(3,4)	(11,12,12,15)	1
58	(0,0),(0,1),(3,0),(4,1)	(12,12,14,14)	1
59	(0,0),(0,1),(0,2),(2,1)	(1,11,11,15)	2
60	(0,0),(0,1),(0,2),(2,0)	(1,11,11,16)	2
61	(0,0),(0,1),(0,5),(2,0)	(4,11,11,14)	2
62	(0,0),(0,1),(0,5),(2,1)	(4,11,11,20)	2
63	(0,0),(0,1),(0,6),(2,1)	(5,5,11,11)	2
64	(0,0),(0,1),(0,6),(2,0)	(5,22,11,11)	2
65	(0,0),(0,1),(0,3),(2,0)	(2,11,15,18)	2
66	(0,0),(0,1),(0,3),(2,1)	(2,11,16,18)	2
67	(0,0),(0,1),(0,4),(2,1)	(3,11,14,18)	2
68	(0,0),(0,1),(0,4),(2,0)	(3,11,20,18)	2
69	(0,0),(0,1),(2,0),(4,6)	(23,11,11,15)	2
70	(0,0),(0,1),(2,0),(4,0)	(6,11,11,14)	2
71	(0,0),(0,1),(2,0),(6,6)	(22,11,15,17)	2
72	(0,0),(0,1),(2,0),(6,0)	(7,11,14,17)	2
73*	(0,0),(0,1),(2,0),(2,1)	(11,11,11,11)	2

| \multicolumn{4}{c}{**Four-Element Motif Classes**—Continued} |
Class Nr.	Representative	Class Weight	Volume
74*	(0,0),(0,1),(2,0),(2,5)	(11,11,11,11)	2
75*	(0,0),(0,1),(2,0),(2,7)	(11,11,11,11)	2
76*	(0,0),(0,1),(2,0),(2,11)	(11,11,11,11)	2
77*	(0,0),(0,1),(2,0),(6,5)	(11,11,11,19)	2
78*	(0,0),(0,1),(2,0),(6,11)	(11,11,11,19)	2
79*	(0,0),(0,1),(2,0),(4,7)	(11,11,15,15)	2
80*	(0,0),(0,1),(2,0),(8,11)	(11,11,15,15)	2
81*	(0,0),(0,1),(2,0),(2,2)	(11,11,15,16)	2
82*	(0,0),(0,1),(2,0),(8,10)	(11,11,15,16)	2
83*	(0,0),(0,1),(2,0),(4,1)	(11,11,14,14)	2
84*	(0,0),(0,1),(2,0),(8,5)	(11,11,14,14)	2
85*	(0,0),(0,1),(2,0),(8,4)	(11,11,14,20)	2
86*	(0,0),(0,1),(2,0),(2,4)	(11,11,14,20)	2
87*	(0,0),(0,1),(2,0),(6,7)	(11,11,17,17)	2
88*	(0,0),(0,1),(2,0),(6,1)	(11,11,17,17)	2
89*	(0,0),(0,1),(2,0),(2,9)	(11,11,18,18)	2
90*	(0,0),(0,1),(2,0),(2,3)	(11,11,18,18)	2
91	(0,0),(0,1),(2,0),(4,3)	(11,15,15,19)	2
92	(0,0),(0,1),(2,0),(4,2)	(11,15,16,19)	2
93	(0,0),(0,1),(2,0),(4,9)	(11,14,14,19)	2
94	(0,0),(0,1),(2,0),(4,8)	(11,14,20,19)	2
95	(0,0),(0,1),(4,2),(6,1)	(15,15,17,17)	2
96	(0,0),(0,1),(4,2),(6,4)	(15,16,18,18)	2
97	(0,0),(0,1),(4,0),(6,1)	(14,14,17,17)	2
98	(0,0),(0,1),(4,0),(6,4)	(14,20,18,18)	2
99	(0,0),(0,1),(0,2),(3,0)	(1,13,12,18)	3
100	(0,0),(0,1),(0,2),(3,1)	(1,12,12,19)	3
101	(0,0),(0,1),(0,5),(3,0)	(4,4,12,12)	3
102	(0,0),(0,1),(0,4),(3,2)	(4,3,13,12)	3
103	(0,0),(0,1),(0,6),(3,2)	(5,13,13,17)	3
104	(0,0),(0,1),(0,6),(3,1)	(5,12,12,17)	3
105	(0,0),(0,1),(0,6),(3,0)	(5,12,12,25)	3
106	(0,0),(0,1),(0,3),(3,2)	(2,13,12,19)	3
107	(0,0),(0,1),(0,3),(3,1)	(2,12,12,18)	3
108	(0,0),(0,1),(0,3),(3,0)	(2,12,21,18)	3
109	(0,0),(0,1),(0,4),(3,0)	(3,3,12,12)	3
110	(0,0),(0,1),(0,4),(3,1)	(3,3,12,21)	3
111	(0,0),(0,1),(3,0),(6,0)	(8,12,12,17)	3
112	(0,0),(0,3),(0,6),(3,0)	(8,21,21,25)	3

Four-Element Motif Classes—Continued			
Class Nr.	Representative	Class Weight	Volume
113*	(0,0),(0,1),(3,0),(3,5)	(13,13,12,12)	3
114*	(0,0),(0,1),(3,0),(3,11)	(13,13,12,12)	3
115*	(0,0),(0,1),(3,0),(9,11)	(13,13,12,12)	3
116	(0,0),(0,1),(3,2),(3,8)	(13,13,17,17)	3
117	(0,0),(0,1),(3,0),(3,2)	(13,12,19,18)	3
118*	(0,0),(0,1),(3,0),(9,7)	(12,12,12,12)	3
119*	(0,0),(0,1),(3,0),(3,7)	(12,12,12,12)	3
120*	(0,0),(0,1),(3,0),(3,1)	(12,12,12,12)	3
121*	(0,0),(0,1),(3,0),(9,3)	(12,12,12,21)	3
122*	(0,0),(0,1),(3,0),(3,3)	(12,12,12,21)	3
123	(0,0),(0,1),(3,0),(6,5)	(12,12,19,19)	3
124	(0,0),(0,1),(3,0),(6,1)	(12,12,17,17)	3
125	(0,0),(0,1),(3,0),(3,6)	(12,12,17,25)	3
126	(0,0),(0,1),(3,0),(3,10)	(12,12,18,18)	3
127	(0,0),(0,1),(3,0),(6,9)	(12,21,18,18)	3
128*	(0,0),(0,3),(3,0),(3,3)	(21,21,21,21)	3
129*	(0,0),(0,3),(3,0),(9,9)	(21,21,21,21)	3
130	(0,0),(0,3),(3,0),(3,6)	(21,21,25,25)	3
131	(0,0),(0,1),(0,2),(4,3)	(1,15,15,15)	4
132	(0,0),(0,1),(0,2),(4,1)	(1,15,14,14)	4
133	(0,0),(0,1),(0,2),(4,0)	(1,15,14,20)	4
134	(0,0),(0,1),(0,5),(4,2)	(4,15,15,14)	4
135	(0,0),(0,1),(0,5),(4,3)	(4,15,15,20)	4
136	(0,0),(0,1),(0,5),(4,0)	(4,14,14,14)	4
137	(0,0),(0,1),(0,5),(4,1)	(4,14,14,24)	4
138	(0,0),(0,1),(0,6),(4,3)	(5,5,15,15)	4
139	(0,0),(0,1),(0,6),(4,1)	(5,5,14,14)	4
140	(0,0),(0,1),(0,6),(4,0)	(5,7,15,14)	4
141	(0,0),(0,1),(0,3),(4,2)	(2,2,15,15)	4
142	(0,0),(0,1),(0,3),(4,1)	(2,2,14,20)	4
143	(0,0),(0,1),(0,3),(4,0)	(2,3,15,14)	4
144	(0,0),(0,1),(0,3),(4,3)	(2,3,15,20)	4
145	(0,0),(0,1),(0,4),(4,1)	(3,3,14,14)	4
146	(0,0),(0,1),(0,4),(4,0)	(3,3,14,24)	4
147	(0,0),(0,2),(2,0),(6,10)	(23,16,16,16)	4
148	(0,0),(0,2),(2,0),(4,4)	(23,16,20,20)	4
149	(0,0),(0,1),(4,0),(8,6)	(6,15,15,14)	4
150	(0,0),(0,2),(0,4),(2,0)	(6,16,16,20)	4
151	(0,0),(0,2),(0,4),(4,2)	(6,20,20,20)	4

Four-Element Motif Classes—Continued			
Class Nr.	Representative	Class Weight	Volume
152	(0,0),(0,2),(0,4),(4,0)	(6,20,20,24)	4
153	(0,0),(0,2),(2,0),(2,6)	(22,22,16,16)	4
154	(0,0),(0,2),(4,0),(6,2)	(22,22,20,20)	4
155	(0,0),(0,2),(0,6),(2,0)	(22,7,16,20)	4
156	(0,0),(0,2),(0,6),(4,0)	(7,7,20,20)	4
157	(0,0),(0,2),(0,6),(4,2)	(7,7,20,24)	4
158	(0,0),(0,1),(4,0),(8,0)	(9,14,14,14)	4
159	(0,0),(0,2),(4,0),(8,0)	(9,20,20,20)	4
160	(0,0),(0,4),(0,8),(4,0)	(9,24,24,24)	4
161*	(0,0),(0,1),(4,2),(4,7)	(15,15,15,15)	4
162*	(0,0),(0,1),(4,2),(4,3)	(15,15,15,15)	4
163*	(0,0),(0,1),(4,0),(4,7)	(15,15,14,14)	4
164*	(0,0),(0,1),(4,0),(4,11)	(15,15,14,14)	4
165*	(0,0),(0,1),(4,0),(4,2)	(15,15,14,20)	4
166*	(0,0),(0,1),(4,0),(4,10)	(15,15,14,20)	4
167*	(0,0),(0,1),(4,0),(4,1)	(14,14,14,14)	4
168*	(0,0),(0,1),(4,0),(4,5)	(14,14,14,14)	4
169	(0,0),(0,1),(4,0),(4,4)	(14,14,14,24)	4
170	(0,0),(0,2),(2,0),(2,2)	(16,16,16,16)	4
171*	(0,0),(0,2),(2,0),(8,8)	(16,16,20,20)	4
172*	(0,0),(0,2),(2,0),(2,4)	(16,16,20,20)	4
173*	(0,0),(0,2),(4,0),(4,2)	(20,20,20,20)	4
174*	(0,0),(0,2),(4,0),(4,10)	(20,20,20,20)	4
175	(0,0),(0,2),(4,0),(4,4)	(20,20,20,24)	4
176	(0,0),(0,4),(4,0),(4,4)	(24,24,24,24)	4
177	(0,0),(0,1),(0,2),(6,1)	(1,1,17,17)	6
178	(0,0),(0,1),(0,2),(6,3)	(1,2,19,18)	6
179	(0,0),(0,1),(0,2),(6,4)	(1,23,18,18)	6
180	(0,0),(0,1),(0,2),(6,0)	(1,22,19,17)	6
181	(0,0),(0,1),(0,5),(6,0)	(4,4,17,17)	6
182	(0,0),(0,1),(0,4),(6,5)	(4,3,19,18)	6
183	(0,0),(0,1),(0,5),(6,3)	(4,6,18,18)	6
184	(0,0),(0,1),(0,5),(6,1)	(4,7,19,17)	6
185	(0,0),(0,1),(0,6),(6,5)	(5,5,19,19)	6
186	(0,0),(0,1),(0,6),(6,1)	(5,5,17,17)	6
187	(0,0),(0,1),(0,6),(6,2)	(5,22,19,19)	6
188	(0,0),(0,1),(0,6),(6,4)	(5,22,18,18)	6
189	(0,0),(0,1),(0,6),(6,3)	(5,8,18,18)	6
190	(0,0),(0,1),(0,6),(6,0)	(5,26,17,17)	6

Four-Element Motif Classes—Continued			
Class Nr.	Representative	Class Weight	Volume
191	(0,0),(0,1),(0,3),(6,0)	(2,2,17,25)	6
192	(0,0),(0,1),(0,3),(6,4)	(2,2,18,18)	6
193	(0,0),(0,1),(0,3),(6,5)	(2,23,19,18)	6
194	(0,0),(0,1),(0,3),(6,1)	(2,22,17,18)	6
195	(0,0),(0,1),(0,3),(6,3)	(2,22,18,25)	6
196	(0,0),(0,1),(0,4),(6,1)	(3,3,17,25)	6
197	(0,0),(0,1),(0,4),(6,3)	(3,3,18,18)	6
198	(0,0),(0,1),(0,4),(6,2)	(3,6,19,18)	6
199	(0,0),(0,1),(0,4),(6,0)	(3,7,17,18)	6
200	(0,0),(0,1),(0,4),(6,4)	(3,7,18,25)	6
201	(0,0),(0,1),(6,3),(6,9)	(22,8,18,18)	6
202	(0,0),(0,3),(0,6),(6,3)	(8,8,25,25)	6
203	(0,0),(0,3),(0,6),(6,0)	(8,26,25,25)	6
204*	(0,0),(0,1),(6,0),(6,5)	(19,19,17,17)	6
205*	(0,0),(0,1),(6,0),(6,11)	(19,19,17,17)	6
206*	(0,0),(0,1),(6,2),(6,3)	(19,19,18,18)	6
207*	(0,0),(0,1),(6,2),(6,5)	(19,19,18,18)	6
208*	(0,0),(0,1),(6,0),(6,1)	(17,17,17,17)	6
209*	(0,0),(0,1),(6,0),(6,7)	(17,17,17,17)	6
210*	(0,0),(0,1),(6,0),(6,3)	(17,18,18,25)	6
211*	(0,0),(0,1),(6,0),(6,9)	(17,18,18,25)	6
212*	(0,0),(0,1),(6,3),(6,4)	(18,18,18,18)	6
213*	(0,0),(0,1),(6,3),(6,10)	(18,18,18,18)	6
214*	(0,0),(0,3),(6,0),(6,3)	(25,25,25,25)	6
215*	(0,0),(0,3),(6,0),(6,9)	(25,25,25,25)	6

O.5 Three Tone Motifs in $OnPiMod_{5,12}$

Refer to the discussion in Section 11.3.8 for the entries of this table. The order of these representatives is a historical one.

	Three-Element Motif Classes in $OnPiMod_{5,12}$	
Cl. Nr.	Representative in $\mathbb{Z}_5 \times \mathbb{Z}_{12}$	Kernel
1	(0,0),(1,0),(0,4)	$\mathbb{Z}_4^2 \times (\mathbb{Z}_3 \times 0) \times (0 \times \mathbb{Z}_5)$
2	(0,0),(1,0),(4,4)	$\mathbb{Z}_4^2 \times (\mathbb{Z}_3 \times 0) \times \mathbb{Z}_5.(1,1)$
3	(0,0),(1,0),(3,4)	$\mathbb{Z}_4^2 \times (\mathbb{Z}_3 \times 0) \times \mathbb{Z}_5.(2,1)$
4	(0,0),(0,4),(1,8)	$\mathbb{Z}_4^2 \times \mathbb{Z}_3.(1,1) \times (\mathbb{Z}_5 \times 0)$
5	(0,0),(1,4),(4,8)	$\mathbb{Z}_4^2 \times \mathbb{Z}_3.(1,1) \times \mathbb{Z}_5.(1,1)$
6	(0,0),(1,0),(0,10)	$(\mathbb{Z}_4 \times 2\mathbb{Z}_4) \times (\mathbb{Z}_3 \times 0) \times (0 \times \mathbb{Z}_5)$
7	(0,0),(1,0),(4,10)	$(\mathbb{Z}_4 \times 2\mathbb{Z}_4) \times (\mathbb{Z}_3 \times 0) \times \mathbb{Z}_5.(1,1)$
8	(0,0),(1,0),(3,10)	$(\mathbb{Z}_4 \times 2\mathbb{Z}_4) \times (\mathbb{Z}_3 \times 0) \times \mathbb{Z}_5.(2,1)$
9	(0,0),(0,4),(1,2)	$(\mathbb{Z}_4 \times 2\mathbb{Z}_4) \times \mathbb{Z}_3.(1,1) \times (\mathbb{Z}_5 \times 0)$
10	(0,0),(1,4),(0,2)	$(\mathbb{Z}_4 \times 2\mathbb{Z}_4) \times \mathbb{Z}_3.(1,1) \times (0 \times \mathbb{Z}_5)$
11	(0,0),(1,4),(4,2)	$(\mathbb{Z}_4 \times 2\mathbb{Z}_4) \times \mathbb{Z}_3.(1,1) \times \mathbb{Z}_5.(1,1)$
12	(0,0),(1,4),(3,2)	$(\mathbb{Z}_4 \times 2\mathbb{Z}_4) \times \mathbb{Z}_3.(1,1) \times \mathbb{Z}_5.(2,1)$
13	(0,0),(0,4),(1,6)	$(\mathbb{Z}_4 \times 2\mathbb{Z}_4) \times (0 \times \mathbb{Z}_3) \times (\mathbb{Z}_5 \times 0)$
14	(0,0),(1,4),(0,6)	$(\mathbb{Z}_4 \times 2\mathbb{Z}_4) \times (0 \times \mathbb{Z}_3) \times (0 \times \mathbb{Z}_5)$
15	(0,0),(1,4),(4,6)	$(\mathbb{Z}_4 \times 2\mathbb{Z}_4) \times (0 \times \mathbb{Z}_3) \times \mathbb{Z}_5.(1,1)$
16	(0,0),(1,4),(3,6)	$(\mathbb{Z}_4 \times 2\mathbb{Z}_4) \times (0 \times \mathbb{Z}_3) \times \mathbb{Z}_5.(2,1)$
17	(0,0),(3,4),(1,6)	$(\mathbb{Z}_4 \times 2\mathbb{Z}_4) \times (0 \times \mathbb{Z}_3) \times \mathbb{Z}_5.(1,2)$
18	(0,0),(1,4),(1,6)	$(\mathbb{Z}_4 \times 2\mathbb{Z}_4) \times (0 \times \mathbb{Z}_3) \times \mathbb{Z}_5.(4,1)$
19	(0,0),(1,0),(0,1)	$(\mathbb{Z}_4 \times 0) \times (\mathbb{Z}_3 \times 0) \times (0 \times \mathbb{Z}_5)$
20	(0,0),(1,0),(4,1)	$(\mathbb{Z}_4 \times 0) \times (\mathbb{Z}_3 \times 0) \times \mathbb{Z}_5.(1,1)$
21	(0,0),(1,0),(3,1)	$(\mathbb{Z}_4 \times 0) \times (\mathbb{Z}_3 \times 0) \times \mathbb{Z}_5.(2,1)$
22	(0,0),(0,4),(1,5)	$(\mathbb{Z}_4 \times 0) \times \mathbb{Z}_3.(1,1) \times (\mathbb{Z}_5 \times 0)$
23	(0,0),(1,4),(0,5)	$(\mathbb{Z}_4 \times 0) \times \mathbb{Z}_3.(1,1) \times (0 \times \mathbb{Z}_5)$
24	(0,0),(1,4),(4,5)	$(\mathbb{Z}_4 \times 0) \times \mathbb{Z}_3.(1,1) \times \mathbb{Z}_5.(1,1)$
25	(0,0),(1,4),(3,5)	$(\mathbb{Z}_4 \times 0) \times \mathbb{Z}_3.(1,1) \times \mathbb{Z}_5.(2,1)$
26	(0,0),(0,4),(1,9)	$(\mathbb{Z}_4 \times 0) \times (0 \times \mathbb{Z}_3) \times (\mathbb{Z}_5 \times 0)$
27	(0,0),(1,4),(0,9)	$(\mathbb{Z}_4 \times 0) \times (0 \times \mathbb{Z}_3) \times (0 \times \mathbb{Z}_5)$
28	(0,0),(1,4),(4,9)	$(\mathbb{Z}_4 \times 0) \times (0 \times \mathbb{Z}_3) \times \mathbb{Z}_5.(1,1)$
29	(0,0),(1,4),(3,9)	$(\mathbb{Z}_4 \times 0) \times (0 \times \mathbb{Z}_3) \times \mathbb{Z}_5.(2,1)$
30	(0,0),(3,4),(1,9)	$(\mathbb{Z}_4 \times 0) \times (0 \times \mathbb{Z}_3) \times \mathbb{Z}_5.(1,2)$
31	(0,0),(1,4),(1,9)	$(\mathbb{Z}_4 \times 0) \times (0 \times \mathbb{Z}_3) \times \mathbb{Z}_5.(4,1)$
32	(0,0),(0,6),(1,1)	$\mathbb{Z}_4.(1,2) \times (\mathbb{Z}_3 \times 0) \times (\mathbb{Z}_5 \times 0)$

	Three-Element Motif Classes in $OnPiMod_{5,12}$—continued	
Cl. Nr.	Representative in $\mathbb{Z}_5 \times \mathbb{Z}_{12}$	Kernel
33	(0,0),(1,6),(0,1)	$\mathbb{Z}_4.(1,2) \times (\mathbb{Z}_3 \times 0) \times (0 \times \mathbb{Z}_5)$
34	(0,0),(1,6),(4,1)	$\mathbb{Z}_4.(1,2) \times (\mathbb{Z}_3 \times 0) \times \mathbb{Z}_5.(1,1)$
35	(0,0),(1,6),(3,1)	$\mathbb{Z}_4.(1,2) \times (\mathbb{Z}_3 \times 0) \times \mathbb{Z}_5.(2,1)$
36	(0,0),(0,10),(1,5)	$\mathbb{Z}_4.(1,2) \times \mathbb{Z}_3.(1,1) \times (\mathbb{Z}_5 \times 0)$
37	(0,0),(1,10),(0,5)	$\mathbb{Z}_4.(1,2) \times \mathbb{Z}_3.(1,1) \times (0 \times \mathbb{Z}_5)$
38	(0,0),(1,10),(4,5)	$\mathbb{Z}_4.(1,2) \times \mathbb{Z}_3.(1,1) \times \mathbb{Z}_5.(1,1)$
39	(0,0),(1,10),(3,5)	$\mathbb{Z}_4.(1,2) \times \mathbb{Z}_3.(1,1) \times \mathbb{Z}_5.(2,1)$
40	(0,0),(0,10),(1,9)	$\mathbb{Z}_4.(1,2) \times (0 \times \mathbb{Z}_3) \times (\mathbb{Z}_5 \times 0)$
41	(0,0),(1,10),(0,9)	$\mathbb{Z}_4.(1,2) \times (0 \times \mathbb{Z}_3) \times (0 \times \mathbb{Z}_5)$
42	(0,0),(4,10),(1,9)	$\mathbb{Z}_4.(1,2) \times (0 \times \mathbb{Z}_3) \times \mathbb{Z}_5.(1,1)$
43	(0,0),(3,10),(1,9)	$\mathbb{Z}_4.(1,2) \times (0 \times \mathbb{Z}_3) \times \mathbb{Z}_5.(1,2)$
44	(0,0),(1,10),(3,9)	$\mathbb{Z}_4.(1,2) \times (0 \times \mathbb{Z}_3) \times \mathbb{Z}_5.(2,1)$
45	(0,0),(1,10),(1,9)	$\mathbb{Z}_4.(1,2) \times (0 \times \mathbb{Z}_3) \times \mathbb{Z}_5.(4,1)$

P

Well-Tempered and Just Modulation Steps

P.1 12-Tempered Modulation Steps

P.1.1 Scale Orbits and Number of Quantized Modulations

In the following table, the exclamation sign (!) in column 6 means that quantization is not possible for every translation quantity p in the notation of Theorem 30.

Orbits and Number of Quantized Modulations			
Class	*# Min. Cadence Sets*	*# Quanta*	*# Quant. Mod.*
38	9	42	54 (!)
38.1	5	20	26
47	6	28	30
47.1	15	66	114
50	7	34	42
50.1	6	36	46
52	5	24	24 (!)
55	6	30	32 (!)
61	10	38	62
62	5	24	24 (!)
39	9	29	93
39.1	6	23	55
40	10	24	108
40.1	7	26	72
41	7	25	75
41.1	6	21	53
42	6	22	54
42.1	7	28	74
43	6	22	57
43.1	7	26	72
44	9	23	89
45	7	21	63

© Springer International Publishing AG, part of Springer Nature 2017
G. Mazzola, *The Topos of Music IV: Roots*, Computational Music Science,
https://doi.org/10.1007/978-3-319-64495-0_16

Orbits and Number of Quant. Mod.—Continued			
Class	# Min. Cadence Sets	# Quanta	# Quant. Mod.
45.1	10	21	105
46	6	26	56
48	10	23	109
48.1	7	28	68
49	7	21	71
49.1	7	26	74
51	9	13	86
53	7	27	67
53.1	9	25	91
54	7	32	71
54.1	21	32	226
56	7	24	70
57	8	21	71
58	18	17	185
59	11	22	101
60	6	21	60

P.1.2 Quanta and Pivots for the Modulations Between Diatonic Major Scales (No.38.1)

Quanta and Pivots for the Modulations Between Diatonic Major Scales				
Transl. p	Cadence	Quantum	Modulator	Pivots
1	$\{II,V\}$	●○●●○●●●●●●	$e^5 11$	$\{II,III,V,VII\}$
1	$\{II,III\}$	●○●●○●●●●●●	$e^5 11$	$\{II,III,V,VII\}$
2	$\{VII\}$	○●●○●●○●○○●●	$e^6 11$	$\{II,IV,VII\}$
2	$\{II,V\}$	○●●○●●○●○●○●	$e^6 11$	$\{II,IV,V,VII\}$
2	$\{IV,V\}$	○●●○●●○●○●○●	$e^6 11$	$\{II,IV,V,VII\}$
3	$\{II,V\}$	●○●○○●○●●●●●	$e^7 11$	$\{II,III,V,VII\}$
3	$\{II,III\}$	●○●○○●○●●●●●	$e^7 11$	$\{II,III,V,VII\}$
4	$\{VII\}$	○○●●○●●○○●○●	$e^8 11$	$\{II,IV,V,VII\}$
4	$\{IV,V\}$	○●●●●●●●○●○●	$e^8 11$	$\{II,III,V,VII\}$
4	$\{II,III\}$	●●●●○●●●●●○●	$e^8 11$	$\{V,VII\}$
5	$\{VII\}$	○○●○●●○●○○●●	$e^9 11$	$\{II,IV,VII\}$
6	$\{II,III\}$	○●●●●●○●●●●	e^6	$\{II,III,V,VII\}$
6	$\{IV,V\}$	○●●●●●●●●●○●	$e^{10} 11$	$\{II,IV,V,VII\}$
6	$\{IV,V\}$	●●●●○●●●●●○●	e^6	$\{II,IV,V,VII\}$
6	$\{II,III\}$	●●●●○●○●●●●●	$e^{10} 11$	$\{II,III,V,VII\}$
7	$\{VII\}$	●○●○○●●○○●○●	$e^{11} 11$	$\{III,V,VII\}$
8	$\{VII\}$	○●●○○●○●○○●●	$e^0 11$	$\{II,VII\}$
8	$\{IV,V\}$	○●●●●●○●●●●●	$e^0 11$	$\{II,IV,V,VII\}$
8	$\{II,III\}$	●●●●○●○●○●●●	$e^0 11$	$\{II,III,V,VII\}$
9	$\{II,V\}$	○○●○●●●●●●○●	$e^1 11$	$\{II,IV,V,VII\}$
9	$\{IV,V\}$	○○●○●●●●●●○●	$e^1 11$	$\{II,IV,V,VII\}$
10	$\{VII\}$	●○●●○●○○○●○●	$e^2 11$	$\{III,V,VII\}$
10	$\{II,V\}$	●○●●○●○●○●○●	$e^2 11$	$\{II,III,V,VII\}$
10	$\{II,III\}$	●○●●○●○●○●○●	$e^2 11$	$\{II,III,V,VII\}$
11	$\{II,V\}$	○●●○●●●●●●●	$e^3 11$	$\{II,IV,V,VII\}$
11	$\{IV,V\}$	○●●○●●●●●●●	$e^3 11$	$\{II,IV,V,VII\}$

P.1.3 Quanta and Pivots for the Modulations Between Melodic Minor Scales (No.47.1)

The symbol p stands for "translation p"; M stands for "Modulator".

Quanta and Pivots for the Modulations Between Melodic Minor Scales

p	Cadence	Quantum	M	Pivots
1	$\{II,IV\},\{IV,VII\}$	●●●●○●●○○●●○	$e^3 11$	$\{II,IV,VII\}$
1	$\{III,VI\},\{V,VI\}$	●●●●●●○●●○●●	$e^3 11$	$\{I,III,V,VI\}$
1	$\{III,VII\}$	●○○●●○●●●●○●	$e^3 11$	$\{III,V,VII\}$
1	$\{IV,V\}$	●●●●○●●●●●●○	$e^3 11$	$\{II,IV,V,VII\}$
1	$\{II,III\}$	●○○●●●●●●●●●	$e^3 11$	$\{II,III,V,VII\}$
1	$\{I,VII\}$	●●●●●○●●●●○●	$e^3 11$	$\{I,III,V,VII\}$
2	$\{III,V\},\{III,VII\},\{II,III\}$	●●○●●●●○●○●●	$e^4 11$	$\{II,III,V,VII\}$
2	$\{II,IV\},\{II,VI\},\{I,II\}$	●○●○●○●○●●○●	$e^4 11$	$\{I,II,IV,VI\}$
2	$\{I,III\},\{III,VI\},\{III,IV\}$	○●●●○●○●○●○●	$e^4 11$	$\{I,III,IV,VI\}$
3	$\{III,V\},\{III,VI\},\{V,VI\},\{I,V\}$	●○●●○●●●○○●●	$e^5 11$	$\{I,III,V,VI\}$
3	$\{III,V\},\{III,VI\},\{V,VI\},\{I,V\}$	●○●●○●●●○○●●	$e^5 11$	$\{I,III,V,VI\}$
3	$\{III,V\},\{III,VI\},\{V,VI\},\{I,V\}$	●○●●○●●●○○●●	$e^5 11$	$\{I,III,V,VI\}$
3	$\{III,V\},\{III,VI\},\{V,VI\},\{I,V\}$	●○●●○●●●○○●●	$e^5 11$	$\{I,III,V,VI\}$
4	$\{III,V\}$	○●●○●●○○●●●○	$e^6 11$	$\{III,V\}$
4	$\{I,III\}$	○○●●●○○●○○○●	$e^6 11$	$\{I,III\}$
4	$\{II,VI\},\{I,II\}$	●●●○●●●○●○●●	$e^6 11$	$\{I,II,IV,VI\}$
4	$\{IV,VII\}$	●●●●●●●○○●○○	$e^6 11$	$\{II,IV,VII\}$
4	$\{III,VI\}$	○●●●●●○●○○○●	$e^6 11$	$\{I,III,VI\}$
4	$\{III,VII\}$	●○○●○○●●○●○●	$e^6 11$	$\{III,V,VII\}$
4	$\{V,VI\}$	●●●●●●●●○○○●	$e^6 11$	$\{I,III,V,VI\}$
4	$\{III,IV\}$	○●●●●○●○●○●●	$e^6 11$	$\{I,III,IV,VI\}$
4	$\{II,III\}$	●●○●○●●●○●○●	$e^6 11$	$\{II,III,V,VII\}$
4	$\{I,VII\}$	●○●●●○●●○●○●	$e^6 11$	$\{I,III,V,VII\}$
4	$\{I,III\}$	●○○○●●○○●●○●	e^4	$\{I,III\}$
4	$\{III,V\}$	○○●●○○●●●○●●	e^4	$\{III,V\}$
4	$\{III,VII\},\{II,III\}$	○●●●○●●●○●●●	e^4	$\{II,III,V,VII\}$
4	$\{III,VI\},\{III,IV\}$	●●●○●●●○●●○●	e^4	$\{I,III,IV,VI\}$
5	$\{I,II\},\{I,V\},\{III,VI\},\{V,VI\}$	●○●●●●○●○●○●	$e^7 11$	$\{I,III,V,VI\}$
5	$\{II,VI\},\{I,II\}$	●○●○○●○●●●●●	$e^7 11$	$\{I,II,IV,VI\}$
5	$\{IV,VII\},\{IV,V\}$	●○●●●●○●○○●○	$e^7 11$	$\{II,IV,V,VII\}$
5	$\{III,VII\}$	●○○●●○○●●●●●	$e^7 11$	$\{III,V,VII\}$
6	$\{III,V\},\{III,VII\},\{II,III\}$	●●○●○●○●●●○●	$e^8 11$	$\{II,III,V,VII\}$
6	$\{I,III\},\{III,VI\},\{III,IV\}$	○●●●○●●●●○●●	$e^8 11$	$\{I,III,IV,VI\}$
6	$\{I,III\},\{III,VI\},\{III,IV\}$	●●○●○●●●○●○●	e^6	$\{I,III,IV,VI\}$

\multicolumn{5}{c}{**Quanta and Pivots for Melodic Minor Scales**—Continued}
p	*Cadence*	*Quantum*	M	*Pivots*
6	$\{III,V\},\{III,VII\},\{II,III\}$	○●●●○●○●●●○●	e^6	$\{II,III,V,VII\}$
7	$\{III,V\},\{I,V\},\{III,VII\},\{I,VII\}$	●○○●●○○●●○●●	$e^9 11$	$\{I,III,V,VII\}$
7	$\{II,VI\},\{I,II\}$	●○●○●●○●○●●	$e^9 11$	$\{I,II,IV,VI\}$
7	$\{IV,VII\},\{IV,V\}$	●○●●●●●●○●○○	$e^9 11$	$\{II,IV,V,VII\}$
7	$\{III,VI\}$	○○●●●●●●○○●●	$e^9 11$	$\{I,III,VI\}$
8	$\{III,V\}$	●○○●●○○○●○○●●	$e^{10} 11$	$\{III,V\}$
8	$\{I,III\}$	○○●●○○○●●○○●	$e^{10} 11$	$\{I,III\}$
8	$\{II,VI\}$	●●●○○●○○●●●●	$e^{10} 11$	$\{II,IV,VI\}$
8	$\{IV,VII\},\{IV,V\}$	●●●●○●○●●●●○	$e^{10} 11$	$\{II,IV,V,VII\}$
8	$\{III,VI\}$	○○●●○●○●●○○●	$e^{10} 11$	$\{I,III,VI\}$
8	$\{III,VII\}$	●●○●○○○●○●●●	$e^{10} 11$	$\{III,V,VII\}$
8	$\{V,VI\}$	●○●●○●○●●●●●	$e^{10} 11$	$\{I,III,V,VI\}$
8	$\{III,IV\}$	○●●●○●○●○●●	$e^{10} 11$	$\{I,III,IV,VI\}$
8	$\{II,III\}$	●●○●○●○●○●●●	$e^{10} 11$	$\{II,III,V,VII\}$
8	$\{I,VII\}$	●●●●○○○●●●●●	$e^{10} 11$	$\{I,III,V,VII\}$
8	$\{I,III\}$	●○○○●●○○●●○●	e^8	$\{I,III\}$
8	$\{III,V\}$	○○●●○○●●○○●●	e^8	$\{III,V\}$
8	$\{III,VII\},\{II,III\}$	○●●●●○●●●○●●	e^8	$\{II,III,V,VII\}$
8	$\{III,VI\},\{II,IV\}$	●●○●●●○●●●○●	e^8	$\{I,III,IV,VI\}$
9	$\{III,V\}$	●○○●●○○●●○○●	$e^{11} 11$	$\{III,V\}$
9	$\{II,IV\},\{II,VI\}$	●○●○○●●○○●○●	$e^{11} 11$	$\{II,IV,VI\}$
9	$\{I,III\},\{I,V\},\{III,VII\},\{I,VII\}$	●○●●●○○●●●○●	$e^{11} 11$	$\{I,III,V,VII\}$
9	$\{I,II\}$	●○●○●●●●○●○●	$e^{11} 11$	$\{I,III,IV,VI\}$
10	$\{II,V\},\{III,VII\},\{II,III\}$	●●○●○●○●○●○●	$e^0 11$	$\{II,III,V,VII\}$
10	$\{II,IV\},\{IV,VII\},\{IV,V\}$	●○●●●○●○●○●○	$e^0 11$	$\{II,IV,V,VII\}$
10	$\{I,III\},\{III,VI\},\{III,IV\}$	○●●●○●○●○●●●	$e^0 11$	$\{I,III,IV,VI\}$
11	$\{II,IV\},\{II,VI\}$	●●●○●●○○●●○●	$e^1 11$	$\{I,IV,VI\}$
11	$\{III,VI\}$	○○●●○●○●●●○●●	$e^1 11$	$\{I,III,VI\}$
11	$\{III,VII\},\{I,VII\}$	●●●●○●○●●●●	$e^1 11$	$\{I,III,V,VII\}$
11	$\{V,VI\}$	●●●●○●●●●○●●	$e^1 11$	$\{I,III,V,VI\}$
11	$\{III,IV\}$	○○●●●●●●●●●	$e^1 11$	$\{I,III,IV,VI\}$
11	$\{I,II\}$	●●●○●●●●●●○●	$e^1 11$	$\{I,II,IV,VI\}$

P.1.4 Quanta and Pivots for the Modulations Between Harmonic Minor Scales (No.54.1)

For this table, we need a numbering of the 21 minimal cadence sets:

$$0 = \{II, VII\} \quad 1 = \{I, III\} \quad 2 = \{II, IV\} \quad 3 = \{III, V\} \quad 4 = \{IV, VI\}$$
$$5 = \{V, VII\} \quad 6 = \{I, VI\} \quad 7 = \{IV, VII\} \quad 8 = \{I, V\} \quad 9 = \{II, VI\}$$
$$10 = \{III, VII\} \quad 11 = \{I, IV\} \quad 12 = \{II, V\} \quad 13 = \{III, VI\} \quad 14 = \{VI, VII\}$$
$$15 = \{I, VII\} \quad 16 = \{I, II\} \quad 17 = \{II, III\} \quad 18 = \{III, IV\} \quad 19 = \{IV, V\}$$
$$20 = \{V, VI\}$$

Quanta and Pivots for the Modulations Between Harmonic Minor Scales

Transl. p	Cadence Nr.	Quantum	Pivots
3/9	1,3,6,8,10,11,15-20	○○○○○○○○○○○○○○	$\{I, II, III, IV, V, VI, VII\}$
3/9	2,4,7,9,14	●○○●●○●●○●●○●	$\{II, IV, VI, VII\}$
3/9	5,12	○●●○●●○●●○●●	$\{II, V, VII\}$
4/8	0,7,12,14-17,19	○○○○○○○○○○○○○○	$\{I, II, III, IV, V, VI, VII\}$
4/8	1,6,13	●○○●●○○●●○●●	$\{I, II, VI\}$
4/8	2	●●●○●●●○●●●●	$\{II, IV\}$
4/8	3	○○●●○○●●○●●	$\{III, V\}$
4/8	4,11,18	●●○●●●○●●●○●	$\{I, III, IV, VI\}$
4/8	5,10	○●●●○●●●○●●	$\{III, V, VII\}$
4/8	8,20	●○●●●○●●●○●	$\{I, III, V, VI\}$
6	2,7	●○●○○●●○●○●●	$\{II, IV, VII\}$
6	3,10,17	○●●●○●○●●●○●	$\{II, III, V, VII\}$
6	4,9,14	●○●○●○●●○●○●	$\{II, IV, VI, VII\}$
6	5,12	○●●○○●○●●○○●	$\{II, V, VII\}$
6	8,11,13,15,16,18	●●●○●●●●●○●	$\{I, II, II, IV, V, VI, VII\}$
6	19	●●○○○●●●●○○●	$\{II, IV, V, VII\}$
1,2,5,7,10,11	any cadence set	●●●●●●●●●●●	$\{I, II, II, IV, V, VI, VII\}$

P.1.5 Examples of 12-Tempered Modulations for All Fourth Relations

Examples of Modulations for all Fourth Relations

Start → Target	Neutral	Pivots	Cadence
$C \rightsquigarrow F$	I_C	$VII_F \cup II_F$	I_F, IV_F, V_F, I_F
$C \rightsquigarrow B_\flat$	I_C, V_C	$III_{B_\flat}, V_{B_\flat} \cup VII_{B_\flat}$	$VII_{B_\flat}, I_{B_\flat}$
$C \rightsquigarrow E_\flat$	I_C, V_C	$II_{E_\flat}, V_{E_\flat} \cup VII_{E_\flat}$	$V_{E_\flat} \cup VII_{E_\flat}, I_{E_\flat}$
$C \rightsquigarrow A_\flat$	I_C, IV_C	$II_{A_\flat} \cup VII_{A_\flat}$	$IV_{A_\flat}, V_{A_\flat}, I_{A_\flat}$
$C \rightsquigarrow D_\flat$	I_C	$III_{D_\flat}, II_{D_\flat} \cup VII_{D_\flat}$	$II_{D_\flat}, V_{D_\flat}, I_{D_\flat}$
$C \rightsquigarrow G_\flat$	$I_C, IV_C \cup VI_C, VI_C \cup VII_C$	$II_{G_\flat} \cup VII_{G_\flat}, II_{G_\flat}$	$V_{G_\flat} \cup VII_{G_\flat}, I_{G_\flat}$

Examples of Modulations for all Fourth Relations—Continued			
$Start \to Target$	$Neutral$	$Pivots$	$Cadence$
$C \rightsquigarrow B$	I_C, V_C, II_C	$IV_B, II_B \cup VII_B$	IV_B, V_B, I_B
$C \rightsquigarrow E$	I_C, VI_C	$V_E \cup VII_E$	IV_E, V_E, I_E
$C \rightsquigarrow A$	I_C, V_C	IV_A	$IV_A \cup VII_A, V_A, I_A$
$C \rightsquigarrow D$	I_C, VI_C, V_C	$VII_D \cup V_D$	II_D, V_D, I_D
$C \rightsquigarrow G$	I_C, V_C	III_G	II_G, V_G, I_G

P.2 2-3-5-Just Modulation Steps

The following tables show data for modulations from C-tonic.

P.2.1 Modulation Steps Between Just Major Scales

Here, we have the two modulators

$$\Phi_1 = e^b, \tag{P.1}$$
$$\Phi_2 = e^b.A. \tag{P.2}$$

The numbering of minimal cadence sets is the one used in Formula (26.2). The tonics D^* and $B_\flat^*$ are the usual third comma shifted representatives of D and B-flat.

Pivots for the Modulations Between Just Major Scales				
$Translation$	$Target\ Tonic$	$Modulator$	$Cadence$	$Pivots$
$(1,0)$	G	Φ_2	5	$\{V, VII\}$
$(-1,0)$	F	Φ_2	1	$\{II, IV\}$
$(2,0)$	D	Φ_2	1	$\{II, IV\}$
$(-2,0)$	$B_\flat$	Φ_2	5	$\{V, VII\}$
$(0,1)$	E	Φ_2	1	$\{II, V, VII\}$
$(0,1)$	E	Φ_1	5	$\{II, V, VII\}$
$(0,-1)$	$A_\flat$	Φ_2	5	$\{II, IV, VII\}$
$(0,-1)$	$A_\flat$	Φ_1	5	$\{II, V, VII\}$
$(1,1)$	B	Φ_2	1	$\{II, V, VI\}$
$(-1,-1)$	$D_\flat$	Φ_2	5	$\{II, V, VII\}$
$(1,-1)$	$E_\flat$	Φ_2	1	$\{II, V, VII\}$
$(1,-1)$	$E_\flat$	Φ_1	1	$\{II, IV, VII\}$
$(-1,1)$	A	Φ_2	5	$\{II, IV, VII\}$
$(-1,1)$	A	Φ_1	1	$\{II, IV, VII\}$
$(-2,1)$	D^*	Φ_2	5	$\{III, V, VII\}$
$(2,-1)$	$B_\flat^*$	Φ_2	1	$\{II, IV, VI\}$

P.2.2 Modulation Steps Between Natural Minor Scales

We have the two modulators

$$\Phi_1 = e^b,$$
$$\Phi_2 = e^b.A.$$

The minimal cadence sets are these:

$$J_1 = \{VII\}, J_2 = \{III, VI\}, J_3 = \{V, VI\},$$
$$J_4 = \{IV, V\}, J_5 = \{II\}, J_6 = \{III, IV\}.$$

Pivots for the Modulations Between Natural Minor Scales				
Translation	Target Tonic	Modulator	Cadence	Pivots
$(-1,0)$	F	Φ_2	5	$\{II, IV\}$
$(1,0)$	G	Φ_2	1	$\{V, VII\}$
$(-2,0)$	$B_\flat$	Φ_2	1	$\{V, VII\}$
$(2,0)$	D	Φ_2	5	$\{II, IV\}$
$(0,-1)$	$A_\flat$	Φ_2	1	$\{II, IV, VII\}$
$(0,-1)$	$A_\flat$	Φ_1	5	$\{II, IV, VII\}$
$(0,1)$	E	Φ_1	5	$\{II, IV, VII\}$
$(0,1)$	E	Φ_2	5	$\{II, V, VII\}$
$(-1,-1)$	$D_\flat$	Φ_2	1	$\{III, V, VII\}$
$(1,1)$	B	Φ_2	5	$\{II, IV, VI\}$
$(1,-1)$	$E_\flat$	Φ_1	1	$\{II, V, VII\}$
$(1,-1)$	$E_\flat$	Φ_2	5	$\{II, V, VII\}$
$(-1,1)$	A	Φ_1	1	$\{II, V, VII\}$
$(-1,1)$	A	Φ_2	1	$\{II, IV, VII\}$
$(2,-1)$	$B_\flat^*$	Φ_2	5	$\{II, IV, VI\}$
$(-2,1)$	D^*	Φ_2	1	$\{III, V, VII\}$

P.2.3 Modulation Steps from Natural Minor to Major Scales

We have the two modulators

$$\Phi_1 = e^b.A,$$
$$\Phi_2 = e^b.B, B = \begin{pmatrix} 1 & 1 \\ 0 & -1 \end{pmatrix}.$$

The minimal cadence sets are the same as for major scales.

\multicolumn{5}{c}{**Pivots for the Modulations From Natural Minor to Major Scales**}
Translation	*Target Tonic*	*Modulator*	*Cadence*	*Pivots*
$(-2,0)$	$B_\flat$	Φ_1	5	$\{V, VII\}$
$(-1,0)$	F	Φ_1	1	$\{II, IV\}$
$(1,0)$	G	Φ_1	5	$\{V, VII\}$
$(2,0)$	D	Φ_1	1	$\{II, IV\}$
$(-1,-1)$	$D_\flat$	Φ_1	5	$\{II, V, VII\}$
$(0,-1)$	$A_\flat$	Φ_1	5	$\{II, IV, VII\}$
$(1,-1)$	$E_\flat$	Φ_1	1	$\{II, V, VII\}$
$(2,-1)$	$B_\flat^*$	Φ_1	1	$\{II, IV, VI\}$

P.2.4 Modulation Steps from Major to Natural Minor Scales

We have the two modulators (same as above).

$$\Phi_1 = e^b.A,$$

$$\Phi_2 = e^b.B, B = \begin{pmatrix} 1 & 1 \\ 0 & -1 \end{pmatrix}.$$

The minimal cadence sets are those of minor scales.

\multicolumn{5}{c}{**Pivots for the Modulations From Major to Natural Minor Scales**}
Translation	*Target Tonic*	*Modulator*	*Cadence*	*Pivots*
$(-2,0)$	$B_\flat$	Φ_1	1	$\{V, VII\}$
$(-1,0)$	F	Φ_1	5	$\{II, IV\}$
$(1,0)$	G	Φ_1	1	$\{V, VII\}$
$(2,0)$	D	Φ_1	5	$\{II, IV\}$
$(-2,1)$	D	Φ_1	1	$\{III, V, VII\}$
$(-1,1)$	A	Φ_1	1	$\{II, IV, VII\}$
$(0,1)$	E	Φ_1	5	$\{II, V, VII\}$
$(1,1)$	B	Φ_1	5	$\{II, IV, VI\}$

P.2.5 Modulation Steps Between Harmonic Minor Scales

We have the unique translation modulator

$$\Phi = e^b.$$

The minimal cadence sets are these:

$$J_1 = \{III\}, J_2 = \{II\}, J_3 = \{VII\}, J_4 = \{I, IV\}, J_5 = \{I, V\},$$
$$J_6 = \{I, VI\}, J_7 = \{IV, V\}, J_8 = \{IV, VI\}, J_9 = \{V, VI\}.$$

Pivots for the Modulations Between Harmonic Minor Scales				
Translation	Target Tonic	Modulator	Cadence	Pivots
$(3,0)$	A^*	Φ	8	$\{II, IV, VI\}$
$(-3,0)$	$E_\flat^*$	Φ	8	$\{II, IV, VI\}$
$(0,1)$	E	Φ	2	$\{II, IV\}$
$(0,1)$	E	Φ	4	$\{I, III, IV, VI\}$
$(0,1)$	E	Φ	8	$\{I, III, IV, VI\}$
$(0,-1)$	$A_\flat$	Φ	2	$\{II, IV\}$
$(0,-1)$	$A_\flat$	Φ	4	$\{I, III, IV, VI\}$
$(0,-1)$	$A_\flat$	Φ	8	$\{I, III, IV, VI\}$
$(2,1)$	$f_\sharp$	Φ	8	$\{II, IV, VI, VII\}$
$(-2,-1)$	$G_\flat$	Φ	8	$\{II, IV, VI, VII\}$
$(0,2)$	$G_\sharp$	Φ	4	$\{I, III, IV, VI\}$
$(0,-2)$	$F_\flat$	Φ	4	$\{I, III, IV, VI\}$
$(1,2)$	$D_\sharp$	Φ	4	$\{I, III, IV, VI\}$
$(-1,-2)$	$B_{\flat,\flat}$	Φ	4	$\{I, III, IV, VI\}$
$(1,0)$	G	Φ	1,5,7,9	$\{I \ldots VII\}$
$(-1,0)$	F	Φ	1,5,7,9	$\{I \ldots VII\}$
$(2,0)$	D	Φ	9	$\{I \ldots VII\}$
$(-2,0)$	$B_\flat$	Φ	9	$\{I \ldots VII\}$
$(3,0)$	A^*	Φ	9	$\{I \ldots VII\}$
$(-3,0)$	$E_\flat^*$	Φ	9	$\{I \ldots VII\}$
$(-2 \ldots 1, 1)$		Φ	7	$\{I \ldots VII\}$
$(-2 \ldots -1, 1)$		Φ	9	$\{I \ldots VII\}$
$(2,1)$	$F_\sharp$	Φ	5,9	$\{I \ldots VII\}$
$(1 \ldots 2, 1)$		Φ	4	$\{I \ldots VII\}$
$(1,1)$	B	Φ	8	$\{I \ldots VII\}$
$(-1 \ldots 2, 1)$		Φ	7	$\{I \ldots VII\}$
$(1 \ldots 2, -1)$		Φ	9	$\{I \ldots VII\}$
$(-2,-1)$	$G_\flat$	Φ	5,9	$\{I \ldots VII\}$
$(-2 \ldots -1, -1)$		Φ	4	$\{I \ldots VII\}$
$(-1,-1)$	$D_\flat$	Φ	8	$\{II, IV, VI\}$
$(0,2)$	$G_\sharp$	Φ	7	$\{II, IV, VI\}$
$(0,-2)$	$F_\flat$	Φ	7	$\{I \ldots VII\}$

P.2.6 Modulation Steps Between Melodic Minor Scales

We have the two modulators

$$\Phi_1 = e^b,$$
$$\Phi_2 = e^b.A.$$

The minimal cadence sets are these:

$$J_1 = \{I\}, J_2 = \{II\}, J_3 = \{III\},$$
$$J_4 = \{I, IV\}, J_5 = \{VI\}, J_6 = \{VII\}.$$

Pivots for the Modulations Between Melodic Minor Scales				
Translation	*Target Tonic*	*Modulator*	*Cadence*	*Pivots*
$(1,0)$	G	Φ_2	6	$\{V, VII\}$
$(-1,0)$	F	Φ_2	2	$\{II, IV\}$
$(2,0)$	D	Φ_2	2	$\{II, IV\}$
$(2,0)$	D	Φ_1	5	$\{II, IV, VII\}$
$(-2,0)$	$B_\flat$	Φ_1	6	$\{V, VII\}$
$(-2,0)$	$B_\flat$	Φ_1	5	$\{II, IV, VII\}$
$(0,1)$	E	Φ_2	2	$\{II, V, VII\}$
$(0,1)$	E	Φ_2	3	$\{I, III, IV, VI\}$
$(0,1)$	E	Φ_1	5	$\{I, III, IV, VI\}$
$(0,1)$	E	Φ_1	6	$\{II, III, V, VII\}$
$(0,-1)$	$A_\flat$	Φ_1	5	$\{I, III, IV, VI\}$
$(0,-1)$	$A_\flat$	Φ_2	5	$\{I, III, V, VI\}$
$(0,-1)$	$A_\flat$	Φ_1	6	$\{II, III, V, VII\}$
$(0,-1)$	$A_\flat$	Φ_2	6	$\{II, IV, VII\}$
$(-1,1)$	A	Φ_1	2	$\{II, IV, VI, VII\}$
$(-1,1)$	A	Φ_1	3	$\{I, III, V, VI\}$
$(-1,1)$	A	Φ_2	5	$\{I, III, V, VI\}$
$(-1,1)$	A	Φ_2	6	$\{II, IV, VII\}$
$(1,-1)$	$E_\flat$	Φ_1	2	$\{II, IV, VI, VII\}$
$(1,-1)$	$E_\flat$	Φ_1	3	$\{I, III, V, VI\}$
$(-1,1)$	$E_\flat$	Φ_2	3	$\{I, III, IV, VI\}$
$(-1,1)$	$E_\flat$	Φ_2	2	$\{II, V, VII\}$
$(-2,1)$	D^*	Φ_2	5	$\{III, V, VII\}$
$(2,-1)$	$B_\flat^*$	Φ_2	1	$\{II, IV, VI\}$

P.2.7 General Modulation Behaviour for 32 Alterated Scales

The following list refers to the 32 scales as defined in 27.1.6.1. Following Radl [871], we say that a scale type

- *has no modulations* if its modulation domain is empty (always excluding the start tonality!),
- *has infinite modulations* if its modulation domain is infinite,
- *has modulations* if its modulation domain is not empty (always excluding the start tonality!), *has limited modulations* if the transitive closure (all tonics which can be reached by successive modulations from relative modulation domains) of its modulation domain is not total space.

\multicolumn{3}{c}{**Modulation Behaviour in 32 Altered Scale Types**}		
No.	Scale Type	Behaviour
1	c, d, e, f, g, a, b	has modulations: see table P.2.1
2	$c, d, e, f, g, a, b_\flat$	has modulations: $\pm(1, 0)$ with III, $\pm(0, 1)$ with V, VI
3	$c, d, e, f, g, a_\flat, b$	has modulations: corresponds to No. 11
4	$c, d, e, f, g, a_\flat, b_\flat$	has modulations: corresponds to No. 9
5	$c, d, e, f_\sharp, g, a, b$	has modulations: see table P.2.2
6	$c, d, e, f_\sharp, g, a, b_\flat$	has modulations (special table in [871])
7	$c, d, e, f_\sharp, g, a_\flat, b$	has modulations: $\pm(1, 0)$ with II, $\pm(0, 1)$ with III, VII
8	$c, d, e, f_\sharp, g, a_\flat, b_\flat$	has limited modulations: see No. 25
9	$c, d, e_\flat, f, g, a, b$	has modulations: see table P.2.6
10	$c, d, e_\flat, f, g, a, b_\flat$	has modulations: corresponds to No. 2
11	$c, d, e_\flat, f, g, a_\flat, b$	has modulations: see table P.2.5
12	$c, d, e_\flat, f, g, a_\flat, b_\flat$	has modulations: see table P.2.2
13	$c, d, e_\flat, f_\sharp, g, a, b$	has modulations: $\pm(1, 0)$ with II, $\pm(-1, 1)$ with I, VII
14	$c, d, e_\flat, f_\sharp, g, a, b_\flat$	has modulations: $\pm(1, 1)$ with I, V, $\pm(0, 1)$ with I, V
15	$c, d, e_\flat, f_\sharp, g, a_\flat, b$	has no modulations
16	$c, d, e_\flat, f_\sharp, g, a_\flat, b_\flat$	has modulations: corresponds to No. 7
17	$c, d_\flat, e, f, g, a, b$	has modulations: see No. 7
18	$c, d_\flat, e, f, g, a, b_\flat$	has modulations: corresponds to No. 14
19	$c, d_\flat, e, f, g, a_\flat, b$	has no modulations: see No. 15
20	$c, d_\flat, e, f, g, a_\flat, b_\flat$	has modulations: corresponds to No. 13
21	$c, d_\flat, e, f_\sharp, g, a, b$	has infinite modulations
22	$c, d_\flat, e, f_\sharp, g, a, b_\flat$	has infinite modulations
23	$c, d_\flat, e, f_\sharp, g, a_\flat, b$	has modulations: $\pm(1, 0)$ with V, $\pm(0, 1)$ with II
24	$c, d_\flat, e, f_\sharp, g, a_\flat, b_\flat$	has infinite modulations
25	$c, d_\flat, e_\flat, f, g, a, b$	has limited modulations
26	$c, d_\flat, e_\flat, f, g, a, b_\flat$	has modulations: corresponds to No. 6
27	$c, d_\flat, e_\flat, f, g, a_\flat, b$	has modulations: see No. 16
28	$c, d_\flat, e_\flat, f, g, a_\flat, b_\flat$	has modulations: see table P.2.1
29	$c, d_\flat, e_\flat, f_\sharp, g, a, b$	has infinite modulations: corresponds to No. 24
30	$c, d_\flat, e_\flat, f_\sharp, g, a, b_\flat$	has infinite modulations: corresponds to No. 22
31	$c, d_\flat, e_\flat, f_\sharp, g, a_\flat, b$	has modulations: corresponds to No. 23
32	$c, d_\flat, e_\flat, f_\sharp, g, a_\flat, b_\flat$	has infinite modulations: corresponds to No. 21

Q

Counterpoint Steps

Q.1 Contrapuntal Symmetries

All the following tables relate to representatives of strong dichotomies (X/Y) which are indicated in the table after the counterpoint theorem 33 in Section 31.3.3.

Q.1.1 Class No. 64

k	g	$g.X[\varepsilon]$	$g.X[\varepsilon] \cap X[\varepsilon]$	$card(g.X[\varepsilon] \cap X[\varepsilon])$
2	$e^6(1+\varepsilon.6)$	$(1+\varepsilon.6)\mathbb{Z}_{12} + \varepsilon.e^6 X$	z even: $z + \varepsilon.\{5,11\}$	48
			z odd: $z + \varepsilon.X$	
	$e^{\varepsilon.8}5$	$\mathbb{Z}_{12} + \varepsilon.e^8 5X$	$\mathbb{Z}_{12} + \varepsilon.\{4,5,7,9\}$	
4	$e^6(1+\varepsilon.6)$	see $k=2$	see $k=2$	48
5	$e^{\varepsilon.11}11$	$\mathbb{Z}_{12} + \varepsilon.e^{11}11X$	$\mathbb{Z}_{12} + \varepsilon.\{2,4,7,9\}$	48
7	$e^6(1+\varepsilon.6)$	see $k=2$	see $k=2$	48
9	$e^6(1+\varepsilon.6)$	see $k=2$	see $k=2$	48
	$e^{\varepsilon.3}7$	$\mathbb{Z}_{12} + \varepsilon.e^3 7X$	$\mathbb{Z}_{12} + \varepsilon.\{2,4,5,7\}$	
11	$e^{\varepsilon.3}7$	see $k=9$	see $k=9$	48
	$e^{\varepsilon.11}11$	see $k=5$	see $k=5$	
	$e^{\varepsilon.8}5$	see $k=2$	see $k=2$	

© Springer International Publishing AG, part of Springer Nature 2017
G. Mazzola, *The Topos of Music IV: Roots*, Computational Music Science,
https://doi.org/10.1007/978-3-319-64495-0_17

Q.1.2 Class No. 68

k	g	$g.X[\varepsilon]$	$g.X[\varepsilon] \cap X[\varepsilon]$	$card(g.X[\varepsilon] \cap X[\varepsilon])$
0	$e^{\varepsilon.6}7$	$\mathbb{Z}_{12} + \varepsilon.e^6 7X$	$\mathbb{Z}_{12} + \varepsilon.(X - \{0\})$	60
1	$e^{\varepsilon.3}7$	$\mathbb{Z}_{12} + \varepsilon.e^3 7X$	$\mathbb{Z}_{12} + \varepsilon.\{0,2,3,5\}$	48
	$e^{\varepsilon.6}(1+\varepsilon.6)$	$(1+\varepsilon.6)\mathbb{Z}_{12} + \varepsilon.e^6 X$	z even: $z + \varepsilon.\{2,8\}$	
			z odd: $z + \varepsilon.X$	
	$7+\varepsilon.6$	$(1+\varepsilon.6)\mathbb{Z}_{12} + \varepsilon.7X$	z even: $z + \varepsilon.\{0,2,8\}$	
			z odd: $z + \varepsilon.(X - \{0\})$	
2	5	$\mathbb{Z}_{12} + \varepsilon.5X$	$\mathbb{Z}_{12} + \varepsilon.\{0,1,3,5\}$	48
3	$e^{\varepsilon.6}(1+\varepsilon.6)$	$(1+\varepsilon.6)\mathbb{Z}_{12} + \varepsilon.e^6 X$	z even: $z + \varepsilon.\{2,8\}$	48
			z odd: $z + \varepsilon.X$	
	$7+\varepsilon.6$	$(1+\varepsilon.6)\mathbb{Z}_{12} + \varepsilon.7X$	z even: $z + \varepsilon.\{0,2,8\}$	
			z odd: $z + \varepsilon.(X - \{0\})$	
5	$e^{\varepsilon.3}11$	$\mathbb{Z}_{12} + \varepsilon.e^3 X$	$\mathbb{Z}_{12} + \varepsilon.\{0,1,2,3\}$	48
	$e^{\varepsilon.6}(1+\varepsilon.6)$	see $k=1$	see $k=1$	
	$7+\varepsilon.6$	see $k=3$	see $k=3$	
8	$e^{\varepsilon.3}11$	see $k=5$	see $k=5$	48
	$e^{\varepsilon.3}7$	see $k=1$	see $k=1$	
	5	see $k=2$	see $k=2$	

Q.1.3 Class No. 71

k	g	$g.X[\varepsilon]$	$g.X[\varepsilon] \cap X[\varepsilon]$	$card(g.X[\varepsilon] \cap X[\varepsilon])$
0	$e^{\varepsilon.8}5$	$\mathbb{Z}_{12} + \varepsilon.e^8 5X$	$\mathbb{Z}_{12} + \varepsilon.\{1,2,6,7\}$	48
	$e^{\varepsilon.8}(5+\varepsilon.6)$	$(1+\varepsilon.6)\mathbb{Z}_{12} + \varepsilon.e^8 5X$	z even: $z + \varepsilon.\{1,2,6,7\}$	
			z odd: $z + \varepsilon.\{0,1,2,7\}$	
1	$e^{\varepsilon.3}(7+\varepsilon.3)$	$(1+\varepsilon.3)\mathbb{Z}_{12} + \varepsilon.e^3 7X$	$z = 0,4,8 : z + \varepsilon.\{0,3\}$	42
			$z = 1,5,9 : z + \varepsilon.\{0,1,3,6,7\}$	
			$z = 2,6,10 : z + \varepsilon.\{3,6\}$	
			$z = 3,7,11 : z + \varepsilon.\{0,1,2,6,7\}$	
	$e^{\varepsilon.3}(7-\varepsilon.3)$	$(1-\varepsilon.3)\mathbb{Z}_{12} + \varepsilon.e^3 7X$	$z = 0,4,8 : z + \varepsilon.\{0,3\}$	
			$z = 1,5,9 : z + \varepsilon.\{0,1,2,6,7\}$	
			$z = 2,6,10 : z + \varepsilon.\{3,6\}$	
			$z = 3,7,11 : z + \varepsilon.\{0,1,3,6,7\}$	
	$e^{\varepsilon.9}(7+\varepsilon.3)$	$(1+\varepsilon.3)\mathbb{Z}_{12} + \varepsilon.e^9 7X$	$z = 0,4,8 : z + \varepsilon.\{3,6\}$	
			$z = 1,5,9 : z + \varepsilon.\{0,1,2,6,7\}$	
			$z = 2,6,10 : z + \varepsilon.\{0,3\}$	
			$z = 3,7,11 : z + \varepsilon.\{0,1,3,6,7\}$	
	$e^{\varepsilon.9}(7-\varepsilon.3)$	$(1-\varepsilon.3)\mathbb{Z}_{12} + \varepsilon.e^9 7X$	$z = 0,4,8 : z + \varepsilon.\{3,6\}$	
			$z = 1,5,9 : z + \varepsilon.\{0,1,3,6,7\}$	
			$z = 2,6,10 : z + \varepsilon.\{0,3\}$	
			$z = 3,7,11 : z + \varepsilon.\{0,1,2,6,7\}$	
2	$e^{\varepsilon.6}(1+\varepsilon.6)$	$(1+\varepsilon.6)\mathbb{Z}_{12} + \varepsilon.e^6 X$	z even: $z + \varepsilon.\{0,1,6,7\}$	60
			z odd: $z + \varepsilon.X$	
3	$e^{\varepsilon.6}(1+\varepsilon.6)$	see $k=2$	see $k=2$	60
6	$e^{\varepsilon.2}5$	$\mathbb{Z}_{12} + \varepsilon.e^2 5X$	$\mathbb{Z}_{12} + \varepsilon.\{0,1,2,7\}$	48
	$e^{\varepsilon.2}(5+\varepsilon.6)$	$(1+\varepsilon.6)\mathbb{Z}_{12} + \varepsilon.e^2 5X$	z even: $z + \varepsilon.\{0,1,2,7\}$	
			z odd: $z + \varepsilon.\{1,2,6,7\}$	
7	$e^{\varepsilon.9}(7+\varepsilon.3)$	see $k=1$	see $k=1$	42
	$e^{\varepsilon.3}(7+\varepsilon.3)$	see $k=1$	see $k=1$	

Q.1.4 Class No. 75

k	g	$g.X[\varepsilon]$	$g.X[\varepsilon] \cap X[\varepsilon]$	$card(g.X[\varepsilon] \cap X[\varepsilon])$
0	$e^{\varepsilon.9}7$	$\mathbb{Z}_{12} + \varepsilon.e^9 7X$	$\mathbb{Z}_{12} + \varepsilon.\{1,4,5,8\}$	48
	$e^{\varepsilon.9}(7+\varepsilon.4)$	$(1+\varepsilon.4)\mathbb{Z}_{12} + \varepsilon.e^9 7X$	$z = 0,3,6,9 : z + \varepsilon.\{1,4,5,8\}$	
			$z = 1,4,7,10 : z + \varepsilon.\{0,1,5,8\}$	
			$z = 2,5,8,11 : z + \varepsilon.\{0,1,4,5\}$	
	$e^{\varepsilon.9}(7-\varepsilon.4)$	$(1-\varepsilon.4)\mathbb{Z}_{12} + \varepsilon.e^9 7X$	$z = 0,3,6,9 : z + \varepsilon.\{1,4,5,8\}$	
			$z = 1,4,7,10 : z + \varepsilon.\{0,1,4,5\}$	
			$z = 2,5,8,11 : z + \varepsilon.\{0,1,5,8\}$	
	$e^{\varepsilon.6}(1+\varepsilon.6)$	$(1+\varepsilon.6)\mathbb{Z}_{12} + \varepsilon.e^6 X$	z even: $z + \varepsilon.\{2,8\}$	
			z odd: $z + \varepsilon.X$	
1	$e^{\varepsilon.6}(1+\varepsilon.6)$	see $k=0$	see $k=0$	48
2	$e^{\varepsilon.8}(5+\varepsilon.4)$	$(1+\varepsilon.4)\mathbb{Z}_{12} + \varepsilon.e^8 5X$	$z = 0,3,6,9 : z + \varepsilon.\{0,1,4,8\}$	56
			$z = 1,4,7,10 : z + \varepsilon.\{0,1,4,5,8\}$	
			$z = 2,5,8,11 : z + \varepsilon.\{0,2,4,5,8\}$	
	$e^{\varepsilon.8}(5-\varepsilon.4)$	$(1-\varepsilon.4)\mathbb{Z}_{12} + \varepsilon.e^8 5X$	$z = 0,3,6,9 : z + \varepsilon.\{0,1,4,8\}$	
			$z = 1,4,7,10 : z + \varepsilon.\{0,2,4,5,8\}$	
			$z = 2,5,8,11 : z + \varepsilon.\{0,1,4,5,8\}$	
4	$e^{\varepsilon.6}(1+\varepsilon.6)$	see $k=1$	see $k=1$	48
5	$e^{\varepsilon.8}(1\pm\varepsilon.4)$	see $k=2$	see $k=2$	56
8	$e^{\varepsilon.5}11$	$\mathbb{Z}_{12} + \varepsilon.e^5 11X$	$\mathbb{Z}_{12} + \varepsilon.\{0,1,4,5\}$	48
	$e^{\varepsilon.5}(11+\varepsilon.4)$	$(1+\varepsilon.4)\mathbb{Z}_{12} + \varepsilon.e^5 11X$	$z = 0,3,6,9 : z + \varepsilon.\{0,1,4,5\}$	
			$z = 1,4,7,10 : z + \varepsilon.\{1,4,5,8\}$	
			$z = 2,5,8,11 : z + \varepsilon.\{0,1,5,8\}$	
	$e^{\varepsilon.5}(11-\varepsilon.4)$	$(1-\varepsilon.4)\mathbb{Z}_{12} + \varepsilon.e^5 11X$	$z = 0,3,6,9 : z + \varepsilon.\{0,1,4,5\}$	
			$z = 1,4,7,10 : z + \varepsilon.\{0,1,5,8\}$	
			$z = 2,5,8,11 : z + \varepsilon.\{1,4,5,8\}$	

Q.1.5 Class No. 78

k	g	$g.X[\varepsilon]$	$g.X[\varepsilon] \cap X[\varepsilon]$	$card(g.X[\varepsilon] \cap X[\varepsilon])$
0	$e^{\varepsilon.9}(7+\varepsilon.3)$	$(1+\varepsilon.3)\mathbb{Z}_{12} + \varepsilon.e^9 7X$	$z=0,4,8 : z+\varepsilon.\{1,4\}$	42
			$z=1,5,9 : z+\varepsilon.\{0,2,4,6,10\}$	
			$z=2,6,10 : z+\varepsilon.\{1,10\}$	
			$z=3,7,11 : z+\varepsilon.\{0,1,4,6,10\}$	
	$e^{\varepsilon.9}(7-\varepsilon.3)$	$(1-\varepsilon.3)\mathbb{Z}_{12} + \varepsilon.e^9 7X$	$z=0,4,8 : z+\varepsilon.\{1,4\}$	
			$z=1,5,9 : z+\varepsilon.\{0,1,4,6,10\}$	
			$z=2,6,10 : z+\varepsilon.\{1,10\}$	
			$z=3,7,11 : z+\varepsilon.\{0,2,4,6,10\}$	
	$e^{\varepsilon.3}(7+\varepsilon.3)$	$(1+\varepsilon.3)\mathbb{Z}_{12} + \varepsilon.e^3 7X$	$z=0,4,8 : z+\varepsilon.\{1,10\}$	
			$z=1,5,9 : z+\varepsilon.\{0,1,4,6,10\}$	
			$z=2,6,10 : z+\varepsilon.\{1,4\}$	
			$z=3,7,11 : z+\varepsilon.\{0,2,4,6,10\}$	
	$e^{\varepsilon.3}(7-\varepsilon.3)$	$(1-\varepsilon.3)\mathbb{Z}_{12} + \varepsilon.e^3 7X$	$z=0,4,8 : z+\varepsilon.\{1,10\}$	
			$z=1,5,9 : z+\varepsilon.\{0,2,4,6,10\}$	
			$z=2,6,10 : z+\varepsilon.\{1,4\}$	
			$z=3,7,11 : z+\varepsilon.\{0,1,4,6,10\}$	
1	$e^{\varepsilon.6}(1+\varepsilon.6)$	$(1+\varepsilon.6)\mathbb{Z}_{12} + \varepsilon.e^6 X$	z even: $z+\varepsilon.\{0,4,6,10\}$	60
			z odd: $z+\varepsilon.X$	
2	$e^{\varepsilon.6}(1+\varepsilon.6)$	see $k=1$	see $k=1$	60
4	$5+\varepsilon.4$	$(1+\varepsilon.4)\mathbb{Z}_{12} + \varepsilon.5X$	$z=0,3,6,9 : z+\varepsilon.\{0,2,6,10\}$	56
			$z=1,4,7,10 : z+\varepsilon.\{0,2,4,6,10\}$	
			$z=2,5,8,11 : z+\varepsilon.\{1,2,4,6,10\}$	
	$5-\varepsilon.4$	$(1-\varepsilon.4)\mathbb{Z}_{12} + \varepsilon.5X$	$z=0,3,6,9 : z+\varepsilon.\{0,2,6,10\}$	
			$z=1,4,7,10 : z+\varepsilon.\{1,2,4,6,10\}$	
			$z=2,5,8,11 : z+\varepsilon.\{0,2,4,6,10\}$	
6	$e^{\varepsilon.9}(7\pm\varepsilon.3)$	see $k=0$	see $k=0$	42
	$e^{\varepsilon.3}(7\pm\varepsilon.3)$	see $k=0$	see $k=0$	
10	$e^{\varepsilon.6}(5+\varepsilon.2)$	$(1+\varepsilon.2)\mathbb{Z}_{12} + \varepsilon.e^6 5X$	$z=0,6 : z+\varepsilon.\{0,2,4,6\}$	52
			$z=1,7 : z+\varepsilon.\{1,2,4,6,10\}$	
			$z=2,8 : z+\varepsilon.\{0,4,6,10\}$	
			$z=3,9 : z+\varepsilon.\{0,2,6,10\}$	
			$z=4,10 : z+\varepsilon.\{0,2,4,10\}$	
			$z=5,11 : z+\varepsilon.\{0,2,4,6,10\}$	
	$e^{\varepsilon.6}(5-\varepsilon.2)$	$(1-\varepsilon.2)\mathbb{Z}_{12} + \varepsilon.e^6 5X$	$z=0,6 : z+\varepsilon.\{0,2,4,6\}$	
			$z=1,7 : z+\varepsilon.\{0,2,4,6,10\}$	
			$z=2,8 : z+\varepsilon.\{0,2,4,10\}$	
			$z=3,9 : z+\varepsilon.\{0,2,6,10\}$	
			$z=4,10 : z+\varepsilon.\{0,4,6,10\}$	
			$z=5,11 : z+\varepsilon.\{1,2,4,6,10\}$	

Q.1.6 Class No. 82

k	g	$g.X[\varepsilon]$	$g.X[\varepsilon] \cap X[\varepsilon]$	$card(g.X[\varepsilon] \cap X[\varepsilon])$
0	$e^{\varepsilon.6}(1+\varepsilon.6)$	$(1+\varepsilon.6)\mathbb{Z}_{12} + \varepsilon.e^6 X$	z even: $z + \varepsilon.\{3,9\}$	48
			z odd: $z + \varepsilon.X$	
	$e^{\varepsilon.6}(7+\varepsilon.6)$	$(1+\varepsilon.6)\mathbb{Z}_{12} + \varepsilon.e^6 7X$	z even: $z + \varepsilon.\{3,7,9\}$	
			z odd: $z + \varepsilon.(X - \{7\})$	
	$e^{\varepsilon.11}(11 - \varepsilon.4)$	$(1+\varepsilon.4)\mathbb{Z}_{12} + \varepsilon.e^{11}11X$	$z = 0,3,6,9 : z + \varepsilon.\{3,4,7,8\}$	
			$z = 1,4,7,10 : z + \varepsilon.\{0,3,7,8\}$	
			$z = 2,5,8,11 : z + \varepsilon.\{0,3,4,7\}$	
	$e^{\varepsilon.11}(11 + \varepsilon.4)$	$(1-\varepsilon.4)\mathbb{Z}_{12} + \varepsilon.e^{11}11X$	$z = 0,3,6,9 : z + \varepsilon.\{3,4,7,8\}$	
			$z = 1,4,7,10 : z + \varepsilon.\{0,3,4,7\}$	
			$z = 2,5,8,11 : z + \varepsilon.\{0,3,7,8\}$	
	$e^{\varepsilon.11}11$	$\mathbb{Z}_{12} + \varepsilon.e^{11}11X$	$\mathbb{Z}_{12} + \varepsilon.\{3,4,7,8\}$	
3&9	$e^{\varepsilon.8}(5 - \varepsilon.4)$	$(1+\varepsilon.4)\mathbb{Z}_{12} + \varepsilon.e^8 5X$	$z = 0,3,6,9 : z + \varepsilon.\{0,4,7,8\}$	56
			$z = 1,4,7,10 : z + \varepsilon.(X - \{7\})$	
			$z = 2,5,8,11 : z + \varepsilon.(X - \{9\})$	
	$e^{\varepsilon.8}(5 + \varepsilon.4)$	$(1-\varepsilon.4)\mathbb{Z}_{12} + \varepsilon.e^8 5X$	$z = 0,3,6,9 : z + \varepsilon.\{0,4,7,8\}$	
			$z = 1,4,7,10 : z + \varepsilon.(X - \{9\})$	
			$z = 2,5,8,11 : z + \varepsilon.(X - \{7\})$	
4	$e^{\varepsilon.6}(1+\varepsilon.6)$	$(1+\varepsilon.6)\mathbb{Z}_{12} + \varepsilon.e^6 X$	see $k = 0$	48
	$e^{\varepsilon.6}(7+\varepsilon.6)$	$(1+\varepsilon.6)\mathbb{Z}_{12} + \varepsilon.e^6 7X$	see $k = 0$	
7	7	$\mathbb{Z}_{12} + \varepsilon.7X$	$\mathbb{Z}_{12} + \varepsilon.(X - \{7\})$	60
8	$e^{\varepsilon.3}7$	$\mathbb{Z}_{12} + \varepsilon.e^3 7X$	$\mathbb{Z}_{12} + \varepsilon.\{0,3,4,7\}$	48
	$e^{\varepsilon.6}(1+\varepsilon.6)$	$(1+\varepsilon.6)\mathbb{Z}_{12} + \varepsilon.e^6 X$	see $k = 0$	
	$e^{\varepsilon.6}(7+\varepsilon.6)$	$(1+\varepsilon.6)\mathbb{Z}_{12} + \varepsilon.e^6 7X$	see $k = 0$	
	$e^{\varepsilon.3}(7+\varepsilon.4)$	$(1+\varepsilon.4)\mathbb{Z}_{12} + \varepsilon.e^3 7X$	$z = 0,3,6,9 : z + \varepsilon.\{0,3,4,7\}$	
			$z = 1,4,7,10 : z + \varepsilon.\{3,4,7,8\}$	
			$z = 2,5,8,11 : z + \varepsilon.\{0,3,7,8\}$	
	$e^{\varepsilon.3}(7-\varepsilon.4)$	$(1-\varepsilon.4)\mathbb{Z}_{12} + \varepsilon.e^3 7X$	$z = 0,3,6,9 : z + \varepsilon.\{0,3,4,7\}$	
			$z = 1,4,7,10 : z + \varepsilon.\{0,3,7,8\}$	
			$z = 2,5,8,11 : z + \varepsilon.\{3,4,7,8\}$	

Q.2 Permitted Successors for the Major Scale

For the sweeping orientation, given a cantus firmus step $CF : x \mapsto y$, one is allowed to move from a consonance c (i.e., $x + \varepsilon.c$) in the top row to a consonance d (i.e., $y + \varepsilon.d$) in the right column iff there is a $*$ in the corresponding matrix entry.

---

1. Oblique Motion in Cantus Firmus

$CF:0\mapsto0$	$CF:2\mapsto2$	$CF:4\mapsto4$	$CF:5\mapsto5$	$CF:7\mapsto7$	$CF:9\mapsto9$	$CF:11\mapsto11$

```
0 4 7 9      0 3 7 9      0 3 7 8      0 4 7 9      0 4 7 9      0 3 7 8      0 3 8
    * * 0      * * * 0      * * * 0        * * 0        * * 0      * * * 0      * * 0
*   * * 4    *   *   3    *   * * 3    *   * * 4    *   * * 4    *   * * 3    *   * 3
* *   * 7    * *   * 7    * *   * 7    * *   * 7    * *   * 7    * *   * 7    * *   8
* * *   9    *   *   9    * * *   8    * * *   9    * * *   9    * * *   8
```

---

2. Minor Ascending Second in Cantus Firmus

$CF:4\mapsto5$	$CF:11\mapsto0$					

```
0 3 7 8      0 3 8
* * * * 0    * * * 0
* * * * 4    * * * 4
* *   * 7    * * * 7
* * * * 9    * * * 9
```

---

3. Minor Descending Second in Cantus Firmus

$CF:5\mapsto4$	$CF:0\mapsto11$					

```
0 4 7 9      0 4 7 9
* * * * 0    * * * * 0
* * * * 3    * * * * 3
* *   * 7    * * * * 8
* * * * 8
```

---

4. Major Ascending Second in Cantus Firmus

$CF:0\mapsto2$	$CF:2\mapsto4$	$CF:5\mapsto7$	$CF:7\mapsto9$	$CF:9\mapsto11$		

```
0 4 7 9      0 3 7 9      0 4 7 8      0 4 7 9      0 3 7 8
*   * * 0    * * * * 0    *   * * 0    *   * * 0    * * * * 0
* * * * 3    * * * * 3    *   * * 4    * * * * 3    * * * * 3
* *   * 7    * *   * 7    * *   * 7    * *   * 7    * * * * 8
* * * * 9    * * * * 8    * * * * 9    *   * * 8
```

---

5. Major Descending Second in Cantus Firmus

$CF:2\mapsto0$	$CF:4\mapsto2$	$CF:7\mapsto5$	$CF:9\mapsto7$	$CF:11\mapsto9$		

```
0 3 7 9      0 3 7 8      0 4 7 9      0 3 7 8      0 3 8
*   * * 0    * * * * 0    *   * * 0    *   * * 0    * * * 0
* * * * 4    * * * * 3    *   * * 4    * * * * 4    * * * 3
* *   * 7    * *   * 7    * *   * 7    * *   * 7    * * * 7
* * * * 9    * * * * 9    * * * * 9    *   * * 9    * * * 8
```

6. Minor Ascending Third in Cantus Firmus

$CF: 2 \mapsto 5$	$CF: 4 \mapsto 7$	$CF: 9 \mapsto 0$	$CF: 11 \mapsto 2$		
0 3 7 9 * *** 0 * * * * 4 ** ** 7 * * 9	0 3 7 8 * * * * 0 * * * * 4 ** ** 7 * *** 9	0 3 7 8 * ** 0 * * * * 4 ** ** 7 * *** 9	0 3 8 * * * 0 * * 3 * * * 7 * * 9		

7. Minor Descending Third in Cantus Firmus

$CF: 5 \mapsto 2$	$CF: 7 \mapsto 4$	$CF: 0 \mapsto 9$	$CF: 2 \mapsto 11$		
0 4 7 9 * * * * 0 * * * 3 ** * 7 *** 9	0 4 7 9 * * * * 0 * * * 3 ** * 7 * * * * 8	0 4 7 9 * * * * 0 * * * 3 ** * 7 * * * * 8	0 3 7 9 * * * * 0 * * * * 3 * * * * 8		

8. Major Ascending Third in Cantus Firmus

$CF: 0 \mapsto 4$	$CF: 5 \mapsto 9$	$CF: 7 \mapsto 11$			
0 4 7 9 * ** 0 * * * * 3 ** * 7 * ** 8	0 4 7 9 * ** 0 * * * * 3 ** * 7 * ** 8	0 4 7 9 * ** 0 * * * * 3 * ** 8			

9. Major Descending Third in Cantus Firmus

$CF: 4 \mapsto 0$	$CF: 9 \mapsto 5$	$CF: 11 \mapsto 7$			
0 3 7 8 * * * * 0 * * * * 4 ** * 7 * * * * 9	0 3 7 8 * * * * 0 * * * * 4 ** * 7 * * * * 9	0 3 8 * * * 0 * * * 4 * * * 7 * * * 9			

10. Ascending Fourth in Cantus Firmus

$CF: 0 \mapsto 5$	$CF: 2 \mapsto 7$	$CF: 4 \mapsto 9$	$CF: 7 \mapsto 0$	$CF: 9 \mapsto 2$	$CF: 11 \mapsto 4$
0 4 7 9 * * * * 0 * * * * 4 ** * 7 * * * * 9	0 3 7 9 * * * * 0 * * * * 4 ** * 7 * * * * 9	0 3 7 8 * * * * 0 * * * * 3 ** * 7 * * * * 8	0 4 7 9 * * * * 0 * * * * 4 ** * 7 * * * * 9	0 3 7 8 * * * * 0 * * * * 3 ** * 7 * * * * 8	0 3 8 * * * 0 * * * 3 * * * 7 * * * 8

11. Descending Fourth in Cantus Firmus

$CF: 5 \mapsto 0$	$CF: 7 \mapsto 2$	$CF: 9 \mapsto 4$	$CF: 0 \mapsto 7$	$CF: 2 \mapsto 9$	$CF: 4 \mapsto 11$
0 4 7 9 * * * * 0 * * * * 4 ** * 7 * * * * 9	0 4 7 9 * * * * 0 * * * * 3 ** * 7 * * * * 9	0 3 7 8 * * * * 0 * * * * 3 ** * 7 * * * * 8	0 4 7 9 * * * * 0 * * * * 4 ** * 7 * * * * 9	0 3 7 9 * * * * 0 * * * * 3 ** * 7 * * * * 8	0 3 7 8 * * * * 0 * * * * 3 * * * * 8

12. Ascending Tritone in Cantus Firmus

$CF: 5 \mapsto 11$	$CF: 11 \mapsto 5$				
0 4 7 9 * ** 0 * * * 3 * ** 8	0 3 8 * * 0 * * * 4 * * * 7 * * 9				

Part XXIV

References and Index

References

1. Abel V and Reiss P: MUTABOR II - Software Manual. Mutabor Soft, Darmstadt 1991
2. Abraham R and Marsden J: Foundations of Mechanics. Benjamin, New York et al. 1967
3. Abramowitz M and I A Stegun: Handbook of mathematical functions. With formulas, graphs and mathematical tables. Dover, 1965
4. Ackermann Ph: Computer und Musik. Springer, Wien and New York 1991
5. Aczel P: Non-well-founded Sets. No. 14 in CSLI Lecture Notes. Center for the Study of Language and Information, Stanford 1988
6. Ackermann Ph: Developing Object-Oriented Multimedia Software. dpunkt, Heidelberg 1996
7. Adorno Th W: Fragment über Musik und Sprache. Stuttgart, Jahresring 1956
8. Adorno Th W: Zu einer Theorie der musikalischen Reproduktion. Suhrkamp, Frankfurt/M. 2001
9. Adorno Th W: Der getreue Korrepetitor (1963). Gesammelte Schriften, Bd. 15, Suhrkamp, Frankfurt am Main 1976
10. Agawu V K: Playing with Signs. Princeton University Press, Princeton 1991
11. Agmon E: A Mathematical model of the diatonic system. JMT, 33, 1-25, 1989
12. Agmon E: Coherent Tone-Systems: a study in the theory of diatonicism. JMT, 40(1), 39-59, 1996
13. Agon A: OpenMusic: Un langage visuel pour la composition musicale assistée par ordinateur. PhD Dissertation, Université Paris VI, Paris 1998
14. Agustín-Aquino O A: Counterpoint in $2k$-tone equal temperament. Journal of Mathematics and Music, Vol 3, nr. 3, pp. 153-164, 2009
15. Agustín-Aquino O: Extensiones Microtonales de Contrapunto. PhD Thesis, Facultad de Ciencias, UNAM, Mexico City, 2012
16. Agustin-Aquino, O A, J Junod, G Mazzola: Computational Counterpoint Worlds. Springer Series Computational Music Science, Heidelberg 2014
17. Akmajian A et al.: Linguistics. MIT Press, Cambridge MA 1995
18. Alain: Alain Citation on Almada's painting in the Gulbenkian Foundation Center. Lisbon 1968
19. Albert R S and M A Runco: A History of Research on Creativity. In: R J Sternberg (Ed.): Handbook of Creativity. Cambridge University Press, New York 1999
20. Aleinikov A G: On Creative Pedagogy. Higher Education Bulletin, 12, 29-34, 1989
21. D'Alembert J Le Rond: Einleitung zur Enzyklopädie (1751). (German Translation) Fischer, Frankfurt/Main 1989
22. Alforhs L V: Complex Analysis. McGraw-Hill, New York et al. 1966
23. Alpaydin R and G Mazzola: A Harmonic Analysis Network. Musicae Scientiae, Vol. 19(2) 191–213, 2015
24. Alunni Ch: Diagrammes & catégories comme prolégomènes à la question: Qu'est-ce que s'orienter diagrammatiquement dans la pensée? In: Batt, N (éd): Penser par le diagramme. Presses Universitaires de Vincennes no.22-2004, Vincennes 2004
25. Alunni Ch: Relativités et puissances spectrales chez Gaston Bachelard. Revue de synthèse, 1999, 120-1, p. 73-110, 1999
26. Alunni Ch: Diagrammes et catégories comme prolégomènes à la question : qu'est-ce que s'orienter diagrammatiquement dans la pensée ? In: Théorie Littérature Enseignement no 22 - Penser par le diagramme de Gilles Deleuze à Gilles Châtelet, p. 83-93, 2004

© Springer International Publishing AG, part of Springer Nature 2017

G. Mazzola, *The Topos of Music IV: Roots*, Computational Music Science,

https://doi.org/10.1007/978-3-319-64495-0

27. Alunni Ch: Le Lemme de Yoneda : enjeux pour une conjecture philosophique ? (variations sous la forme pro-lemmatique, mais en prose), in Andreatta M, F Nicolas and C Alunni (Eds.): A la lumière des mathématiques et à l'ombre de la philosophie. Coll. Musique/Sciences, ENS, Delatour, Ircam, Centre Pompidou, p. 195-211, 2012

28. American Mathematical Society (AMS): Mathematical Subject Classification 2010, 47 pages
http://www.ams.org/msc/pdfs/classifications2010.pdf

29. Amuedo J: Computational Description of Extended Tonality. Master's Thesis, U Southern California, Los Angeles 1995

30. Andreatta M: Group-theoretical Methods Applied to Music. Independent Study Dissertation, University of Sussex, Sussex 1997

31. Andreatta M: Méthodes algébriques en musique et musicologie du XXème siècle : aspects théoriques, analytiques et compositionnels. PhD thesis, Ircam/EHESS 2003

32. Andreatta M, A Ehresmann, R Guitart, G Mazzola: Towards a Categorical Theory of Creativity for Music, Discourse, and Cognition, in J. Yust, J. Wild, and J.A. Burgoyne (Eds.): MCM 2013, LNAI 7937, pp. 19-37, 2013

33. Andreatta M: La théorie mathématique de la musique de Guerino Mazzola et les canons rythmiques. Mémoire DEA, EHESS Paris IV & IRCAM, Paris 1999

34. Andreatta M and G Mazzola: From a Categorical Point of View: K-nets as Limit Denotators. Perspectives of New Music, volume 44, number 2 (Summer 2006)

35. http://www.staps.uhp-nancy.fr/bernard

36. http://www.staps.uhp-nancy.fr/bernard/revues_corps.htm

37. Ansermet E: Die Grundlagen der Musik im menschlichen Bewusstsein. 4th Ed. Piper, München and Zürich 1986

38. Apfel E: Diskant und Kontrapunkt in der Musiktheorie des 12. bis 15. Jahrhunderts. Heinrichshofen, Wilhelmshafen 1982

39. Apollonius: Les coniques d'Apollonius de Perge, trad. Paul Ver Eecke. Blanchard, Paris, 1959

40. Aristoteles: Topik (Organon V, 345 b.C.). Rolfes E (German transl.), Meiner, Hamburg 1992

41. Artin M: Algebra. Birkhäuser, Basel et al. 1993

42. Artin M, Grothendieck A, Verdier J L: Théorie des Topos et Cohomologie Etale des Schémas (Tomes 1,2,3). Springer Lecture Notes in Mathematics 269, 270, 305, Springer, New York et al. 1972-1973

43. Askenfelt A et al.: Musical Performance. A Synthesis-by-Rule Approach. Computer Music J. 7/1, 1983

44. Assayag G: Du calcul secret au calcul visuel. In: Delalande F and Vinet H (Eds.): Interface homme-machine et création musicale. Hermes, Paris 1999

45. Assayag G, G Bloch, M Chemillier: OMax-Ofon. Sound and Music Computing (SMC), Marseille 2006

46. Assayag G and S Dubnov: Using Factor Oracles for Machine Improvisation. Soft Computing 8, pp. 1432-7643, September 2004

47. Assmann J: Schöpfungsmythen und Kreativitätskonzepte im Alten Ägypten. In: R M Holm-Hadulla (Ed.): Kreativität. Springer, Heidelberg 2000

48. Auroux S: La sémiotique des encyclopédistes. Payot, Paris 1979

49. Babbitt M: Some Aspects of Twelve-Tone Composition. In: Hays W (Ed.): The Score and IMA Magazine, 12, 53-61, 1955 (reprinted in "Twentieth Century Views of Music History", 364-371, Scribner, New York 1972)

50. Babbitt M: Twelve-Tone Invariants as Compositional Determinants. Musical Quarterly, 46, 245-259, 1960

51. Babbitt M: Set Structure as a Compositional Determinant. JMT, 5(2), 72-94, 1961

52. Babbitt M: Twelve-Tone Rhythmic Structure and the Electronic Medium. PNM, 1(1), 49-79, 1962

53. Babbitt M: The Structure and Function of Music Theory. College Music Symposium, Vol.5, 1965 (reprinted in Boretz and Cone, 1972, 10-21)

54. Babbitt M: Words about Music. Dembski S and Straus J N (Eds.), University of Wisconsin Press, Madison 1987

55. Bach J S: Krebskanon. In: *Musikalisches Opfer*, 70. (BWV 1079), Neue Gesamtausg. sämtl. Werke Ser. VIII, Bd.1, Bärenreiter, Kassel 1978

56. Bach J S: Choral No.6, In: *Himmelfahrtsoratorium* (BWV 11), Neue Gesamtausg. sämtl. Werke Ser. II, Bd.8, Bärenreiter, Kassel 1978

57. Bachelard G: La Valeur inductive de la Relativité. 257 p., Vrin, Paris 1929

58. Bacon F: De dignitate et augmentis scientiarum. 1st Ed. 1605, improved 1623

59. Bacon F: Novum Organum, trad. Malherbe et Pousseur. coll. épiméthée, PUF, Paris, 1986

60. Badiou A: Court traité d'ontologie transitoire. Seuil, Paris, 1998

61. Badiou A, F Tarby: La philosophie et l'événement. Entretiens. Suivis d'une courte introduction à la philosophie d'Alain Badiou. Germina, 2010
62. Badiou A: La République de Platon. Pluriel, 2012
63. Badiou A: Le Séminaire – Parménide. L'être 1 - Figure ontologique. 1985-1986. Fayard, 2014
64. Badiou A: Le Séminaire – Heidegger. L'être 3 - Figure du retrait. 1986-1987. Fayard, 2015
65. Badiou A: Éloge des mathématiques. Flammarion, 2015
66. Baer J and J C Kaufman: Creativity Research in English-Speaking Countries. J C Kaufman and R J Sternberg (Eds.): The International Handbook of Creativity. Cambridge University Press, Cambridge, pp. 10-38, 2006
67. Bagchee S N: Understanding Raga Music. BPI, New Delhi 1996
68. Bahle J: Der musikalische Schaffensprozess. Psychologie der schöpferischen Erlebnis- und Antriebsformen. S Hirzel, Leipzig 1936
69. Baker J, Beach D, Bernard J: Music Theory in Concept and Practice. Eastman Studies in Music, University of Rochester Press, 1997
70. Bakhtin M: Toward a Philosophy of the Act. Translated by Vadim Liapunov. University of Texas Press, Austin 1993
71. Balzano G: The group-theoretic description of 12-fold and microtonal pitch systems. CMJ, 4, 66-84, 1980
72. Bandemer H and Gottwald H: Fuzzy Sets, Fuzzy Logic, Fuzzy Methods. Wiley, New York et al. 1995
73. Banter H: Akkord-Lexikon. Schott, Mainz 1982
74. Barbin E: Histoire et enseignement des mathématiques : pourquoi ? comment ?. Bulletin de l'Association Mathématique du Québec, XXXVII-1, 20-25
75. Barbin E: La révolution mathématique du XVIIème siècle. Ellipses, 2006
76. Barbin E et R Guitart: La pulsation entre les conceptions optiques, algébriques, articulées et projectives, des ovales cartésiennes, in: L'Océan indien au carrefour des mathématiques arabes, chinoises, européennes et indiennes. IUFM de La Réunion, 1998, pp. 359-394
77. Barbin E et R Guitart: Algèbre des fonctions elliptiques et géométrie des ovales cartésiennes. Revue d'histoire des mathématiques, SMF, n7, 2001, pp. 161-205
78. Barbin E: Gaston Tarry et la doctrine des combinaisons. Labyrinthes et carrés magiques, conférence au colloque *Les travaux combinatoires entre 1870 et 1914 et leur actualité*, Guéret, 30 Sept. - 2 Oct. 2015
79. Barbin E et J-P Cléro (Eds.): Les mathématiques et l'expérience : ce qu'en ont dit les philosophes et les mathématiciens. Hermann, Paris 2015
80. Barbin E: La révolution mathématique du XVIIème siècle. Ellipses, 2006
81. Bareil H: Analyse de "La pulsation mathématique", *A.P.M.E.P.* 427, March-April 2000
82. Barlow K: Über die Rationalisierung einer harmonisch irrationalen Tonhöhenmenge. Preprint, Köln 1985.
83. Barrow J and F Tipler: The Anthropic Cosmological Principle: Oxford University Press, New York 1986
84. Bars I: Extra Dimensions in Space and Time (Multiversal Journeys). Springer, Heidelberg 2009
85. Bartenieff I: The Roots of Laban Theory: Aesthetics and Beyond. In: Bartenieff et al.: Four Adaptations of Effort Theory. Dance Notation Bureau, New York 1970
86. Barthes R: Eléments de sémiologie. Communications 4/1964
87. Barwise J and L Moss: Hypersets. Mathematical Intelligencer, 13(4):3141, 1991
88. Basten A: Personal e-mail communciaction to G.M. November 22, 1996
89. Bastiani A (= A Ehresmann): Théorie des ensembles. CDU, 328 p., Paris, 1970
90. Bätschmann O: Einführung in die kunstgeschichtliche Hermeneutik. Wissenschaftliche Buchgemeinschaft Darmstadt, Darmstadt 1986
91. Batt N (éd): Penser par le diagramme. Presses Universitaires de Vincennes no.22-2004, Vincennes 2004
92. Bauer M: Die Lieder Franz Schuberts. Breitkopf und Härtel, Leipzig 1915
93. Bazelow A and F Brickel: A Partition Problem Posed by Milton Babbitt. PNM, 14(2), 15(1), 280-293, 1976
94. Becker M: Nineteenth Century Foundations of Creativity Research. Creativity Research Journal, Vol. 8, pp. 219-229, 1995
95. Bednarik R G: A Figurine from the African Acheulian. Current Anthropology, Vol. 44:3, pp. 405-413, 2003
96. Beethoven L van: *Grosse Sonate für das Hammerklavier* op.106 (1817-1818). Ed. Peters, Leipzig 1975
97. Békésy G von: Experiments in Hearing. McGraw-Hill, New York 1960
98. Bell J L: From absolute to local mathematics. *Synthese* 69, 409-426, 1986
99. Bellal A: Il faut sortir des mathématiques (humour) (see also [437]), http:::environ-energie.org/category/humourpastiches/
100. Benveniste É: La nature des pronoms. In: Benveniste É: Problèmes de linguistique générale, I. Gallimard, Paris 1966
101. Beran J: *Cirri*. Centaur Records, 1991

102. Beran J: Statistics in musicology. Chapman & Hall (CRC Press), New York 2003

103. Beran J and G Mazzola: *Immaculate Concept.* SToA music 1002.92, Zürich 1992

104. Beran J and G Mazzola G: Rational composition of performance. In Auhagen W, R Kopiez (Eds.): Regulation of creative processes in music, Staatliches Institut für Musikforschung (Berlin), pp. 37-67, Lang Verlag, Frankfurt/New York, 1997

105. Beran J and G Mazzola: Analyzing Musical Structure and Performance—a Statistical Approach. Statistical Science. Vol. 14, No. 1, 47-79, 1999

106. Beran J and G Mazzola: Visualizing the Relationship Between Two Time Series by Hierarchical Smoothing. Journal of Computational and Graphical Statistics, Vol. 8, No. 2, 213-238, 1999

107. Beran J and G Mazzola: Timing Microstructure in Schumann's "Träumerei" as an Expression of Harmony, Rhythm, and Motivic Structure in Music Performance. Computers and Mathematics with Applications, Vol. 39, Issue 5/6, 99-130, 2000

108. Beran J and G Mazzola: Musical composition and performance - statistical decomposition and interpretation. Student, Vol. 4, No.1, 13-42, 2001

109. Beran J: Maximum likelihood estimation of the differencing parameter for invertible short and long-memory ARIMA models. J.R. Statist. Soc. B, 57, No.4, 659-672, 1995

110. Beran J, R Goswitz, G Mazzola, P Mazzola: On the relationship between tempo and quantitative metric, melodic and harmonic information in Chopin's Prélude op. 28, No. 4: a statistical analysis of 30 performances. Journal of Mathematics and Music, Vol 8., No. 3, p. 225-248, 2014

111. Beranek L L: Acoustics, 1954 -1993, The Acoustical Society of America, ISBN: 0-88318-494-X

112. Berg A: Die musikalische Impotenz der Neuen Ästhetik Hans Pfitzners. Musikblätter des Anbruchs. Jg 02, S. 400, Wien 1920

113. Berger M: Geometry I, II. Springer, Berlin et al. 1987

114. Bernard J: Chord, Collection, and Set in Twentieth-Century Theory. In: Baker J et al. [69], 11-52

115. Béziau J-Y: La Théorie des Ensembles et la Théorie des Catégories : Présentation de deux Sœurs Ennemies du Point de Vue de leurs Relations avec les Fondements des mathématiques. Boletin de la Asociación Matemática Venezolana, Vol. IX, No 1, pp. 45-52, 2002

116. Bilkey D and K M Douglas: Amusia Is Associated with Deficits in Spatial Processing. Nature Neuroscience 10, pp. 915 - 921, 2007

117. Biok H-R: Zur Intonationsbeurteilung kontextbezogener sukzessiver Intervalle. Bosse, Regensburg 1975

118. Biot J-B: Essai de géométrie analytique, 8 ième éd. (1ère éd. 1802), Bachelier, Paris 1834

119. Birkhoff G: Universal Algebra. In: Rota G-C and J.S. Oliveira J S (Eds.): Collected papers of G. Birkhoff, pp. 111-115, Birkhäuser, Basel

120. Blind Boy Fuller: ttp://www.allaboutbluesmusic.com/blind-boy-fuller/ 2015

121. Block I: Oscar Peterson Lays Down the Story of a Life Lived in the Swing of Things. The Gazette (Montreal), August 31, 2002

122. Blumenthal L M: Distance Geometry. Chelsea 1953

123. Boden M A: Conceptual Spaces. In [753], pp. 235-244

124. Boden M A: The Creative Mind: Myths and Mechanisms. Routledge, London 2004

125. Boden M A: Creativity & Art. Oxford University Press, Oxford 2010

126. Blumenthal L M: Distance Geometry. Chelsea 1953

127. Blood A, J Zatorre, P Bermudez, A C Evans: Emotional response to pleasant and unpleasant music correlates with activity in paralimbic brain regions. Nature Neuroscience, Vol. 2, No. 4, 382-387, April 1999

128. Boi L: Le problème mathématique de l'espace. Une quête de l'intelligible. Springer, 1995

129. Boirel R: Comment résoudre aisément les problèmes de mathématiques. Initiation au dynamisme opératoire mathématique. éd. Culture humaine, Sermaise, 1964

130. Boissière A: Geste, interprétation, invention selon Pierre Boulez. Revue DEMéter, Dec. 2002, Univ. Lille-3

131. Borceux F: Handbook of Categorical Algebra: Volume 1. Cambridge University Press, Cambridge 1994

132. Borceux, F: Handbook of Categorical Algebra: Volume 2. Cambridge University Press, Cambridge 1995

133. Borgo D: Sync or Swarm. Continuum International Publishing Group 2006

134. Boretz B and E T Cone: Perspectives on Contemporary Music Theory. W.W. Norton and Company, New York 1972

135. Bose R C and S S Shrikhande: On the construction of sets of mutually orthogonal Latin squares and the falsity of a conjecture of Euler. Trans. Amer. Math. Soc. 95, 191-209, 1960

136. Bose R C, S S Shrikhande and E T Parker: Further results on the construction of mutually orthogonal Latin squares and the falsity of Euler's conjecture. Canadian J. of Math. 12, 189-203, 1960

137. Bott J, J Crowley, J LaViola Jr: Exploring 3D Gestural Interfaces for Music Creation in Video Games. CFDG 2009 April 2630, 2009, Orlando 2009

138. Boulez P: Musikdenken heute I,II; Darmstädter Beiträge V, VI. Schott, Mainz 1963, 1985

139. Boulez P: Le timbre et l'écriture, le timbre et le langage. In: Bourgeois Chr (Ed.): Le timbre, métaphore pour la composition. IRCAM, Collection Musique/Passé/Présent, Paris 1991

140. Boulez P: Jalons (dix ans d'enseignement au Collège de France). Bourgeois, Paris 1989

141. Boulez P: L'Ecriture du geste. Bourgeois, Paris 2002

142. Boulez P: Structures, premier livre. UE, London 1953

143. Boulez P: Structures, deuxièxime livre. UE, London 1967

144. Bourgeois Chr (Ed.): Le timbre, métaphore pour la composition. IRCAM, Collection Musique/Passé/Présent, Paris 1991

145. Bourbaki N: Eléments de Mathématique, Algèbre, Ch.1-9. Hermann, Paris 1970-1973

146. Bourbaki N: Eléments de Mathématique, Algèbre Commutative, Ch.1-7. Hermann, Paris 1961-65

147. Bourbaki N: Eléments de Mathématique, Topologie Générale, Ch.1-4. Hermann, Paris 1971

148. Bourbaki N: Foundations of mathematics for the working mathematician. Journal of Symbolic Logic, 14, pp. 1-8, 1949

149. Boyer C B: Newton as an Originator of Polar Coordinates. American Mathematical Monthly 56: 73-78, 1949

150. Braem P B and T Braem. A Pilot Study of the Expressive Gestures Used by Classical Orchestra Conductors. In: Emmorey K and H Lane (Eds.): The signs of Language Revisited. Lawrence Erlbaum Associates, 2000

151. Brändle L: Die "Wesentlichen Manieren" (Ornamente in der Musik). Oesterreichischer Bundesverlag, Wien 1987

152. Brandt C and C Roemer: Standardized Chord Symbol Notation: Roerick Music, Sherman Oaks, CA 1976

153. Brecht B et al.: Conductor Follower. In: ICMA (Ed.): Proceedings of the ICMC 95, San Francisco 1995

154. Brockwell P J and R A Davis: Time Series: Theory and Methods. Springer, New York 1987

155. Brofeldt H: Piano Music for the Left Hand Alone. http://www.left-hand-brofeldt.dk Non-dated

156. Bruhn H: Harmonielehre als Grammatik der Musik. Psychologie Verlags Union, München et al. 1988

157. Bruter C-P: Comprendre les mathématiques. Les 10 notions fondamentales. Ed. Odile Jacob, Paris, 1996

158. Bulwer J: Chironomia. Thomas Harper, London 1644

159. Buteau Ch and G Mazzola: From Contour Similarity to Motivic Topologies. Musicae Scientiae 2, pp. 125-149, 2000

160. Buteau Ch and G Mazzola: From Contour Similarity to Motivic Topologies. Musicae Scientiae, Vol. IV, No. 2, 125-149, 2000.

161. Busch H R: Leonhard Eulers Beitrag zur Musiktheorie. Bosse, Regensburg 1970

162. Buser P and M Imbert: Audition. Hermann, Paris 1987

163. Buteau Ch: Motivic Topologies and their Signification in Musical Motivic Analysis. Master's Thesis, U Laval/Québec 1998

164. Buteau Ch: Reciprocity Between Presence and Content Functions on a Gestalt Composition Space. Tatra Mt. Math. Publ. 23, 17-45, 2001

165. Butsch C and H Fischer: Seashore-Test für musikalische Begabung. Huber, Bern/Stuttgart 1966

166. Byers W: How Mathematicians Think. Using Ambiguity, Contradiction and Paradox to Create Mathematics. Princeton U. Press, 2007

167. Cabral B and L Leedom: Imaging vector fields using line integral convolution. Computer Graphics 27 (SIG-GRAPH'93 Proceedings), 263-272, 1993

168. Cadoz C and M M Wanderley: In: Gesture - Music. Trends in Gestural Control of Music, M M Wanderley and M Battier eds., Ircam - Centre Pompidou, 2000

169. Cadoz C: Le geste canal de communication homme-machine. La communication 'instrumentale'. Sciences Informatiques, numéro spécial: Interface homme-machine. 13(1): 31-61, 1994

170. Cage J: ASLSP. http://www.john-cage.halberstadt.de

171. Calvet O et al.: Modal synthesis: compilation of mechanical sub-structure and acoustical sub-systems. In: Arnold S, Hair G, ICMA (Eds.): Proceedings of the 1990 International Computer Music Conference. San Francisco 1990

172. Campos D G: Peirce on the Role of Poietic Creation in Mathematical reasoning, trans. Charles S. Peirce Society: A Quaterly J. in American Philosophy, Vol. 43, No 3, pp. 470-489, Summer 2007

173. Franco Ferrara's conducting technique, Personal conversation with maestro C. Caruso at State Music Conservatory of Palermo, 2014

174. Carnot L: Mémoire sur la Relation qui existe entre les distances respectives de cinq points quelconques dans l'espace, suivi d'un Essai sur la théorie des transversales. 1806

175. Castellengo M and N H Bernardoni: Interplay between harmonics and formants in singing: when vowels become music. Proceedings of ISMA, International Conference on Noise and Vibration Engineering 2014

176. Camuri A et al.: Toward a cognitive model for the representation and reasoning on music and multimedia knowledge. In: Haus G and Pighi I (Eds.): X Colloquio di Informatica Musicale. AIMI, LIM-Dsi, Milano 1993

177. Carey N and D Clampitt: Aspects of well-formed scales. MTS, 11, 187-206, 1989

178. Cardelli L and P Wegner: On Understanding types, data Abstraction and polymorphism. Computing Surveys, Vol. 17,4, 1985

179. Castagna G: Foundations of Object-oriented programming. ETAPS, Lisbon 1998

180. Castine P: Set Theory Objects. Lang, Frankfurt/Main et al. 1994

181. Cavaillès J: Méthode axiomatique et formalisme, p. 178, 1938. In: Oeuvres complètes de Philosophie des sciences. Hermann, 1994

182. Cayley A: On a Theorem in the Geometry of Position. Camb. Math. Jour., t. II, pp. 267-271, 1841

183. Celibidache Video: http://www.youtube.com/watch?v=_BAsS5SmiNU

184. Celibidache at Wikipedia: http://en.wikipedia.org/wiki/Sergiu_Celibidache

185. CERN. Internet Information via http://www.cern.ch 1995

186. Certeau M de: L'étranger ou l'union dans la différence. 1ère éd. 1969, Le Seuil, 2005

187. Châtelet G: Arithmétique et algèbre modernes, II, PUF, Paris 1956

188. Chastel A: Le geste dans l'art. Liana Levi, Paris 2001

189. Châtelet G: Les enjeux du mobile. Seuil, Paris 1993

190. Châtelet G: Figuring Space. Kluwer 2000 (translation of the French original *Les enjeux du mobile*, [189])

191. Chatterjee S and B Price: Regression Analysis by Example. Wiley, 2nd ed., New York 1995

192. Chauviré C: Peirce et la signification. PUF, Paris 1995

193. Chauviré C: L'œil mathématique. Essai sur la philosophie mathématique de Peirce. éd. Kimé, 2008

194. Cherlin M: Memory and Rhetorical Trope in Schoenberg's String Trio. Journal of the American Musicological Society, 51/3, pp. 559-602, fall 1998

195. Cherry D & E Blackwell: Mu (Parts I and II). Actuel 1969

196. Choi I: Cognitive Engineering of Gestural Primitives for Multi-Modal Interaction in a Virtual Environment. In: Proceedings of the 1998 IEEE International Conference on Systems, Man and Cybernetics (SMC'98), 1998

197. Chomsky N and M Halle: The Sound Pattern of English. Harper and Row, New York 1968

198. Chowning J: The Synthesis of Complex Audio Spectra by Means of Frequency Modulation. Journal of the Audio Engineering Society 21 (7), 1985

199. Chowning J: The synthesis of complex audio spectra by means of frequency modulation. Journal of the Audio Engineering Society, 21, 1973.

200. Clarke E: Imitating and Evaluating Real and Transformed Musical Performances. Music Perception 10/3, 317-341, 1993

201. Clayton M and G Fatone, L Leante, and M Rahaim: Imagery, Melody, and Gesture in Cross-Cultural Perspective. In: King E: New Perspectives on Music and Gesture (2016), Routledge 2011

202. Clough J and Myerson G: Variety and multiplicity in diatonic systems. JMT, 29, 249-270, 1985

203. Clough J and J Douthett: Maximally even sets. JMT, 35, 93-173, 1991

204. Clough J: Diatonic Interval Cycles and Hierarchical Structure. PNM, 32(1), 228-253, 1994

205. Clynes M: Sentics. The Touch of Emotions. Anchor Doubleday, New York 1977

206. Clynes M: Secrets of life in music. In: Analytica, Studies in the description and analysis of music in honour of Ingmar Bengtsson, Royal Swedish Academy of Music, No 47 pp. 3-15, 1985

207. Colbourn C J and J H Dinitz: Handbook of Combinatorial Designs. Chapman & Hall/CRC, 2007

208. Cont A, S Dubnov and G Assayag: On the Information Geometry of Audio Streams with Applications to Automatic Structure Discovery. IEEE Transactions on Audio, Speech and Language Processing 2009

209. Cohn R: Introduction to Neo-Riemannian Theory: A Survey and a Historical Perspective. JMT 42(2), 167-180, 1998

210. Conen H: Formel-Komposition. Schott, Mainz et al. 1991

211. Cook P: http://www.cs.princeton.edu/ prc/SingingSynth.html

212. Cook N: Playing God: Creativity, Analysis, and Aesthetic Inclusion. In: I Deliège and G A Wiggins (Eds.): Musical Creativity: Multidisciplinary Research in Theory and Practice. Psychology Press, Hove/New York, pp. 9-24, 2006

213. Coolidge J L: The origin of polar coordinates. American Mathematical Monthly, 59 : 78-85, 1952

214. Cooley J W and J W Tuckey: An Algorithm for the Machine Calculation of Complex Series. Math. Comput. 19, 297301 1965

215. Cope D: Virtual Music: Computer Synthesis of Musical Style. MIT Press, Cambridge, MA 2001

216. Cope D: Computer Models of Musical Creativity. MIT Press, Cambridge, MA 2005

217. Corbin A, J-J Courtine, G Viagarello (Eds.): Histoire du corps, T3, Les mutations du regard. Le XX siècle. Seuil, Paris 2006

218. Cox A: Hearing, Feeling, Grasping Gestures. In: Gritten A and E King (Eds.): Music and Gesture. Aldershot: Ashgate, 2006

219. Cooper K N D et al.: Handwritten Music-Manuscript Recognition. In: ICMA (Ed.): Proceedings of the ICMC 96, San Francisco 1996

220. Couasnon B et al.: Using a Grammar For a Reliable Full Score Recognition System. In: ICMA (Ed.): Proceedings of the ICMC 95, San Francisco 1995

221. Creutzfeldt O D: Cortex Cerebri. Springer, Berlin et al. 1983

222. Cros S C H: Théorie de l'homme intellectuel et moral, vol.2. Bachelier, 1842

223. Csíkszentmihályi M: Beyond Boredom and Anxiety. Jossey-Bass, San Francisco 1975

224. Csíkszentmihályi M: Society, Culture, and Person: A Systems View of Creativity. In: R J Sternberg (Ed.): The Nature of Creativity: Contemporary Psychological Perspectives. Cambridge University Press, New York, pp. 325-339, 1988

225. Csíkszentmihályi M: The Psychology of Optimal Experience. Harper and Row, New York 1990

226. Csíkszentmihályi M: Creativity: Flow and the Psychology of Discovery and Invention. HarperCollins, New York 1996

227. Cusick S: Feminist Theory, Music Theory, and the Mind/Body Problem. Perspectives of New Music 32/1 (Winter 1994): 8-27

228. Czerny C: Pianoforte Schule. 1840

229. Dacey J S and K Lennon: Understanding Creativity: The Interplay of Biological, Psychological, and Social Factors. Josey-Bass, San Francisco, 1998

230. Dahlhaus C: Zur Theorie des klassischen Kontrapunkts. Kirchenmusikalisches Jb 45, 1961

231. Dahlhaus C: Über den Begriff der tonalen Funktion. In: Vogel M (Ed.): Beiträge zur Musiktheorie des 19. Jahrhunderts. Bosse, Regensburg 1966

232. Dahlhaus C and H H Eggebrecht: Was ist Musik? Heinrichshofen, Wilhelmshaven et al. 1985

233. Dahlhaus C: Untersuchung über die Entstehung der harmonischen Tonalität. Bärenreiter, Kassel et al. 1967

234. Dahlhaus C et al.: Neues Handbuch der Musikwissenschaft, Bd. 1-13: Athenaion and Laaber, Laaber 1980-1993

235. Dahlhaus C and G Mayer: Musiksoziologische Reflexionen. In: Dahlhaus, C and de la Motte-Haber H (Eds.): Neues Handbuch der Musikwissenschaft, Bd. 10: Systematische Musikwissenschaft. Laaber, Laaber 1982

236. Dahlhaus C: Ludwig van Beethoven und seine Zeit. Laaber, Laaber 1987

237. D'Alembert J Le Rond: Einleitung zur 'Enzyklopädie'. Fischer, Frankfurt/Main 1989

238. D'Alembert, J Le Rond: Einleitung zur Enzyklopädie (1751). (German Translation) Fischer, Frankfurt/Main 1989

239. Dallos P: The active cochlea. J. Neurosci. Dec;12(12):4575-85, 1992

240. Dannenberg R B: An on-line algorithm for real-time accompaniment. In: ICMA (Ed.): Proceedings of the ICMC 84, San Francisco 1984

241. Dannenberg R B et al.: Automatic Ensemble Performance. In: ICMA (Ed.): Proceedings of the ICMC 94, San Francisco 1994

242. Alighieri, D: La Divina Commedia (1321). CreateSpace Independent Publishing Platform 2014

243. Danuser H et al. (Eds.): Neues Handbuch der Musikwissenschaft, Bd. 11: Interpretation. Laaber, Laaber 1992

244. Darwin Ch: The Origin of Species. London 1859

245. Daval R and G-T Guilbaud: Le raisonnement mathématique. PUF, 1945

246. de Bruijn N G: Pólya's Theory of Counting. In: Beckenbach E F (Ed.): Applied Combinatorial Mathematics, Ch.5. Wiley, New York 1964

247. de Bruijn N G: On the number of partition patterns of a set. Nederl. Akad. Wetensch. Proc. Ser. A 82 = Indag. Math. 41, 1979

248. Debussy C: *Préludes*, Livre I (1907-1910). Henle, München 1986

249. Dechelle F et al.: The Ircam Real-Time Platform and Applications. In: ICMA (Ed.): Proceedings of the ICMC 95, San Francisco 1995

250. Delalande F: La gestique de Gould. In: Guertin G (Ed.): Glenn Gould — pluriel. Corteau, Verdun 1988

251. Degazio B: A Computer-Based Editor For Lerdahl and Jackendoff's Rhythmic Structures. In: ICMA (Ed.): Proceedings of the ICMC 96, San Francisco 1996

252. de la Motte D: Harmonielehre. Bärenreiter/dtv, Kassel 1976

253. de la Motte-Haber H and H Emons: Filmmusik. Eine systematische Beschreibung. München, Hanser 1980

254. de la Motte-Haber H: Handbuch der Musikpsychologie. Laaber-Verlag, 2.Ed., Laaber 1996

255. de la Motte-Haber H: Rationalität und Affekt. In: Götze H and Wille R (Eds.): Musik und Mathematik. Springer, Berlin et al. 1985

256. de la Motte-Haber H: Musikalische Hermeneutik und empirische Forschung. In: Dahlhaus, C and de la Motte-Haber H (Eds.): Neues Handbuch der Musikwissenschaft, Bd. 10: Systematische Musikwissenschaft. Laaber, Laaber 1982

257. de la Motte-Haber H: Die Umwandlung der Interpretationsparameter in Struktureigenschaften. In: "Das Paradox musikalischer Interpretation", Symposion zum 80. Geburtstag von K. von Fischer, Univ. Zürich 1993

258. Deleuze G: Francis Bacon. La logique de la sensation. Éditions de la Différence, Paris 1981

259. Demazure M and P Gabriel: Groupes Algébriques. Masson & Cie./North-Holland, Paris/Amsterdam 1970

260. Dennett, D C: Quining qualia. In Marcel A and Bisiach E (Eds.): Consciousness in Contemporary Science. Oxford University Press, 1988

261. Desain P and Honing H: The Quantization of Musical Time: A Connectionist Approach. Computer Music Journal 13 (3), 56-66, 1989

262. Descartes R: Musicae Compendium. Herausgegeben und ins Deutsche übertragen als "Leitfaden der Musik" von J. Brockt, Wiss. Buchgesellschaft, Darmstadt 1978

263. Descartes R: Principia philosophiae (Principles of Philosophy). Translation with explanatory notes by Valentine Rodger and Reese P. Miller (Reprint ed.). Dordrecht: Reidel

264. Descartes R: Œuvres. Vrin, Paris, 1989. See also in English: The Geometry of René Descartes, transl. by E Smith and M L Latham, Dover Publ., 1954

265. Detienne M: Les Maîtres de Vérité dans la Grèce archaïque. 1ère éd. 1967, Le livre de poche/Références, no 611, Paris, 2012

266. De Tranquelléon: Sur l'intersection de deux cônes. Nouvelles Annales de mathématiques, 2ème série, tome III, pp. 539-540, 1864

267. Deyoung L: Pitch Order and Duration Order in Boulez' Structure 1a. PNM 16/2: 27-34, 1978

268. Dieudonné J: Foundations of Modern Analysis. Academic Press, New York et. al. 1960

269. Dieudonné J: La conception des mathématiques chez Valéry. In: Robinson-Valéry J: Fonctions de l'esprit. Treize savants redécouvrent Paul Valéry. Hermann, p. 183-191, Paris 1983

270. Dieudonné J: Mathematics – Music of Reason. Springer, 1992

271. Discogs: Horace Parlan. http://www.discogs.com/artist/253065-Horace-Parlan Not-dated

272. Döhl F: Webern - Weberns Beitrag zur Stilwende der Neuen Musik. Katzbichler, München et. al. 1976

273. Donald M: Mimesis and the Executive Suite: missing links in language evolution. In: J R Hurford, M Studdert-Kennedy, Ch Knight (Eds.): Approaches to the evolution of language, Cambridge U Press 1998

274. Dominguez M, D Clampitt and Th Noll: Well-formed Scales, Maximally Even Sets, and Christoffel Words. In: Proceedings of the MCM 2007, Springer, Heidelberg et al. 2009

275. Douady R: Jeux de cadres et dialectique outil-objet dans l'enseignement des mathématiques. Une réalisation dans le cursus primaire. Thèse, Paris 7, 1984

276. Douthett J and R Steinbach: Parsimonious Graphs: A Study in Parsimony, Contextual Transformations, and Modes of Limited Transposition. Journal of Music Theory 42.2, pp. 241-263, 1998

277. Doyle P, J Conway, J Gilman, B Thurston: Geometry and the Imagination in Minneapolis, version 0.91 dated 12 April 1994

278. Drake R M: Drake Musical Aptitude Tests. Science Research Associates, Chicago 1954

279. Dreiding A et al.: Classification of Mobile Molecules by Category Theory. In: Symmetries and Properties of Non-Rigid Molecules. Studies in Physical and Theoretical Chemistry, 23, 1983

280. Drossos C A: Sets, Categories and Structuralism. In: Sica, G. (Ed.): What is Category Theory?, Polimetrica, Monza 2005

281. Duffy M and S Tsuda: Shield Your Eyes Sonic Anarchy. The Sonic Alchemists, Minneapolis 2010

282. Dufourt H: Les difficultés d'une prise de conscience théorique. In: Le compositeur et l'ordinateur. 6-12, Ircam, Centre Georges Pompidou, Paris 1981

283. Duvignau K, B Gaume, J-L Nespoulous: Proximité sémantique et stratégies palliatives chez le jeune enfant et l'aphasique. Revue Parole, numéro spécial: 'Handicap langagier et recherches cognitives: Apports mutuels', J-L Nespoulous and J Virbel (Eds.), UMH Belgique, vol. 31-32: 219-255, 2004

284. Eberle G: "Absolute Harmonie" und "Ultrachromatik". In: Kolleritsch O (Ed.): Alexander Skrjabin. Universal Edition, Graz 1980

285. Ebisch S J H, M G Perrucci, A Ferretti, C Del Gratta, G Luca Romani, V Gallese: The Sense of Touch: Embodied Simulation in a Visuotactile Mirroring Mechanism for Observed Animate or Inanimate Touch. Journal of Cognitive Neuroscience 20:1-13, 2008

286. Eco U: Kunst und Schönheit im Mittelalter. Hanser, Wien 1991

287. Eco U: The Search for the Perfect Language. Wiley-Blackwell 1997
288. Eco U: Die Suche nach der vollkommenen Sprache. Beck, München 1994
289. Eco U: Le Signe. biblio essais n4159, éd. Labor 1988
290. Edelsbrunner H and J L Harer: Computational Topology. AMS 2010
291. Eggebrecht H H: Interpretation. In: "Das Paradox musikalischer Interpretation", Symposion zum 80. Geburtstag von K. von Fischer, Univ. Zürich 1993
292. Eggebrecht H H: Musik im Abendland. Piper, München and Zürich 1996
293. Ehrenfels Chr von: Über Gestaltqualitäten. Vierteljahresschrift für wissenschaftliche Philosophie XIV, 1890
294. Ehresmann C: Œuvres complètes et commentées. Part. IV-2 : esquisses et structures monoïdales fermées. ed. par Andrée Ch. Ehresmann, Amiens, 1983
295. Ehresmann A C, J-P Vanbremeersch: Memory Evolutive Systems. Hierarchy, Emergence, Cognition. Elsevier Amsterdam, 2007
296. Ehresmann A and J Gomez-Ramirez: Conciliating Neuroscience and Phenomenology via Category Theory. Progress in Biophysics and Molecular Biology, Vol. 19, issue 2, Elsevier, 2015
297. Eitz C: Das mathematisch reine Tonsystem. Leipzig 1891
298. Eimert H: Grundlagen der musikalischen Reihentechnik. Universal Edition, Wien 1964
299. Eisenbud D: Commutative Algebra with a View Toward Algebraic Geometry. Springer New York et al. 1996
300. Eisenbud D, and J Harris: The Geometry of Schemes. Springer, New York 2000
301. Engström B: Stereocilia of sensory cells in normal and hearing impaired ears. Scand. Audiol. Suppl. 19, 1-34, 1983
302. Engquist B and W Schmid (Eds.): Mathematics Unlimited — 2001 and Beyond. Springer, 2001
303. Ericson K: Heinz Holliger. Spurensuche eines Grenzgängers. Peter Lang, Bern 2004
304. Erlewine S: Movin' & Groovin' - Horace Parlan.
     http://www.allmusic.com/album/movin-groovin-mw0000381525   2015
305. Essl K: Strukturgeneratoren. Beiträge zur elektronischen Musik 5, IEM, Graz 1996
306. Euclid: Euclid's Elements. Translated with introduction and commentary by Sir Th L Heath, vol. 1, 2 and 3, Dover, 1956
307. Euler L: Tentamen novae theoriae musicae (1739). In: Opera Omnia, Ser.III, Vol.1 (Ed. Bernoulli, E et al.). Teubner, Stuttgart 1926
308. Euler L: Conjecture sur la raison de quelques dissonances générales reçues dans la musique (1764). In: Opera Omnia, Ser.III, Vol.1 (Ed. Bernoulli, E et al.). Teubner, Stuttgart 1926
309. Euler L: De harmoniae veris principiis per speculum musicum representatis (1773). In: Opera Omnia, Ser.III, Vol.1 (Ed. Bernoulli, E et al.). Teubner, Stuttgart 1926
310. Euler L: Recherches sur une nouvelle espèce de quarrés magiques. Opera Omnia, Ser. I, Vol. 7: 291-392, 1782
311. Fauré G: Requiem in D minor, Op. 48. Edited by P. Legge 2000
312. Feldman J et al.: Force Dynamics of Tempo Change in Music. Music Perception, *10*, 1992
313. Felver Chr: Cecil Taylor - All The Notes. A Chris Felver Movie, 2003
314. Ferretti R and G Mazzola: Algebraic Varieties of Musical Performances. Tatra Mt. Math. Publ. 23, 59-69, 2001
315. Feulner J et al.: MELONET: Neural Networks that Learn Harmony-Based Melodic Variations. In: ICMA (Ed.): Proceedings of the ICMC 94, San Francisco 1994
316. Fichte J G: Doctrine de la science. Nova methodo (1799), Livre de poche 4621, LGF, 2000
317. Fichtner R: Die verborgene Geometrie in Raffaels "Schule von Athen". Oldenburg 1984
318. Fink E: Grundlagen der Quantenmechanik. Akademische Verlagsgesellschaft, Leipzig 1968
319. Finke R A, Th B Ward, S M Smith: Creative Cognition: Theory, Research, and Applications. MIT Press, Cambridge, MA 1992
320. Finscher L: Studien zur Geschichte des Streichquartetts. Bärenreiter, Kassel 1974
321. Finsler P: Über die Grundlegung der Mengenlehre. Erster Teil. Die Mengen und ihre Axiome. Math. Z. 25, 683-713, 1926
322. Finsler P: Aufsätze zur Mengenlehre. Unger G (Ed.), Wiss. Buchgesellschaft, Darmstadt 1975
323. Fisher G and J Lochhead: Analyzing from the Body. In: Theory and Practice, Journal of the Music Theory Society of New York State, Vol. 27
324. Fleischer A: Eine Analyse theoretischer Konzepte der Harmonielehre mit Hilfe des Computers. Magisterarbeit, MWS, HU Berlin 1996
325. Fleischer A, G Mazzola, Th Noll: Zur Konzeption der Software RUBATO für musikalische Analyse und Performance. Musiktheorie, Heft 4, 314-325, 2000
326. Florida R: The Rise of the Creative Class. Basic Books, 2003

327. Floris E: La rupture cartésienne et la naissance d'une philosophie de la culture dans les œuvres juvéniles de J.-B. Vico. Thèse de 3ème cycle, dir. P. Ricœur, Université Paris X-Nanterre, 22 juin 1974

328. Flusser V: Gesten — Versuch einer Phänomenologie. Fischer, Frankfurt am Main 1994

329. Forster M: Technik modaler Komposition bei Olivier Messiaen. Hänssler, Neuhausen-Stuttgart 1976

330. Forte A: Structure of Atonal Music: Practical Aspects of a Computer-Oriented Research Project. In: Musicology and the Computer. Musicology 1966-2000. A Practical Program. Three Symposia. American Musicological Society, NY 1970

331. Forte A: Structure of Atonal Music. Yale University Press, New Haven 1973

332. Forte A: La Set-complex theory: Elevons les enjeux! Analyse musicale, 4e trimestre, 80-86, 1989

333. Foster W C: Singing Redefined. Recital Publications 1998

334. Frank H: RUBATO® Broadcast. ORF2: Modern Times, Jan. 10, 1997

335. Freedman D Z and P van Nieuwenhuizen: Supergravitation und die Einheit der Naturgesetze. In: Dosch H G (Ed.): Teilchen, Felder und Symmetrien. Spektrum der Wissenschaft, Heidelberg 1984

336. Friberg A: Generative Rules for Music Performance: A Formal Description of a Rule System. Computer Music Journal, Vol. 15, No. 2, 1991

337. Friberg A et al.: Performance Rules for Computer-Controlled Contemporary Keyboard Music. Computer Music Journal, Vol. 15, No. 2, 1991

338. Friberg A et al.: Recent Musical Performance Research at KTH. In: Sundberg J, (Ed.): Generative Grammars for Music Performance. KTH, Stockholm 1994

339. Friberg A: A Quantitative Rule System for Musical Performance. KTH PhD-Thesis, Stockholm 1995

340. Fripertinger H: Enumeration in Musical Theory. Beiträge zur elektronischen Musik 1, Hochschule für Musik und Darstellende Kunst, Graz 1991

341. Fripertinger H: Die Abzähltheorie von Pólya. Diplomarbeit, Univ. Graz 1991

342. Fripertinger H: Endliche Gruppenaktionen in Funktionenmengen—Das Lemma von Burnside—Repräsentantenkonstruktionen—Anwendungen in der Musiktheorie. Doctoral Thesis, Univ. Graz 1993

343. Fripertinger H: Untersuchungen über die Anzahl verschiedener Intervalle, Akkorde, Tonreihen und anderer musikalischer Objekte in n-Ton Musik. Magisterarbeit, Hochschule für Musik und Darstellende Kunst, Graz 1993

344. Fripertinger H: Anwendungen der Kombinatorik unter Gruppenaktionen zur Bestimmung der Anzahl "wesentlich" verschiedener Intervalle, Chorde, Tonreihen usw. Referat an der Univ. Innsbruck, Math. Institut d. Karl-Franzens-Univ., Graz 1996

345. Fripertinger H: Enumeration of Mosaics. Discrete Mathematics, 199, 49-60, 1999

346. Fripertinger H: Enumeration of Non-isomorphic Canons. Tatra Mt. Math. Publ. 23, 47-57, 2001

347. Frova A: Fisica nella musica. Zanichelli, Bologna 1999

348. Fussi F and S J Ghiotti: Il canto fra terminologia e didattica, 2015
    http://www.voceartistica.it/it-IT/index-/?Item=ghiotti

349. Fux J J: Gradus ad Parnassum (1725). Dt. und kommentiert von L. Mitzler, Leipzig 1742

350. Gabriel P: Personal Communication. Zürich 1979

351. Gabriel P: Des catégorie abéliennes. Bull. Soc. Math. France 90, 1962

352. Gabriel P: Representations of Finite-Dimensional Algebras. In: Kostrikin, A.I.; Shafarevich, I.R. (Eds.): Encyclopaedia of Mathematical Sciences, Vol. 73, Springer Berlin 1992

353. Gabrielsson A: Music Performance. In: Deutsch D (Ed.): The Psychology of Music (2nd ed.). Academic Press, New York

354. Gabrielsson A: Expressive Intention and Performance. In: Steinberg R (Ed.): Music and the Mind Machine. Springer, Berlin et al. 1995

355. Gaither C C and A E Cavazos-Gaither: Mathematically Speaking. A Dictionary of Quotations. IOP Publishing Ltd, 1998

356. Galenson D W: Old Masters and Young Geniuses: The Two Life Cycles of Creativity. Princeton University Press, Princeton, NJ 2006

357. Gamer C: Some combinatorial resources of equal-tempered systems. JMT, 11, 32-59, 1967

358. Gamma E, R Helm, R Johnson, J Vlissides: Design Patterns, Elements of Reusable Object-Oriented Software. Addison-Wesley, Reading, MA et al., 1994

359. Gardner H: Kreative Intelligenz: Was wir mit Mozart, Freud, Woolf und Gandhi gemeinsam haben (Translated by Andreas Simon). Campus Verlag, Frankfurt/New York 1999

360. Gasquet S: L'illusion mathématique. Le malentendu des maths scolaires. Syros, 1997

361. Geisser H, G Mazzola, S Onuma: Imaginary Time. Video, Vimeo 2015

362. Gentilucci M & and M C Corballis: From manual gesture to speech: A gradual transition. Neuroscience & Biobehavioral Reviews Vol. 30, Issue 7, pp 949-960 2006

363. Gerwin Th: IDEAMA, Zentrum fur Kunst und Medientechnologie, Karlsruhe 1996

364. Geweke J: A comparison of tests of independence of two covariance-stationary time series. J. Am. Statist. Assoc., 76, 363-373, 1981

365. Giannitrapani D: The Electrophysiology of Intellectual Functions. Karger, Basel 1985

366. Gilson E: Introduction aux arts du beau. Vrin, Paris 1963

367. Ginzburg C: Leçon de Montaigne. Sud-Ouest, 17 octobre 2015

368. Ginzburg C: L'estrangement. In à distance, Neuf essais sur le point de vue en histoire. Traduction française par Pierre Antoine Fabre, éditions Gallimard NRF, Paris 2001

369. Giocanti S: L'art sceptique de l'estrangement dans les *Essais* de Montaigne, in *L'estrangement, Retour sur un thème de Carlo Ginzburg* [573, p. 19-35]

370. Godement R: Topologie algébrique et théorie des faisceaux. Hermann, Paris 1964

371. Godøy R I and M Leman (Eds.): Musical Gestures—Sound, Movement, and Meaning. Routledge, New York and London 2010

372. Goeller S: Object Oriented Rendering of Complex Abstract Data. PhD. thesis, Universität Zürich, 2004

373. Goethe J W von: Brief an Zelter. 9. Nov. 1829

374. Goffman E: The Presentation of Self in Everyday Life. University of Edinburgh, Edinburgh 1956

375. Goffman E: Behavior in Public Places. The Free Press, New York 1963

376. Goldblatt R: Topoi. North-Holland, Amsterdam et al. 1984

377. Goldin-Meadows S: Hearing Gesture: How Our Hands Help Us Think. Harvard University Press 2003

378. Goldstein J L: An Optimum Processor Theory for the Central Formation of the Pitch of Complex Tones. J. Acoust. Soc. Am. 54, 1496 1973

379. Gomes Teixeira F: Traité des courbes spéciales remarquables planes et gauches. 3 vol., (Coïmbre, Impr. de l'Université, 1908-1915), éd. Jacques Gabay, 1995. Traduction de *Tratado de las curvas especiales notables*, Ac. Sc. Madrid, 1899

380. Interview with Emilia Gómez: "PHENICX is a new way of experiencing a classical music concert", http://www.upf.edu/enoticies/en/entrevistes/1232.html#.VbfU7s63QUU

381. Gorenstein D: Classifying the finite simple groups. Bull. A.M.S. 14, 1-98, 1986

382. Gottschewski H: Tempohierarchien. Musiktheorie, Heft 2, 1993

383. Götze H and R Wille R (Eds.): Musik und Mathematik. Springer, Berlin et al. 1985

384. Gould G: The Glenn Gould Reader. Alfred A. Knopf, New York 1984

385. Goupillaud P, A Grossmann, J Morlet: Cycle-octave and related transforms in seismic signal analysis. Geoexploration, 23, 85-102, 1984-1985

386. Grabusow N: Vielfalt akustischer Grundlagen der Tonarten und Zusammenklänge - Theorie der Polybasiertheit. Musiksektion des Staatsverlags, Moskau 1929

387. Graeser W: Bachs "Kunst der Fuge". In: Bach-Jahrbuch 1924

388. Graeser W: Der Körpersinn. Beck, München 1927

389. Gramain A: Géométrie élémentaire. Hermann, 1997

390. Greimas A J: Les actants, les acteurs et les figures. In: Chabrol C (Ed.): Sémiotique narrative et textuelle. Larousse, Paris 1974

391. Greub W: Linear Algebra. Springer, Berlin et al. 1967

392. Grisey G: Tempus ex machina: a Composer's Reflections on Musical Time. Contemporary Music Review 2, no. 1:238-275, 1987

393. Grondin J: L'herméneutique. Coll. Que sais-je ? no 3758, PUF, Paris 2006

394. Gross D: A Set of Computer Programs to Aid in Music Analysis. Ph. Diss. Indiana Univ. 1975

395. Grothendieck A and J Dieudonné: Eléments de Géométrie Algébrique I. Springer, Berlin et al. 1971

396. Grothendieck A and J Dieudonné: Eléments de Géométrie Algébrique I-IV. Publ. Math IHES no. 4, 8, 11, 17, 20, 24, 28, 32, Bures-sur-Yvette 1960-1967

397. Grothendieck A: Correspondence with G. Mazzola. April 1, 1990

398. Grothendieck A: Récoltes et Semailles. Université Univ. Sci. et Tech. Languedoc et CNRS, Montpellier, 1st edn. 1985

399. Guevara R C L et al.: A Modal Distribution Approach to Piano Analysis and Synthesis. In: ICMA (Ed.): Proceedings of the ICMC 96, San Francisco 1996

400. Guillaume P: La psychologie de la forme. Journal de Psychologie, XXII, p. 768-800, 1925

401. Guillaume P: La psychologie de la forme. Flammarion, 1979

402. Guitart R: Sur l'ébauche des structures. Proc. 3d Congress of Bulgarian math., summaries part II, p. 354, Varna, Sept. 6-15, 1972

403. Guitart R: Sur le foncteur diagrammes, CTGD XIV-2, pp. 181-182, 1973

404. Guitart R: Remarques sur les machines et les structures, CTGD XV-2, pp. 113-144, 1974

405. Guitart R: Remarques sur les machines et les structures. CTGD XV-2, pp. 113-144, 1974

406. Guitart R and L Van den Bril: Décompositions et lax-complétions. CTGD XVIII-2, pp. 333-407, 1977

407. Guitart R: Des machines aux bimodules. polycopié, Univ. Paris 7, 24 p., 1978

408. Guitart R: Relations et carrés exacts. Ann. Sc. Math. Qué., vol. IV-2, p. 103-125, juillet 1980

409. Guitart R and C Lair: Calcul syntaxique des modèles et calcul des formules internes, *Diagrammes*, vol. 4, déc. 1980, 106 p.

410. Guitart R: On the geometry of computations, I and II. Cahiers Top. Géo. Diff. Cat. XXVII,4, 107-136, 1986; XXIX,4, p. 297-326, 1988

411. Guitart R: Mathématiques. 1ère année. Polycopié U. Paris 7; 27 chap., 432 p. 1992-1993

412. Guitart R: La pulsation mathématique. L'Harmattan, Paris, 1999

413. Guitart R: évidence et étrangeté. PUF, Paris, 2000

414. Guitart R: Sur les places du sujet et de l'objet dans la pulsation mathématique. Questions éducatives, revue du Centre de Recherche en éducation de l'Université Jean Monnet de Saint-Étienne, no. 22-23, pp. 49-81, décembre 2002

415. Guitart R: Toute théorie est algébrique et topologique. Cahiers Top. Géo. Diff. Cat. vol. XLIX-2 (2008), p. 83-128. 2008 (first partial version: "Toute théorie est algébrique", in Journée Mathématique en l'Honneur d'Albert Burroni, vendredi 20 septembre 2002, Université Paris 7, Institut de Mathématiques de Jussieu, Prépublication 368, p. 79-102, avril 2004

416. Guitart R and Ch Ehresmann: Au carrefour des structures locales et algébriques. Colloque International Ch Ehresmann : 100 ans, Amiens 7-9 octobre 2005; Cahiers Top. Géo. Diff. Cat., vol. XLVI-3, p. 172-175, 2005

417. Guitart R: L'évidemment des objets et le dehors comme substance. Colloque Le lemme de Yoneda, "Enjeux mathématiques et philosophie", ENS, Paris 18 juin 2007

418. Guitart R: An anabelian definition of abelian homology. Cahiers Top Géo Diff Cat, XXXXVIII, 4, pp. 261-269, 2007

419. Guitart R: Mathematical Pulsation at the root of invention. Anais do Congresso Htem 4, Rio de Janeiro, de 5 a 9 de maio de 2008

420. Guitart R: Sur la représentation de l'herméneutique comme commerce gratuit des interprétations, conférence au *CERCI*, Université de Nantes, 7 mars 2009

421. Guitart R: Cohomological Emergence of Sense in Discourses (As Living Systems Following Ehresmann and Vanbremeersch). Axiomathes, 19(3), 245-270, 2009

422. Guitart R: Les coordonnées curvilignes de Gabriel Lamé, représentations des situations physiques et nouveaux objets mathématiques, 12 p. In: Actes du Colloque International Gabriel Lamé. Les pérégrinations d'un ingénieur au XIXème siècle, 15-16-17 janvier 2009, Nantes. SABIX, no 44, p. 119-129, octobre 2009

423. Guitart R: Sense and Signs in the Mathematical Invention of Coordination (in the Perspective of Arrows, Relations and Sketches). Talk at the colloquium *Semiotic Approaches to Mathematics, the History of Mathematics and Mathematics Education (SemMHistEd)*, 3rd Meeting Aristotle University of Thessaloniki, July 16-17, 2009

424. Guitart R: Modélisation qualitative catégoricienne : modèles, signes et formes. In: Andreatta M, F Nicolas and Ch Alunni (Eds.): à la lumière des mathématiques et à l'ombre de la philosophie, Coll. Musique/Sciences, ENS, Delatour, Ircam, Centre Pompidou, pp. 133-147, Paris 2012

425. Guitart R: Bachelard et la pulsation mathématique. Revue de synthèse, tome 136, 6ème série, n. 1-2, p. 33-74, 2015

426. Guitart R: Note sur deux problèmes pour une épistémologie transitive des mathématiques, *Revue de synthèse*, tome 136, 6ème série, n. 1-2, p. 237-279, 2015

427. Guitart R: Nietzsche face à l'expérience mathématique, in: Barbin E and J-P Cléro éds, *Les mathématiques et l'expérience : ce qu'en ont dit les philosophes et les mathématiciens*, Hermann, p. 247-278, 2015

428. Guitart R: History in mathematics according to André Weil. ESU7, Copenhagen, July 2014

429. Guitart R: L'évidement et la nécessité de la contingence. Conférence au Colloque *Les fins de cure : la poursuite du désêtre*, 7/8, AECF Lille, novembre 2015

430. Gurlitt et al. (Eds.): Riemann Musiklexikon/Sachteil. Schott, Mainz 1967

431. Gusdorf G: La révolution galiléenne, tomes I et II. Payot, Paris, 1969

432. Hába A: Neue Harmonielehre. (Reprint from 1927) Universal Edition, Wien 1978

433. Habermann R: Elementary Applied PDEs, Prentice Hall 1983

434. Hadamard J: An Essay on the Psychology of Invention in the Mathematical Field. Princeton University Press, 1945, Dover, 1954

435. Halsey D and E Hewitt: Eine gruppentheoretische Methode in der Musiktheorie. Jahresber. d. Dt. Math.-Vereinigung 80, 1978

436. Handschin J: Der Toncharakter. Zürich 1948

437. Hansen P: Pourquoi il faut sortir des mathématiques. Mercredi 1er avril 2015 (see also [99]). http://www.energie-crise.fr/spip.php?article176

438. Hanslick E: Vom Musikalisch-Schönen. Breitkopf und Härtel (1854), Wiesbaden 1980

439. Hardy G H: A Mathematician's Apology. Cambridge U. Press, 1967

440. Hashimoto K: D-Brane. Springer, Heidelberg et al. 2012

441. Hardt M: Zur Zahlenpoetik Dantes. In: Baum R. and Hirdt W (Eds.): Dante Alighieri 1985. Stauffenburg, Tübingen 1985

442. Harkness R: Songs of Seoul. University of California Press 2014

443. Harris C and A R Brinkman: A unified set of software tools for computer-assisted set-theoretic and serial analysis of contemporary music. Proc. ICMC 1986, ICA, San Francisco 1986

444. Hartshorne R: Algebraic Geometry. Springer, New York et al. 1977

445. Hashimoto S and H Sawada: Musical Performance Control Using Gesture: Towards Kansei Technology for Art. In: Kopiez R and Auhagen W (Eds.): Controlling Creative Processes in Music. Peter Lang, Frankfurt am Main et al. 1998

446. Hatten R: Interpreting Musical Gestures, Topics, and Tropes. Indiana University Press 2004

447. Hawking S: A Brief History of Time: From the Big Bang to Black Holes. Bantam Books, New York 1988

448. Haugh L D: Checking for independence of two covariance stationary time series: a univariate residual cross correlation approach. J. Am. Statist. Assoc., 71, 378-385, 1976

449. Haus G et al.: Stazione di Lavoro Musicale Intelligente. In: Haus G and Pighi I (Eds.): X Colloquio di Informatica Musicale. AIMI, LIM-Dsi, Milano 1993

450. Hayes J R: Cognitive Processes in Creativity. In: J A Glover, R R Roning, C R Reynolds (Eds.): Handbook of Creativity. Plenum, New York 1989

451. Hebb D O: Essay on mind. Hillsdale, New Jersey, Lawrence Erlbaum Associates, 1980

452. Hegel G W F: Wissenschaft der Logik I (1812). Felix Meiner, Hamburg 1963

453. Hegel G W F: Wissenschaft der Logik, Erster Band, Erstes Buch. Schrag, Nürnberg 1812
Online http://www.deutschestextarchiv.de/book/show/hegel_logik0101_1812
Modern German online version:
http://www.zeno.org/Philosophie/M/Hegel,+Georg+Wilhelm+Friedrich/Wissenschaft+der+Logik

454. Hegel G W F: Hegel's Science of Logic.
Online http://www.marxists.org/reference/archive/hegel/works/hl/hlconten.htm

455. Heiberg J L and H Menge (Eds.): Euclidis opera omnia. 8 vol. & supplement, in Greek. Teubner, Leipzig 1883-1916.

456. Heijink H, P Desain, H Honing, L Windsor: Make me a match: An evaluation of different approaches to score-performance matching. Computer Music Journal, 24(1), 43-56, 2000

457. Helmholtz H von: Die Lehre von den Tonempfindungen als physiologische Grundlage der Musik (1863). Nachdr. Darmstadt 1968

458. Henck H: Karlheinz Stockhausens Klavierstück IX. Verlag für systematische Musikwissenschaft, Bonn-Bad Godesberg 1978

459. Hentoff N: Liner notes to Coltrane's last album "Expression". Impulse AS-9120, New York 1967

460. Hecquet S and A Prokhoris: Fabriques de la danse. PUF, Paris 2007

461. Herbort H J: Keine Ausweispflicht für cis. Die Zeit Nr. 43, 21. Oktober 1988

462. Hermann J: Consideratio curvarum in punctum positione datum projectarum, et de affectionibus earum inde pendentibus. Commentarii Academiae Petropolitanae, IV, 37-46, 1729

463. Hesse H: Das Glasperlenspiel (1943). Fretz und Wasmuth, Zürich 1943

464. Hesse H: Materialien zu H. Hesses "Das Glasperlenspiel", Vol. I. : Vom Wesen und von der Herkunft des Glasperlenspiels (1934). Suhrkamp, Frankfurt am Main 1981

465. Heussenstamm G: Norton Manual of Music Notation. Norton & Comp., New York 1987

466. Hichert J: Verallgemeinerung des Kontrapunkttheorems für die Hierarchie aller starken Dichotomien in temperierter Stimmung. Diplomarbeit, TU Ilmenau 1993

467. Hindemith P: Unterweisung im Tonsatz. Schott, Mainz 1940

468. Hilbert D and S Cohn-Vossen: Geometry and the Imagination. AMS Chelsea Pub. (transl. by Paul Nemenyi from *Anschauliche Geometrie*, 1932), 1952

469. Hiller L and P Ruiz: Synthesizing sounds by solving the wave equation for vibration objects. J. of the Audio Engineering Soc. 19: 463-470, 542-551, 1971

470. Hilton P J and S Wylie: Homology Theory. Cambridge University Press, London et al. 1967

471. Hjelmslev L: La Stratification du Langage. Minuit, Paris 1954

472. Hjelmslev L: Prolégomènes à une théorie du langage. Minuit, Paris 1968-71

473. Hjelmslev L: Nouveaux essays. PUF, Paris 1985

474. Hofstädter D: Gödel, Escher, Bach. New York: Basic Books, New York 1979

475. Hofstadter D R: Staring Emmy Straight in the Eye—and Doing my Best not to Flinch. In: D Cope (Ed.): Virtual Music: Computer Synthesis of Musical Style. MIT Press, Cambridge, MA 2001

476. Hodgson J: Mastering Movement: the life and work of Rudolf Laban. Routledge, New York 2001

477. Holden S P: Right-Handed and Right On. The New York Times, No. 3, June 2, 1999

478. Holm-Hadulla R M: Kreativität. Konzept und Lebensstil. Vandenhoeck & Ruprecht, Göttingen 2005

479. Hong Y: Testing for independence between two covariance stationary time series. Biometrika, 83, No.3, 615-626, 1996

480. Honing H: Expresso, a strong and small editor for expression. In: ICMA (Ed.): Proceedings of the ICMC 92, San Francisco 1992

481. Hooft G 't: Symmetrien in der Physik der Elementarteilchen. In: Dosch H G (Ed.): Teilchen, Felder und Symmetrien. Spektrum der Wissenschaft, Heidelberg 1984

482. Hörnl D et al.: Learning Musical Structure and Style by Recognition, Prediction and Evolution. In: ICMA (Ed.): Proceedings of the ICMC 96, San Francisco 1996

483. Horry Y: A Graphical User Interface for MIDI Signal Generation and Sound Synthesis. In: ICMA (Ed.): Proceedings of the ICMC 94, San Francisco 1994

484. Hosken D: An Introduction to Music Technology. Routledge New York 2011

485. Howe H: Some combinatorial properties of Pitch-Structures. PNM, 4(1), 45-61, 1965

486. Howe M J A: Prodigies and Creativity. In: R J Sternberg (Ed.): Handbook of Creativity. Cambridge University Press, New York 1999

487. Hu S-T: Mathematical Theory of Switching Circuits and Automata. University of California Press, Berkeley and Los Angeles 1968

488. Hudak P et al.: Haskore Music Notation - An Algebra of Music. J. Functional Programming Vol. 6 (3) 1996

489. Hudspeth A J and D P Corey: Sensitivity, Polarity, and Conductance Change in the Response of Vertebrate Hair Cells to Controlled Mechanical Stimuli. Proc. Nat. Acad. Sci. Am. 74(6), 2407-2411, 1977

490. Hull P et al.: The Music Business and Recording Industry. Third Edition. Routledge, New York and London 2011

491. Humphreys J E: Introduction to Lie Algebras and Representation Theory. Springer, New York et al. 1972

492. Hung R et al.: The Analysis and Resynthesis of Sustained Musical Signals in the Time Domain. In: ICMA (Ed.): Proceedings of the ICMC 96, San Francisco 1996

493. Hunziker E and G Mazzola: Ansichten eines Hirns. Birkhäuser, Basel 1990

494. Husmann H: Einführung in die Musikwissenschaft. Heinrichshofen, Wilhelmshaven 1975

495. Ilchmann A: Kritik der Übergänge zu den ersten Kategorien in Hegels Wissenschaft der Logik. Hegel-Studien, Bd. 27, Bouvier, Bonn 1992

496. Jackendoff R and F Lerdahl: A Generative Theory of Tonal Music. MIT Press, Cambridge MA, 1983

497. Jackson J D: Classical Electrodynamics. Wiley, New York 1998

498. Jairazbhoy N A: What Happened to Indian Music Theory? Indo-Occidentalism? Ethnomusicology 52 (3): 34977, 2008

499. Jakobson R and M Halle: Fundamentals of Language. Mouton, Le Hague 1957

500. Jakobson R: Linguistics and Poetics. In: Seboek, TA (Ed.): Style in Language. Wiley, New York 1960

501. Jakobson R: Language in relation to other communication systems. In: Linguaggi nella società e nella tecnica. Edizioni die Communità, Milano 1960

502. Jakobson R: Six leçons sur le son et le sens. Ed. de Minuit, Paris 1976

503. Jakobson R: Hölderlin, Klee, Brecht. Suhrkamp, Frankfurt/Main 1976

504. Jakobson R and K Pomorska: Poesie und Grammatik. Dialoge. Suhrkamp, Frankfurt/Main 1982

505. Jakobson R: Semiotik - Ausgewählte Texte 1919-1982. Holenstein E (Ed.), Suhrkamp, Frankfurt/Main 1988

506. Jauss H R: Rückschau auf die Rezeptionstheorie—Ad usum Musicae Scientiae. In: Danuser H and Krummacher F (Hsg.): Rezeptionsästhetik und Rezeptionsgeschichte in der Musikwissenschaft. Laaber, Laaber 1991

507. Jeans J H: Science and Music. 1938, reprinted by Dover, 1968

508. Jeppesen K: Kontrapunkt. Breitkopf und Härtel, Wiesbaden 1952

509. John F: Partial Differential Equations. Springer, Heidelberg et al. 1978

510. Johnson T: See his web page at http://www.tom.johnson.org
511. Johnson-Laird P N: Freedom and Constraint in Creativity. In: R J Sternberg (Ed.): The Nature of Creativity: Contemporary Psychological Perspectives. Cambridge University Press, New York, pp. 202-219, 1988
512. Junod J: Counterpoint Worlds and Morphisms. Dissertation, University of Zurich, Zurich 2010
513. Julia G: Sur l'itération des fonctions rationnelles. J. de Math. Pure et Appl. 8, 1918
514. Jurek T: The Complete Blue Note Horace Parlan Sessions. AllMusic 2015.
    `http://www.allmusic.com/album/the-complete-blue-note-horace-parlan-sessions-mw0001007267`
515. Kagel M: Translation - Rotation. Die Reihe Bd.7, Universal Edition, Wien 1960
516. Kahle W: Taschenatlas der Anatomie, Bd.3: Nervensystem und Sinnesorgane. Thieme/dtv, Stuttgart 1979
517. Kaiser J: Beethovens 32 Klaviersonaten und ihre Interpreten. Fischer, Frankfurt/Main 1979
518. Kan D: Adjoint functors. Transactions of the American Mathematical Society, 87: 294-329, 1958
519. Kant I: Kritik der reinen Vernunft. Meiner, Hamburg, 2nd edn. 1957
520. Kantor J-M (Ed.): Ou en sont les mathématiques?. Vuibert/SMF, 2002
521. Karg-Elert S: Polaristische Klang- und Tonalitätslehre (1931). Out of print, cf.: Schenk P.: Karg-Elerts polaristische Harmonielehre. In: Vogel, M (Ed.) Beiträge zur Musiktheorie des 19. Jahrhunderts. Bosse, Regensburg 1966
522. Kasner E and J Newman: Mathematics and the imagination. Simon and Schuster, New York, 1940
523. Katayose H et al.: Demonstration of Gesture Sensors for the Shakuhachi. In: ICMA (Ed.): Proceedings of the ICMC 94, San Francisco 1994
524. Katsman R: Anthropoetic Gesture: A Key to Milorad Pavić's Poetics (Landscape Painted With Tea). Academic Electronic Journal in Slavic Studies No. 16, Univ. Toronto, 2006
525. Katsman R: Gestures Accompanying Torah Learning/recital among Yemenite Jews. Gesture 7 (1): 119, 2007
526. Kayser H: Akróasis. Die Lehre von der Harmonik der Welt. Schwabe, Basel 1946
527. Kelley J L: General Topology. Van Nostrand, Princeton et al. 1955
528. Kempe A B: A memoir on the theory of mathematical form. Philosophical Transactions of the Royal Society of London, 177, 1-70, 1886
529. Kendon A: How Gestures Can Become like Words. Crosscultural Perspectives in Nonverbal Communication. 1988
530. Kendon A: Gesture: Visible Action as Utterance. Cambridge University Press, Cambridge 2004
531. Keyser C J: The Meaning of Mathematics. Scripta Mathematica, 1933, p. 15-28
532. Kiczales G, J Rivieres, D G Bobrow: The Art of the Metaobject Protocol. The MIT Press, Boston 1991
533. Kinderman W: Artaria 195. University of Illinois Press, Urbana and Chicago 2003
534. King E: New Perspectives on Music and Gesture. Routledge 2016
535. Kirsch E: Wesen und Aufbau der Lehre von den harmonischen Funktionen. Leipzig 1928
536. Klemm M: Symmetrien von Ornamenten und Kristallen. Springer, Berlin et al. 1982
537. Klumpenhouwer H: Deep Structure in K-net Analysis with special reference to Webern's opus 16,4. Unpublished manuscript
538. Klumpenhouwer H: A Generalized Model of Voice-Leading for Atonal Music. Ph.D. Thesis, Harvard University 1991
539. Knapp J: Franz Liszt. Berlin 1909
540. Köhler E: Brief an Guerino Mazzola. Hamburg 1988
541. Kollmann A: An Essay on Musical Harmony. London 1796
542. Komparu K: The Noh Theatre. Weatherhill/Tankosha, New York et al. 1983
543. Koenig Th: Robert Schumanns *Kinderszenen* op.15. In: Metzger H-K und Riehn R (Hrg.): Robert Schumann II. edition text+kritik, München 1982
544. Kononenko N O: Ukrainian minstrels and the blind shall sing. Armonk, N.Y.: M E Sharpe, 1998
545. Kopiez R and J Langner: Entwurf einer neuen Methode der Performanceanalyse auf Grundlage einer Theorie oszillierender Systeme. In: Behne K-E and de la Motte H (Eds.): J.buch der D. Ges. für Musikpsychologie. **12** (1995), Wilhelmshaven 1996
546. Kopiez R: Aspekte der Performanceforschung. In: de la Motte H: Handbuch der Musikpsychologie. Laaber, Laaber 1996
547. Kopiez R: Mensch - Musik - Maschine. Musica, 50 (1), 1996
548. Kopiez R: "The most wanted song/The most unwanted song" – Klangfarbe als wahrnehmungsästhetische Kategorie. Musicology Conference, Halle 1998
549. Kopiez R, J Langner, Ch Stoffel: Realtime analysis of dynamic shaping. Talk at the 6th International Conference on Music Perception and Cognition (ICMPC), Keele, England, August 5-10, 2000

550. Kopiez R, J Langner and B Feiten: Perception ad Representation of Multiple Tempo Hierarchies in Musical Performance ad Composition: Perspectives from a New Theoretical Approach. In: Kopiez R and W Auhagen (Eds.): Controlling Creative Processes in Music. Peter Lang, Frankfurt am Main et al. 1998

551. Körner T W: Fourier Analysis. Cambridge University Press 1988, Cambridge

552. Kostelanetz R: John Cage. Praeger, New York 1968

553. Kouneiher J, D Flament, P Nabonnand, J-J Szczeciniarz (Eds.): Geometrie au XXème siècle, 1930-2000. Histoire et horizons. Hermann, Paris 2005

554. Kouzes R T et al.: Collaboratories: Doing Science on the Internet. Computer, August 1996

555. Kozbelt A, R A Beghetto, M A Runco: Theories of Creativity. In: J C Kaufman and R J Sternberg (Eds.): The Cambridge Handbook of Creativity. Cambridge University Press, New York, pp. 20-47, 2010

556. Krömer R: La théorie des catégories : un outil de musicologie scientifique aux yeux de la critique philosophique. In: Andreatta M, F Nicolas et C Alunni (Eds.): Á la lumière des mathématiques et à l'ombre de la philosophie. Coll. Musique/Sciences, ENS, Delatour, Ircam, Centre Pompidou, p. 213-222, Paris 2012

557. Kronland-Martinet R: The Wavelet Transform for Analysis, Synthesis, and Processing of Speech and Music Sounds. Computer Music Journal, 12 (4), 1988

558. Kreiman J and D Sidtis: Foundations of Voice Studies. Wiley-Blackwell 2013

559. Kronland-Martinet R: The Wavelet Transform for Analysis, Synthesis, and Processing of Speech and Music Sounds. Computer Music Journal, 12 (4), 1988

560. Kronman U and J Sundberg: Is the Musical Ritard an Allusion to Physical Motion? In: Gabrielsson A (Ed.): Action and Perception in Rhythm and Meter. Bibl. of the Royal Swedish Acad. of Sci. 55, Stockholm 1987

561. Kuratowski K and A Mostowski: Set Theory. North Holland, Amsterdam 1968

562. Kühner H: Virtual Table. CG TOPICS 3/97

563. Laban R: Language of Movement: A guidebook to Choreutics, ed. Lisa Ullmann, Plays, Boston 1974

564. Laban R: Die Welt des Tänzers. W. Seifert, Stuttgart 1920

565. Laban R: The Laban Sourcebook. McCaw D (Ed.), Routledge, New York 2011

566. Laban R: The Mastery of Movement, 4th ed. Macdonald & Evans, Plymouth 1980

567. Laban R: The Rhythm of Effort and Recovery I. LAMG Magazine, in Hodgson, Mastering Movement, November 1959

568. Lachenmann H: Mouvement (- vor der Erstarrung) für Ensemble. Breitkopf und Härtel, 1984

569. Lages Lima E: Fundamental Groups and Covering Spaces. Peters, Natick, MA 2003

570. Lakoff G and R Núñez: Where mathematics comes from: How the embodied mind brings mathematics into being. Basic Books, New York 2000

571. Lamé G: Examen des différentes méthodes employées pour résoudre les problèmes de géométrie (1903). Hermann, Paris 1918

572. Lamé G: Leçons sur les coordonnées curvilignes, et leurs diverses applications. Mallet-Bachelier, 1859

573. Landi S (Ed.): L'estrangement, Retour sur un thème de Carlo Ginzburg. Essais, revue interdisciplinaire d'humanités, No Hors série, école doctorale Montaigne-Humanités, Univ. Michel de Montaigne, Bordeaux 3, 2013

574. Landi S (Ed.): Avant-propos. In: L'estrangement, Retour sur un thème de Carlo Ginzburg [573, p. 9-17]

575. Landry, E and J-P Marquis: Categories in Context: Historical, Foundational, and Philosophical. Philosophia Mathematica 13.1: 1-43, 2005

576. Lang S: Introduction to Differentiable Manifolds. Interscience, New York et al. 1962

577. Lang S: Elliptic Functions. Addison-Wesley, Reading, Mass. 1973

578. Lang S: $SL_2(\mathbb{R})$. Addison-Wesley, Reading, Mass. et al. 1975

579. Langley P, H A Simon, G L Bradshaw, J M Zytkow: Scientific Discovery: Computational Explorations of the Creative Process. MIT Press, Cambridge, MA, 1987

580. Langner G: Evidence for Neuronal Periodicity Detection in the Auditory System of Guinea Fowl: Implications for Pitch Analysis in the Time Domain. Exp. Brain Res. 52, 333-355 1983

581. Langner G: Periodicity coding in the auditory system. Hear Res. Jul;60(2):115-42, 1992

582. Langer S: Feeling and Form, Routledge and Kegan Paul, London 1953

583. Lawson R A: The First Century of Blues: One Hundred Years of Hearing and Interpreting the Music and the Musicians. Southern Cultures, pp. 39-61, 2007

584. Langner J, R Kopiez, Ch Stoffel, M Wilz: Realtime analysis of dynamic shaping. In: Woods C et al. (Eds.): Proceedings of the Sixth International Conference on Music Perception and Cognition, Keele, UK, 2000

585. Lawvere F W: An elementary theory of the category of sets. Proc. Natl. Acad. Sci. 52, 1506-1511, 1964

586. Lawvere F W and R Rosebrugh: Sets for mathematics, Cambridge U. Press, 2003

587. Lazier R: Making Schoenberg Dance. In: Sound Moves, Proceedings of the Int. Conf. on Music and Dance, Roehampton University 2005

588. Leach J L: Towards a Universal Algorithmic System for Composition of Music and Audio-Visual Works. In: ICMA (Ed.): Proceedings of the ICMC 96, San Francisco 1996

589. Ledderose L: Kreativität und Schrift in China. In: R M Holm-Hadulla (Ed.): Kreativität. Springer, Heidelberg 2000

590. Lee, J M: Introduction to Topological Manifolds. Springer, Heidelberg et al. 2000

591. Legrand M: Revue critique de "La pulsation mathématique". Repères-IREM, 39, avril 2000

592. Le Guin E: Boccherini's Body. University of California Press, Berkeley and Los Angeles 2006

593. Leman M: Schema Theory. Springer, Berlin et al. 1996

594. Lerdahl F: Timbral Hierarchies. Contemporary Music Review, vol.2, no.1, 1987

595. Lerdahl F: Cognitive Constraints on Compositional Systems. In: ed. J Sloboda (Ed.): Generative Processes in Music. Oxford University Press, Oxford 1988. Reprinted in Contemporary Music Review 6, no. 2 (1992):97-121

596. Levy M: Intonation in North Indian Music. Biblia Impex, 1982

597. Lévi-Strauss C: Le cru et le cuit: Mythologies I. Plon, Paris 1964

598. Lewin D: Re: Intervallic Relations between Two Collections of Notes. JMT, 3(2), 298-301, 1959.

599. Lewin D: Klumpenhouwer Networks and Some Isographies that Involve Them. Music Theory Spectrum 12/1, 1990

600. Lewin D: The Intervallic Content of a Collection of Notes, Intervallic Relations between a Collection of Notes and Its Complement: An Application to Schoenberg's Hexachordal Pieces. JMT, 4(1), 98-101, 1960

601. Lewin D: Forte's Interval Vector, My Interval Function, and Regener's Common-Note Function. JMT, 21(2), 194-237, 1977

602. Lewin D: A Formal Theory of Generalized Tonal Functions. JMT 26(1), 32-60, 1982

603. Lewin D: On Formal Intervals between Time-Spans. Music Perception, 1(4), 414-423, 1984

604. Lewin D: Musical Form and Transformation: 4 Analytic Essays. Yale University Press, New Haven and London 1993

605. Lewin D: Generalized Musical Intervals and Transformations (1987). Cambridge University Press 2007

606. Lewis C I: Mind and Word Order. Dover, New York 1956

607. Leyton M: Symmetry, Causality, Mind. MIT Press, Cambridge/MA and London 1992

608. Leyton M: A Generative Theory of Shape. Springer, Berlin et al. 2001

609. Lichtenhahn E: Romantik: Aussen- und Innenseiten der Musik. In: "Das Paradox musikalischer Interpretation", Symposion zum 80. Geburtstag von K. von Fischer, Univ. Zürich 1993

610. Lidov D: Mind and Body in Music. Semiotica 66/1 (1987), pp. 77-78

611. Ligeti G: Pierre Boulez: Entscheidung und Automatik in der Structure Ia. Die Reihe IV, UE, Wien 1958

612. Ligeti G and G Neuweiler: Motorische Intelligenz. Wagenbach, Berlin 2007

613. Lippe E and A H M ter Hofstede: A Category Theory Approach to Conceptual Data Modeling. Informatique Theorique et Applications, Vol 30, 1, 31-79, 1996

614. Liszka J: A general introduction to the Semeiotic of Charles S. Peirce. Bloomington M.A. : Harvard University Press, 1996

615. Liszt F: Mephisto-Walzer I. Ed. Peters, No. 7203, (Edited by E. Von Sauer) 1913-17

616. Locke J: An essay concerning human understanding (first publication 1690). Pennsylvania State University 1999

617. Loomis L H and S Sternberg: Advanced Calculus. Addison-Wesley, Reading, MA et al. 1968

618. Loria G: Perfectionnements, Evolution, Métamorphoses du concept de "coordonnées" Contribution à l'histoire de la géométrie analytique. Osiris, Vol. 8, pp. 218-288, 1948

619. Loria G: A. L. Cauchy in the History of Analytic Geometry. Scripta Mathematica, p. 123-128, 1933

620. Losada C: Isography and Structure in the Music of Boulez. Journal of Mathematics and Music 2008, Vol. 2, no. 3, pp. 135-155, 2008

621. Luck G, P Toiviainen, M Thompson: Perception of Expression in Conductors' Gestures: A Continuous Response Study. Music Perception, 20(1), 47-57, 2010

622. Louis A K et al.: Wavelets. Teubner, Stuttgart 1994

623. Lubet A: (Paralyzed on One) Sideman: Disability Studies Meets Jazz, through the Hands of Horace Parlan. Critical Studies in Improvisation/Études critiques en improvisation 6.2. 2010
    http://www.criticalimprov.com/article/view/1268

624. Lubet A: Music, Disability, and Society. Temple University Press, Philadelphia 2011

625. Lubet A: Social Confluence and Citizenship: A View from the Intersection of Music and Disability. In: Hirschmann N J, and B Linker (Eds.): Civil Disabilities: Citizenship, Membership, and Belonging. University of Pennsylvania Press, Philadelphia 2015

626. Lubet A: Oscar Peterson's piano prostheses: Strategies of performance and publicity in the post-stroke phase of his career. Jazz Research Journal, 7(2), pp. 151-182, 2015

627. Lubet A: A Case of the Blues (misc). Unpublished, 2014

628. Lucas E: Récréations mathématiques. Gauthier-Villars, Paris, 1882

629. Lüdi W: Fax to G Mazzola. Malans 01/23/1991

630. Ludwig H: Marin Mersenne und seine Musiklehre. Olms, Hildesheim 1971

631. Lussy M: Traité de l'expression musicale. Paris 1874

632. Lutz R: Le poète dans son oeuvre. Seminar Jung, Univ. Zürich, 1980

633. Lyttleton H: The Best of Jazz. Robson Books, London 1998

634. Machado A: Chant XXIX, Proverbios y cantarès, Campos de castilla, (1917), Ed. Cátedra, Madrid, 1989, 2015

635. Mac Lane S: Homology. Springer, Heidelberg et al. 1963

636. Mac Lane S: The development of mathematical ideas by collision: The case of Categories and Topos Theory. In: Categorical Topology and its relation to analysis, algebra and combinatorics, 1989

637. Mac Lane S: Categories for the Working Mathematician. Springer, New York et al. 1971

638. Mac Lane S: Mathematics, Form and Function. Springer, New York 1986

639. Mac Lane S and Moerdijk I: Sheaves in Geometry and Logic. Springer, New York et al. 1994

640. MacLean P D: The triune brain, emotion, and scientific bias. In: Schmitt F O (Ed.): The Neurosciences: Second Study Program, 336-348, Rockefeller Univ. Press, New York 1970

641. Maeder R: Programming Mathematica. Addison-Wesely, Reading, Mass. 1991

642. Manin Y: Interview in the Notices of the American Mathematical Society, November 2009, p. 1268

643. Mannone M and G Mazzola: Hegel's Conceptual Group Action on Creative Dynamics in Music. Gli spazi della musica, University of Torino 2014

644. Mannone M and G Compagno: Characterization of the degree of Musical non-Markovianity. Unpublished

645. Mannone M and S Lo Franco and G Compagno: Comparison of non-Markovianity criteria in a qubit system under random external fields. Phys. Scr. T, 2013

646. Mannone M: Segmentation des séries temporelles pour l'orchestration automatique. IRCAM - UPMC, Paris 2013 http://www.atiam.ircam.fr/Archives/Stages1213/MANNONE_Maria.pdf

647. Mannone M: Dalla Musica all'Immagine, dall'Immagine alla Musica - Relazioni matematiche fra composizione musicale e arte figurativa. Compostampa 2011

648. Mannone M and G Mazzola: Hypergestures in Complex Time: Creative Performance Between Symbolic and Physical Reality. Proceedings of the MCM 2015 conference, Springer, Heidelberg 2015

649. Mannone M and G Mazzola: Hypergestures in Complex Time: Creative Performance Between Symbolic and Physical Reality. Talk at MCM conference 2015, Queen Mary U, London 2015

650. Mansouri B: Normes internationales élargies et dépassement du calcul classique de la valeur. Thèse, Doctorat ès Sciences de gestion, Université de Bordeaux, 321 p., 3 juillet 2015

651. Marek C: Lehre des Klavierspiels. Atlantis 1977

652. Marino G, M-H Serra, J-M Raczinski: The UPIC system: Origins and innovations. Perspectives of New Music, 31(1):258–69, 1993

653. Marquis J-P: Category theory and the foundations of mathematics: philosophical excavations, *Synthese*, 103, pp. 229-254, 1995

654. Marrou H-I: De la connaissance historique. 1954, nouv. éd. coll. Points/Histoire, H21, Paris, 1975

655. Martinet A: Eléments de Linguistique Générale. Colin, Paris 1960

656. Martin B: CeDell Davis: The Last Man Standing. Huffington Post, 18 February 2015 http://www.huffingtonpost.com/barrett-martin/cedell-davis-the-last-man_b_6696710.html

657. Martino D: The Source Set and Its Aggregate Formations. JMT, 5(2), 224-273, 1961

658. De Martino E: La terra del rimorso. Bellaterra edizioni, 1999

659. Marx A B: Lehre von der musikalischen Komposition. Breitkopf & Härtel, Leipzig 1837-1847

660. Mason R M: Enumeration of Synthetic Musical Scales (...). J. of Music Theory 14, 1970

661. Massey W S: A Basic Course in Algebraic Topology. Springer, New York et al. 1991

662. Mathiesen Th J: Transmitting Text and Graphics in Online Database: The Thesaurus Musicarum Latinarum Model. In: Hewlett W B and Selfridge-Field E (Eds.): Computing in Musicology Vol.9, CCARH, Menlo Park 1993-94

663. Matisoff S: The legend of Semimaru, blind musician of Japan. Columbia University Press, New York 1978

664. Mattheson J: Der vollkommene Kapellmeister. Hamburg 1739

665. Maxwell H J: An Expert System for Harmonizing Analysis of Tonal Music. In: Balaban M et al. (Eds.): Understanding Music with AI: Perspectives on Music Cognition, MIT Press. Cambridge, MA 1992

666. May J P: A Concise Course in Algebraic Topology. The University of Chicago Press, Chicago 2007

667. Mazzola G: *Akroasis—Beethoven's Hammerklavier-Sonate in Drehung (für Cecil Taylor)*. LP, recorded August 30, 1979, WERGO/Schott, Mainz 1980

668. Mazzola G: Musique et théorie des groupes. Conference at Institut de Mathématique, U Genève, January 31, 1980

669. Mazzola G: Die gruppentheoretische Methode in der Musik. Lecture Notes, Notices by H. Gross, SS 1981, Mathematisches Institut der Universität, Zürich 1981

670. Mazzola G: Gruppen und Kategorien in der Musik. Heldermann, Berlin 1985

671. Mazzola G et al.: Rasterbild-Bildraster. Springer, Berlin et al. 1986

672. Mazzola G: Die Rolle des Symmetriedenkens für die Entwicklungsgeschichte der europäischen Musik. In: Symmetrie, Katalogband Vol.1 zur Symmetrieausstellung, Mathildenhöhe, Darmstadt, 1986

673. Mazzola G: Obertöne oder Symmetrie: Was ist Harmonie?. In: Herf F R (Ed.): Mikrotöne, Helbling, Innsbruck 1986

674. Mazzola G: Mathematische Betrachtungen in der Musik I,II. Lecture Notes, Univ. Zürich 1986/87

675. Mazzola G: Der Kontrapunkt und die K/D-Dichotomie. Manuscript, University of Zürich 1987

676. Mazzola G: Die Wahl der Zahl - eine systematische Betrachtung zum Streichquartett. In: Dissonanz 17, 1988

677. Mazzola G and G R Hofmann: Der Music Designer MD-Z71 - Hardware und Software für die Mathematische Musiktheorie. In: Petsche H (Ed.): Musik - Gehirn - Spiel, Beiträge zum 4. Herbert-von-Karajan-Symposion. Birkhäuser, Basel 1989

678. Mazzola G et al.: A Symmetry-Oriented Mathematical Model of Classical Counterpoint and Related Neurophysiological Investigations by Depth-EEG. In: Hargittai I (Ed.): Symmetry II, CAMWA, Pergamon, New York 1989

679. Mazzola G et al.: Hirnelektrische Vorgänge im limbischen System bei konsonanten und dissonanten Klängen. In: Petsche H (Ed.): Musik - Gehirn - Spiel, Beiträge zum 4. Herbert-von-Karajan-Symposion. Birkhäuser, Basel 1989

680. Mazzola G: *presto* Software Manual. SToA music, Zürich 1989-1994

681. Mazzola G: *Synthesis*. SToA 1001.90, Zürich 1990

682. Mazzola G: Geometrie der Töne. Birkhäuser, Basel 1990

683. Mazzola G and Muzzulini D: Tempo- und Stimmungsfelder: Perspektiven künftiger Musikcomputer. In: Hesse H P (Ed.): Mikrotöne III. Edition Helbling, Innsbruck 1990

684. Mazzola G and Muzzulini D: Deduktion des Quintparallelenverbots aus der Konsonanz-Dissonanz-Dichotomie. Accepted for publication in: Musiktheorie, Laaber 1990, but unpublished (!)

685. Mazzola G: Mathematische Musiktheorie: Status quo 1990. Jber. d.Dt. Math.-Verein. 93, 6-29, 1991

686. Mazzola G: Mathematical Music Theory—An Informal Survey. Edizioni Cerfim, Locarno 1993

687. Mazzola G: RUBATO at SMAC KTH, Stockholm 1993

688. Mazzola G and O Zahorka: Tempo Curves Revisited: Hierarchies of Performance Fields. Computer Music Journal **18**, No. 1, 1994

689. Mazzola G and O Zahorka: The RUBATO Performance Workstation on NeXTSTEP. In: ICMA (Ed.): Proceedings of the ICMC 94, San Francisco 1994

690. Mazzola G and O Zahorka: Geometry and Logic of Musical Performance I, II, III. SNSF Research Reports (469pp.), Universität Zürich, Zürich 1993-1995

691. Mazzola G et al.: Analysis and Performance of a Dream. In: Sundberg J (Ed.): Proceedings of the 1995 Symposium on Musical Performance. KTH, Stockholm 1995

692. Mazzola G et al.: The RUBATO Platform. In: Hewlett W B and E Selfridge-Field (Eds.): Computing in Musicology 10. CCARH, Menlo Park 1995

693. Mazzola G and O Zahorka: The PrediBase Data Base System of RUBATO on NEXTSTEP. In: Selfridge-Field E (Ed.): Handbook of Musical Codes. CCARH, Menlo Park 1995

694. Mazzola G: Inverse Performance Theory. In: ICMA (Ed.): Proceedings of the ICMC 95, San Francisco 1995

695. Mazzola G et al.: Neuronal Response in Limbic and Neocortical Structures During Perception of Consonances and Dissonances. In: Steinberg R (Ed.): Music and the Mind Machine. Springer, Berlin et al. 1995

696. Mazzola G and O Zahorka: RUBATO und der Einsatz von Hypermedien in der Musikforschung. Zeitschrift des Deutsches Bibliotheksinstituts, Berlin Jan. 1996

697. Mazzola G and O Zahorka: Topologien gestalteter Motive in Kompositionen. In: Auhagen W et al. (Eds.): Festschrift zum 65. Geburtstag J.P. Fricke. Köln 1995
http://www.uni-koeln.de/phil-fak/muwi/publ/fs_fricke

698. Mazzola G: Towards Big Science. Geometry and Logic of Music and its Technology. In: Enders B and Knolle N (Eds.): Symposionsband Klangart '95, Rasch, Osnabrück 1998

699. Mazzola G et al.: The RUBATO Homepage.
http://www.rubato.org, Univ. Zürich, since 1996

700. Mazzola G: Objective C and Category Theory. Seminar Notes, Multimedia Lab, CS Department, U Zurich, Zurich 1996

701. Mazzola G et al.: *Orbit*. Music & Arts CD-1015, Berkeley 1997

702. Mazzola G and J Beran: Rational Composition of Performance. In: Kopiez R and W Auhagen (Eds.): Proceedings of the Conference "Controlling Creative Processes in Music". Lang, Frankfurt and New York 1998

703. Mazzola G: Semiotic Aspects of Musicology: Semiotics of Music. In: Posner R et al. (Eds.): A Handbook on the Sign-Theoretic Foundations of Nature and Culture. Walter de Gruyter, Berlin and New York 1998

704. Mazzola G: Semiotic Aspects of Musicology: Semiotics of Music. - Art. 152, In: Posner et al. (eds): Semiotics (Handbook, Vol.III). W. de Gruyter, Berlin 1999

705. Mazzola G: Music@EncycloSpace. In: Enders B and J Stange-Elbe (Eds.): Musik im virtuellen Raum (Proceedings of the Klangart congress'98). Rasch, Osnabrück 2000

706. Mazzola G: Humanities@EncycloSpace. FER-Studie Nr.XX. Schweizerischer Wissenschaftrat, Bern 1998
http://www.swr.ch

707. Mazzola G: Die schöne Gefangene – Metaphorik und Komplexität in der Musikwissenschaft. In: Kopiez R et al. (Eds.): Musikwissenschaft zwischen Kunst, Aesthetik und Experiment. Königshausen & Neumann, Würzburg 1998

708. Mazzola G: L'Essence du Bleu (sonate pour piano). Acanthus, Rüttenen 2002

709. Mazzola G: Les mathématiques de l'interprétation musicale. Séminaire MaMuX, IRCAM, Paris, May 18, 2002

710. Mazzola G, G Milmeister, J Weissmann: Comprehensive Mathematics for Computer Scientists, Vols. I, II; 700 pp. Springer, Heidelberg et al. 2004

711. Mazzola G: The Topos Geometry of Musical Logic. Proceedings of the Fourth Diderot Symposium of the European Math. Soc., Springer, Heidelberg et al. 2002

712. Mazzola G: Degenerative Theory of Tonal Music. In: Proceedings of the Klangart congress'99. Universität Osnabrück 2002

713. Mazzola G: Classifying Algebraic Schemes for Musical Manifolds. Tatra Mt. Math. Publ. 23, 71-90, 2001

714. Mazzola G et al.: The Topos of Music. Birkhäuser, Basel 2002

715. Mazzola G, G Milmeister, K Morsy, F Thalmann: Functors for music: The Rubato Composer System. In: Digital Art Weeks Proceedings, ETH Zürich, 2006

716. Mazzola G: Global Networks in Computer Science? Invited Talk at the Computer Science Department Colloquium, ETH Zurich, January 2006

717. Mazzola G: Elemente der Musikinformatik. Birkhäuser, Basel et al. 2006

718. Mazzola G: La vérité du beau dans la musique. Delatour/IRCAM, Paris 2007

719. Mazzola G and M Andreatta: From a Categorical Point of View: K-nets as Limit Denotators. Perspectives of New Music, Vol. 44, No. 2, 2006

720. Mazzola G and M Andreatta: Diagrams, Gestures, and Formulas in Music. Journal of Mathematics and Music 2007, Vol. 1, no. 1, 2007

721. Mazzola G and P B Cherlin: Flow, Gesture, and Spaces in Free Jazz—Towards a Theory of Collaboration. Springer Series Computational Music Science, Heidelberg et al. 2009

722. Mazzola G et al.: Topos Theory for a Creative Analysis of Boulez's Structures. In: Naimpally S, G Di Maio (Eds.): Quaderni di Matematica, Theory And Applications of Proximity, Nearness and Uniformity, Vol. 23, 2009

723. Mazzola G: Categorical Gestures, the Diamond Conjecture, Lewin's Question, and the Hammerklavier Sonata. Journal of Mathematics and Music Vol. 3, no. 1, 2009

724. Mazzola G, C Losada, F Thalmann, S Tsuda: Topos Theory for a Creative Analysis of Boulez's Structures. In: Somashekhar Naimpally, Giuseppe Di Maio (Eds.): Quaderni di Matematica, Theory And Applications of Proximity, Nearness and Uniformity, Vol 23, 2009

725. Mazzola G: Musical Performance. Springer, Heidelberg et al. 2011

726. Mazzola G, J Park, F Thalmann: Musical Creativity. Springer, Heidelberg 2011

727. Mazzola G: Singular Homology on Hypergestures. Journal of Mathematics and Music Vol. 6, Nr. 1, 2012

728. Mazzola G: Manifolds and Singular Homology of Compositions, Networks, and Gestures: Classification, Connecting Functors, Examples. Unpublished

729. Mazzola G and F Thalmann: The BigBang Rubette: Gestural Music Comoposition With Rubato Composer. Proceedings of the ICMC 2008, ICMA, Ann Arbor 2008

730. Mazzola G and F Thalmann: Musical Composition and Gestural Diagrams. In Agon C et al. (Eds.): Proceedings of the Third Conference on Mathematics and Computation in Music (MCM), Springer, Heidelberg, 2011

731. Mazzola G, M Mannone, Y Pang: Cool Math for Hot Music. Springer, Heidelberg 2016
732. Mesnage M: La Set-Complex Theory: de quels enjeux s'agit-il? Analyse musicale, 4e trimestre, 87-90, 1989
733. Messiaen O: Technique de mon langage musical. Leduc, Paris 1944
734. Meyer J: Akustik und musikalische Aufführungspraxis. Verlag Das Musikinstrument, Frankfurt/Main 1980
735. Meyer-Eppler W: Grundlagen und Anwendungen der Informationstheorie. Springer, Berlin 1959
736. Meyer-Eppler W: Zur Systematik der elektrischen Klangtransformationen. In: Darmstädter Beiträge III, Schott, Mainz 1960
737. Michels U: dtv-Atlas zur Musik I,II. dtv/Bärenreiter, München/Kassel 1977, 1985
738. Miklaszewski K: A case Study of a Pianist Preparing a Musical Performance. Psychology of Music, 17, 95-109, 1989
739. Milmeister G: The Rubato Composer Music Software: Component-Based Implementation of a Functorial Concept Architecture. Springer Series Computational Music Science, Heidelberg 2009
740. McKay B D and I M Wanless: On the number of Latin squares. arXiv:0909.2101v1 [math.CO] 11 Sep 2009
741. McNeill D: Gesture and Thought. University of Chicago Press, Chicago 2005
742. McPartland M: Marian McPartland's Piano Jazz. National Public Radio, 9 November 1997
     http://www.npr.org/player/v2/mediaPlayer.html?action=1&t=1&islist=false&id=17622315&m=90469170
743. Mead V H: More than Mere Movement: Dalcroze Eurhythmics. Music Educators Journal, Feb. 1986
744. Mendes R S et al.: Universal patterns in sound amplitudes of songs and musical genres. Phys. Rev. E, Vol. 83, pp. 017101 ff., 2011
745. Menez M: L'expérience mathématique. In: Dorier J-L, S Coutat (Eds.): Enseignement des mathématiques et contrat social : enjeux et défis pour le 21ème siècle — Actes du Colloque EMF20123 (GT4, pp. 632-640)
746. Meredith D, G Wiggins, K Lemstrom: Pattern induction and matching in polyphonic music and other multidimensional datasets. In: Callaos N et al. (Eds.): Proceedings of the 5th World Multiconference on Systemics, Cybernetics and Informatics. Orlando 2001
747. Meredith D, K Lemstrom, G Wiggins: Algorithms for discovering repeated patterns in multidimensional representation of polyphonic music. Journal of New Music Research, Vol 4, 2002
748. Merker B H: Layered Constraints on the Multiple Creativities of Music. In: I Deliège and G A Wiggins (Eds.): Musical Creativity: Multidisciplinary Research in Theory and Practice. Psychology Press, Hove/New York, pp. 25-41, 2006
749. Merleau-Ponty M: The Phenomenology of Perception in The Body: Classic and Contemporary Readings, (ed. Donn Welton), Blackwell, Oxford 1999
750. Merleau-Ponty M: Phenomenology of Perception. Trans. Colin Smith, Routledge & Kegan Paul, London 1978 [1962]
751. Merleau-Ponty M: Phénoménologie de la perception. Gallimard 1945
752. Merleau-Ponty M: The prose of the world, translation by J O'Neill. Northwestern University Press, Evanston 1973 (La prose du monde, Gallimard, 1969)
753. Meusburger P, J Funke, E Wunder (Eds.): Milieus of Creativity: An Interdisciplinary Approach to Spatiality of Creativity. Springer, 2009
754. Miller M: A Jazz Legend Handles his Own Legacy with Care. Globe and Mail (Canada), 23 February: C3. 1999
755. Misch C and Wille R: Stimmungslogiken auf MUTABOR: Eine Programmiersprache. In: Herf F R (Ed.): Mikrotöne II. Edition Helbling, Innsbruck 1988
756. http://forumnet.ircam.fr/701.html
757. Mokkapati V: Modal Action: An Indo-European Perpective on Gregorian Chant. Phd Thesis, Wesleyan University 1996
758. Molino J: Fait Musical et Sémiologie de la Musique. Musique en Jeu 17 1975
759. Monti G: Théorie des catégories et méthode conditionnelle appliquées aux comptes en normes internationales élargies. Thèse, Doctorat ès Sciences de gestion, Université de Bordeaux 2017
760. Montiel-Hernandez M: El denotador: Su estructura, construcción y papel en la teoría matemática de la musica. Master's thesis, UNAM, Mexico City, 1999
761. Moog R: MIDI, Musical Instrument Digital Interface. Audio Eng. Soc. 34, Nr.5, 1986
762. Moon T: A Master of Improv is Renewed by Melody. Philadelphia Inquirer, 29 September: H02, 2002
763. Moore C-L: The Harmonic Structure of Movement, Music and Dance According to Rudolf Laban. Edwin Meller Press, Lewiston 2009
764. Moritz R E: On Mathematics. A collection of witty, profound, amusing passages about mathematics and mathematicians. Copyright 1914 Robert Edouard Moritz, 410 p., Dover 1958
765. Morris R D: Composition with Pitch-Classes. Yale University Press, New Haven et al. 1987
766. Morris R D: Compositional Spaces and Other Territories. PNM 33, 328-358, 1995

767. Morris R D: K, Kh, and Beyond. In: Baker J et al. [69], 275-306

768. Morse Ph M and H Feshbach: Methods of Theoretical Physics, Part I. McGraw-Hill, New York et al. 1953

769. Mozart W A: *Die Zauberflöte*. (Klavierauszug) Ed. Peters 1932

770. Müller W: Darstellungstheorie von endlichen Gruppen. Teubner, Stuttgart 1980

771. Müller S: Pianist's Hands: Synthesis of Musical Gestures. PhD Thesis, University of Zurich, Zurich 2004

772. Müller S and G Mazzola: Constraint-based Shaping of Gestural Performance. Proceedings of the ICMC 03, ICMA, Ann Arbor 2003

773. Müller S and G Mazzola: The Extraction of Expressive Shaping in Performance. Computer Music Journal, Vol. 27(1), MIT Press, 2003

774. Mumford D: Lectures on Curves on an Algebraic Surface. Princeton University Press, Princeton 1966

775. Mumford D and K Suominen: Introduction to the theory of moduli. In: Oort F (Ed.): Algebraic Geometry Oslo 1970. Wolters-Noordhoff, Groningen 1972

776. Murenzi R: Wavelets. Combes J M, Grossman A, Tchmitchian P (Eds.), Springer Berlin et al., 1988

777. MusicKit Online-Documentation. Version 4.0 1994

778. Muzzulini D: Konsonanz und Dissonanz in Musiktheorie und Psychoakustik. Lizenziatsarbeit, MWS Univ. Zürich 1990

779. Muzzulini D: Musical Modulation by Symmetries. J. for Music Theory, Vol. 39, No. 2, pp. 311-328, 1995

780. Muzzulini D: Tempo Modifications and Spline Functions. NF-Report 1993, Univ. Zürich 1993

781. Narayan S S, Temchin A N, Recio A, Ruggero M A: Frequency tuning of basilar membrane and auditory nerve fibers in the same cochleae. Science. 282(5395):1882-4, Dec 4 1998

782. Nattiez J-J: Fondements d'une Sémiologie de la Musique. Edition 10/18 Paris 1975

783. Negro G Del: The Passeggiata and Popular Culture in an Italian Town: Folklore and the Performance of Modernity. Montréal; McGill-Queen's University Press 2004

784. Neumaier W: Was ist ein Tonsystem? Lang, Frankfurt/Main et al. 1986

785. Neuman D A: A House of Music: The Hindustani Musician and the Crafting of Traditions. PhD Thesis, Columbia University 2004

786. Neuwirth G: Josquin Desprez, "Erzählen von Zahlen". In: Musik-Konzepte 26/27, edition text+kritik, München 1982

787. Newlove J and J Dalby: Laban for All. Routledge, New York 2004

788. NeXTSTEP Online-Documentation; Version 3.3. NeXt Inc., Redwood City 1995

789. Nicolas F: René Guitart : la pulsation d'une intellectualité mathématique étrangement évidente. Conférence à la *Journée en l'honneur de René Guitart pour ses 65 ans*, 9 novembre 2012, Université Paris 7 Denis Diderot, 8 p, inédit. See a summary in: `http://www-lmpa.univ-littoral.fr/ sic`

790. Nicolas F: Le Monde-Musique III. Le musicien et son intellectualité musicale. Aedam Musicae, 384 p., juin 2015

791. Nicotra E: Introduzione alla tecnica della direzione d'orchestra. Edizioni Curci 2007

792. Nieberle R: IRCAM Signal Processing Workstation. Keyboards, Dec. 1992

793. `http://www.nime.org`

794. Noether E: Hyperkomplexe Grössen und Darstellungstheorie. Math. Zeitschr., Vol. XXX 1929

795. Noland C: Agency and Embodiment: Performing Gestures/Producing Culture. Harvard University Press, Cambridge 2009

796. Noll J: Musik-Programmierung. Addison-Wesley, Bonn 1994

797. Noll Th: Morphologische Grundlagen der abendländischen Harmonik. Doctoral Thesis, TU Berlin 1995

798. Noll Th: Fractal Depth Structure of Tonal Harmony. In: ICMA (Ed.): Proceedings of the ICMC 95, San Francisco 1995

799. Noll Th: `http://www.cs.tu-berlin.de/ noll/ChordDictionary.sea.hqx`, TU Berlin 1996

800. Noll Th: Harmonische Morpheme. Musikometrika 8, 7-32, 1997

801. Noll Th: The Consonance/Dissonance-Dichotomy Considered from a Morphological Point of View. In: Zannos I (Ed.): Music and Signs. ASCO Publ., Bratislava 1999

802. Noll Th: Morphologische Grundlagen der abendländischen Harmonik. Doctoral Thesis, TU Berlin 1995

803. Noll Th: Wie blättert man in einem hypermedialen Gebärdenlexikon? In: Schmauser C, Th Noll (Eds.): Körperbewegungen und ihre Bedeutungen. Berlin Verlag Arno Spitz, Berlin 1998

804. Noll Th et al.: Computer-aided transformational analysis with tone sieves. Contribution au congrés: SMC 2006

805. Nunn D et al.: Acoustic Quanta. In: ICMA (Ed.): Proceedings of the ICMC 96, San Francisco 1996

806. Oettingen A von: Das duale Harmoniesystem. Leipzig 1913

807. Okopenko A: Lexikon-Roman. Residenz Verlag, Salzburg 1970

808. Opcode: MAX. http://www.opcode.com/products/max/, Opcode Systems Inc. 1997

809. Marco Orsini conducting Le Nozze di Figaro 2014 https://www.youtube.com/watch?v=M7QuLXy_61U

810. Osgood C E et al.: The Nature and Measurement of Meaning. Psychological Bulletin 49, 1952

811. Osher S and R Fedkiw: Level Set Methods and Dynamic Implicit Surfaces. Springer, 2003

812. Owens Th: Bebop: The Music and the Players. Oxford University Press, New York 1995

813. Pakkan M and V Akman: Hypersets. Information Sciences: An International Journal, 85(1):4361, 1996

814. Pallasmaa J: The Thinking Hand. Wiley 2009

815. Palmer R: Liner notes, Oscar Peterson: A Night in Vienna. DVD. Verve 02498 6253-4, 2004

816. Pappas T: The Music of Reason: Experience the Beauty of Mathematics Through Quotations. Wide World Publishing/ Tetra, 1995

817. Parker W: Conversations. Rogueart, 2011

818. Parlan H: The Maestro. CD. SteepleChase SCCD 31167, 1995

819. Parlan H: Horace Parlan by Horace Parlan. DVD. Image ID1505R5DVD, 2000

820. Parlan H: In Copenhagen. CD. Storyville 101 8521, 2008

821. Parlan H: Musically Yours. CD SteepleChase SCS 31141, 2010

822. Parmentier M: Connaissances, vérités et idées mathématiques dans l'*Essai sur l'entendement humain* de Locke. In: Barbin E et J-P Cléro (éds): Les mathématiques et l'expérience : ce qu'en ont dit les philosophes et les mathématiciens, p. 95-114, Hermann, Paris 2015

823. Parncutt R: The Perception of Pulse in Musical Rhythm. In: Gabrielsson A (Ed.): Action and Perception in Rhythm and Music. Royal Swedish Adademy of Music, No.55

824. Parncutt R: Recording Piano Fingering in Live Performance. In: Enders B, Knolle N (Eds.): KlangArt-Kongress 1995, Rasch, Osnabrück 1998

825. Parncutt R et al.: Interdependence of Right and Left Hands in Sight-read, Written, and Rehearsed Fingerings. Proc. Euro. Sco. Cog. Sci. Music, Uppsala 1997

826. Parncutt R: Modeling Piano Performance: Physics and Cognition of a Virtual Pianist. In: ICMA (Ed.): Proceedings of the ICMC 97, San Francisco 1997

827. Patras F: La pensée mathématique contemporaine. PUF, Paris, 2001

828. Patras F: La possibilité des nombres. PUF, Paris, octobre 2014

829. Peck R: Klein-Bottle Tonnetze. MTO, Vol. 9, no. 3, August 2003

830. Penrose R: The Road to Reality. Vintage, London 2002

831. Perle G: Serial Composition and Atonality: An Introduction to the Music of Schoenberg, Berg and Webern. 5th ed., revised, University of California Press, Berkeley 1981

832. Peterson O: An Oscar Peterson Christmas. CD, Telarc 515956K, 1995

833. Peterson O: Oscar in Paris. Telarc 2CD-83414, 1997

834. Peterson O: A Tribute to Oscar Peterson: Live at the Town Hall. Telarc 2CD-83401

835. Peterson O: A Summer Night in Munich. CD. Telarc CD-83450, 1999

836. Peterson O: A Night in Vienna. DVD, Verve 02498 6253-4, 2004

837. Peterson O, R Brown, M Jackson: The Very Tall Band: Live at the Blue Note. CD. Telarc CD-83443, 1999

838. Peterson O, R Brown, M Jackson: What's Up? The Very Tall Band. CD. Telarc CD-83663, 2007

839. Peterson O and B Green: Oscar and Benny. CD. Telarc CD-83406, 1998

840. Peterson O, R Hargrove, R Moore: Oscar Peterson Meets Roy Hargrove and Ralph Moore. CD. Telarc, CD-83399, 1996

841. Peterson O, and M LeGrand: Trail of Dreams: A Canadian Suite. CD. Telarc CD-83500, 2000

842. Peterson O, and I Perlman: Side by Side. CD. Telarc, CD-83341, 1994

843. Petsche H et al.: EEG in Music Psychological Studies. In: Steinberg R (Ed.): Music and the Mind Machine. Springer, Berlin et al. 1995

844. Petsche H: Private correspondence. Vienna, March 2001

845. Piaget J: Morphismes et categories. Comparer et transformer, Delachaux et Niestlé, Neuchâtel, 1990

846. Pfeifer R and J Bongard: How the Body Shapes the Way We Think: A New View of Intelligence. MIT Press 2006

847. Peirce C S: Peirce, *Collected Papers*, vol. III, vol. IV The simplest mathematics, 1960

848. Pierce A: Deepening Musical Performance through Movement: The Theory and Practice of Embodied Interpretation (Musical Meaning and Interpretation). Indiana University Press, Bloomington 2010

849. Plato: Phaidron.

850. Plomp R and W Levelt: Tonal Consonance and Critical Bandwidth. J. Acoust. Soc. Am. 38, 548, 1965

851. Plotinus: The Six Enneads, in *Great Books of the Western World*, Vol. 17. Encyclopædia Britannica, Inc., Chicago, 1952

852. Poincaré H: La valeur de la science. Flammarion, Paris 1905

853. Pólya G: Mathematical Discovery. On Understanding, Learning and Teaching Problem Solving (Combined Edition). Wiley, New York, 1962, 1965, 1981
854. Pólya G: Comment poser et résoudre un problème. Dunod, 1965
855. Polyanin A D: Handbook of Linear Partial Differential Equations for Engineers and Scientists. Chapman & Hall/CRC, 2002; see also: http://eqworld.ipmnet.ru, section 3.2.
856. Pope S T: Music Notations and the Representation of Musical Structure and Knowledge. Perspectives of New Music, Spring-Summer 1986
857. Popelard M-D and A Wall: Des Faits et gestes. éditions Bréal, Rosny-sous-Bois 2003
858. Popper K R: Conjectures and Refutations. Routledge & Kegan Paul, London 1963
859. Popper K: The open society and its enemies. Vol. 2, Routledge & Kegan, London, 1945
860. Posner R: Strukturalismus in der Gedichtinterpretation. In: Blumensadth H (Ed.): Strukturalismus in der Literaturwissenschaft. Kiepenheuer & Witsch, Köln 1972
861. Prame E: Measurements of the Vibrato Rate of 10 Singers. In: Friberg A et al. (Eds.): Proceedings of the 1993 Stockholm Music Acoustic Conference. KTH, Stockholm 1994
862. Pribram K H: Brain and Perception. L Erlbaum, Hillsdale, NJ 1991
863. Priestley M B: Spectral Analysis of Time Series. Academic Press, London 1981
864. Prinz W: Modes of linkage between perception and action. In W Prinz & A-F Sanders (Eds.): cognition and motor processes (pp. 185–193), Springer, Berlin 1984
865. Promies W: Symmetrie in der Literatur. In: Symmetrie, Katalogband Vol.1 zur Symmetrieausstellung. Mathildenhöhe, Darmstadt, 1986
866. Promies W: Stolbergs Gedicht im Poesiegarten In: Symmetrie, Katalogband Vol.3 zur Symmetrieausstellung. Mathildenhöhe, Darmstadt, 1986
867. Puckette M and Lippe C: Score Following in Practice. In: ICMA (Ed.): Proceedings of the ICMC 92, San Francisco 1992
868. Pulkki V et al.: DSP Approach to Multichannel Audio Mixing. In: ICMA (Ed.): Proceedings of the ICMC 96, San Francisco 1996
869. Puttke M: Learning to Dance Means Learning to Think! In: Bläsing B, Puttke M, Schack Th. (Eds.): The Neurocognition of Dance. Psychology Press, Hove and New York 2010
870. Quinn I: Fuzzy Extensions to the Theory of Contour. Music Theory Spectrum. Vol. 19/2, 1997
871. Radl H: Versuch über die Modulationstheorie Mazzolas in reiner Stimmung. Diploma Thesis, U Augsburg, Augsburg 1998
872. Rahaim M: Musicking Bodies. Wesleyan U P, Middletown, CT 2012
873. Rahn J: Basic Atonal Theory. Longman, New York 1980
874. Rahn J: Review of D. Lewin's "Generalized Musical Intervals and Transformations". JMT, 31, 305-318, 1987
875. Raffman D: Language, Music, and Mind. MIT Press, Cambridge et al. 1993
876. Rameau J-Ph: Traité de l'Harmonie, Réduite à ses Principes Naturels. Paris 1722
877. Ramachandran V S: Mirror Neurons and imitation learning as the driving force behind the "great leap forward" in human evolution, 2015. See: http://www.edge.org/3rd.culture/ramachandran/ramachandran p6.html
878. Ramstein C: Analyse, Représentation et Traitement du Geste Instrumental. PhD thesis, Institut National Polytechnique de Grenoble, December 1991
879. Ratz E: Einführung in die musikalische Formenlehre. Universal Edition, Wien 1973
880. Rauchfleisch U: Musik schöpfen, Musik hören. Ein psychologischer Zugang. Vandenhoeck und Ruprecht, Göttingen and Zürich 1996
881. Rav Y: Why do we prove theorems?. Philosophia mathematica, 7 (1), p. 5-41, 1999
882. Read G: Music Notation. Crescendo Publ., Boston 1969
883. Reeves H: Patience dans l'azur. L'évolution cosmique. Seuil, Paris 1981
884. Reichardt J F: Vermischte Musikalien. Riga 1777
885. Reidmeister K: Elementare Begründung der Knotentheorie. Abh. Math. Sem. Univ. Hamburg, 5, 24-32, 1926
886. Repp B: Diversity and Commonality in Music Performance: An Analysis of Timing Microstructure in Schumann's "Träumerei". J. Acoustic Soc. Am. 92, 1992
887. Repp B: e-mail communication of tempo data. Haskins Laboratories, New Haven, June 2, 1995
888. Repp B: Patterns of note onset asynchronies in expressive piano performance. J. Acoustic Soc. Am. 100, 1996
889. Repp B: Pedal Timing and Tempo in Expressive Piano Performance: A Preliminary Investigation. Psychology of Music 24, 1996
890. Repp B: The Art of Inaccuracy: Why Pianists' Errors Are Difficult to Hear. Music Perception, 14, 2 1997
891. Repp B: Expressive Timing in a Debussy Prélude: A Comparison of Student and Expert Pianists. Musicae Scientiae 1, 1997

892. Reti R: The Thematic Process in Music (1951). Greenwood Press, Westport, 2nd ed. 1978

893. Reti R, commented by Kopfermann M: Schumanns Kinderszenen: quasi Thema mit Variationen. In: Musik-Konzepte Sonderband Robert Schumann II, edition text + kritik, München 1982

894. Riguet J: Systèmes de coordonnées relationnels. CRAS, t. 236, p. 2369-2371, séance du 22 juin 1953

895. Riguet J: Systèmes de coordonnées relationnels II. Applications à la théorie des groupes de Kaloujnine. CRAS t. 238, p. 435-437, séance du 25 janvier 1954

896. Riguet J: Systèmes de coordonnées relationnels III. Fermetures et systèmes symétriques. CRAS, t. 238, p. 1763-1765, séance du 3 mai 1954

897. Rhode W S and L Robles: Evidence from Mössbauer Experiments for Nonlinear Vibration in the Cochlea. J. Acoust. Soc. Am. 48, 988, 1970

898. Richter P et al.: How Consistent are Changes in EEG Coherence Patterns Elicited by Music Perception? In: Steinberg R (Ed.): Music and the Mind Machine. Springer, Berlin et al. 1995

899. Riemann B: Über die Hypothesen, welche der Geometrie zugrunde liegen (1854). Gött. Abh. No.13 (published in 1867)

900. Riemann H: Musikalische Logik. Leipzig 1873

901. Riemann H: Musikalische Syntaxis. Leipzig 1877

902. Riemann H: Vereinfachte Harmonielehre oder die Lehre von den tonalen Funktionen der Akkorde. London 1893

903. Riemann H: Über Agogik. In: Präludien und Studien II. Leipzig 1900

904. Riemann H: System der musikalischen Rhythmik und Metrik. Breitkopf und Härtel, Leipzig 1903

905. Riemann H: Handbuch der Harmonielehre. Leipzig 6/1912

906. Riemann H: Grundriss der Musikwissenschaft. Leipzig 1928

907. Riemann H and H H Eggebrecht: Riemann Musiklexikon, Sachteil, 12. Auflage. Schott, Mainz 1967

908. Ries F: Biographische Notizen über L. van Beethoven (1838). New edition by Kalischer A Ch, 1906

909. Risset J-C: Real-time: Composition or performance? Reservations about real-time control in computer music and demonstration of a virtual piano partner. In: Kopiez R and W Auhagen (Eds.): Controlling creative processes in music. Peter Lang, Frankfurt/Main 1998

910. Rizzolati G and M Fabbri-Destro: Mirror neurons: from discovery to autism. Exp. Brain Res. 200:223-237, 2010

911. Roads C: The Computer Music Tutorial. MIT Press, Cambridge, Mass. and London 1998

912. Rodet X et al.: Xspect: a New X/Motif Signal Visualization, Analysis and Editing Program. In: ICMA (Ed.): Proceedings of the ICMC 96, San Francisco 1996

913. Rodet X: Synthesis and Processing of the Singing Voice. Proc.1st IEEE Benelux Workshop on Model based Processing and Coding of Audio (MPCA-2002), Leuven, Belgium, November 15, 2002

914. Roederer J G: Physikalische und psychoakustische Grundlagen der Musik. Springer, Berlin 1977

915. Rogers C R: Toward a theory of creativity, ETC: A review of General Semantics, Vol 11, 1954, 249-260

916. Rossi A: Die Geburt der modernen Wissenschaft in Europa. Beck, München 1997

917. Rostand F: Souci d'exactitude et scrupules des mathématiciens. Vrin, Paris 1960

918. Rüetschi U-J: Denotative Geographical Modelling—an attempt at modelling geographical information with the Denotator system. Diploma thesis, University of Zürich, 2001

919. Rufer J: Die Komposition mit zwölf Tönen. Bärenreiter, Kassel 1966

920. Runco M A and I Chand: Cognition and Creativity. Educational Psychology Review, Vol. 7:3, pp. 243-267, 1995

921. Ryan R A: Introduction to Tensor Products of Banach Spaces. Springer, Heidelberg 2002

922. Ruwet N: Langage, Musique, Poésie. Seuil, Paris 1972

923. S-Plus. MathSoft Inc., Seattle, Washington 1995

924. Sachs K-J: Der Contrapunctus im 14. und 15. Jahrhundert. AMW, Franz Steiner, Wiesbaden 1974

925. Sacks O W: Musicophilia: tales of music and the brain. (Rev. and expanded 1st ed.) Vintage Books, New York 2008

926. Saiber A and H S Turner (Eds.): Mathematics and the Imagination. Configurations, 17, 195 pp. 2009

927. Sandall R: Miraculously, Magically, the Years Fell Away. Daily Telegraph (London), 1 July: 21 2006

928. Sander E: Notre pensée progresse par analogies. La conscience. La Recherche Hors-série, no 15, pp. 54-57, octobre-novembre 2015

929. Sattler D E (Ed.): Friedrich Hölderlin – Einhundert Gedichte. Luchterhand, Frankfurt/Main 1989

930. Salzer F: Structural Hearing: Tonal Coherence in Music. (German: Strukturelles Hören. Noetzel, Wilhelmshaven 1960) Dover, New York 1962

931. Sarasúa A and E Guaus: Dynamics in Music Conducting: A computational Comparative Study Among Subjects. Proceedings of the International Conference on New Interfaces for Musical Expression, NIME'14, Goldsmiths London, UK, 2014

932. Sauvageot F: Pierre Deligne. *Images des mathématiques*, CNRS, 12 mai 2013
http://images.math.cnrs.fr/Pierre-Deligne.html

933. Saussure F de: Cours de Linguistique Générale. Payot, Paris 1922

934. Saussure F de: Linguistik und Semiotik - Notizen aus dem Nachlass. Fehr J (Ed.), Suhrkamp, Frankfurt/Main 1997

935. Savan D: An introduction to C.S. Peirce's Full System of Semeiotic. Toronto Semiotic Circle, Toronto 1988

936. Sawyer R K: Improvisation and the creative process: Dewey, Collingwood, and the aesthetics of spontaneity. The Journal of Aesthetics and Art Criticism, Vp;. 58, pp. 149-61, 2000

937. Sawyer R K: Group Creativity: Music, Theater, Collaboration. Lawrence Erlbaum Associates, Mahwah 2003

938. Scarborough D L et al.: Connectionist Models for Tonal Analysis. Computer Music Journal Vol. 13, 1989

939. Schenker H: Fünf Urlinien-Tafeln. Universal Edition, Wien 1932

940. Schenker H: Theorien und Phantasien III: Der freie Satz. Universal Edition, Wien 1935

941. Scherchen H: Manuale del direttore d'orchestra. Edizioni Curci 2009

942. Schewe K-D: Type concept in OODB Modelling and its Logical Implications. In Kawaguchi E et al. (Ed.), Information Modelling and Knowledge Bases XI, pp. 256–274. IOS Press, 2000

943. Schmalz R: Out of the mouth of mathematicians. A quotation book for philomaths. Mathematical Association of America, 1993

944. Schmidt J (Ed.): Celibicache. Book and Movie PARS (ISBN 3-9803265-1-9), München 1992

945. Schmidt-Biggemann W: Topica universalis. Meiner,Hamburg 1983

946. Schmitt, J-C: La raison des gestes dans l'occident médiéval. Gallimard, Paris 1990

947. Schneps L and P Lochak (Eds.): Geometric Galois Actions. Cambridge University Press, Cambridge, 1997

948. Schönberg A: Harmonielehre (1911). Universal Edition, Wien 1966

949. Schönberg A: Die Komposition mit zwölf Tönen. In: Style and Idea, New York 1950

950. Schönberg A: Fundamentals of Musical Composition. Edited by Gerald Strang, with an introduction by Leonard Stein. St. Martin's Press, New York 1967

951. Schopenhauer A: De la quadruple racine du principe de raison suffisante. éd. complète (1813-1847), Vrin, 1991

952. Schubert F: *Auf dem Wasser zu singen*, op. 72 (1823). Bärenreiter/Henle München/Kassel 1982

953. Schubert H: Kategorien I, II. Springer, Berlin et al. 1970

954. Schumann R: Kinderszenen, op. 15 (1839). Henle, München 1977

955. Schumann R: Dritter Quartett-Morgen. NZ f. Musik 1838

956. Schweitzer A: Johann Sebastian Bach. (1908) Breitkopf & Härtel, Wiesbaden 1979

957. Scriba C J and P Schreiber: 5000 Years of Geometry: Mathematics in History and Culture. Birkhäuser, 2015

958. Seashore C E: Psychology of Music. McGraw-Hill, New York 1938

959. Shepard R N and L A Cooper: Mental Images and Their Transformations. MIT Press 1982

960. Shepp A: The Geneva Concert. DVD, inakustic LC13921, 2006

961. Shepp A and H Parlan: Goin' Home. CD. SteepleChase (D) SCCD-31079, 1985

962. Shepp A and H Parlan: Trouble in Mind. CD. SteepleChase ? SCCD 31139, 1986

963. Shepp A and H Parlan: En Concert: First Set. CD. 52e Rue Est (F) RECD 015, 1987

964. Shepp A and H Parlan: En Concert: Second Set. CD. 52e Rue Est (F) RECD 016, 1987

965. Shepp A and H Parlan: Reunion. CD. L+R CDLR 45003, 1987

966. Shepp A and H Parlan: Goin' Home. CD. SteepleChase (D) SCCD-31079, 2010

967. Shepp A and H Parlan: Trouble in Mind. CD. SteepleChase ? SCCD 31139, 2010

968. Siebers T: Disability Aesthetics (pp. 2-3). University of Michigan Press, Ann Arbor 2010

969. Serre J-P: Faisceaux Algébriques Cohérents. Annals of Mathematics, 61 (2): 197-278, 1955

970. Serre J-P: Trees. Springer, New York 1980

971. Serres M (Ed.): Eléments d'histoire des sciences. Bordas, Paris 1989

972. Sessions R: Questions About Music. W. W. Norton and Company, 1971

973. Shaffer L H: Musical Performance as Interpretation. Psychology of Music, Vol. 23, 1995

974. http://www.fau.edu/bodymindculture

975. Shusterman R: Somaesthetics: A Disiplinary Proposa. The Journal of Aesthetics and Art Criticism. Vol.57, No.3, pp. 299-313, 1999

976. Simondon G: Du mode d'existence des objets techniques. Aubier, Paris 1989

977. Simonton D K: Thematic fame, melodic originality, and musical zeitgeist: A biographical and transhistorical content analysis. Journal of Personality and Social Psychology, Vol. 38, pp. 972-83, 1980

978. Simonton D K: Thesaurus of scales and melodic patterns. Coleman-Ross Company, New York 1947

979. Siron J: La partition intérieure. Outre Mesure, Paris 1992

980. Slawson W: The Musical Control of Sound Color. Canadian Univ. Music Review, No.3, 1982

981. Slawson W: Sound Color. U. Cal. Press, Berkeley 1985

982. Sloan D: From DARMS to SMDL, and back again. In: Haus G and I Pighi (Eds.): X Colloquio di Informatica Musicale. AIMI, LIM-Dsi, Milano 1993

983. Sloboda J: The Musical Mind: An Introduction to the Cognitive Psychology of Music. Calderon Press, Oxford 1985

984. Sloboda, A S and A C Lehmann: Tracking Performance Correlates of Changes in Perceived Intensity of Emotion During Different Interpretations of a Chopin Piano Prelude. Music Perception, Summer 2001, Vol 19, No. 1, pp. 87-120, 2001

985. Sloterdijk P: Selbstexperiment. Hanser, München & Wien 1996

986. Sluming V et al.: Broca's Area Supports Enhanced Visuospatial Cognition in Orchestral Musicians. The Journal of Neuroscience, Vol. 27(14), pp. 3799-3806, April 4, 2007

987. Smart M A: Mimomania: Music and Gesture in Nineteenth-Century Opera. New Ed edition, University of California Press, Berkeley 2005

988. SMDL committee: Standard Music Description Language Draft, Pittsford, NY 1995

989. Smith III J O: Synthesis of bowed strings. In: Strawn J, Blum T, ICMA: Proceedings of the International Computer Music Conference, San Francisco 1982

990. Smith III J O: Physical Modeling of Musical Instruments, part I. Computer Music Journal, 16 (4), 1992

991. Smith III J O: Physical Modeling Synthesis Update. Computer Music Journal, 20 (2), 1996

992. Sodomka A: KUNSTRADIO at documentaX.
http://www.culture.net/orfkunstradio/BIOS/sodomkabio.html, Vienna 1997

993. Spanier E H: Algebraic Topology. McGraw-Hill, New York et al. 1966

994. Späth H: Eindimensionale Spline-Interpolations-Algorithmen. Oldenburg, München et al. 1990

995. Späth H: Zweidimensionale Spline-Interpolations-Algorithmen. Oldenburg, München et al. 1991

996. Spellman A B: Four Lives in the Bebop Business. Pantheon 1966

997. Sperry R W: Neurology and the mind-body problem. American Scientist, 40, 291-312, 1952

998. Spicker V: Das Abstraktmotiv bei Cecil Taylor. MusikTexte, 73/74 March 1998

999. Spivak M: Calculus on Manifolds. Benjamin, New York 1965

1000. Squire R L and Butters N (Eds.): Neuropsychology of Memory. D. Guildford Press, NewYork and London 1984

1001. Stahnke M: Struktur und Ästhetik bei Boulez. Wagner, Hamburg 1979

1002. Stange-Elbe J, G Mazzola: Cooking a Canon with RubatoPerformance Aspects of J.S. Bach's "Kunst der Fuge". In: ICMA (Ed.): Proceedings of the ICMC 98, 179-186, San Francisco 1998.

1003. Stange-Elbe J: Analyse- und Interpretationsperspektiven zu J.S. Bachs "Kunst der Fuge" mit Werkzeugen der objektorientierten Informationstechnologie. Habilitationsschrift, Osnabrück 2000

1004. Stange-Elbe J: Instrumentaltechnische Voraussetzungen für eine computergestützte Interpretation. In: Enders B (Ed.): Proceedings of the Klangart congress '01, Osnabrück

1005. Starr D and R Morris: A General Theory of Combinatoriality and the Aggregate. PNM, 16(1), 364-389, 16(2), 50-84, 1977-78

1006. Starr D: Sets, Invariance and Partitions. JMT, 22(1), 1-42, 1978

1007. Stein S and S Sandor: Algebra and Tiling: Homomorphisms in the Service of Geometry.: Math. Assoc. Amer., Washington, DC 1994

1008. Steinberg R (Ed.): Music and the Mind Machine. Springer, Heidelberg 1995

1009. Steinberg R and A Fairweather (Eds.): Blues-Philosophy for Everyone: Thinking Deep About Feeling Low. John Wiley 2012

1010. Sternberg R J and T I Lubart: The Concept of Creativity: Prospects and Paradigms. In: Sternberg R J (Ed.): Handbook of Creativity. Cambridge University Press, New York 1999

1011. Stinson D R: Combinatorial designs: constructions and analysis. Springer, 2004

1012. Stollberg A, J Weissenfeld, F H Besthorn (Eds.): DirigentenBilder—Musikalische Gesten — verkörperte Musik. Schwabe, Basel 2015

1013. Stravinsky I: The Poetics of Music. Harvard University Press 1947

1014. Steele G L: Common Lisp: The Language, 2nd Edition. Digital Press, 1990

1015. Stone P: Symbolic Composer. http:/www.xs4all.nl/~psto/

1016. Stopper B: Gleichstufig temperierte Skalen unter Berücksichtigung von Oktavstreckungen. In: Hosp I (Hrsg.): Bozener Treffen 1991 — Skalen und Harmonien. Südtiroler Kulturinstitut, Bozen 1992

1017. Straub H: Beiträge zur modultheoretischen Klassifikation musikalischer Motive. Diplomarbeit ETH-Zürich, Zürich 1989

1018. Straub H: Kadenzielle Mengen beliebiger Stufendefinitionen. Unpublished manuscript, Zürich 1999/2000

1019. Straus J N: Voice Leading in Atonal Music. In: Baker J et al. [69], 237-274

1020. Stumpf C: Tonpsychologie. Leipzig 1883-1890

1021. Sundberg J, A Askenfelt, L Frydén: Musical Performance: A Synthesis-by-Rule Approach, Computer Music Journal, 7, 37-43, 1983

1022. Sundberg J and V Verillo: On the Anatomy of the Retard: A Study of Timing in Music. J. Acoust. Soc. Am. 68, 772-779, 1980

1023. Sundberg J: Music Performance Research. An Overview. In: Sundberg J, L Nord, R Carlson (Eds.): Music, Language, Speech, and Brain. London 1991

1024. Sundberg J (Ed.): Generative Grammars for Music Performance. KTH, Stockholm 1994

1025. Sundin N-G: Musical Interpretation in Performance. Mirage, Stockholm 1984

1026. Suzuki T et al.: Musical Instrument Database with Multimedia. In: ICMA (Ed.): Proceedings of the ICMC 96, San Francisco 1996

1027. Suppan W: Zur Verwendung der Begriffe Gestalt, Struktur, Modell und Typus in der Musikethnologie. In: Stockmann D and Steszwski J (Eds.): Analyse und Klassifikation von Volksmelodien. PWM Edition Krakau 1973

1028. Swan, R G: Vector Bundles and Projective Modules, *Transactions of the American Mathematical Society* 105 (2): 264-277

1029. Sylvester D: Interviews with Francis Bacon. Pantheon Books, New York 1975

1030. Szczeciniarz J-J: Réflexion sur Riemann, philosophe mathématicien, in [79], p. 195-221

1031. Symmetrie. Katalogband Vol.1 zur Symmetrieausstellung, Mathildenhöhe, Darmstadt, 1986

1032. Takemitsu T: Confronting Silence. Fallen Leaf Press, Berkeley 1995

1033. Takhtajan L A: Quantum Mechanics for Mathematicians. American Mathematical Society, Providence 2008

1034. Teschl G: Mathematical Methods in Quantum Mechanics. American Mathematical Society, Providence 2009

1035. Tanaka A: Interaction, Experience, and the Future of Music. In: O'Hara K and B Brown (Eds.): Consuming Music Together. Springer, Dordrecht 2006

1036. Tarry G: Le problème des 36 officiers. AFAS, Compt 1900–1901

1037. Tatarkiewicz W: A History of Six Ideas: An Essay in Aesthetics. Polish Scientific Publishers, Warszawa 1980

1038. Taylor Y and H Barker: What's It Like to Be Blind? Faking It. `fakingit.typepad.com` 2007

1039. Taylor C: Burning Poles. VHS, Mystic Fire 1991

1040. Taylor C: Silent Tongues. Freedom 1974

1041. Thalmann F: Musical composition with Grid Diagrams of Transformations. Master's Thesis, University of Bern, Bern 2007

1042. Thalmann F: Gestural Composition with Arbitrary Musical Objects and Dynamic Transformation Networks. PhD thesis, University of Minnesota, 2014

1043. Thalmann F and G Mazzola: The Bigbang Rubette: Gestural music composition with Rubato Composer. In: Proceedings of the International Computer Music Conference, International Computer Music Association, Belfast 2008

1044. Thalmann F and G Mazzola: Affine musical transformations using multi-touch gestures. Ninad, 24:58–69, 2010

1045. Thalmann F and G Mazzola: Gestural shaping and transformation in a universal space of structure and sound. In: Proceedings of the International Computer Music Conference, International Computer Music Association, New York City 2010

1046. Thalmann F and G Mazzola: Poietical music scores: Facts, processes, and gestures. In: Proceedings of the Second International Symposium on Music and Sonic Art, MuSA, Baden-Baden 2011

1047. Thalmann F and G Mazzola: Using the creative process for sound design based on generic sound forms. In: MUME 2013 proceedings, AAAI Press, Boston 2013

1048. Thalmann F and G Mazzola: Visualization and transformation in general musical and music-theoretical spaces. In: Proceedings of the Music Encoding Conference 2013, MEI, Mainz 2013

1049. Tormoen D, F Thalmann, G Mazzola: The composing hand: Musical creation with leap motion and the Bigbang Rubette. In: Proceedings of 14th International Conference on New Interfaces for Musical Expression (NIME), London 2014

1050. Terhardt E: Zur Tonhöhenwahrnehmung von Klängen. II. Ein Funktionsschema. Acustica 26, 187-199, 1972

1051. Thomas d'Aquin: Première question disputée. La vérité (De Veritate). (teaching in the years 1256-1259), Paris, Vrin, 2002

1052. Thom R: Stabilité structurelle et morphogénèse. Benjamin, Reading, MA 1972

1053. Tinctoris J: Opera Theoretica, Vol. 2: Liber de arte contrapuncti. Seay A (Ed.): Amer. Institute of Musicology 1975. gopher://IUBVM.UCS.INDIANA.EDU/00/tml/15th/tincon2.text

1054. Tittel E: Der neue Gradus. Doblinger, Wien-München 1959

1055. Titze I R and D T Talkin: A theoretical study of the effects of various laryngeal configurations on the acoustics of phonation. Acoustical Society of America Journal, Vol. 66, No. 60, 1979

1056. Titze I R: Speed, accuracy, and stability of laryngeal movement in singing. Acoustical Society of America Journal, Vol. 115, p. 2590 ff., 2004

1057. Todd N P M: A Model of Expressive Timing in Tonal Music. Music Percep. **3**, pp. 33-58, 1985

1058. Todd N P M: Towards a Cognitive Theory of Expression: The Performance and Perception of Rubato. Contemporary Music Review **4**, 1989

1059. Todd N P M: A Computational Model of Rubato. Contemporary Music Review **3**, 1989

1060. Todd N P M: The Dynamics of Dynamics: A Model of Musical Expression. J. Acoustic Soc. Am. **91**, pp. 3540-3550, 1992

1061. Todoroff T: A Real-Time Analysis and Resynthesis Instrument for Transformation of Sounds in the Frequency Domain. In: ICMA (Ed.): Proceedings of the ICMC 96, San Francisco 1996

1062. Tomasello M: Why Don't Apes Point? In: Enfield N J & S C Levinson (Eds.): Roots of Human Sociality: Culture, Cognition and Interaction. Berg, pp. 506-524, Oxford & New York

1063. Tomasello M, M Carpenter, U Liszkowski: A New Look at Infant Pointing. Child Development, Vol. 78, No. 3. pp. 705-722, May/June 207

1064. Tymoczko D: The Geometry of Musical Chords. Science, Vol 313, 2006

1065. Tymoczko D: Mazzola's Model of Fuxian Counterpoint. In: Agon, C, G. Assayag, E. Amiot, J. Bresson, J. Mandereau (Eds.): Mathematics and Computation in Music. LNAI 6726, Springer, Heidelberg 2011

1066. Uhde J: Beethovens Klaviermusik III. Reclam, Stuttgart 1974

1067. Uhde J and R Wieland: Forschendes Üben. Bärenreiter 2002

1068. Uhde J and R Wieland: Denken und Spielen. Bärenreiter, Kassel et al. 1988

1069. Ungvary T and R Vertegaal: The Sentograph: Input Devices and the Communication of Bodily Expression. ICMA 1995

1070. Utgoff P E and P B Kirlin: Detecting Motives and Recurring Patterns in Polyphonic Music. Proceedings of the International Computer Music Conference, ICMA 2006

1071. van Dalen D: Logic and Structure. Springer, Berlin et al. 1997

1072. Valéry P: Leçon Inaugurale du Cours de Poétique au Collège de France. Gallimard, Paris 1945

1073. Valéry P: Cahiers I-IV (1894-1914). Celeyrette-Pietri N and J Robinson-Valéry (Eds.), Gallimard, Paris 1987

1074. Valéry P: Philosophie de la danse. Oeuvres, tome I, Variété, "Théorie poétique et esthétique", pp. 1390-1403, Gallimard, 1956

1075. Valéry P: Analecta (1926), in *Tel quel*, tome II, Collection Idées, no 241, Gallimard, 1971

1076. Vallès J: "Penser c'est oublier" *Images des mathématiques*. CNRS, http://images.math.cnrs.fr/Penser-c-est-oublier.html, 10 juillet 2015

1077. Valtieri S: La Scuola d'Athene. Mitteilungen des Konsthistorischen Instituts in Florenz 16, 1972

1078. van der Waerden B L: Algebra I, II. Springer, Berlin et al. 1966

1079. van der Waerden B L: Algebra I. (English tranlsation) Springer, New York 2003

1080. van der Waerden B L: Die Pythagoreer. Artemis, Zürich 1979

1081. Varèse E: Erinnerungen und Gedanken. In: Darmstädter Beiträge III. Schott, Mainz 1960

1082. Varèse E: Rückblick auf die Zukunft. edition text+kritik, München 1978

1083. Vercoe B: The Synthetic Performer in the Context of Live Performance. In: ICMA (Ed.): Proceedings of the ICMC 84, San Francisco 1984

1084. Veyne P: Comment on écrit l'histoire, 1959. nouv. ed., Points/Histoire 226, Paris 1996

1085. Vico J-B: La méthode des études de notre temps. Présentation et traduction de A. Pons, (De nostri temporis studiorum ratione, 1708), online, site du Réseau Intelligence de la Complexité, 1981

1086. Vieru A: The Book of Modes. Editura Muzicală, Bucarest 1993

1087. Vinci A C: Die Notenschrift. Bärenreiter, Kassel 1988

1088. Violi P: Beyond the body: towards a full embodied semiosis. In: Dirven R, and R Frank (Eds.): Body, Language and Mind. Mouton de Gruyter, Berlin 2006

1089. Vogel M: Die Lehre von den Tonbeziehungen. Verlag für systematische Musikwissenschaft, Bonn-Bad Godesberg 1975

1090. Vogel M: Arthur v. Oettingen und der harmonische Dualismus. In: Vogel M (Ed.): Beiträge zu Musiktheorie des 19. Jahrhunderts. Bosse, Regensburg 1966

1091. Vogel M: Berechnung emmelischer und ekmelischer Mehrklänge. In: Herf F R (Ed.): Mikrotöne II. Helbling, Innsbruck 1988

1092. Vogt H: Neue Musik seit 1945. Reclam, Stuttgart 1972

1093. Vossen G: Datenbankmodelle, Datenbanksprachen und Datenbank-Management-Systeme. Addison-Wesley, Bonn et al. 1994

1094. Vuza D T: Sur le rythme périodique. Revue Roumaine de Linguistique—Cahiers de Linguistique Théorique et Appliquée 22, no. 1 1985

1095. Vuza D T: Some Mathematical Aspects of David Lewin's Book *Generalized Musical Intervals and Transformations*. Perspectives of New Music, vol. 26, no. 1, 1988

1096. Vuza D T: Supplementary Sets and Regular Complementary Unending Canons (Part One). Perspectives of New Music, vol. 29, no. 2, 1991

1097. Vuza D T: Supplementary Sets and Regular Complementary Unending Canons (Part Two). Perspectives of New Music, vol. 30, no. 1, 1992

1098. Vuza D T: Supplementary Sets and Regular Complementary Unending Canons (Part Three). Perspectives of New Music, vol. 30, no. 2, 1992

1099. Vuza D T: Supplementary Sets and Regular Complementary Unending Canons (Part Four). Perspectives of New Music, vol. 31, no. 1, 1993

1100. Vygotsky L S: Thought and Language (1934). Edited and translated by E Hanfmann and G Vakar, revised and edited by A Kozulin. MIT Press, Cambridge 1986

1101. Wanske H: Musiknotation. Schott, Mainz 1988

1102. Waldvogel M. et al.: *presto* source code. TRIMAX, Zürich 1986-1993

1103. Wallas G: The Art of Thought. Harcourt, New York 1926

1104. Wallis J: Tractatus de Sectionibus Conicis. 1655

1105. Walker J S: Fourier Analysis and Wavelet Analysis. Notices of the AMS, vol.44, No.6, July 1997

1106. Ward J: Gregorian Chant according to the Principles of André Mocquereau of Solesmes. The Catholic Education Press 1923 (Reprint)

1107. Waugh W A O'N: Music, probability, and statistics. In: Encyclopedia of Statistical Sciences, **6**, 134-137, 1985

1108. Webern A: *Variationen op.27* (mit Interpretationsanweisungen). UE 16845, Wien 1980

1109. Wegner P: Interactive Foundations of Object-Oriented Programming. Computer, IEEE Computer Soc., October 1995

1110. Weidman A: Singing the Classical, Voicing the Modern: The Postcolonial Politics of Music in South India. Duke University Press 2006

1111. Weil A: Œuvres scientifiques. Collected papers, Vol. I, Springer, New York, 1979

1112. Weisberg R W: Creativity and Knowledge: A Challenge to Theories. In: R J Sternberg (Ed.): Handbook of Creativity. Cambridge University Press, New York 1999

1113. Weyl H: Symmetrie. Birkhäuser, Basel 1955

1114. Weinberger N M: Musica Research Notes. Music and Science Information Computer Archive, Center for Neurobiology of Learning and Memory, U California Irvine, Vol. III, Issue II, Fall 1996

1115. Whiteside A: Indispensables of Piano Playing. Coleman-Ross, New York 1955

1116. Wicinski A A: Psichologyiceskii analiz processa raboty pianista-ispolnitiela nad muzykalnym proizviedieniem. Izviestia Akademii Piedagogiceskich Nauc Vyp., 25 [Moscow], 171-215, 1950

1117. Widmer G: Learning Expression at Multiple Structural Levels. In: ICMA (Ed.): Proceedings of the ICMC 94, San Francisco 1994

1118. Widmer G: Modeling the Rational Basis of Musical Expression. Computer Music J. 18, 1994

1119. Widmer G and W Goebl: Computational Models of Expressive Music Performance: The State of the Art. Journal of New Music Research 33(3), 203-216, 2004

1120. Wiedemann M and E Zimmermann: Kreativität und Form. Springer Vieweg, Berlin Heidelberg 2012

1121. Wieser H-G: Electroclinical features of the psychomotor seizure. Fischer, Stuttgart and Butterworth, London 1983

1122. Wieser H-G and G Mazzola: Musical consonances and dissonances: are they distinguished independently by the right and left hippocampi? Neuropsychologia 24 (6):805-812, 1986

1123. Wieser H-G and G Mazzola: EEG responses to music in limbic and auditory cortices. In: Engel J Jr, Ojemann G A, Lüders H O, Williamson P D (Eds.): Fundamental mechanisms of human function. Raven, New York 1987

1124. Wieser H-G: Musik und Gehirn. Revue Suisse Méd. 7, 153-162, 1987

1125. Wieser H-G and S Moser: Improved multipolar foramen ovale electrode monitoring. J Epilepsy 1: 13-22, 1986

1126. Wiggins G et al.: A Framework for Evaluation of Music Representation Systems. Computer Music Journal 17:3, 1993

1127. Wiil U K (Ed.): Computer Music Modeling and Retrieval (CMMR 2003). Springer, Heidelberg 2004

1128. Wiil U K (Ed.): Computer Music Modeling and Retrieval (CMMR 2004). Springer, Heidelberg 2005

1129. Wille R: Symmetrie — Versuch einer Begriffsbestimmung. In: Symmetrie, Katalogband Vol.1 zur Symmetrieausstellung, Mathildenhöhe, Darmstadt, 1986

1130. Wille R: Personal Communication. Darmstadt 1982

1131. Wille R: Musiktheorie und Mathematik. In: Götze H and R Wille (Eds.): Musik und Mathematik. Springer, Berlin et al. 1985

1132. Wille R: Bedeutungen von Begriffsverbänden. Preprint Nr. 1058, TH Darmstadt 1987

1133. Wille R: Personal Communication. Darmstadt 1985

1134. Winson I: Brain and Psyche. The Biology of the Unconscious. Anchor Press/Doubleday, Garden City, NY 1985

1135. Wilson F R: The Hand: How Its Use Shapes the Brain, Language, and Human culture. Pantheon Books, New York 1998

1136. Witten E: Magic, Mystery, and Matrix. Notices of the AMS, Vol. 45, Nr. 9, 1998

1137. Wittgenstein L: Tractatus Logico Philosophicus – Logisch-Philosophische Abhandlung, Wilhelm Ostwald (Ed.), Annalen der Naturphilosophie, 14, 1921

1138. Wittgenstein L: Tractatus Logico-Philosophicus (1918). Suhrkamp, Frankfurt/Main 1969

1139. Wolfram S: Software für Mathematik und Naturwissenschaften. In: Chaos und Fraktale. Spektrum der Wissenschaft Verlagsgesellschaft, Heidelberg 1989

1140. Xenakis I: Formalized Music. Indiana Univ. Press, Bloomington 1972

1141. Yaglom A M and I M Yaglom: Wahrscheinlichkeit und Information. Deutscher Verlag der Wissenschaften, Berlin, 1967

1142. Yoneda N: On the homology theory of modules. J. Fac. Sci. Univ. Tokyo, Sct. I,7, 1954

1143. Ystad S et al.: Of Parameters Corresponding to a Propagative Synthesis Model Through the Analysis of Real Sounds. In: ICMA (Ed.): Proceedings of the ICMC 96, San Francisco 1996

1144. Yusuke A et al.: Hyperscore: A Design of a Hypertext Model for Musical Expression and Structure. J. of New Music Research, Vol. 24, 1995

1145. Zahorka O: Versuch einer Charakterisierung des altrömischen Melodiestils. Seminar Lütolf, MWS Univ. Zürich 1993

1146. Zahorka O: From Sign to Sound—Analysis and Performance of Musical Scores on RUBATO. In: Enders B (Ed.): Symposionsband Klangart '95, Schott, Mainz 1997

1147. Zahorka O: PrediBase—Controlling Semantics of Symbolic Structures in Music. In: ICMA (Ed.): Proceedings of the ICMC 95, San Francisco 1995

1148. Zahorka O: RUBATO – Deep Blue in der Musik? Animato 97/3, 9-10, Zürich 1997

1149. Zalamea F: Synthetic Philosophy of Contemporary Mathematics. Urbanomic, Falmouth, UK, & Sequence Press, New York, USA 2011

1150. Zarca B: L'univers des mathématiciens. L'ethos professionnel des plus rigoureux des scientistes. Presses Univ. Rennes, 2012

1151. Zarlino G: Istitutioni harmoniche. Venezia 1558

1152. Zbikowski Z: Conceptualizing Music: Cognitive Structure, Theory, and Analysis. Oxford University Press 2005

1153. Zbikowski Z: Music, Emotion, Analysis. Music Analysis 29/1-3(2010): 37-60

1154. Zekl G: Topos. Meiner, Hamburg 1990

1155. Zlotin B and A Susman: Creative Pedagogy. Journal of TRIZ. no.4, pp. 9-17 (in Russian) 1991

1156. Zweifel P F: Generalized Diatonic and Pentatonic Scales: A Group-theoretic Approach. PNM, 34(1), 140-161, 1996

1157. Zwicker E and H Fastl: Psychoacoustics. Facts and Models. Springer, Berlin, et al. 1999

1158. Zwicker E and M Zollner: Elektroakustik (2nd edition). Springer, New York et al. 1987

1159. Zwiebach B: A First Course in String Theory. Cambridge University Press, Cambridge et al. 2004

Index

2^{Fu}, 56
$B_\epsilon x$, 229
$Dia(Names/\mathcal{E})$, 1437
$H^i(\Delta/G^I)$, 355
$Mono(\mathcal{E})$, 1437
$R^n \dot{@} B \tilde{@} M$, 169
$T_2(R)$, 186
$\coprod$, 44
$\mathfrak{C}^{(3)}$, 266
$\mathcal{A}_2(U, V; W)$, 1399
$\prod$, 44
$\{\}$, 44
!, 1424, 1432
$!/\alpha$, 331
(G), 1386
$(G : H)$, 1379
$(G^I)^\vee$, 275
(I_1/J_1), 519
(I_7/J_7), 519
(K_1/D_1), 519
(K_7/D_7), 519
$(X/C(X))$, 518
$(X[\varepsilon]/Y[\varepsilon])$, 521
$(X|C(X))$, 518
(f/α), 334
$(r_{i,j})$, 1392
(x, y), 1372
$/N$, 296
0_R, 1385
12-$Temp$, 422
1_R, 1385
2, 1375
$3CH$, 448
$3Chains$, 260
$< N$, 340
$<_{\mathbf{Mod}}$, 76, 78
$?^s$, 1414
$?_e$, 429
$?_{\beta\sigma\mu}$, 122
$A(f, \zeta, \eta, J(\zeta), U(\eta))$, 1451
AK_ξ-topology, 1237

$ALLSER_n$, 195
$ASCII$, 335
AX, 1434
$A@F$, 1423
$A@MOT$, 384
$A@MOT_F$, 384
$A@MOT_{F,n}$, 384
$A@ <_M$, 79
$A \times B$, 1372
A^B, 1375
$A^\star$, 1398
A_q, 1406
A_s, 1406
A_Γ, 1457
$A_{t,y,G}$, 788
$A_{threshold}$, 1348
$AbsDyn$, 68, 641
$AbsDynamicEvents$, 641
$AbsTpo$, 341
$AbsTpoEvt$, 341
$AbsTpoInComp$, 341
$Accent_{U,S}$, 632
Ad, 141
$AddTo(c1, c2, at)$, 807
$Alg_{\mathbb{R}}$, 925
$AllExt_B(m)$, 443
$AllExt_B(n, m)$, 443
$AllSet$, 1372
$Alphabet$, 1376
$Alt(s)$, 1468
$AltMaj(i)$, 475
$AltMaj(i)^{(3)}$, 475
$An(B, M)$, 1400
$ArchaicForm$, 84, 85
Arg, 616
$ArtiSlur_{S,U}$, 632
Aut, 1378
$Aut(K)$, 144
$Aut(c)$, 1420
$Aut_R(M)$, 1391
$Aut_\Phi(S)$, 1093

© Springer International Publishing AG, part of Springer Nature 2017
G. Mazzola, *The Topos of Music IV: Roots*, Computational Music Science,
https://doi.org/10.1007/978-3-319-64495-0

$B(\Delta)$, 1402
$B(f, \zeta, \eta, J(\zeta), U(\eta))$, 1451
$BA \models \alpha$, 1434
$BD_V^{mean}(w)$, 648
$BD_V(w)$, 648
BIT, 44, 60
$\tilde{B}@F$, 168
B_*, 946
B_N, 290
B_n, 966
$B_r(x)$, 1449
$B_{/M}$, 1400
$BarLine$, 68
$Basic$, 343
$BeetMotChordFiber$, 347
$BeetMotChordFiberObject$, 348
$BetterForm$, 84
C, 67, 439, 1349
$C(F)$, 55, 1087
$C(G, F, P, w)$, 191
$C(X)$, 518
$C(i)$, 475
$C+$, 439
$C/OnBeetSon$, 347
$C0$, 439
CD, 342, 676
$CINT(S)$, 208
$CINT(Ser)$, 207
$CINT_m(S)$, 208
$CN(G^I)$, 283, 988
$COM(Cont)$, 206
$CONT_{n,k}(X)$, 206
CP, 391
$CT(D)$, 57
C^*-algebra, 1492
$C^*(K; M)$, 1447
$C^0[a, b]$, 1450
C^5, 420
$C^i(X)$, 518
$C^n(K; M)$, 1447
$C^\star(G^I)$, 307
$C_n(R, X)$, 966
C_{G^I}, 288
Cad, 454
Cc, 518
$CcFourier$, 72
CcM, 518
$CellHierarchy_{BP}$, 596
$Cell_U$, 596
$Cell_{BP}$, 596
$Ch/PianoChord$, 340
$Chain_R$, 1476
$Chrono - Fourier$, 235
$ChronoFourierSound$, 235
Ci, 518
CiM, 518

$Cl_H(U)$, 588
Cl_n, 177
$ClassZChord$, 420
$Closed_X$, 1443
Cm, 439
$Cochain_R$, 1477
$ColimCirc$, 65
$CommaZModule$, 422
$Consonant$, 234
$Conv(D)$, 1462
$Count$, 343
$Covens$, 1445
$Crescendo$, 67
$Ct(GestSem)$, 1183
D, 67, 263, 592, 1347
DEG, 457
DF_0, 1451
DNR, 210
$D^r f$, 1451
$D_\epsilon^\mu(M)$, 400
D_K, 1415
D_f, 1412
$D_x(T_0)$, 660
$D_\epsilon(M)$, 397
$D_{k,n}$, 125
$D_{n,m}$, 1213
$Daughters$, 596
Dc, 518
DcM, 518
$DenOrb(R, n, B, M)$, 170
$Der(L)$, 1408
$Der(X)$, 1469
Df, 1451
Df_x, 1451
Dg, 457
Dg^*, 457
Di, 518
DiM, 518
Dia, 387
$Dia(Names)$, 1437
$Dia\pi$, 392
$Dia^*(Y/\mathcal{E})$, 1089
Dia_k, 387
Dil, 1392
Dir, 656
$Dir(Q)$, 1376
$Dom(X)$, 380
Dp, 457
$Dup(T)$, 335
$Duration$, 67
$DurationValue$, 809
$DynSymb$, 68
E, 67, 1347
$E2M$, 62
$EHLD$, 631, 787
$EHLDGC$, 658

$EMB(A, B)$, 208
ENH, 422
EX, 1432
$E \rightsquigarrow F$, 109
E_N, 1209
$E_{n,m}$, 1213
Ed_{Rg}, 389
$Ed_{\Delta Rg}$, 389
$El(G)$, 403
$El(M)$, 403
$Elast$, 388
$End(A)$, 141
$End(c)$, 1420
$End_R(M)$, 1391
$Envelope$, 71
$EulerChord$, 92
$EulerChordEvent$, 93
$EulerModule$, 62, 90, 92
$EulerModule_q[\varepsilon]$, 508
$EulerPlane$, 93, 96
$EulerZChord$, 420
$Euler\mathbb{R}Module$, 563
$Evt_S(U)$, 630
Ex, 422
$Ex(D, L_{\alpha,\beta})$, 306
$Ex(GestSem)$, 1183
$Ex(L/K)$, 1461
$ExTop_A(F)$, 430
$Ext_R^n(Z, W)$, 945
$Ext_A(M)$, 427
$F \dashv G$, 1423
$F(D)$, 57
$FCM(Alphabet)$, 1376
FD, 676
$FG(Alphabet)$, 1379
$FM(Alphabet)$, 1376
$FM\text{-}Object$, 73
$FO(PEX)$, 1435
$FORMS$, 1438
FSH, 343
F_0, 1451
F_x, 1411
F_π, 554
F_{sym}, 554
$Fermata$, 643
$Fermata_{E,S}$, 632
$Fib(e_.)$, 758
$Field_U$, 597
$Field_{U,R}$, 597
Fin, 1426
$Fin(Onset)$, 81
$Fin(Pitch)$, 81
$Fin(S)$, 1374
$Flatten$, 75
$Flatten^n$, 75
$Fold$, 1103

$ForSem$, 1092
$FormList(S)$, 60
$Fourier$, 70, 72
$FourierSound$, 71
$Frame_U$, 596
$Func(\mathbf{C}, \mathbf{D})$, 1422
$Fund(H)$, 588
$Fushi$, 343
$FushiOrnament$, 343
$FushiPic$, 342
$FushiSTRG$, 343
G, 67, 913, 1350
$G(K)$, 941
$G.x$, 257
G/H, 1379
$G/\Omega \pitchfork \Omega^*$, 633
$GL(n, p)$, 1382
$GL_n(R)$, 1382
$G[f]$, 945
$G \wr H$, 1380
G^I, 252, 987
G^{I*}, 289
G^{opp}, 1378
G_m, 1378
$G_{v,k,l}$, 185
$Gal(L/K)$, 1460
$Ges_t^P(M)$, 391
$Ges_t^\mu(M)$, 400
$Ges_t(M)$, 391
$GestSem$, 1183
GlC^t, 557
$Glissando$, 67
$GlobPerfScore_{BP.}$, 599
$GlobPerfScore_{BP}$, 598
$Grass_{r,n}$, 1416
H, 67, 588, 1339, 1350
$HA \models \alpha$, 1434
$H \backslash G$, 1379
$H \lhd G$, 1379
H^*-emergent property, 1062
$H_\Delta^i(G^I)$, 355
$H^n(K; M)$, 1447
$H^q(C)$, 1477
$H^\star(G^I)$, 307
H_X, 922
$H_X(x, y)$, 922
$H_l^\star(X^{MaxMet})$, 379
$H_n(K.c.)$, 970
$H_n(R, X)$, 966
$H_q(C)$, 1475
$H_k(f)$, 986
$H_{n-1}^{net}(R, \mathcal{C}, \Gamma_*)$, 986
H_{prime}, 1351
$HarMin$, 474
$HarMin^{(3)}$, 474
Hom, 1378, 1385, 1419

Hom^{loc}, 281
I, 573, 592
$I(F)$, 55, 1087
$I(H)$, 577
$ICV(A, B)$, 207
ID, 593
III_X, 262
II_X, 262
$INT(Ser)$, 207
$INT_m(Ser)$, 207
$IV(A, B)$, 208
IV_X, 262
$I|i$, 269
I^n, 966
I^q, 1476
I_U, 596
I_X, 262
I_x, 511
Id_A, 141, 1372
Id_e, 1420
$Idempot(M)$, 1376
$Im(f)$, 1375
$Importance$, 342
$InTop(F)$, 429
$InitPerf_U$, 597
$InitSet_U$, 596
$InstruName$, 69
$Int(E)$, 427
$Int(M)$, 427
$IntMod_{12,q}[\varepsilon]$, 511
$IntMod_{12}$, 510
$IntThirds_{3,4,q}[\varepsilon]$, 511
$IntThirds_{3,4}$, 510
$Int^e(M)$, 427
Int_A, 430
Int_g, 1379
$Intonation$, 563
$Item(F)$, 809
J-manifold, 1062
J-shape, 1062
$J(X)$, 1247
$J(\mathbf{\Sigma}_E, w)$, 654
JCK, 342
$JCKFU$, 344
$J_K(X)$, 1431
J_{Kt}, 266
$K©L$, 939
$K(X)$, 1430
$K.c.$, 967
$KU_{s,t}$, 120
$K_1 \wedge K_2$, 178
K_4, 206
K_E, 575
K_b, 708
K_{Top_S}, 631
$Ker(f)$, 1379

$Kernel_U$, 597
$Knot$, 73
$Knot_{Basic}$, 74, 270
$Knot_{PerOns}$, 503
Kq, 62
Kt, 62
$K|k$, 1445
L, 67, 1348
$LARM$, 974
LB, 804
LCP, 391
$LPS^{BP,k}_{Instrument}$, 599
$LPS^{BP.}_{Instrument}$, 599
$LPS^{BP}_{Instrument}$, 599
$L_F f$, 1455
$L_\psi f(a, b)$, 236, 1344
$L_{n,1}$, 922
$Label$, 342
$LegatoSlur_U$, 632
$LegatoSlur_{U,S}$, 632
Lev, 269
$LimCirc$, 65
$Limint(F)$, 259
Lin_R, 1391
$Lin_R(G^I, M)$, 289
$List(F)$, 809
$ListEntry_F$, 343
$List_F$, 343
$LocC^t$, 552
$LocPerfScore_{BP}$, 595
Loc^F, 91
$Loudness$, 67
M, 913
M/N, 289
MOT, 384
MOT_F, 384
MOT_n, 384
$MOT_{F,n}$, 384
$M[\varepsilon]$, 107
$M@F$, 1437
$M@$, 1397, 1422
$M|f/Id_A$, 289
$M^\star$, 1376
M_n, 207
M_q, 1406
M_s, 1406
M_t, 449
$M_x(T_0)$, 660
M_{12}, 260
M_{OP}, 384
M_O, 384
$M_{[\varphi]}$, 1392
$M_\mathbb{C}$, 994
$M_{d,a}$, 444
$M_{i,j}$, 787
$Maelzel$, 341

Maj, 474
$Makro_{Basic}$, 74, 270
$Makro_{PerOns}$, 503
Man^r, 1466
$Marcato_{U,S}$, 632
$Match_P$, 1432
$MathPitch$, 61
$Max(X)$, 268
$MaxMet(X)$, 268
$Media$, 342
$MelMin$, 474
$MelMin^{(3)}$, 474
$Min(L,S)$, 581
$Min(X,S)$, 581
Mod, 1111
$ModelForm$, 84
$Modulator$, 1115
$MonEnd(F)$, 427
$Mono(\mathcal{E})$, 1089
$Mor(\mathbf{C})$, 1420
Mos_n^G, 310
$Mos_{n,\lambda}^G$, 311
$Mos_{n,k}^G$, 310
$Mother$, 596
$Mt/Motif$, 340
$N(D)$, 57
$N(F)$, 54, 1086
$N(U)$, 1446
$NComp$, 341
NF, 209
$N \rtimes_\phi H$, 1380
N_e, 1376
$N_k(U)$, 1446
N_{red}, 300
$Nat(F,G)$, 1422
$NatMin$, 474
$NatMin^{(3)}$, 474
$NewContainer$, 807
$Note$, 1111
$NoteGroup$, 340
$Number$, 342
OPD, 120
$OPLD_{\mathbb{Z}}$, 437
OS, 552
$Ob(\mathbf{C})$, 1420
$ObExTop_B(F)$, 433
$ObGl$, 987
$ObLoc$, 987
$OnMod_m$, 63
$OnPiMod_{m,n}$, 63
$Onset$, 67, 93
$Onset??_S$, 67
$OnsetProj$, 347
Op, 616
$Open(G)$, 335
$Open(\mathcal{S})$, 1443

$Open_X$, 1411, 1443
$Open_f$, 1411
$Open_{X,x}$, 1411
$Open_{f,x}$, 1411
$OrchSet$, 69
$Overtone$, 1115
$P(E)$, 525
$P(Q)$, 1388
$P(Q)/\sim$, 1420
PEX, 1435
PF, 210
PO, 502
$P \wedge Q$, 346
P^+, 1432
$P_{Träumerei}$, 720
$Para$, 99, 100
$Part(I)$, 310
$Part(a.,b.)$, 1462
$Partial$, 72, 1122
$Path(\Gamma)$, 1458
$Pause$, 69
$Per(R)$, 312
$PerCell$, 1254
$PerOns$, 502
$PerOns_{12}$, 503
$Percussion$, 502
$Periods_w$, 1340
$PhysCrescendo$, 68
$PhysDuration$, 68
$PhysGlissando$, 68
$PhysInstr_i$, 555
$PhysLoudness$, 68
$PhysOnset$, 68
$PhysOrchestra$, 555
$PhysPitch$, 68
$PhysicalBruteForceOperator$, 651
$PiMod_{12,(7)}$, 508
$PiMod_n$, 63
$PiThirds_{3,4}$, 510
$PianoSelector$, 47
$Piano\text{-}Note$, 44
$Pitch$, 67
$PitchChange$, 343
$PitchSymb$, 68
$PowerCirc$, 65
$Ptch$, 117
$PythagorasLine$, 460
$QNormalize(c)$, 807
$QReduce(c,[n])$, 807
Q_w, 653
$Q_{\beta\sigma\mu}$, 122
$Q_\theta(E_1,E_2)$, 526
$Q_\theta(K/D)$, 526
$R(Para)$, 267
$R.S$, 106
$R.f$, 143

REd, 389
RT, 809
RTC, 370
$RULES$, 1434
$R[\varepsilon]$, 1387
R-Gestoid, 923
$R\langle M\rangle$, 1386
$R\langle Q\rangle$, 1388
RC, 923
R-H_X, 923
R^*, 503
R^*_{Ons}, 503
R^2, 727
R^C, 1392
$R^\times$, 1385
R_R, 1391
R_{G^I}, 289
R_{Vuza}, 312
R_{max}, 596
R_{min}, 596
$Rad(M)$, 1395
$Rad(f)$, 926
$Rad(x)$, 925
Rat, 616
$Rate$, 341
$RelDyn$, 68, 641
$RelDynamicEvents$, 641
$Remainder_{Split} \propto \mu$, 649
$RemoveFrom(c1, c2)$, 807
$Review$, 342
Rg, 386
$Rg_o(M)$, 386
$Rg_p(M)$, 386
$Rhythm(Para)$, 267
$S!$, 344
S, 263
$S(E)$, 525
$S(EX)$, 1432
$S(p, u.)$, 1381
$S/\sim$, 1373
$SEG_k[M^{1/n}]$, 207
$SERM_n$, 125
$SERM_{k,n}$, 125
SER_n, 125
$SER_{k,n}$, 125
$SO_2(71)$, 786
SPE, 234
$S[Alphabet]$, 1386
$S\langle Alphabet\rangle$, 1386
$S\otimes_R?$, 1397
S^1, 921, 950, 971
S^3, 922
$S^{-1}A$, 1406
$S^{-1}M$, 1406
$S^\diamond$, 101
$S_n(K)$, 1447

S_q, 508
$S_q[\varepsilon]$, 508
$S_t^{M_t}$, 449
Sat, 370
Sat_B^F, 434
$Satellite$, 668
$SatelliteLevel$, 1108
$ScalarOperator_w$, 654
$ScoreForm$, 596
$ScoreInstr_i$, 555
$ScoreOrchestra$, 555
Sem, 1092
$SemInTop(F)$, 429
$Sema(\mathcal{E})$, 1439
$SemiEnd(F)$, 427
$Semitone$, 1350
Sg, 457
$Sg(GestSem)$, 1183
$Sh(\mathbf{C}, J)$, 1431
$Simplex_U$, 596
$Simplexes_U$, 596
$Sound$, 121
Sp, 457
$Sp(X)$, 378
$Sp(x)$, 269
$Sp_\mu(x)$, 385
$Spec$, 1412
$Spec(A)$, 1412
$Spec(R)$, 148
$Spec(f)$, 1412
$Special$, 343
$Split \propto \mu$, 649
$Split_{U,\nu}$, 649
$Staccatissimo_{U,S}$, 632
$Staccato_{U,S}$, 632
$StepTune$, 644
$Sub(X)$, 1375
$Sub(BP)$, 588
$SubGal(L/K)$, 1461
$Sub_{M*}(N)$, 395
$Support$, 1336
$SupportForm(E)$, 339
$Switch$, 60
$Syllabic$, 234
$Sym(K)$, 144
$Sym_i(A, B)$, 208
$SymbolicBrueForceOperator$, 650
$SynCirc$, 64
T, 263, 592, 1338
$T(E)$, 575
$T(F)$, 55, 1086
$T(G)$, 1381
$T(p., u.\dot{})$, 1381
TF, 447
$TF_{f,t}$, 447
TG^I, 558

TID, 593
TI_n, 125
TK, 551
TO, 1451
$TRUTH(F)$, 436
$TRUTH(I)$, 335
$TRUTH(h)$, 336
TTO, 207
TX, 1467
T_s^r, 1468
$T^r f$, 1451
$T^t K$, 551
T_0, 660, 1445
T_1, 1445
T_2, 1445
T_S, 927
$T_S(x,y)$, 927
T_f, 927
$T_k K$, 551
$T_{2,\mathbb{R}}$, 391
$T_{X,x}$, 927, 1415
T_Λ, 660
T_{TON}, 484
T_ε, 107
$T_{\zeta,\eta,f}$, 1452
T_{red}, 107
$Tempo$, 562
$TempoOperator_w$, 653
$Tension(\lambda^\cdot, \mu_\cdot, \omega, \phi_{min})$, 484
$Tenuto_{U,S}$, 632
$Terminal$, 809
Tf, 551, 1451
Tf_k, 552
Tg, 457
$TimeSig$, 69
$TimeSig(p/q)_S$, 632
$Title$, 342
$Top(H)$, 588
Top_S, 631
$TorSeq$, 388
$Toroid_\gamma^{m,l}$, 388
Tp, 457
$Tr(D,T)$, 522
$Trans$, 312
$Trans(D,T)$, 522
$Triv(K)$, 144
$Tune$, 644
$Types$, 1437
$UNICODE$, 335
$U \boxtimes V$, 1400
U^n, 597
U^o, 1443
U_g, 118
U_n, 207
U_s, 118
U_x, 262

U_{n+1}, 597
$U_{x/x+1}$, 262
$Utai$, 344
$V(E)$, 1412
$VF(O)$, 1455
VII_X, 262
VI_X, 262
V^*, 1212
V_X, 262
V_Γ, 1457
$V_{n,x}$, 175
$Val(I)$, 335
$ValCh(p/q = r/s)_S$, 632
$VerbAbsTpo$, 341
$Vowel$, 234
$WP_n(U, \mathbb{T}_\mathbb{R})$, 597
W^*, 498
$W_{n,i}^R$, 631
$Weight(U)$, 597
$WeightList_{BP}$, 597
$Weight_n(U)$, 597
$Weight_{BP}$, 597
$X(G, M)$, 913
$X(R, n, B, M)$, 170
$X(harmo)$, 725
$X(melod)$, 725
$X(metric)$, 725
$X@F$, 1397
X-chromatic, 420
X-harm, 420
X-major, 420
X-mel, 420
X^*, 1488
$X^{(3)}$, 262
X^{MaxMet}, 268
$X^{MetLg[L]}$, 268
$X^{MetPer[P]}$, 268
$X^\star$, 1397
$X_\cdot^{cyc(g)}$, 1382
X_6, 459
X_n^r, 175
X_{add}, 725
$Y \,☞\, Z$, 946
$Y_{\mathbf{Aff}}$, 1414
$Y_{\mathbf{ComRings}}$, 1414
$Year$, 342
$Yoneda$, 982
$Z(R)$, 1385
$Z(harmo)$, 725
$Z(metric)$, 725
ZNF, 210
Z_n, 966
$[C/C(X)]$, 518
$[C|C(X)]$, 518
$[K)$, 231
$[K]$, 193

$[L:K]$, 1460

$[M]$, 402

$[M^{1/n}]$, 206

$[R]$, 335

$[S]$, 1422

$[X, A; Y, B]_{X'}$, 1474

$[\frac{n}{i}]$, 311

$[\mathcal{X}_C, \mathcal{X}_F]$, 1246

$[a, b]$, 457

$[b, p, g]$, 267

$[f]$, 943

$[l|x|k]$, 110

$[s]$, 1373

$[x)$, 402

$[x]$, 1454

$[xy]$, 444

$\&$, 1432

$\mathbb{A}^n$, 1415

$\overrightarrow{?}$, 944

$\overrightarrow{F}$, 942

$\overrightarrow{X}$, 914, 1457

$\overrightarrow{@}$, 916, 941

$\overrightarrow{f}$, 914, 1458

Ab, 1411

Aff, 1412

$\|f\|_1$, 1450

$\|f\|_2$, 1450

$\|f\|_\infty$, 1450

$\|x\|_1$, 1450

$\|x\|_2$, 1450

$\|x\|_\infty$, 1450

$\| \|$, 1450

$@M$, 1397, 1422

$@$, 53, 1396

$@_R^{red}M$, 1401

$@_R$, 1396

$@_{loc}X$, 280

$\mathbb{C}$, 1385

$\mathbb{C}_m$, 1112

CHR, 44

CatMan, 993

Colimit, 1437

$ComGlob_A$, 285

$ComLoClass_{n+1,End(B)}^{gen,M}$, 170

$ComLoClass_{n+1,0_R}^{gen,lf,sp}$, 177

$ComLoClass_{n+1,O_R}^{gen}$, 173

$ComLoClass_{n+1,0_R}$, 177

$ComLoc$, 979

$ComLoc_A$, 142

$ComLoc_A^{emb}$, 143

$ComLoc_A^{gen}$, 143

$ComLoc_A^{in}$, 143

ComMod, 1412

ComRings, 1411

ComRings$_@$, 1415

ΔRg, 386

$\Delta n \Gamma(G^I)$, 294

Δ, 565, 628, 1432

Δ-formula, 926

$\Delta(G^I)$, 989

Δ_d, 1461

Δ_n, 173, 293

Δ_q, 1476

Δ_{G^I}, 294

$\Delta_{i,j}$, 1403

Den, 331

$Den(x, y)$, 331

$Den_{\textbf{Colimit}}$, 331

$Den_{\textbf{Limit}}$, 331

$Den_{\textbf{Power}}$, 331

$Den_{\textbf{Simple}}$, 331

$Den_{\textbf{Syn}}$, 331

Den_∞, 335

Den_∞/sig, 338

ⓓ, 975

Digraph, 914, 938, 1457

$FlatLocNet(\mathcal{C})$, 981

$\mathcal{F}_\psi$, 928

$Formula$, 926

$ForSem$, 1440

ΓB_f, 291

$\Gamma \Lambda_f^n$, 291

Γ, 290, 1364, 1412

Γ-machine, 1038

$\Gamma(G^I)$, 288

$\Gamma(U, F)$, 1412

$\Gamma(f/Id_A)$, 290

$\Gamma(myFMObject)$, 73

$\Gamma \overrightarrow{@} K$, 941

$\Gamma \overrightarrow{@} X$, 937

$\Gamma \overrightarrow{@}_A \Sigma^I$, 998

$\Gamma \circlearrowright \mathcal{C}$, 984

Γ^*, 1457

$\Gamma^G \overrightarrow{@}_A \Sigma^I$, 998

Γ^w, 239

$\Gamma_t, _k$386

Γ_t, 385

$\Gamma_t(M, j)$, 309

Γ_{Dir}, 660

$\Gamma_{RedIndia,c}$, 387

$\Gamma_{t,n}$, 385

$Gesture$, 937

$Gesture(F)$, 942

$Gesture(K)$, 941

$_\mu Glob$, 352

$GlobalDigraph$, 996

Grp, 1411, 1421

$Graph$, 1458

$\mathbb{H}$, 1385

$Heyting$, 930

LCTop, 1255

$\Lambda \downarrow$, 628

$\Lambda\uparrow$, 628
$\Lambda\updownarrow$, 628
$\Lambda^s T$, 1468
Λ^∞, 628
Lie$_R$, 1408
Limit, 1437
LinMod, 1421
LinMod$_R$, 1421
Loc, 136
$Loc_{End(A)}$, 141
$Loc_{@A}$, 137
$_\mu Loc$, 352
$LocNet$, 981
$LocNet(\mathcal{C})$, 981
LocRgSpaces, 1411
$\mathbb{M}_R$, 1420
$\mathbb{M}_{m,n}(R)$, 1392
Mod$^@$, 961, 979, 1397
Mod, 1397, 1411, 1421
Mod$_R$, 1397, 1421
Mon, 1411, 1421
$\mathbb{N}$, 1371, 1420
$_\mu ObGlob$, 352
$ObLoc$, 132, 979
$_\mu ObLoc$, 352
$ObLoc_{End(A)}$, 141
$ObLoc_{@A}$, 137
$\Omega\propto\mu$, 633
Ω, 598, 628, 720, 1428, 1459
$\Omega(p)$, 483
Ω_{Sh}, 1431
Ω_ω^X, 723
Φ, 1432
Π, 610, 1435
$\Pi_1(X)$, 920
Π_w, 191
Power, 1437
Ψ, 610
$\mathbb{Q}$, 61, 62, 1371, 1385
$\mathbb{Q}_m$, 1112
$\mathbb{R}$, 44, 61, 1371, 1385
$\mathbb{R}\Delta$, 926
$\mathbb{R}_m$, 1112
$\mathbb{R}_{[\mathbb{Q}]}$, 61
$RadicalDigraph$, 926
$\Rightarrow$, 1342, 1433
$\circledcirc\mathbb{R}$, 83
Rings, 1411, 1421
$BC(A,W)$, 1453
BP, 588
$B(A,W)$, 1452
$C_1\times C_2|K$, 587
$GlDiff^t$, 557
$Glob$, 275
$LocDiff^t$, 552
$ObGlob$, 275

Tan_R^t, 552
Schémas, 1414
Ens, 1054
Ens$^\sigma$, 1028
Ens$_U$, 1420
$\Sigma_I M_i$, 1391
Simpl, 1445
$SinLoc$, 141
$SinLoc_{End(A)}$, 141
$SinLoc_{@A}$, 141
Sob, 1414
$SpaceDigraph$, 914
Simple, 1437
    denotator, 1112
Syn, 1437
TopCat, 939, 961, 1085
$\overrightarrow{\textbf{TopCat}}^{©}$, 962, 1086
TON, 447, 448, 483, 485
$\mathbb{T}(h)$, 337
$\mathbb{T}_I$, 335
$\mathbb{T}_I^A$, 335
Tex, 334
$Texig(Den)(h)$, 337
$Texig(Den)_I$, 334
Top2, 1474
Top, 993, 1085, 1421, 1444
$©$, 939
$UNICODE$, 1090
Υ, 1435
VAL, 447, 448, 483, 485
VAL$_{mode}$, 483
VAL$_{type}$, 483
Ξ, 1432, 1435
Ξ_D, 419
Ξ_K, 419
Ƨ, 568, 1254
$\mathbb{Z}$, 44, 1371, 1385
$\mathbb{Z}_m$, 1112
$\mathbb{Z}_n$, 1381, 1386
$\mathbb{Z}_2$, 60
$\mathbb{Z}_{71}^4$, 784
$\text{Ƨ}_{(x,y)}^t$, 1191
$\text{Ƨ}_{\Lambda,Dir}$, 656
$\text{Ƨ}_{\Lambda,U}$, 659
Ƨ_Λ, 656
Ƨ_{ex}, 591
α, 524
α_+, 508, 567
α_-, 508
α_+, 108
α_-, 108
$\bar{X}$, 402
$\bar{z}$, 1385
β, 524
$\beta(x)$, 291
$\bigcap V$, 1372

$\bigcup V$, 1372

$\bigoplus_I M_i$, 1392

$\bigwedge^{\oplus k} \mathcal{L}$, 975

$\perp$, 336, 1433

$\boxtimes$, 1400

$\bullet$, 804

$\cap$, 283

$\overset{?}{@A}$, 137

$\check{S}$, 101

χ, 418, 419

$\chi(Y)$, 1375

$\chi^\star$, 418

χ_σ, 1428, 1459

$\coprod_I M_i$, 1375

δY, 264

δx, 1481

δ, 524, 610

$\delta(X/Y)$, 520

$\delta(Y|X)$, 520

$\delta[X|Y]$, 520

$\delta_{@A}$, 294

δ_{ij}, 1392

$\dot{f}/Id_A$, 295

$\dot{s}$, 295

$\emptyset$, 146, 1371

$\emptyset_R$, 1397

$\exists x$, 1436

$\exists_x P(x,y)$, 346

$\forall x$, 1436

$\forall_x P(x,y)$, 346

$\mathfrak{D}$, 1412

$\mathfrak{H}$, 593

$\mathfrak{MusGen}$, 133

$\mathfrak{M}$, 1436

$\mathfrak{M} \models \alpha[x]$, 1436

$\mathfrak{M}_0$, 459, 952

$\mathfrak{M}_1$, 459

$\mathfrak{M}_2$, 459

$\mathfrak{S}_M$, 1378

$\mathfrak{T}$, 399

$\mathfrak{T}/Ges$, 399

$\mathfrak{T}^{sp}_{Toroid}$, 403

$\mathfrak{T}^{sp}_{Toroid}/Ges$, 403

$\mathfrak{T}_{\mu/Ges}$, 399

$\mathfrak{T}_\mu$, 399, 400

$\mathfrak{T}_\mu/Ges$, 400

$\mathfrak{T}_{t,P,d}$, 397

$\mathfrak{T}_{t,P,d}/Ges_t$, 398

$\mathfrak{W}$, 598

$\mathfrak{gl}(L)$, 1408

$\mathfrak{gl}(n,R)$, 1408

$\mathfrak{h}$, 590, 628

$\mathfrak{h}|U$, 590

$\mathfrak{sl}(L)$, 1408

$\mathfrak{sl}(n,R)$, 1408

γ, 610

$\gamma_i(t)$, 911, 915

$\overline{GL}(n,p)$, 1382

$\hat{?}$, 56, 131

$\hat{b}$, 721

$\hat{f}$, 131, 1344

$->$, 1432

$\in$, 1371

$\int_C F$, 1061

$\int_c \omega$, 975

$\int_{\mathbf{C}} F$, 1424

$\int_x f$, 1454

ι, 628

ι_j, 1375, 1392

κ, 454

$\kappa(x)$, 1412

κ_J, 455

κ_{RelDyn}, 642

κ_{orb}, 460

$\langle S \rangle_e$, 1376

$\langle S \rangle$, 423, 1376

$\langle\!\langle S \rangle\!\rangle$, 1422

$\vdash_{CL} \alpha$, 1434

$\mathbb{D}$, 1021

$\mathbb{L}_4$, 1026

$\mathbb{T}^2$, 966

globfact, 992

Frame$_n$, 983

Frame$_{X.}$, 1254

Sel, 989

$\mathcal{B}$, 1067, 1070, 1443

$\mathcal{B}^n$, 1488

$\mathcal{C}^r$, 1451

$\mathcal{C}^{\text{☞}}$, 946

$\mathcal{C}^\tau$, 982

$\mathcal{C}_{Im}$, 1289

$\mathcal{C}_{Ph}$, 1290

$\mathcal{D}(\mathcal{C})$, 1031

$\mathcal{D}_\Gamma \mathcal{X}$, 1037

$\mathcal{E}$, 1437

$\mathcal{E}(x,y)$, 1186

$\mathcal{F}(O)$, 1455

$\mathcal{F}X$, 1469

$\mathcal{G}$, 1067, 1071

$\mathcal{L}$, 1482

$\mathcal{L}(X,Y)$, 1487

$\mathcal{L}(x,y)$, 1185

$\mathcal{L}^p(X,d\mu)$, 1489

$\mathcal{M}$, 172

$\mathcal{M}_r$, 175

$\mathcal{P}$, 1290

$\mathcal{R}$, 1437

$\mathcal{T}^r_s(X)$, 1468

$\mathcal{V}(X)$, 1468

μ, 628

∇, 939

$\neg$, 1433

ω, 349
$\overline{Y}$, 1443
$\overline{f}$, 1447
$\partial Dynamics$, 567
$\partial Intonation$, 567
$\partial Tempo$, 567
∂U, 590, 1443
∂Z, 568
$\partial \mathbf{\Sigma}_{\lambda,ex}$, 591
$\partial f_i/\partial x_j$, 1451
$\partial^Y R$, 580
$\partial_Y R$, 580
∂_q, 1475
∂_n, 966
$\phi(n)$, 1381
πY, 264
$\pi_1(X)$, 920
$\pi_1(X,x)$, 921
π_j, 1392
$\pi_{m,l}$, 393
$\prod_I M_i$, 1374
$\psi_{a,b}$, 236
ρ_D, 1337
$\searrow(g)$, 944
σ-algebra, 1488
$\sigma(X/Y)$, 520
$\sigma(X|Y)$, 520
$\sigma[X|Y]$, 520
$\sim$, 1373
$\sim_P s$, 1362
$\sqrt{x}$, 1376
$\sqsubset$, 398, 1388
$\sqsubset_i^{\lambda^i}$, 482
$\sqsubset_i^{\lambda^i}$, 1388
$\square$, 1400
$\subset$, 1371
$\subseteq$, 1371
$\tau(X,D,S)$, 581
$|Ser|_1$, 229
$|Ser|_2$, 229
θ, 524
$\tilde{A}$, 1412
$\tilde{X}_n^r$, 175
$\times$, 177
$\top$, 336, 1433
$\vec{T}_2(R)$, 185
$\underline{U}$, 591
$\underline{\Lambda}$, 628
$\underline{\mathbf{\Sigma}}$, 591
$\uparrow$, 915, 1457
$\uparrow^n$, 1457
$\vDash \alpha$, 1433
φ_x, 661
$\vee$, 1433
$\wedge$, 1433
$\wedge^r$, 1407

$\widehat{M}$, 434
$\wp_{score}$, 912
$]x[$, 265
$^+\tau_x$, 263
$^0\tau_x$, 263
$^G X$, 965
$^R Glob$, 354
$^R Glob_A$, 354
$^{loc} Loc^@$, 281
$_0\mathbf{C}$, 1420
$_1\mathbf{C}$, 1420
$_A\Delta_n$, 293
$_R R$, 1391
$_R Sat_A^F$, 434
$_\mu^R Glob$, 354
$_\mu^R Glob_A$, 354
$|m|_\alpha$433
$a \sqcup_c b$, 1424
$a \times_c b$, 1424
$abcard_t(M)$, 386
$abcard_t(m)$, 386
$ad(x)$, 1408
ad, 1375
add, 335
at, 448
$b \diamond f$, 721
b_*, 945, 946
$book$, 65, 72
$bottom(x)$, 1390
c, 1348
$c\text{-}space_n(X)$, 205
c^5, 420
c^m, 766
$c^{(3)}$, 461
$c_h^{(3)}$, 461
$c_m^{(3)}$, 461
c_{har}, 474
c_{mel}, 474
c_{nat}, 474
$card(A)$, 1373
$causalEnd$, 767
$causalStart$, 767
$char(F)$, 1386
$codom(f)$, 1419
$colim(\Delta)$, 1424
$coord(F)$, 1438
$ct_{\mu,\epsilon}(M)$, 409
$cyc(g)$, 1382
d, 389, 1347
dB, 1348
$d\omega$, 973
d^*, 1449
$d_E^2 x$, 721
d_P^*, 394
d_n, 1447
d_t, 389

$d_{1,c}$, 390
$d_{2,c}$, 390
$d_E x$, 721
$d_{P,n}^*$, 394
$d_{\infty,c}$, 390
$deformLiGr$, 766
$den(E)$, 339
df, 1455, 1469
$dim(K)$, 1445
$dim(M)$, 1392
$dom(f)$, 1419
$drap(k)$, 170
dw, 1470
e, 1347
e^M, 1396
e^m, 1396
e_z, 756
enh, 425
ev_p, 148
ex_z, 431
$exp(F, G)$, 334
$exp(ad(x))$, 1409
$ext(E)$, 337
f/α, 129
$f: a \to b$, 1419
$f@S$, 102
$f \circ g$, 1419
$f \overset{\iota}{/} \alpha$, 274
$f \overset{\iota}{/} \alpha \star M$, 288
$f^{-1}(C)$, 1375
$f_* F$, 1411
$f_{S,T}$, 1406
$finalEnd$, 767
$finalStart$, 767
$frame(F)$, 1438
$fun(F)$, 1438
g, 457, 1350
$g \circ f$, 1372
g_{U,w_s,w_p}, 1225
glb, 347
$grad(f)$, 1455
$groundclass(C^I)$, 313
h, 1349, 1350
h_x, 1411
i, 1432
$i<x,y>$, 207
$i\{x, y\}$, 207
$ic<k>$, 207
$ic\{k\}$, 207
$id(F)$, 1438
$intex_{z,B,G}(Ch)$, 459
$intex_{z,B}$, 432
$ip<a,b>$, 207
$ip\{a, b\}$, 207
is_X, 378

$iso\cap$, 283
j, 1414
jak, 213
jak_π, 218
k-spectroid, 925
k-Spectroid, 925
k-Contra, 387
k_s, 119
key, 1350
l, 1375
$l(Cont)$, 206
$l(M)$, 268, 1395
$l^2(\mathbb{N})$, 1486
$lambda_w(x)$, 259
lev, 1374
$lev(x)$, 269
lev_i, 269
$lim(\Delta)$, 1424
$lim(\mathbf{D})$, 1387
loc, 280
lub, 347
m^{-1}, 1378
m_x, 1411
m_{12}, 260
mod_o, 420
mod_p, 95
$monex_z$, 431
$myFM^w$, 239
$myFM_w$, 239
$n(U)$, 1445
nW, 409
$nW_\epsilon(M)$, 409
$n\Gamma(\sigma)$, 288
$n_k(U)$, 1445
$n_\infty(K^I)$, 259
$newset$, 146
o, 62
o-TempClass, 93
$obintex_{z,B}$, 433
p, 457
$p(M)$, 268
p-ClassChord, 93
p-EulerClass, 93
p-Scale, 95
p_x^Q, 1335
$p_{Instrument}^{BP}$, 599
p_B, 567
p_X, 1467
p_j, 1374
p_{Onset}, 100
$p_{U,V}$, 588
p_Δ^x, 523
$p_{\beta\sigma\mu}$, 121
p_{cf}, 508
p_{int}, 508
p_{meter}, 267

pct, 642

$pr_{\mu,\epsilon}(M)$, 409

$prof$, 379

pv, 62

q, 62

q_X, 1414

$r(\mathbf{S}_E, w)$, 654

r_H, 1337

r_t, 207

$rel(x, a_i)$, 648

res_{G^I}/Id_A, 294

$res_{@A}$, 294

res_{f/Id_A}, 294

$resclass(C^I)$, 313

ret, 298

rev_k, 125

$round(x)$, 1390

$s' \leqslant s$, 1445

$s(\mathbf{S}_E, w, t)$, 654

$s(x)$, 1412

$s(x, y)$, 1185

sG, 457

sG^*, 457

sP, 457

$scalemod_p$, 95

sem, 1438

set, 146

sig_{Den}, 334

$sing$, 1426

$sp(x)$, 269

$span$, 384

$st^i(x)$, 259

$supp(\chi)$, 337

$sym(A)$, 208

t, 62, 385

$t(F)$, 1438

$t(x, y)$, 1185

t_n, 385

$top(x)$, 1390

$true$, 1428, 1459

$v \wedge w$, 1468

w, 628

w-$Pitch$, 133

w-$PitchClass$, 94

w-$TemperedScale$, 96

w_{harmo}^S, 647

w_{metro}^S, 646

w_{motif}^S, 647

$w_{Evt., RelEvt..}$, 642

$w_{GrpArti}$, 645

w_{loc}, 763

w_{orth}, 763

x_{hmax}, 720

x_{hmean}, 720

$x_{melodic}$, 720

x_{metric}, 720

x/E, 338

$x < y$, 1373

$x > y$, 1413

$x@\hat{G}$, 257

$x^{\sim}$, 332

$x \succ z$, 227

$x \twoheadrightarrow y$, 402

x_h, 261

x_m, 261

$x_{U,V}$, 445

x_{alt}, 108

$x_{j,s}$, 710

x_{red}, 108

$y(t_i, j)$, 724

$\mathbf{C}^{/opp}b$, 1424

$\mathbf{C}/b$, 1424

$\mathbf{C} \models \alpha$, 1434

$\mathbf{C}^@$, 1422

$\mathbf{C}^{opp}$, 1421

$\mathbf{C}^{spaces}$, 1411

$\mathbf{C}_?^{spaces}$, 1411

$\mathbf{C}_X^{spaces}$, 1411

$\mathbf{a}P$, 1432

$|$, 1432

$|?|$, 1446

$|?|_d$, 1446

$|K|$, 1446

$|f|$, 1446

$|s|$, 1446

ébauche, 1038

Colimit, 59

   denotator, 1113

Limit, 58

   denotator, 1113

List

   denotator, 1113

Loc, 89

Mod$^@$, 54

Power, 58, 90

Simple, 58

Syn, 58

T^2, 917

head, 1375

tail, 1375

2-category, 939, 946

2D Fourier series, 1212

A

A-addressed

   function, 288

   gesture, 961

A-parametrized gesture, 1317

A2C, 870

abbreviation, 1053

abelian, 1378

abelian group, finitely generated -, 306

Ableton Live, 1158
absolute
    dynamical sign, 640
    dynamics, 685
    logic, 145
    music, 774
    novelty, 1062
    pitch, 577
    symbolic dynamics, 68
    tempo, 561, 643
absolutely relative, 1028
absorbing point, 432
absorption, 1337
    coefficient, 1337
abstract
    cardinality, 386
    complement, 208
    gestalt, 391
        specialization, 402
    identity, 14
    inclusion, 208
    motif, 386
    onset, 125
    specialization, 402
abstraction, 405, 407, 425, 1026
    concept framework, 385
    textual -, 362
Abstraktmotiv, 909
accelerando, 608, 631, 644
accelerated motion, 607
accentuation, 592
accessory parameter, 831
accumulation point, 1443
acoustical wave, 963
acoustics, virtual -, 701
action
    bodily -, 1305
    complement -, 518
    faithful -, 1378
    free -, 1378
    gestural -, 912
    Lagrangian -, 1188, 1481
    left -, 1378
    local -, 1038
    motor -, 608, 897
    Nambu-Goto -, 1185, 1484
    right -, 1378
    transitive -, 1378
action-homunculus, 869
action-to-cognition layer, 870
activities, fundamental -, 3, 6
activity
    combinatorial -, 198
    construction -, 162
    instinctive -, 623
    interpretative -, 246, 251, 252

actology, 1001
acuteness, 237
Aczel, Peter, 1094
ad-hoc polymorphism, 802
adapted tempo curve, 576
Add-Element, 814
AddObjects, 1132
address, 53, 54, 1179, 1397, 1423
    change, 54, 70
        technique, 70
    faithful, 431
    fixed vs. variable, 89
    for a chord, 94
    full -, 431
    fully faithful -, 431
    functor, 141
    killing, 168
    navigation, 140
    subcategory, 1089
    variable, 53
    zero -, 53
addressed
    adjointness, 138
    comma category, 137
adic representation, 1389
adjoint
    functors, 907
    left -, 1423
    right -, 1423
adjointness, addressed, 138
adjunction, 1375
admissible parallel hypergesture, 1250
admitted
    successor, 533
    tonalities, 465
Adorno, Theodor Wiesengrund, v, 154, 245, 248, 548, 569, 574, 609, 623, 652, 889–891, 893, 894, 1182
Adrien, Jean-Marie, 1346
affect, 890, 893
    categorical -, 1266
    vitality -, 1266
affine
    counterpoint group, 391
    dual, 1398
    Lie bracket, 444
    simplex, 1461
    tensor product, 1400
    transformation, modular -, 786
affine functions
    complex of -, 288
    on functorial global compositions, 354
after qualifier, 815
Agawu, Kofi, 329
Age of Enlightenment, 36
aggregate, 207
Agmon, Eytan, 203, 205

agogical
   architecture, 799
   operator, 720
agogics, 249, 643
   global -, 629
   primavista -, 629
AgoLogic, 576, 623, 789, 799
Agon, Carlos, xi, 210, 314
Air, 1283
Alain, 145
Alberti, Leon Battista, 856
aleatoric component, 198
aleatorics, 59
Alexander Technique, 868
Alfohrs, Lars V., 1465
algebra, 1385
   Boolean -, 103, 1433
   general linear -, 1408
   group -, 925
   Heyting -, 103, 930, 934, 1433
   Lie -, 1408
   logical -, 1433
   monoid -, 1386
   quiver -, 1388
   Riemann -, 481
   tensor -, 1469
algebra of gestures, 1002
algebraic
   element, 1460
   field extension, 1460
   geometry, 550, 1009
   topology, 164, 919
   universe, 1029
algorithm, 1342
   Euclidean -, 1386, 1389
   in FM synthesis, 73
   off-line -, 761
   real-time -, 761
   TX802, 236
algorithmic extraction of performance fields, 759
Alhaiya Bilwal, 1306
Alighieri, Dante, 116, 161, 1218
aliquid pro aliquo, 15
all-interval
   n-phonic series, 195
   series, 200
allomorph, 443
allomorphic extension, 443
allowed successor pairing, 531
almost regular manifold, 974
Alpaydin, Ruhan, 678, 1098, 1099
α-restriction, 433
alphabet
   of creativity, 198
   of music, 90
alphabetic ordering, 36, 38, 50

alterated note, 107
alteration, 107, 108, 162, 226, 466, 507, 1148
   as tangent, 108
   direction of -, 788
   elementary -, 108
   force field, 788
   in pitch, 53
   pitch -, 788
   successively increased -, 788
   two-dimensional -, 787
Alteration, 1135
altered scale, 480
alternating tensor, 1468
Alunni, Charles, 859, 861, 910, 1036, 1059, 1060, 1177
ambient
   space, 90
      coproduct -, 105
      product -, 104
ambient space
   dual -, 107
ambiguity, 245, 251
   theory of -, 245
   tonal -, 493
ambitus, 261
American
   (musical) set theory, 116, 180, 202–211, 979
   contour
      theory, 385
   jazz, 442
      notation, 438
      theory, 439
amplitude, 238, 1340
   modulation, 236, 833
   spectrum, 1340
AMS, 1034
Amuedo's decimal normal rotation, 210
Amuedo, John, 209, 211, 439
amusia, 872
analyse créatrice, 1139
analysis, 1338
   -by-synthesis, 609, 621
   chord -, 438
   coherent -, 634
   comparative -, 273
   complex -, 1465
   FM -, 236
   gestural -, 1274
   immanent -, 378, 383
   metrical -, 688
   motivic -, 214, 404
   musical -, 612
   neutral -, 222, 250
   normative -, 377
   notic -, 1306
   principal component -, 739
   regression -, 711, 724, 726

situs, 164, 227
  sonic -, 694
  spectral -, 524
  text -, 609
analytic, 1465
analytical
  discourse, 12
  method, 1012
  vector, 723
  weight, 549, 553, 646
Anaxagoras, 855
anchor note, 625
Andersen, Anne Sophie, 1277
Andreatta, Moreno, vi, xi, 210, 314, 1036, 1173
Andrieu, Bernard, 868
ANSI-C, 783
anthropic principle, 465, 466, 543
anthropocentricity, 1177
anthropology, computer-aided -, 765
anti homomorphism, ring -, 1385
antisymmetric, 1373
antiworld, 460, 497, 498, 773
anvil, 1354
apophatic theology, 1004
Applebaum, Mark, 1264
application framework, 665
apposition, 16
approach
  bigeneric -, 444
  categorical -, 801
  historical -, 465, 472
  nonparametric -, 708
  statistical -, 707
  systematic -, 472
  transformational -, 204
approximate, 683
Aprea, Bruno, 1292
arbitrary, 16
archetypal gesture, 1269
archicortex, 528
Archimedes, 864
architectural principle, 718
architecture
  agogical -, 799
  modulatory -, 495
Arduino, 1277
area
  Broca's -, 872
  Wernicke's -, 872
Argerich, Martha, 735, 736, 768
argument, 616
Aristotle, ix, 28, 38, 43, 773, 885
arm-based gesture, 1274
Arnold, Troy, 1276
aroh, 1306
arpeggio, 74, 134, 575, 592, 625

field, 575
arrow, 508, 898, 908, 1037, 1046, 1050, 1375, 1457
  self-addressed -, 514
arsis, 1192, 1287, 1325
art, scientific -, 1045
articulated listening, 249
articulating notes, 1303
articulation, 249, 562, 632, 645, 685
  double -, 17
  field, 565–568
  initial -, 578
  operator, 591
Artin, Emil, 1087
artistic
  difficulty, 1218
  fantasy, 570
  operator, 1204
artistry, combinatorial -, 199
arts, 5
Ashkenazy, Vladimir, 730, 735, 737
Assayag, Gérard, xi, 791
Assisi, Francis of, 847
associated
  metric, 1450
  metrical rhythm, 267
  topology, 1450
AST, 202, 271, 387, 409, 439
  global -, 314–316
  software for -, 209
asymmetries of communication, 754
Atari®, 624
  Mega ST4, 791
atlas, 252, 557, 987, 991, 1466
  A-addressed, 252
  projective -, 296
  standard -, 293
atlases, equivalent -, 257
atom, semantic -, 442
atomic formula, 1435
atomism, ontological -, 24
atonal music, 204
attack, 1339
attarantate, 874
audience, 1278
auditory
  cortex, 524, 527
  gestalt, 396
  nerve, 1355
  representation, 197
augmentation, 134
augmented, 444
Augustinus, 463, 502, 1322
Auroux, Sylvain, 36
auto, 1420
autocomplementarity, 181, 427
  function, 419, 519

autocomplementary, marked - dichotomy, 518
autocorrelation, 766
autocorrelogram, 715
autograph, free -, 1052
automorphism, 1378, 1385, 1420
   group, 144, 1391
     of interpretable compositions, 306
   relative -, 1460
autonomy, 6
autoreferential, 849
avaroh, 1306
Avison, Charles, 248
awareness, kinesthetic -, 1281
axiom, 1434
   of choice, 1374
axioms for QM, 1492
axis
   third -, 96
   diachronic -, 328, 472
   fifth -, 96
   of combination, 116, 212
   of selection, 116, 212
   paradigmatic -, 160
   synchronic -, 328, 472
   syntagmatic -, 16, 160
Ayler, Albert, 933

B

Bätschmann's Bezugssystem, 11
Bätschmann, Oskar, 11, 155
Bénabou, Jean, 1038
Béziau, Jean-Yves, 1028
Békésy, Georg von, 1358
Babbitt, Milton, 203, 204
Bach, Johann Sebastian, 114, 115, 118, 121, 161, 189,
     199, 203, 248, 249, 325, 488, 571, 608, 688, 709,
     711, 752, 753, 890
Bachelard, Gaston, 1059, 1060
background, 415
Bacon, Francis, 5, 859, 860, 1005
Badiou, Alain, 1007, 1016
Bagatelle op.126,2, Beethoven, 894
Bageshri, 1305
Bakhtin, Michail, 856
ball, open -, 1449
Ballade op. 23, 1161
Banach space, 1450, 1486
Banach, Stefan, 1450
band, frequency -, 526
bandwidth, 708, 721
bankruptcy, scientific -, 22
bar grouping, 713
bar-line, 592, 631
bar-lines, 68
Barbin, Évelyne, 1040
Barker, Hugh, 886

barline meter, 97
Barlow, Klarenz, 1368
Barthes, Roland, 15, 160
barycentric coordinate, 1446
base, 1393
   for a topology, 1443
   sheaf on -, 1412
Basic, 74
basic
   extension, 427
   intension, 427
   series, 125
   theme, 202
basilar membrane, 1356
basis, 1430
   calculation, 761
   coordinate, 1347
   deformation, 656
   of a tangent composition, 551
   of disciplines, 5
   parameter, 67, 654
   Schauder -, 1209, 1486
   space, 567, 588, 628
   specialization, 656
basis-pianola operator, 655
Batt-Rawden, Andrew, 1277
Baudelaire, Charles, 218, 493, 799
Bauer, Moritz, 215
Baumgarten, Alexander, 868
beat, 377, 1365
   extraction, 1292
   frequency, 1365
   meter, 97
   strong -, 377
   weak -, 377
beauty, 345
becoming, 1069
Beethoven, Ludwig van, 100, 120, 134, 201, 248, 267,
     276, 325, 347, 405, 407, 459, 460, 463, 466, 488,
     495, 571, 774, 780, 937, 1178, 1353
before qualifier, 815
behave well, 398
being, 1067, 1068
Bell, John L., 1058
Benjamin, Walter, 548, 569, 652
Benveniste, Émile, 856, 903
Beran operator, 723, 744
Beran, Jan, vi, xi, 201, 202, 488, 612, 707, 723
Berg, Alban, 125, 204, 247, 1260
Berger, Hans, 525
Bernard, Jonathan W., 204
Bernoulli, Jakob, 1007, 1048
Bernoulli, Johann, 1007
Beschler, Edwin, xii
Bessel function, 1344
Bevilacqua, Frédéric, 902

Bezugssystem, Bätschmann's, 11

Bezuoli, Giuseppe, 28

biaffine, 1399

bicategory, 946

   gesture -, 937, 946

bidirectional dialog, 31

big bang, 329

big science, 196

BigBang, 1103, 1187

   rubette, 947, 1097, 1127, 1157

BigBangObject, 1103, 1114

bigeneric

   approach, 444

   major tonality, 449

bigeneric morphemes, construction of -, 445

bijective, 1372

bilinear form, 291

Binnenstruktur, 247

biological inheritance principle, 627

bipolar recording, 524

Bissonanz, 424

Blake, William, 417, 755

block, 589, 784

   type, 311

blues, 885

Bobillier, Etienne, 1027

Boccherini, Luigi, 241, 826, 1264

Boden, Margaret A., 1034

bodily

   action, 1305

   technique, 1280

body, 870, 910, 1271

   Hegel -, 1066, 1070

   performance -, 586

   performative -, 1283

   philosophy of the -, 868

Boesendorfer, 577, 629, 687, 700

boiling down method, 648

book concept, 48

BOOLE, 44

Boole, George, 1433

Boolean

   algebra, 103, 1433

   combination

     of (class) chords, 94

   operation, 784

   topos, 1434

Boquiren, Sidney, 1277

Borel

   function, 1488

   set, 1488

boson, 1184, 1485

bottle, Möbius -, 558

bottom wall, 631

Boulez, Pierre, 30, 35, 36, 40, 89, 90, 114, 127, 245, 246, 253, 287, 303, 779, 781, 831, 839, 859, 1179, 1181, 1363

bound

   Chopin rubato, 625

   variable -, 1436

boundary, 1443

   homomorphism, 966

   manifold with -, 1467

   operator, 969

bounded operator, 1487

Bourbaki, Nicolas, 1028, 1041

bow

   angle, 235

   application, 235

   parameter, 833

   pressure, 235, 833

   velocity, 235, 833

box, 802, 1066

   factory -, 805

   flow -, 1455

   Hegel's -, 1072

   temporal -, 811

Box-Value, 814

bracket, Lie -, 1246, 1408, 1455

Brahms, Johannes, 1264

brain, emotional -, 527

brane, 1184

breaking, symmetry -, 775

breath, 1286

breathe together, 1317

breathing, 1315, 1316

Brendel, Alfred, 729, 730

Bresin, Roberto, 902

brilliance, 237

Brinkman, Alexander, 209

Broca's area, 872

Brofeldt, Hans, 883

Bruhat, standardized - transformation, 1154

Bruijn, Nicolass Govert de, 190, 194, 311

Bruter, Claude-Paul, 1035

Bulwer, John, 855

bundle

   of ontologies, 142

   tangent -, 1451

Bunin, Stanislav, 735

Buteau, Chantal, xi, 376, 403

C

C-major, inner symmetry of -, 123

C-major, 457

c-motif, 100

C-scale frame, 475

C-scheme, 1415

C-space, 1411

c-space, 205

CAC, 801
cadence, 414, 453–462, 465, 817, 840
  parameter, 454
  Rameau's -, 455
cadential, 454, 465
  family, minimal -, 455
  formula, 453
  set, 455
Cadoz, Claude, 901, 1182
Cage, John, 59, 250, 571
Calbris, Genevieve, 1301
calculation, 1002, 1029
  basis -, 761
  field -, 761
  precision, 639
calculus, 570
  of variations, 1471
camera obscura, 603
camino, 1005
Campanella, Tommaso, 849, 856, 1037, 1177
Campos, Daniel G., 1053
Camurri, Antonio, 902
canon, 117, 159, 268
canonical
  curve, 720, 733, 739
  operator, 207
  Para-meter, 268
  program, 325
canons, classification of -, 312
Cantor, Georg, 1006
cantus
  durus, 261
  firmus, 509, 931
  mollis, 261
Capova, Sylvia, 735
cardinality, 672, 1373
  abstract, 386
  of a gestalt, 391
  of a local composition, 90
Carey, Mariah, 1303
Carnatic vocalist, 1301
Carnot, Lazare, 1049
Carpitella, Diego, 874
carpus, 912
carrier, 73, 1342
Cartan, Élie, 1055
cartesian, 938
  closed, 1429
  coordinates, 1007
  doubt, 1012
  product, 1372, 1424
  thinking, 898, 1182
case
  linear -, 205
  cyclic -, 205
Casella, Alfredo, 183

Castine, Peter, 209
catastrophe, 497–499, 501, 957
  theory, 226, 497
categorical
  affect, 1266
  approach, 801
  digraph, 940
  manifold, 993
  manifolds, category of -, 993
  modeling, 1001, 1003, 1045, 1046
categorically natural gesture, 918
categories, equivalent -, 1422
categorization, 1004
category
  cocomma -, 1424
  cocomplete -, 1425
  comma -, 1424
  complete -, 1425
  finitely
    cocomplete -, 1425
    complete -, 1425
  graph -, 939
  homotopy -, 1474
  internal -, 1464
  isomorphic -, 1421
  linear -, 923
  matrix -, 1420
  mimetic -, 892
  of categorical manifolds, 993
  of cellular hierarchies, 593
  of commutative global composition, 285
  of coverings of sets, 293
  of denotators, 331–334
  of elements, 132, 1061, 1424
  of factorizations, 943
  of forms, 57
  of formulas, 926
  of functorial global compositions, 275
  of gestures, 1289
  of Heyting algebras, 930
  of local compositions, 89
  of locally almost regular manifolds, 974
  of objective global compositions, 275
  of performance cells, 586
  of points, 981, 1255
  of sheaves, 1431
  of sketches, 1029
  of textual semioses, 337
  opposite -, 1421
  path -, 1388, 1458
  product -, 1422
  quotient -, 1420
  simplex -, 939
  skeleton -, 1421
  tangent -, 927
  theory, 908, 1002, 1028

topological -, 939, 961, 1085, 1464
category of form semiotics, 1092
category of known simple models, 1062
Cauchy
  problem, 1456
  sequence, 1450
Cauchy, Augustin, 1450, 1456
Cauchy-Riemann equation, 1465
causal
  coherence, 769
  depth, 676
  relation, 816
causal-final variable, 767
causality, 766
Cavaillès, Jean, v, 859, 861, 873, 909, 924, 1035, 1177,
    1181, 1300
Cavalieri, Bonaventura, 1048
CDC Cyber, 525
Čech cohomology, 354, 2131
Celibidache, Sergiu, 117, 574, 1285
cell, 306
  complex, 325
  Deiters' -, 1356
  hair -, 1356
  outer
    hair -, 1356
    pillar -, 1356
  performance -, 585, 1254
cellular
  hierarchies
    category of, 593
    classification of -, 591
  hierarchy, 590, 596, 1254
    product -, 590
    restriction -, 590
    type of a -, 590
  organism, 325
Cent, 1349
center, 1385
central
  pitch detector, 1361
CERN, 196
Certeau, Michel de, 1040
Châtelet, Gilles, 859, 862, 872, 894, 898, 1036, 1060,
    1177, 1180
chain, 1395
  complex, 1475
  proof -, 1434
  third -, 675
chamber pitch, 563, 1349
change
  of frame, 1058
  of material, 813
  of orientation, 509, 515, 531
  of perspective, 324
  program -, 784

value -, 632
CHANT, 238, 1314
chant, Gregorian -, 509
character string, 60
characteristic, 1386
  function, 335, 1375
  gesture, 956
  map, 1428
characteristics, method of -, 1456
charge, semantic -, 403
chart, 251, 252
  of level j, 269
cheap design, 873
checkboxes
  matrix of -, 1110
Cherlin, Michael, 934
chest voice, 1315
Chin, Unsuk, 1295
Chinese philosophy, 1278
chironomic
  game, 1327
Choi, Insook, 901
choice, axiom of -, 1374
Chomsky, Noam, 234
Chopin rubato, 550, 562, 576, 624, 765
  bound -, 625
  free -, 625
Chopin, Frédéric, 82, 625, 743, 1161
chord, 92, 180, 187, 249, 413
  structural -, 415
  a scale's -, 95
  addresses, 94
  analysis, 438
  circle -, 424
  class -, 94
  classification, 159, 414
  closure, 431
  complement, 94
  core -, 485
  dictionary, 180
  difference, 94
  diminished seventh -, 463, 497, 499, 501
  event, 93
  foundation -, 439
  fundamental -, 439
  inspector, 675
  intersection, 94
  inversion, 420
  isomorphism classes, 180
  just - class, 94
  n-, 93
  pivotal -, 463
  progression, 413, 414
  prolongational -, 415
  self-addressed -, 185
  sequence, 486

coherent -, 486
  standard -, 436
  symbol, 438, 439
  tempered - class, 94
  tesselating -, 309
  union, 94
CHORD-CLASSIFIER, 210, 441
Chowning, John, 1342
Christian religion, 847
christianization, 889
  of music, 891
chromatic
  (tempered) class chord, 94
  Michel -, 478
  octave, just -, 96
  Roederer -, 478
  scale, 418
  Vogel -, 478
chronocentricity, 1175, 1177
chronospectrum, 235
Cicero, 856
circle
  chord, 424
  of fifths, 423
  of fourths, 423
circular
  colimit, 65
  definition, 48, 146
  denotator, 72–75
  denotators
  folding - denotators, 369
  form, 64
  gesture, 1269
  limit, 65
  set, 66
  synonymy, 64
circularity, 849, 851
  conceptual -, 146
  of forms, 48
CL, 1434
class, 156, 802
  contour -, 206
  resultant -, 313
  chord, 93
    just -, 94
    tempered -, 94
  contiguity -, 1447
  counterpoint dichotomy -, 518
  dichotomy -, 518
  equivalence -, 1373
  ground -, 313
  marked
    counterpoint dichotomy -, 518
    dichotomy -, 518
  meta-object -, 814
  nerve, 283, 309, 322, 988

number, 180
precedence
  list, 804
segment -, 206, 208
set -, 207, 208
third comma -, 265
Vuza -, 312
weight, 188, 283
classical logic, 1434
classification, 828
  epistemological -, 5
  semiotics of sound -, 240
  chord -, 414
  geometric -, 177, 989
  in musicology, 157
  local musical interpretation of -, 173
  local theory of -, 157
  of canons, 312
  of cellular hierarchies, 591
  of chords, 159
  of motives, 187–190
  of music-related activities, 4
  of rhythms, 312
  of sounds, 10
  recursive -, 177
  sound -, 232
  technique, 169
  theory, 831
classifier, subobject -, 1428, 1459
CLOS, 210, 801, 802
closed
  cartesian -, 1429
  locally -, 1445
  path, 1375
  point, 228
  set, 1443
  sieve, 1431
  simplex, 1446
closure, 1443
  hierarchy -, 588
  objective -, 432
closure operator, Kuratowski -, 1237, 1443
Clough, John, 203
cluster
  Cortot -, 735, 739
  Horowitz -, 735, 739
Clynes, Manfred, 604, 607, 889, 891, 895, 1182
CMAP, 209
coarser, 1443
coboundary map, 1447
cochain, 1477
  complex, 1477
    of a global composition, 307
  singular -, 1447
cochlear Fourier analysis, 1365
cockpit, 1104

cocomma category, 1424
cocomplete category, 1425
cocone, 1423
coconut parable, 1089
coda, 249, 496
code, 212
codification of a symmetry, 128
codomain, 1372, 1419
coefficient
    absorption -, 1337
    largest -, 733
    system, 1447
coefficients, signs of -, 731
cognitive
    dimension, 179
    effort, 179
    embryology, 870
    independence, 179
    layer, 869
    musicology, 21
    psychology, 179, 226
    science, 611
    stratum, 869
coherence, 415, 550, 634, 688
    causal -, 769
    final -, 769
    harmonic -, 447
    inter-period -, 769
coherent
    analysis, 634
    chord sequence, 486
    topology, 1444
Cohn, Richard, 315
cohomological technique, 1061
cohomology, 307, 1477
    Čech -, 354, 1231
    de Rham -, 973
    group, 1447
    l-adic -, 379
    module, 1477
        of a global composition, 307
    resolution -, 355
coinduced
    topology, 1444
Coleman, Ornette, 794, 900, 1183
Coleman, Steve, 377
colimit, 252, 1290, 1424
    circular -, 65
    form, 57
    topology, 1444
colinear, 230
collaborative environment, 197
collaboratory, 32, 667
collective responsibility, 633
color, 1108
    coordinate, 1341

encoding, 765
parameter, 835
sound -, 159
space, 832
coloring, 784
Coltrane, John, 571, 603, 1295
COM matrix, 387
combination
    axis of -, 116, 212
    linear -, 1391
    weight -, 681
combinatorial
    activity, 198
    artistry, 199
    design, 1025
    topology, 254
combinatoriality, 210
combinatorics, creative -, 246
comes, 159, 199, 688
comma
    category, 1424
        addressed -, 137
    fifth -, 62
    Pythagorean -, 62
    syntonic -, 62, 97
    third -, 62, 265
common
    language, 22
    taste, 752
common-note function, 204
communication, 3, 4, 9, 11, 24
    asymmetries of -, 754
    coordinates, 14
    process, 14
communicative dimension, 22
commutative, 1376
    local composition
        module of a -, 105
    diagram, 1421
    global compositions, 285
    homotopy -, 1474
    local composition, 105
    polynomials, 1386
commutativity relation, 1420
compact, 1445
    locally -, 1464
compact-open topology, 914, 937, 1464
comparative
    analysis, 273
    discourse, 493
comparative criticism, 756
comparison matrix, 206
competence, 331
    historical, 349
    stylistic, 349
complement, 1371

abstract -, 208, 210
action, 518
literal -, 210
theorem, 208
complete
  category, 1425
  harmony, 827
  quiver, 1375
  uniform space, 1450
completeness, 36, 42, 49
  finite -, 138
completion, semantic -, 49
complex
  analysis, 1465
  cell -, 325
  chain -, 1475
  cochain -, 1477
  derivative, 1465
  gesture, 1042
  module -, 288
  number, 948
  numbers, 1385
  of affine functions, 288
  quotient -, 289
  set -, 314
  simplicial -, 779, 1445
  simplicial cochain -, 1447
  time, 1186, 1218
complexity, 165
  degree of -, 162
  formal -, 383
  measure, 254
  of performance, 548
component
  aleatoric -, 198
  alteration, 107
  idempotent -, 1377
  irreducible -, 269
  metrical -, 267
  reduced -, 107
  semiotic -, 854
composed frame, 802
composer, 12, 895, 1263
  perspective of the -, 246
composition, 162, 1372
  t-fold tangent -, 551
  global -
    standard -, 293
  commutative
    local -, 105
  computer
    assisted -, 801
  computer-aided -, 774, 791
  concept, 571
  dodecaphonic -, 124
  functorial -, 1180

functorial local -, 101
generic, 174
global -, 41, 140, 831, 987, 1466
  N-formed -, 291
  oriented -, 291
  resolution of a -, 323
global functorial -, 256
global objective -, 252
interpretable -, 304
local -, 41, 75, 89, 90
  embedded -, 106
  dimension of a -, 178
  generating -, 106
  projecting -, 178
  standard -, 293
local objective -, 90
locally free local -, 175
modular -, 251
musical -, 30
non-interpretable -, 305, 308
non-interpretable global -, 990
tangent -, 551
tools, fractal -, 115
compositional
  design, 209
  idea, 322
  space, 204
compositions
  commutative global -, 285
  local -
    category of -, 89
computation
  symbolic -, 801
computational musicology, 21
computer assisted composition, 801
  science, 6, 156
computer-aided
  anthropology, 765
  composition, 774, 791
computer-assisted performance research, 629, 701
conativity, 212
concatenation, 132, 1372
  principle, 133, 512
concept, 9, 35
  architecture, 152
  composition -, 571
  construction history, 47
  critical -, 1072
  denotator -, 665
  form -, 665
  format, 42
  framework, 3, 5, 9
    abstraction -, 385
    dynamic -, 329
  fuzzy -, 164
  grouping -, 250

human - construction, 48
leafing, 50, 52
of a book, 48
of instantaneous velocity, 27
of music, 21
paradigmatic -, 229
poietical -, 14
point -, 145
RUBATO®-, 665
score -, 250, 571, 754, 811
set -, 146
space, 21, 31, 32, 1034
spatial -, 894
stable -, 226
surgery, 83–85, 633
concepts
standard of basic musicological -, 92
void pointer -, 32
conceptual
circularity, 146
explicitness, 21
failure, 23
genealogy, 64
identification, 229
laboratory, 30
navigation, 35
precision, 32
profoundness, 92
universality, 92
zoom-in, 19
conceptualization
dynamic -, 67
fuzzy -, 375
human -, 145
precise -, 211
process, 201
concert
form, 793
master, 626
pitch, 576
condition
Dirichlet -, 1483
initial -, 1452
instrumental -, 701
Neumann -, 1483
conducting, mechanism of -, 1291
conductor, 551, 562, 626
gestures of the -, 1285
primitive gesture of -, 1287
role of -, 1288
cone, 1423
configuration
counting series, 191
view -, 1108
conformal, 1466
conjecture, diamond -, 929, 937

conjugation, 1379, 1385
conjugation class
of endomorphisms, 180
of symmetry group, 180
conjunction, 346, 1432
connective, predicate -, 1435
connotation, 16
connotator, 328, 1440
conservation, energy -, 973
consonance, 463, 469, 509, 1364
deformed -, 532, 1236
imperfect -, 522, 532, 541
perfect -, 522, 532, 541
consonance-dissonance, 1353
dichotomy, 419, 519, 543
consonant, 234
interval, 414, 527
mode, 450
constant
functor, 1422
module complex, 288
part, 433
shift, 109
structural -, 1408
constraint
geometric -, 913
gestural -, 617
musical -, 913
physical -, 913
programming, 774, 801
constraints
geometric -, 912
mechanical -, 912
semiotic -, 232
construction
activity, 162
of bigeneric morphemes, 445
recursive -, 42
construction history of concept, 47
construction using compass and straightedge, 1461
contact, 212
point, 235
container, 806, 811
content, 15, 337, 409, 647, 673
interval -, 207
mathematical -, 15
maximal structure -, 1362
musical -, 15, 1286
textual -, 1286
context, 212, 736, 1072
problem, 674
real-time -, 760
contiguity, 16
class, 1447
contiguous simplicial maps, 1447
continuous, 1444

gesture, 817
   method, 639
   presheaf, 982
   stemma, 661
   weight, 639
continuum, Kendon -, 854
contour, 159, 387
   class, 206
   space, 205
   theory, 271
      American -, 385
contra, 532
contractible, 1474
contraction, 1452, 1474
contrapuntal
   form, 249
   group, 115
   interval, oriented -, 509
   meaning of $\mathbb{Z}$-addressed motives, 100
   motion shape type, 387
   sequence, 531
   symmetry, 532, 1235
      local character of a -, 533
   technique, 159
   tension, 531
   tradition, 199, 1367
contravariant functor, 1421
contravariant-covariant
   rule, 805
control
   group, 775
   interactive -, 813
   of transformation, 200
conversation, topos of -, 826
convex, 1462
   hull, 1462
Cooper, Lynn, 872
coordinate
   barycentric -, 1446
   basis -, 1347
   color -, 1341
   fifth -, 1350
   function, 174
   geometric -, 1341
   gestural -, 893
   octave -, 1350
   ontological -, 10
   pianola -, 1347
   third -, 1350
coordinates, 45
   cartesian -, 1007
   of existence, 578
coordinator, 43, 1438
   form -, 55
   gestural form -, 1087
   of a form, 55

coproduct, 1375
   ambient space, 105
   of local compositions, 105
   type, 47
Corballis, Michael C., 871
Corbin, Alain, 874
core chord, 485
corporeal, 1264
correlate
   electrophysiological -, 524
   physiological -, 897
correspondence
   exponential -, 1464
   Galois -, 1093
cortex, auditory -, 524, 527
Corti, organ of -, 1356
Cortot cluster, 735, 739
Cortot, Alfred, 734, 735, 737
coset
   left -, 1379
   right -, 1379
cosmology, 465
counterpoint, 134, 419, 508, 523, 826, 966, 1235
   dichotomy, 518
      class, 518
   double -, 513
   rubette for -, 1097
   theorem, 534, 538
   theory, 775, 839
counterpoint dichotomy
   class, marked -, 518
countersubject, 689
counting series, configuration -, 191
coupling
   monogamic -, 632
   polygamic -, 632
Courtine, Jean-Jacques, 874
covariant functor, 1421
covering, 252
   equivalent -, 252
   family, 1430
   motif, 384
   sieve, 1430
Cox, Arnie, 1264
cp, 205
CPL, 804
cpset, 206
creation, melodic -, 1077
creative
   combinatorics, 246
   extension, 201
   mathematics, 1033
   process, 1066, 1072
   space, 1034
creativity, 198, 329, 1001, 1031, 1035, 1038, 1044
   alphabet of -, 198

essence of -, 1287
    gestural -, 1183
    mathematical -, 1002
    musical -, 1227
creator, 11, 12
crescendo, 67, 551, 594, 607, 1348
    wedge, 641
critical
    concept, 1072
    distance, 1337
    fiber, 755, 758
criticism
    comparative -, 756
    journalistic -, 730
    music -, 634
critique, 755
    music -, 751
cross-correlation, stemmatic -, 634
cross-modal, 1274
cross-semantical relation, 612
cube, 1476
    n-dimensional -, 966
    topographic -, 17, 32
cul-de-sac, 541
    interval, 538
cultural mimesis, 874
culture of performance, 623
curve, 230, 1451
    canonical -, 720, 733, 739
    exact -, 945
    integral -, 1245, 1453
    intonation -, 563
    tempo -, 562, 607, 623, 723, 784
curvilinear reduction, 776
Cusick, Susan, 1264
cutting, 1032
CX5M, Yamaha -, 525
cycle, 1375, 1382, 1454
    index, 191, 1382
    of variations, 791
    pitch -, 207
cyclic
    case, 205
    extension, 207
    group, 1381
    interval succession, 207, 208
Czerny, Carl, 624, 764, 911, 914

D

d'Alembert, Jean Le Rond, 5, 36, 50
D-brane, 1482, 1483
da capo, 117
dactylus, 213
    grid, 216, 217
Dahlhaus, Carl, 23, 123, 245, 264, 265, 447, 472, 488,
    674, 826, 1293, 1367

Damasio, Antonio, 854
dance, 604, 874, 951, 1301
Dannenberg, Roger, 761
darkening, 1315
data, ethnomusicological -, 83
dataglove, 607
daughter, 619
    tempo, 562
daughter performance, 984
daughters, 596
Davin, Patrick, 817
Davis, CeDell, 876
DBMS, 665
de la Motte, Helga, 22, 571
De Martino, Ernesto, 874
De Morgan, Augustus, 1002
De Novellis, Romina, 874
de Rham cohomology, 973
Deacon, the Deacon, 1321
Debussy, Claude, 183, 492, 622, 1279
decay, 1339
decomposition
    hierarchical -, 710, 720
    natural -, 707
    orthonormal -, 10
    spectral -, 708
    Sylow -, 445, 510
Deep Purple, 189
deep-frozen gesture, 912
default weight function, 482
definition
    circular, 48, 146
    of music, 5
    of musical concepts, 97
definitive performance, 895
deformation, 226, 591
    basis -, 656
    degree of -, 787
    hierarchy -, 658
    non-linear, 681
    non-linear -, 640, 733
    of a tempo curve, 576
    pianola -, 591, 656
deformed
    consonance, 532, 1236
    dichotomy, 532
    dissonance, 532, 1236
degree, 249, 262, 441, 442, 465
    different -, 264
    modulation -, 465
    of complexity, 162
    of deformation, 787
    of freedom, 179
    of organization, 718
    of symmetry, 208
    parallel -, 264

system, irreducible -, 457
 theory, 436
 triadic -, 1245
Deiters' cell, 1356
Delalande, Francois, 607, 608
delay, 833
 relative -, 236
Deleuze, Gilles, 859, 860
Deligne, Pierre, 351, 1022, 1179
delta, Kronecker -, 1392
denaturation, sacred -, 891
Dennett, Daniel, 149
denotator, 41, 57–59, 1074
 Simple -, 1112
 $\mathbb{Z}$-addressed -, 54
 Colimit-, 1113
 Limit-, 1113
 List-, 1113
 attributes, 42
 circular -, 72–75
 concept, 665
  genealogy, 41
 flow chart, 42
 image, 59
 language, 595
 morphism, 91
 name, 45
 non-zero-addressed -, 69
 ontology, 328
 orchestra instrumentation -, 69
 philosophy, 153
 Pow -, 1113
 reference -, 332
 regular -, 66–71
 self-addressed -, 69
 truth -, 335
denotators
 circular - folding, 369
 linear ordering among -, 50
 ordering on -, 75–83
 ordering principle on -, 49
 relations among -, 89
Denotex, 668, 1441
DenotexRUBETTE®, 668
dense, 229
densification, 817
density
 Lagrange -, 1482
 Lagrangian -, 1185
deocentricity, 1175
deontology, 848
depiction, 1308
depth, 21, 23, 198
 causal -, 676
 EEG, stereotactic -, 524
 electrode, 525

final -, 676
 in mathematics, 23
 in musicology, 23
 in the humanities, 485
 semantic -, 383
derivation, 1408, 1455, 1469
 inner -, 1408
 outer -, 1408
derivative, 1451
 complex -, 1465
 exterior -, 973
 Lie -, 1455, 1469
derived serial motif, 194
Desain, Peter, 547
Descartes' duality, 1187
Descartes, René, 11, 148, 1005, 1015, 1031, 1364
description
 object -, 200
 verbal -, 622
design
 combinatorial -, 1025
 compositional -, 209
 matrix, 724
 sound -, 1121
Desmond, Paul, 179
development, 249, 496
 history, 612
 software -, 595
 syn- and diachronic of music, 199
Dezibel, 1348
Di, 1103
di-alteration, 108
diachronic, 16
 axis, 328, 472
 index, 223
 normalization, 754
diaffine homomorphism, 1396
diagonal
 embedding, 1403
 field, 565
diagram, 898, 907, 1017, 1032, 1053, 1421
 scheme, 1420
 commutative -, 1421
 filtered -, 1411
 gestural -, 1036
 Hasse -, 218, 1374
 of forms, 57
diagrammatician, 859
dialog, 827
 bidirectional, 31
 experimental navigation -, 32
dialogical principle, 828
diameter, 520
diamond conjecture, 929, 937
diastematic, 672
 index shape type, 387

notation, 1322
shape type, 387
diatonic scale, 543
dichotomy class, 518
    marked -, 518
    consonance-dissonance -, 419, 519, 543
    counterpoint -, 518
    deformed -, 532
    interval -, 518
    major -, 519, 543
    marked counterpoint -, 518
    marked interval -, 518
    Riemann -, 522
    Saussurean -, 15
dictionary of expressive rules, 615
Diderot, Denis, 5, 36, 50
Dieudonné, Jean, 1044
difference, 1371
    genealogical -, 756
    phenomenological -, 756
different degree, 264
differentiable, 1451
    n-chain, 975
    manifold, 1466
differential, 1455, 1469
    equation, 1452
    form, 975
    operator, 1470
    semantic -, 163
differentiation rules, 610
difficulty, artistic -, 1218
digital age, 35
digraph, 914
    categorical -, 940
    dual -, 1457
    global -, 996
    radical -, 926
    spatial -, 914, 940, 1457
    topological -, 993
digraph (directed graph), 1457
diinjective, 1405
dilatation, 133, 1402
    time -, 70
dilinear
    homomorphism, 1392
    part, 1397
dimension, 1392, 1445
    cognitive -, 179
    communicative -, 22
    of a local composition, 178
    of a simplex, 1445
    ontological -, 17
diminished, 444
diminished seventh chord, 463, 497, 499, 501
Ding an sich, 21
direct

image, 1411
    sum module, 1392
directed graph (digraph), 907, 914, 1375, 1457
direction of alteration, 788
directional endomorphism, 656
Dirichlet condition, 1483
disabled gesture, 875
discantus, 509, 931
disciplinarity, dynamic -, 666
discipline, basic -, 5
discourse
    analytical -, 12
    comparative -, 493
    esthesic -, 14
discoursivity, 36, 42, 49
discrete, 638
    interpretation, 254
    field, 760
    gesture, 817, 944
        functor, 944
    nerve, 254
    topology, 1443
disease, psychosomatic -, 1288
disjoint, 1372
    sum, 1424
disjunction, 346, 1432
dissonance, 463, 469, 509, 1364
    deformed -, 532, 1236
    emancipation of -, 31
dissonant interval, 527
    mode, 450
distance, 226, 228
    critical -, 1337
    Euclidean - for diastematic types, 389
    Euclidean - for rigid types, 389
    for toroidal types, 390
    function, 389
    natural -, 362
    on toroidal sequences, 390
    relative Euclidean - for rigid types, 389
    third -, 511
    to an initial set, 580
Distributed RUBATO®, 764
distributed laboratory, 32
distributive, 1433
distributor, 689
division
    of pitch distances, 61
    of time
        regular -, 376
divisor, resulting -, 313
Dockery, Wayne, 877
documentation, 3–6
dodecaphonic
    composition, 124
    composition principle, 115

method, 775, 780
  paradigm, 125
  series, 125, 162, 194, 247, 253, 325
  vocabulary, 199
dodecaphonism, 134, 206
  communicative problem of -, 135
  esthetic principles of -, 134
domain, 1372, 1419
  fundamental scientific -, 5
  modulation -, 476
dominance, 218, 269
  topology, 231, 402
dominant, 263, 414, 445, 448
  role of major scale, 541
  seventh, 419
dominate, 227, 1413
Donald, Merlin, 874
double
  articulation, 17
  counterpoint, 513
doubt
  cartesian -, 1012
  sceptical -, 1012
drama, musical -, 753
dramaturgy, emotional -, 893
drawing, 1147
Dreiding, André, 291
Dress, Andreas, 291
drill, 1305
driving grid, 787
drum, ear -, 1354
dual
  affine -, 1398
  ambient space, 107
  digraph, 1457
  gesture, 940
  linear -, 1397
  numbers, 508, 1387
  space, 1488
dual numbers, 107
dualism between major and minor, 122
duality, Descartes' -, 1187
Dufourt, Hugues, 801
duration, 44, 67
  period, 97
dux, 159, 199
DX7, 1342
Dylan, Bob, 885
dynamic
  concept framework, 329
  conceptualization, 67
  disciplinarity, 666
  navigation, 40
dynamical
  initialization, 578
  knowledge management, 329

modularity -, 666
  sign
    absolute -, 640
    relative local -, 640
    relative punctual -, 640
dynamically loadable module, 665
dynamics, 248, 249, 562, 564
  absolute -, 685
  historical -, 221, 223
  mechanical -, 608
  of performance, 658
  primavista -, 629
  relative -, 685
  symbolic
    absolute, 68
    relative, 68

E

ϵ-ball, 229
E-MU Xboard, 1151
ϵ-neighborhood, 397
ϵ-paradigm, 229
ear
  drum, 1354
  inner -, 1355
  middle -, 1354
  outer -, 1354
ecclesiastical mode, 261, 540, 541
Eco, Umberto, 1050
edge, 1458
editing, geometric -, 783
editor, 802
EEG
  depth -, 524
  response, 524
  semantic charge of -, 524, 525
  test, 524
effect, groove -, 788
effective, 1378
Eggebrecht, Hans Heinrich, 21–23, 743
Ego, poetic -, 214, 219
Ehrenfels' transpositional invariance criterion, 91
Ehrenfels, Christian von, 91, 167, 226, 246, 271, 273, 383
Ehresmann, Andrée, 1013, 1029, 1060
Ehresmann, Charles, 934
eigenvector, 899
Eilenberg, Samuel, 1178
Eimert, Herbert, 127, 211
Eitz, Carl, 1350
elastic, 672
  shape type, 388
electrode, depth -, 525
electrophysiological correlate, 524
element, 1371
  algebraic -, 1460
  neutral -, 1376

primitive -, 1460
transcendental -, 1460
varying -, 1023
elementary
    alteration, 108
    gesture, 817
    neighborhood, 403
    shift, 109
elementary gesture of the pianist, 1192
elements, category of -, 132, 1061, 1424
Ellington, Duke, 1297
emancipation of dissonance, 31
embedded local composition, 106
embedding, 1422
    diagonal -, 1403
    number, 208
    Yoneda -, 1397, 1423
embodied
    gesture, 1274, 1279
    musical gesture, 1263, 1264
embodiment, 846, 864, 909, 952, 1176
    musical -, 1314
    science, 868
embryology, cognitive -, 870
emergence, 1061, 1063
emotion, 528, 604–606, 894
emotional
    brain, 527
    dramaturgy, 893
    function of music, 528
    landscape, 241
emotive state, 1266
emotivity, 212
empty
    form name, 47
    set, 146
    string, 45
encapsulated history, 556
encapsulation, 23, 156, 806
    speculative -, 28
encoding
    color -, 765
    formula, rubato -, 618
Encyclopédie, 5, 36, 38, 50
encyclopedia, 36, 362
encyclopedic ordering, 50
encyclopedism, 49
encyclospace, 36, 38, 50
endo, 1420
endolymph, 1355
endomorphism, 1420
    directional -, 656
    enharmonic -, 426
    right-absorbing -, 432
    ring, 1391
energy, 608

conservation, 973
spectrum, 1340
enharmonic, 425
    endomorphism, 426
    group, 426
    identification, 425
ensemble rules, 610
enumeration
    musical - theory, 190
    of motives, 195
    theory, global -, 309
envelope, 71, 1339
environment
    collaborative -, 197
    experimental -, 681
epi, 1420
epilepsy therapy, surgical -, 524
epileptiform potential, 524
epimorphism, 1420
epistemology
    of musicology, 27
    transitive -, 1059
epsilon
    gestalt topology, 398
    topology, 397
Epstein, David, 607
equation, 898, 907
    Cauchy-Riemann -, 1465
    differential -, 1452
    Euler-Lagrange -, 1188, 1189, 1227, 1482
    Poisson -, 1189, 1197, 1471
    Schrödinger -, 1493
    spring -, 1341
equations, Maxwell's -, 973
equivalence
    phonological -, 215
    class, 1373
    homotopy -, 1474
    paradigmatic transformation -, 212
    perceptual -, 229
    relation, 250, 1373
    syntagmatic -, 215
equivalent
    atlases, 257
    categories, 1422
    covering, 252
    norms, 1450
equivariant, 1378
Erlewine, Stephen Thomas, 877
Escher Theorem, 916, 931, 941, 961, 965, 966, 985, 995, 998, 1184
Escher, Maurits Cornelis, 161, 916
EspressoRUBETTE®, 759, 764
essence of creativity, 1287
essential parameter, 831
essentic form, 895

esthesic, 11, 1342
    identification, 248
esthesis, 12, 13, 211
esthetic, 211
esthetics, 13, 211
    of music, 323
estrangement, 1003
ethnological form, 49
ethnology
    inverse -, 754
ethnomusicological data, 83
ethnomusicology, 754
Euclid, 147, 148, 507, 1021, 1047, 1178
Euclidean
    algorithm, 1386, 1389
    geometry, 290
    metric, 229
Euler
    function, 1381
    module, 62, 179
    plane, 93
    point, 62, 1349
    space, 1349
Euler's identity, 1341
Euler, Leonhard, 62, 478, 509, 1350, 1364, 1497
Euler-Lagrange equation, 1188, 1189, 1227, 1482
Eulerian square, 1025
European score notation, 67
Eustachian tube, 1355
evaluation, 295, 1433
Evans, Bill, 880
event
    percussion -, 503
    time -, 556
evolution, 628
ex movere, 894
exact
    curve, 945
    sequence, split -, 1380
exactness, 1002
exchange
    of pitch and onset, 127
    parameter -, 133, 134
existence, 57, 327
    mathematical -, 145, 328, 340
    musical -, 340
experiment
    mental -, 550
    musicological -, 30, 31
    physical -, 29
experimental
    environment, 681
    humanities, 27
    material, 330
    natural sciences, 27
    strategy, 694, 701

experimentation, 813
experiments of the mind, 31
explanatory variable, 724
explicitness conceptual -, 21
exponentiable, 1429
exponential
    correspondence, 1464
    law, 1464
exposition, 249, 496, 794
expression, 335, 603, 759
    human -, 570
    instrumental -, 825
    of expression, 892
    rhetorical -, 570
expressive rules, dictionary of -, 615
expressivity
    pure -, 606
    rhetorical -, 556
extension, 306, 330, 337, 552
    allomorphic -, 443
    basic -, 427
    creative -, 201
    cyclic -, 207
    Galois -, 1460
    separable -, 1460
    strict -, 443
    topology, 430
exterior
    derivative, 973
    score, 571
extraction
    beat -, 1292
    monotone -, 1019
extraterritorial part, 591
extroversive semiosis, 329

F

f-morphism, 1411
F-to-enter level, 727
face, 969, 1445
    operator, 968
fact, 1067
facticity, 327, 345, 464
    finite - support, 338
    full -, 338
factor
    pressure decrease -, 1337
    strength -, 610
factory box, 805
faithful
    action, 1378
    address, 431
    functor, 1421
    local network, 980
    point, 431
False, 1433

'false' voice, 1315
family, 239
  covering -, 1430
  minimal cadential -, 455
  of violins, 828
  violin -, 241, 840
fanfare, 958, 1075
fantasy, artistic -, 570
faster uphill, 610
father, 619
Fauré, Gabriel, 1285
Feldman, Jacob, 607
Feldman, Morton, 250
Felver, Christopher, 1182
Fermat, Pierre de, 23, 1007
fermata, 550, 630, 632, 643
fermion, 1485
Ferrara, Franco, 1292
Ferretti, Roberto, xi
feuilleton, 635
feuilletonism, 751
FFT, 524
fiber, 611, 1375
  critical -, 755, 758
  group, 775
  product, 1387, 1424
    of local compositions, 138
  structure, 756
  sum, 1424
    of local compositions, 140
Fibonacci sequence, 340
Fibonacci, Leonardo, 59, 340
fibration, linear -, 758
fiction, 327, 465
fictitious performance history, 627
field, 597
  arpeggio -, 575
  calculation, 761
  diagonal -, 565
  discrete -, 760
  finite -, 786
  fundamental -, 591
  interpolation, 761, 764
  intonation -, 563
  of equivalence, 157
  of fractions, 1406
  operator, 652
  paradigmatic -, 125
  parallel articulation -, 567
  parallel crescendo -, 567
  parallel glissando -, 567
  performance -, 564, 568, 585, 983
  prime -, 1386
  selection, 803
  skew -, 1385
  tempo -, 562

  tempo-intonation -, 565
  tensor -, 1468
  vector -, 1245, 1451, 1468
  writing, 803
field extension, algebraic -, 1460
fifth, 62, 1350
  axis, 96
  coordinate, 1350
  sequence, 262
figuratio, 847
film music, 603
filtered diagram, 1411
filtering, input -, 760
final
  coherence, 769
  depth, 676
  retard, 607
  ritard, 896
  vertex, 660
finale, 791
finalis, 261
finality, 766
fine arts, 13, 153
finer, 1443
finger, 912
  space, 915
finger-based gesture, 1280
fingering, 248, 607, 622
finite, 1371
  completeness, 138
  cover topology, 353
  field, 786
  locally -, 1446
  monoid, 1376
  multigraph, 1375
finitely
  cocomplete category, 1425
  complete category, 1425
  generated, 1381, 1391
    abelian group, 306
Finscher, Ludwig, 825, 826
Finsler's principle, 145
Finsler, Paul, x, 145, 1094
Fiordilino, Emilio, 1471
first representative, 180
FIS, 204
Fisher, George, 1264
Fitting's lemma, 1396
Fitting, Hans, 1396
fixpoint, 1452
  group, 1378
flasque module complex, 304
flat, 109
  local network, 981
flatten, 74, 1111
flattening operation, 75, 270

Fleischer, Anja, xi, 484
Fleisher, Leon, 883
FLOAT, 44
Floris, Ennio, 1012
flow, 886, 1299, 1300
  box, 1455
  interpolation, 582
Flusser, Vilém, 848
flying carpet, 767
FM, 236, 1342
  -object, generalized, 239
  analysis, 236
  synthesis, 73
folding, 363
  circular denotators, 369
  colimit denotators, 367
  limit denotators, 366
foramen ovale recording, 524
force
  field, alteration -, 788
  modulation -, 466, 469
forces in physics, 534
foreground, 415
form, 43, 52–57, 1004, 1074, 1086
  bilinear -, 291
  circular -, 64
  circularity, 48
  colimit -, 57
  concept, 665
  concert -, 793
  contrapuntal -, 249
  coordinator, 55
  differential -, 975
  essentic -, 895
  ethnological -, 49
  Forte's prime -, 210
  functor, 55
  gestural -, 1086
  identifier, 55, 56
  large -, 1298
  limit -, 57
  list -, 809
  morphisms, wrap -, 331
  musical -, 6
  name, 43, 44
    empty -, 47
  normal -, 209
  of a symmetry, 113
  pointer character, 48
  powerset -, 56
  prime -, 210
  Rahn's normal -, 209
  regular -, 64
  semiotic, global -, 1440
  simple -, 56
    simplify to a -, 63

sonata -, 249, 496, 791
space, 55
Straus' zero normal -, 210
synonym -, 56
type, 55
typology, 55
form semiotics
  morphism of -, 1439
  category of -, 1092
formal
  complexity, 383
  structure, 801
formalism, Lie -, 658
formant, 238
  manifold, 238
  open - set, 238
formoid, radical -, 928
forms
  category of -, 57
  diagram of -, 57
  ordering on -, 75–83
formula, 907, 1435
  Δ- -, 926
  atomic -, 1435
  cadential -, 453
  generalized -, 927
  propositional -, 1435
  quantifier -, 1435
formulas, category of -, 926
Forte's prime form, 210
Forte, Allen, 203, 204, 209, 314
foundation chord, 439
foundations, 1028
four part texture, 826
Fourier analysis
  cochlear -, 1365
  decomposition, 70
  ideology, 233
  paradigm, 232
  representation, 740, 832
  theorem, 10
  theory, 925
  transform, 1344
Fourier series, 2D -, 1212
Fourier's theorem, 1340
Fourier, Jean Baptiste Joseph, 422
fractal, 59, 162, 782
  composition tools, 115
  principle, 799
fractions, field of -, 1406
frame, 585, 596, 802, 1346, 1438
  change of -, 1058
  composed -, 802
  simple -, 802
  space, 55, 1087
  structure, 591

wavelet -, 236
framework, 806
    application, 665
    concept -, 3, 5, 9
    hermeneutical -, 11
free
    action, 1378
    autograph, 1052
    Chopin rubato, 625
    commutative monoid, 1376
    group, 1379
    jazz, 548, 1177
    locally -, 1413
    module, 1392
    monoid, 1376
    variable -, 1436
free jazz, v, 13, 900
freedom of choice, 543
Frege, Gottlob, 849, 859, 867, 898, 937, 1180
French gesture theory, 894
frequency, 61, 70, 1340, 1341
    band, 526
    beat -, 1365
    fundamental -, 1340
    modulation, 236, 1111, 1342
    modulation -, 236, 833
    of variable inclusion, 732
Freud, Sigmund, 528
Friberg, Anders, 610
Fripertinger, Harald, xi, 167, 190, 211, 309, 310, 1382
Frost, Robert, 273
Frydén, Lars, 610, 621
Frye, Roger, 1024
fugue, 159, 199
full
    address, 431
    functor, 1421
    model, 726
    point, 431
    subcategory, 1422
    subcomplex, 1445
fully faithful
    address, 431
    functor, 1421
    point, 431
function, 859, 898, 1181, 1372
    A-addressed -, 288
    autocomplementarity -, 419, 519
    Bessel -, 1344
    Borel -, 1488
    characteristic -, 335, 1375
    common-note -, 204
    Euler -, 1381
    generic -, 804
    gradus suavitatis -, 1364, 1497
    Green -, 1189, 1471

horizontal poetical -, 781
index -, 1390
interval -, 204
inverse -, 1373
level -, 269, 1374
of a symmetry, 113
poetical -, 16, 116, 212, 241, 248, 774, 781
theory, 265, 437
tonal -, 249, 263, 447
value, tonal -, 447
vertical poetical -, 782
weight -, 1253
function harmony, 32
functional, 1372
    gesture, 1034
    object, 1036
    programming, 801
    relation, 1196
    semantics, 445
functor, 1179, 1421
    address -, 141
    constant -, 1422
    contravariant -, 1421
    covariant -, 1421
    faithful -, 1421
    form -, 55
    full -, 1421
    fully faithful -, 1421
    gestural form -, 1087
    global section -, 1412
    God -, 1179
    homology -, 1476
    limiting -, 962
    local -, 926
    logical -, 946
    module -, 143
    nerve -, 988, 1445
    of orbits, 1417
    open -, 1416
    open covering of -, 1416
    representable -, 1423
    resolution -, 294
    simplicial -, 988
    support -, 257
    topological -, 939
functorial
    composition, 1180
    global composition, 256
    local composition, 101
functors, adjoint -, 907
fundamental
    activities, 3, 6
    chord, 439
    field, 591
    group, 920, 1475
    groupoid, 920, 1475

note, 441
period, 1340
pitch, 437
scientific domain, 5
series, 115
space, 588
Fundamental Lemma of Calculus of Variations, 1471
fushi, 13, 342
Fux rules, 541, 1235
Fux, Johann Joseph, 523, 541, 840, 1367
fuzziness, 437
fuzzy
    concept, 164
    conceptualization, 375
    logic, 336
    set, 163, 1029
    theory, 159

G

G-prime form, 208
Gabriel, Peter, 153, 925
Gabrielsson, Alf, 604, 607
gait, 1305
Galilei, Galileo, 27, 29, 32, 548
Galois
    correspondence, 1093
    extension, 1460
    group, 1460
    theory, 1087, 1460
Galois, Evariste, x
game, chironomic -, 1327
Garbers, Jörg, xi, 665, 1088, 1441
Garbusow, Nikolai, 181
Gast, Peter, 1033
gate function, hippocampal -, 528
Geary, James, 1022
Gegenklang, 265, 457
Geisser, Heinz, 1176, 1295
Gell-Mann, Murray, 145
genealogical difference, 756
genealogy
    conceptual, 64
    of denotator concept, 41
    poietic, 128
general linear algebra, 1408
    pause, 644
    position, 322
General Midi, 235
general position, 106, 174
    musical meaning of -, 106
generalization, 1026
generalized formula, 927
generated, finitely -, 1381, 1391
generating local composition, 106
Generative Theory of Tonal Music (=GTTM), 255, 376
generator

sound -, 700
time -, 775
generators for a topology, 1463
generic
    composition, 174
    function, 804
    linear visualization, 362
    point, 228, 269, 1413
    score, 548
genotype, 782
Gentilucci, Maurizio, 871
geocentricity, 1175, 1177
geodesic, 241
geographic
    information system (GIS), 667
    orientation, 38
geometric
    classification, 177, 989
    constraint, 913
    constraints, 912
    coordinate, 1341
    editing, 783
    intuition, 1036
    language, 893
    morphism, 946
    parameter, 832
    realization, 988, 1446
    representation, 783, 892
geometry
    analytical -, 148
    algebraic -, 148, 165, 550, 1009
    Euclidean -, 290
germ, 267, 1451
    rhythmic -, 127, 266
Germain, Sophie, 1017, 1033
germinal melody, 220, 221, 792
gestalt, 90, 167, 271, 383, 405
    abstract -, 391
    auditory -, 396
    cardinality of a -, 391
    global -, 251
    musical -, 90, 127
    paradigm, 672
    psychology, 90
    small -, 398
    specialization, 402
        category, 403
    stability, 226
gestoid, 923
gestural
    action, 912
    analysis, 1274
    constraint, 617
    coordinate, 893
    creativity, 1183
    diagram, 1036

form, 1086
groove, 1303
morphism, 1287
performance, 893
philosophy, 893
presheaf, 962, 1086
primitive, 901
rationale, 753
semantics, 753
similarity, 1289
strategy, 1270
symbol, 912
theory, 1182
transformation, 1272
vibration, 893
gestural form
    coordinator, 1087
    functor, 1087
    identifier, 1087
    space, 1087
    type, 1086
gestural interaction of both hands, 1299
gesturalize, 1154, 1162
gesturally representable, 962
gesture, 604, 607–609, 891, 895, 898, 909, 914, 1001,
        1005, 1032, 1036, 1067, 1176, 1180, 1245, 1300
    A-addressed -, 961
    A-parametrized -, 1317
    archetypal -, 1269
    arm-based -, 1274
    bicategory, 937, 946
    categorically natural -, 918
    characteristic -, 956
    circular -, 1269
    complex -, 1042
    continuous -, 817
    deep-frozen -, 912
    disabled -, 875
    discrete -, 817, 944
    dual -, 940
    elementary -, 817
    embodied -, 1274, 1279
    embodied musical -, 1263, 1264
    finger-based -, 1280
    functional -, 1034
    functor, discrete -, 944
    global -, 993
    high level -, 1041
    human -, 1314
    imaginary -, 1042, 1314, 1315
    instrumental -, 817, 902, 1318
    invisible -, 1302
    mathematical -, 1002, 1036, 1037, 1040
    melodic -, 1302
    mimic -, 893
    mirror -, 1270

normative -, 1305
objective -, 1042
of gestures, 894
of locally compact points, 983
of silence, 1327
of stemmata, 1256
orchestral -, 817
performative -, 1283
phenomenology, 848
physical -, 911, 1186, 1229, 1289
physical - curve, 912
semiotic, 1183
sketch of -, 1030, 1038
smartphone -, 1279
spectral -, 1318
speech -, 1311
subjective -, 1042
symbolic -, 911, 912, 1186, 1228, 1229, 1289
symbolic - curve, 912
technology, 1283
theory, 893, 1173
    French -, 894
topos, 945
vocal -, 1301, 1313, 1315
gesture disabled, 875
gesture-sound mapping, 1279
gestures
    of the conductor, 1285
    algebra of -, 1002
    category of -, 1289
    types of -, 1286
Get-Editor, 814
Get-View, 814
Giannitrapani, Duilio, 525
Gianoli, Reine, 735
Gilels, Emil, 622
Gilson, Etienne, 12
Ginsburg, Carlo, 1040
GIS, 204, 315
    structure, 203
Glarean, 261
glide reflection, 1402
glissando, 67, 551, 567, 594, 817, 1350
global, 245
    affine functions, module of -, 355
    agogics, 629
    AST, 314–316
    composition, 140, 831, 987, 1466
        cochain complex of a -, 307
    digraph, 996
    enumeration theory, 309
    form semiotic, 1440
    functorial composition, 256
        morphism, 274
    gestalt, 251
    gesture, 993

hypergesture, 1223
molecule, 292
molecules, morphism of -, 292
morphisms, 246
network, 990
object, 245
objective composition, 252
  morphism, 274
performance score, 598
predicate, 454
score, 250, 783
section, 288, 1424
  functor, 1412
skeleton, 996
slope, 677
solution, 1454
standard composition, 293
tangent composition, 557
technical parameter, 835
tension, 677
theory, 220
threshold, 675
variational principle, 1231
world-sheet, 1224
Zarlino network, 992
globalization
  metrical -, 99
  orchestral -, 555
gluing, 1032
God, 1004, 1016, 1177, 1186
  functor, 1179
Godøy, Rolf Inge, 902
Goethe, Johann Wolfgang von, 123, 163, 325, 748, 828
Goffman, Ervin, 853
Goldbach conjecture, 30
Goldbach, Christian, 30
golden section, 59
Goldin-Meadow, Susan, 899
Goldstein, Julius, 1361
Göller, Stefan, xi, 362, 1103, 1088, 1441
Goswitz, Robert, vi
Gottschewski, Hermann, 29
Gould, Glenn, 550, 608, 694, 702, 751, 752, 899, 909
GPL, xi
GPS, 598
gradus suavitatis function, 1364, 1497
Graeser, Wolfgang, 113, 115, 203, 249, 890, 1182
Graffman, Gary, 883
Gram identity, 292
grammar
  locally linear -, 660
  performance -, 615
  rule-based -, 615
grand unification, 463
granddaughter, 660
grandmother, 660

graph, 1372, 1375, 1458
  category, 939
  directed -, 907, 914, 1375
  of a FM-denotator, 73
  operation -, 1138
  Riemann -, 677
  unordered -, 1458
  weighted -, 239
graphical
  interface design, 361
  MOP, 813
Grassmann scheme, 1416
Greek prosody, 1324
greeking, 363, 381, 1103
Green function, 1189, 1471
Gregorian chant, 509 1321
Gregorian musical figures, 1328
Greimas, Algirdas Julien, 774–776
grid
  dactylus -, 216, 217
  driving -, 787
  vector
    horizontal -, 788
    vertical -, 788
Grondin, Jean, 1050
groove
  effect, 788
  gestural -, 1303
Grothendieck
  topology, 149, 353, 1430
  topos, 1431
Grothendieck, Alexander, x, xii, 145, 149, 153, 351, 353,
    358, 850, 908, 1031, 1038, 1173, 1177, 1179, 1430
ground class, 313
group, 1378
  affine counterpoint -, 391
  algebra, 925
  automorphism -, 144, 1391
  cohomology -, 1447
  contrapuntal -, 115
  control -, 775
  cyclic -, 1381
  enharmonic -, 426
  fiber -, 775
  fixpoint -, 1378
  free -, 1379
  fundamental -, 920, 1475
  Galois -, 1460
  Hegel -, 1066
  homomorphism, 1378
  isomorphism, 1378
  isotropy -, 1378
  Klein -, 206, 451
  linear counterpoint -, 391
  opposite -, 1378
  p-Sylow -, 1380

paradigmatic -, 390
  of isometries, 394
product -, 1380
quotient -, 1379
rhythmical -, 810
simple -, 1379
Sylow -, 179
symmetric -, 1378
symmetry -, 144, 180, 470, 672
theory, 157, 212
torsion -, 1381
group-theoretical method, 198, 205
grouping, 376, 415, 608, 629
  bar -, 713
  concept, 250
  hierarchical -, 611
  instrumental -, 599
  metrical -, 248
  of sounds, 74
  rules, 610
  stemmatic -, 633
  structure (=G), 377
  time -, 99
groupoid, 922
  fundamental -, 920 1475
Growth Point, 854
GTTM, 255, 376, 619
Guérin, Michel, 850
GUI, network -, 1096
Guitart lemma, 1006
Guitart, René, 1006
gyri, Heschl's -, 1360

H

Hölderlin, Friedrich, 9, 116
Hüllakkord, 431
Haegi, Hans, 291
hair cell, 1356
Hajós group, 310, 314
Hajós, Gyorgy, 310
Halle, Morris, 234
Halsey, George, 309, 313
Hameer, 1305
Hamilton operator, 1493
Hamilton's variational principle, 1243, 1481
Hamilton, William, 1385
Hamiltonian, 658
hammer, 1354
Han singing, 1321
hand, 855, 912, 915
  impairment, 881
  left -, 1289
  pianist's -, 911, 912
hand's position, 1192
hands
  gestural interaction of both -, 1299

shapes of the -, 1310
handshape, 1311
hanging orientation, 108, 509
Hanslick, Eduard, 15, 248, 251, 775, 828, 1181
harkat, 1306
HarmoRUBETTE®, 744
harmolodic, 794
harmonic
  coherence, 447
  knowledge, 485
  logic, 449
  minor, 479
  morpheme, 449
  morphology, 908
  motion, 415
  path, 482
  progression, 128
  semantics, 436
  strip, 253, 262, 442
  tension, 481, 482
  topology, 443
  weight, 482, 647
harmonic minor
  scale, 472, 474
  tonality, 460
harmonical-rhythmical scale, 794
harmony, 90, 181, 523
  complete -, 827
  jazz -, 276
  Keplerian -, 30
  Riemann -, 263
  rubette for -, 1098
HarmoRUBETTE®, 448, 481, 648, 674, 714
Harnoncourt, Nicolas, 751, 754
Harris, Craig, 209
Hashimoto, Shuji, 607, 613
Hasse diagram, 218, 1374
  specialization -, 220
hat, Mexican -, 1345
Hatten, Robert S., 871, 889, 898, 899, 909, 950, 979, 1265
Hauptmann, Moritz, 436
Hausdorff
  locally compact -, 983
  topology, 1445
Hausdorff, Felix, 1445
Hausegger, Friedrich von, 1293
Hawking, Stephen, 1187
hayashi, 13
Haydn, Joseph, 241, 826, 827
Hazlitt, William, 685
head, 914, 1457
head voice, 1315
heartbeat, 607
Hebb, Donald O, 719
Hegel

body, 1066, 1070
box, 1072
group, 1066
Hegel, Georg Wilhelm Friedrich, 35, 748, 781, 1040,
    1066, 1174
Heidegger, Martin, 854
Heijink, Hank, 761
helicotrema, 1355
Helmholtz, Hermann von, 509, 1364, 1365
Hemmert, Werner, xi
Hentoff, Nat, 603
Hermann, Jakob, 1048
hermeneutics, 1051
    unicorn of -, 13
Hertz, 1340
Hervé, Jean-Luc, 817
Heschl's gyrus, 524, 1360
Hess system, 524
Hesse, Hermann, v, 164, 573, 1177
Hewitt, Edwin, 309, 313
hexagram of Pascal, 1023
hexameter, 213
Heyting
    algebra, 103, 335, 930, 934, 1433
    category of - algebras, 930
    logic, 436, 1460
Heyting, Arend, 335, 436, 1433
Hichert, Jens, 517, 534, 536, 538, 543
hidden symmetry, 114
hierarchical
    decomposition, 710, 720
    grouping, 611
    organism, 249
    smoothing, 708
hierarchy, 250, 555, 631
    cellular -, 590, 596, 1254
    closure, 588
    deformation, 658
    metrical -, 375
    of performance development, 622
    parallel -, 591
    performance -, 556
    phrasal -, 1311
    piano -, 593
    space -, 588
    standard -, 590
    tempo -, 624
    violin -, 593
high level gesture, 1041
Hilbert space, 1486
Hindemith, Paul, 123, 414, 422
Hindustani
    musician, 1305
    vocalist, 1301
Hintergrund, 415
hippocampal

gate function, 528
    memory function, 528
hippocampus, 524, 528
histogram, 716
historical
    approach, 465, 472
    dimension of music, 91
    dynamics, 221, 223
    instrumentation, 324
    localization, 223
    musicology, 328
    process, 627, 634
    rationale, 826
    reality, 487
historicity in music, 220
history, 5
    development -, 612
    encapsulated -, 556
    of mathematics, 1040
    of music, 6
hit point, 582
    problem, 580
Hjelmslev, Louis, 14, 16, 328, 849, 1183
Ho, Jocelyn, vi, vii
Hofmann, Ernst Theodor Amadeus, 248
Hofstadter, Douglas, 1022
holomorphic, 1465
homeomorphism, 1444
homology, 1475
    functor, 1476
    module, 966, 1475
    network -, 986
    singular -, 965, 1476
    theory, 967
homology module, hypergestural -, 970
homomorphism
    boundary -, 966
    diaffine -, 1396
    dilinear -, 1392
    group -, 1378
    Lie algebra -, 1408
    linear module -, 1391
    monoid -, 1376
    ring -, 1385
    structural -, 1385
homotopy, 916, 920, 957, 1181, 1185, 1289, 1447, 1474
    relative -, 1447
    category, 1474
    commutative, 1474
    equivalence, 1474
    inverse, 1474
    theory, 1474
    type, 1474
Honing, Henkian, 547
Horace, 621
horizontal

grid vector, 788
  poetical function, 781
  poeticity, 214
Horowitz cluster, 735, 739
Horowitz, Vladimir, 730, 735, 736, 738, 768, 899, 909
hue, 1115
Hul, Bopha, 1276
human
  conceptualization, 145
  expression, 570
  gesture, 1314
  precision, 623
human phonatory system, 1314
humanism, 828
humanities, 164, 225
  experience in the -, 31
  experimental -, 27
Husmann, Heinrich, 1366
hypergestural
  homology module, 970
  singular homology, 1236
hypergesture, 870, 894, 915, 937, 959, 967, 985, 1036,
    1184, 1185
  global -, 1223
  non-singular -, 1247
  parallel -, 1245
  spatial -, 917
  vocal -, 1316
hypermedia, 38
hypernetwork, 985
hyperouranios topos, 21, 38
hyperset theory, 1094
hypothesis, mimetic -, 1264, 1276

I

I, 22
I, the, 892
ICMC, 197, 612
icon, instrumental -, 784
Ide, Takefumi, 1274
idea
  compositional -, 322
  musical -, 774
ideal, 1386
  left -, 1386
  right -, 1386
ideas, intermediate -, 1010
idempotent, 1376
  component, 1377
identical reproduction, 891
identification
  conceptual -, 229
  enharmonic -, 425
  esthesic -, 248
identifier, 1438
  form -, 55, 56

gestural form -, 1087
identity, 405, 408, 1419
  of a point, 147
  abstract, 14
  Euler's -, 1341
  Jacobi -, 1408
  of a work, 14
  resolution of the -, 1494
  slice, 275
ideology, Fourier -, 233
IL, 1434
image, 1375
  denotator -, 59
  direct -, 1411
  inverse -, 1375
imaginary
  gesture, 1042, 1314, 1315
  time, 957, 1189, 1297, 1300
  world, 1058
imagination, 5, 1032
imitation, 405, 407
immanent analysis, 383
impairment, hand -, 881
imperfect, consonance, 522, 532, 541
implementation, 627
implication, 346, 1432, 1433
importance, relative -, 648
improvisation, 40, 1278, 1295
  jazz -, 179
  melismatic -, 1303
improvised music, listen to -, 1295
in absentia, 16, 789
in praesentia, 16, 789
in-time music, 817
inbuilt performance grammar, 753
incarnation, 847
included, literally -, 208
inclusion
  abstract -, 208, 210
  literal -, 210
incomplete semiosis, 330
incorrect, politically -, 752
indecomposable, 1395
  space, 588
independence, cognitive -, 179
index
  cycle -, 1382
  diachronic -, 223
  function, 1390
  set, 1373
indiscrete
  interpretation, 254
  topology, 1443
individual variable, 1435
ineffability, 22, 570
ineffable, 1285

infinite, 1371
  interpretation, 259
  message, 751
  performance, 549
infinitely small, 569
infinitesimal, 638
information, 35
  paratextual -, 685
  system, geographic -, 667
InfoRUBETTE®, 668
ingenium, 1012
ingenuity, 1015
inharmonic, 675
inharmonicity, 237
inheritance, 628, 802
  principle, biological -, 627
  property, 395
initial, 1424
  articulation, 578
  condition, 1452
  design matrix, 724
  moment, 574
  performance, 586, 597
  set, 573, 586, 596, 1254
    polyhedral -, 580
  value, 562
initial set, distance to an -, 580
initialization, dynamical -, 578
injective, 1372
  module, 1407
inlet, 805
inner
  derivation, 1408
  ear, 1355
  logic, 623
  score, 548, 571, 1295
innervation, mimic -, 891
input
  filtering, 760
  real-time -, 783
inspector, chord -, 675
instance, 802
instantiation, 802
instinctive activity, 623
instrument
  makeshift -, 1279
  name, 69
  space, 597
instrumental
  condition, 701
  expression, 825
  gesture, 817, 902, 1318
  grouping, 599
  icon, 784
  parameter, 832, 835
  technique, 832

variety, 555
  vector, 835
  voice, 220
instrumentation
  historical -, 324
  orchestra - denotator, 69
instrumentum, 38
INTEGER, 44
integer, 1385
integral
  curve, 1245, 1453
  of perspectives, 325
  surface, 1456
integrated serial motif, 195
integration
  Lebesque -, 1488
  method, 683
intellectuality
  mathematical -, 1006
  musical -, 1006
intelligence, 873
intensification, 799
intension, 330, 427
  basic -, 427
  topology, 429
intensity, 608
inter-corporeal, 1274
inter-period coherence, 769
inter-sensoriality, 1274
inter-sensory, 1264
interaction
  interpretative -, 720
  matrix, 709
interactive control, 813
interface design, graphical -, 361
interictal period, 524
interior, 1443
interiorization, 893
interlude, 689
intermediate
  ideas, 1010
  performance, 622
internal
  category, 1464
  movement, 897
  structure, 254
interpolation, 570, 760
  field -, 761, 764
  flow -, 582
interpretable, 989
  composition, 304
    automorphism group of -, 306
  molecule, 292, 316
interpretation, 220, 222, 549, 989
  discrete -, 254
  indiscrete -, 254

infinite -, 259
iterated -, 258
just triadic degree -, 265
metrical, 268
motivic -, 271, 384
of a local composition, 258
of weights, 659
rhythmical -, 268
semantic -, 490
silly -, 254
singleton -, 275
sketchy -, 622
tangent -, 558
tetradic -, 276
third chain -, 260
triadic -, 276, 450, 455, 466
triadic degree -, 261
interpretative
    activity, 246, 251, 252
    interaction, 720
interpreter, 892, 893
interspace, 569
    sequence, 192
    structure, 192
interval
    unordered p-space -, 207
    unordered pc -, 207
    class content vector, 207
    consonant -, 414, 527
    content, 207
    contrapuntal -, oriented -, 509
    cul-de-sac -, 538
    cyclic - succession, 207
    dichotomy, 518
    dissonant -, 527
    function, 204
    multiplication, 512
    ordered p-space -, 207
    ordered pc -, 207
    succession, 207
        cyclic -, 208
        m^{th} -, 208
    successive -, 526
    time -, 70
    vector, 208, 210
interval-class vector, 204
intonation, 562, 563
    curve, 563
    field, 563
intratextual, 329
introversive semiosis, 329
intuition
    geometric -, 1036
    mathematical, 1027
    musical -, 202
intuitionistic logic, 443, 934, 1434

invariance
    transformational -, 226, 271
    vector, 208
invariant pcset, 208
invent problems, 1042
invention, 1001
    method of -, 1039, 1044
inversa, 692
inverse, 1378
    ethnology, 754
    function, 1373
    homotopy -, 1474
    image, 1375
    left -, 1378
    performance theory, 611, 651, 756
    right -, 1378
inversion, 117, 247, 262
    chord -, 420
    real, 124
    retrograde -, 61, 120
    tonal, 124
    tonal -, 788
inverted weight, 681
invertible, 1385
invisible gesture, 1302
IRCAM, v, 238, 801, 1314
irreducible, 589
    component, 269
    degree system, 457
    topological space, 1413
iso, 1420
isometry, 250, 1449
isomorphic, 1378
    category, 1421
isomorphism, 1420
    group -, 1378
    monoid -, 1376
    ring -, 1385
isomorphism classes
    of local rhythms, 181
    of chords, 180
isotropy group, 1378
isotypic tesselation, 309
ISPW, 761
Issigonis, Sir Alec, 1044
istesso tempo, 555
Italian conducting school, 1292
iterated interpretation, 258

J

Jackendoff, Ray, 250, 254, 376, 379, 619, 721
Jacobi identity, 1408
Jacobi, Carl, 1408
Jacobian, 1190, 1247, 1451
Jacobson, Nathan, 1395

Jakobson, Roman, 16, 116, 157, 212, 222, 234, 241, 250, 329, 774, 781, 1050
James, William, 873
Jandl, Ernst, 799
Japanese philosophy, 1266
Jaques-Dalcroze, Émile, 875
Jauss, Hans Robert, 155
Java, 666
Java2D, 764
jazz, 12, 13, 40, 179, 571, 909
    American -, 442
    CD review, 342
    free -, 13, 548, 900, 1177
    harmony, 276
    improvisation, 179
    lead-sheet notation, 438
JCK, 342
Jefferson, Blind Lemon, 885
jnd, 229
Johnson, Tom, 488, 789
join, 1433
journalistic criticism, 730
joystick, 895
Józef Marja Hoene-Wroński, 319
JSynNote, 1111
Julia set, 162
Julia, Gaston, 162
Junod, Julien, 543, 1097
Jurek, Thom, 877
just, 93
    chromatic octave, 96
    class chord, 94
    modulation, 475
    scale, 96
    triadic degree interpretation, 265
    tuning, 1350
just-tempered tuning, 1351
justest
    scale, 266
    tuning, 460
juxtaposition, 61

K

K-net, 898, 924, 1182
k-partition, 310
Köhler, Egmont, 189
Körpersinn, 1182
Kagel, Maurizio, 127, 325
kairos, 826
Kaiser, Joachim, 248, 495, 752
Kan, Daniel, 907
kansei, 607, 613
Kant, Immanuel, ix, 9, 21, 29, 38, 145, 1043
Karajan, Herbert von, ix, 577, 608, 783, 1292
Karg-Elert, Sigfrid, 115, 415
Kashalkar, Ulhas, 1306

Katsman, Roman, 847
Kayser, Hans, 1177
Kendon continuum, 854
Kendon, Adam, 853, 899, 909, 1180, 1301
Kepler, Johannes, 114
Keplerian harmony, 30
kernel, 597, 1379
    Naradaya-Watson -, 709
    Priestley-Chao -, 744
    smoothing, 708, 721
    smoothing -, 708
    symbolic -, 585
    view, 680
kernel (of a performance cell), 1254
key, 784
    function of music, 529
    musicogenic -, 529
    signature, 631
keyboard, 915
    piano -, 912
Keyser, Cassius, 1052
khyal, 1309
killing, address -, 168
kindred, 240
kinesthetic awareness, 1281
kinetic model, 1306
Kircher, Athanasius, 198
Klangrede, 17, 827
Klein group, 206, 391, 451
Klumpenhouwer, Henry, 898, 980, 1182
Klumpenhouwer network, 924
knot in FM synthesis, 73
knowledge, 35, 361
    crash, 339
    harmonic -, 485
    hiding, 198, 362
    human -, 5
    management, dynamical -, 329
    ontology, 345
    private -, 27
    space, 9, 27
    theory of -, 1005
Koenig, Thomas, 687
Kollmann, August, 827
Kopiez, Reinhard, 237, 604, 605
KORG, 1346
Kronecker delta, 1392
Kronecker, Leopold, 1392
Kronman, Ulf, 607, 896
Krull, Wolfgang, 1396
KTH school, 617, 621
Kubalek, Antonin, 735
Kunst der Fuge, 115, 203, 249, 608, 688, 699
Kuriose Geschichte, 628, 629, 634, 701, 711
Kuratowski closure operator, 1237, 1443
Kurth, Ernst, 1293

Kurzweil, 701

L

l, 1348
λ-*abstraction*, 803
l-adic cohomology, 379
L-system, 782
Lüdi, Werner, 548, 791
Lévi-Strauss, Claude, 487
Laban, Rudolf, 874
laboratory
  conceptual -, 30
  distributed -, 32
Lagrange
  density, 1482
  potential, 1218
Lagrangian, 658, 1481
  action, 1188, 1481
  density, 1185
Lakoff, George, 859, 1181
Lamé, Gabriel, 1027
Landry, Elaine, 908
landscape, emotional -, 241
Langer, Jörg, 605
Langer, Susan, 604
language, 17, 853, 893
  common -, 22
  denotator -, 595
  geometric -, 893
  natural -, 1034
langue, 17
Laplace operator, 1471
large
  form, 1298
  performance of a - orchestra, 626
largest coefficient, 733
laryngeal movements, 1316
larynx, 1316
Lasker, Emanuel, 1027
Latin square, 1024
lattice, 1433
law
  exponential -, 1464
  Weber-Fechner -, 1348
Lawrence, David Herbert, 751
Lawvere, William, 149, 357, 934
layer
  action-to-cognition -, 870
  cognitive -, 869
  RUBATO®-, 668
layers
  of reality, 10
  time -, 1299
Lazier, Rebecca, 933
lazy path, 1375
LCA, 525

Le Guin, Elizabeth, 1264
lead-sheet notation, 441, 571
leaf, 949
Leap Motion, 1143
learning
  by doing, 32
  process, 556
leaves of a stemma, 628
Lebesque integration, 1488
left
  action, 1378
  adjoint, 1423
  coset, 1379
  hand, 1289
  ideal, 1386
  inverse, 1378
legato, 645
LEGO, 782
Lehmann, Andreas, 743
Leibnitz, Gottfried Wilhelm, 465, 743, 869, 1178
Leitfaden, vii
λεκτον, 16
Leman, Marc, 902, 1182
lemma
  Fitting's -, 1396
  Guitart -, 1006
  Yoneda's -, 938
length, 268, 1395
  minimal -, 689
  of a local meter, 97
  path -, 1375
lens space, 922
LEP, 196
Lerdahl, Fred, 237, 238, 240, 250, 254, 376, 379, 619, 721
Les fleurs du mal, 219, 799
level, 269
  connotative -, 16
  denotative -, 16
  F-to-enter -, 727
  function, 269, 1374
  meta -, 17
  metrical -, 377
  neutral -, 211, 956
  object-, 17
  sound pressure -, 1348
levels of reality, 10
Levelt, Wilhelm, 1364, 1367
Levi, Beppo, 1040
Lewin, David, 70, 203, 204, 309, 315, 409, 889, 898, 908, 937, 938, 951, 956, 980, 1177, 1182, 1236
Lewis, Clarence Irving, 570
lexical, 344
lexicographic ordering, 50, 76, 1373
Leyton, Michael, xi, 773, 775
LH, 629
library, 51, 367

Lidov, David, 1265
Lie algebra, 1408
    homomorphism, 1408
    linear -, 1408
    bracket, 1246, 1408, 1455
        affine -, 444
    derivative, 1455, 1469
    formalism, 658
    operator, 638
Lie type operator, 1255
Lie, Sophus, 1408, 1455
Ligeti, György, 30, 1179, 1295
limbic
    structure, 524
    system, 528, 606, 1361
limit, 1424
    circular -, 65
    form, 57
    ring, 1387
    topology, 1444
limited
    modulations, 481
    transposition, 126
        mode with -, 126
limiting functor, 962
line, 1375
linear
    (in)dependence, 1393
    algebra, special -, 1408
    case, 205
    category, 923
    combination, 1391
    counterpoint group, 391
    dual, 1397
    fibration, 758
    Lie algebra, 1408
    module homomorphism, 1391
    ordering, 1373
        on a colimit, 77
        on a limit, 77
        on finite subsets, 77
    representation, 1393
    visualization
        generic -, 362
        metrical -, 362
linear ordering among denotators, 50
linearization, 923
linguistics, 160
    structuralist -, 250
Lipschitz, locally -, 1452
Lipschitz, Rudolf, 1452
LISP, 439
list form, 809
listen to improvised music, 1295
listener, 11, 13
listening

articulated -, 249
modes of -, 1281
music -, 1353
procedure, 610
Liszt, Franz, 16, 18, 495, 1076
literally included, 208
Lloyd, Sam, 1025
Lluis Puebla, Emilio, xi
local, 245
    action, 1038
    character of a contrapuntal symmetry, 533
    composition, 75, 89, 90
        commutative -, 105
        embedded -, 106
        functorial -, 101
        generating -, 106
        morphism, 105
        objective -, 90, 979
        sequence of a -, 192
        wrapped as -, 91
    compositions
        coproduct of -, 105
        fiber sum of -, 140
        product of -, 104
    functor, 926
    meter, 97
        length of a -, 97
        period of a -, 97
    meters, simultaneous -, 500
    morphism, 1411
    network, 980, 1253
        faithful -, 980
        flat -, 981
    optimization, 676
    orientation, 264
    Para-meter, 267
    performance score, 595, 1254
    rhythm, 99, 106
    ring, 1395
    score, 250, 783
    solution, 1452
    standard composition, 293
    symmetry, 533, 534
    technical parameter, 835
    threshold, 675
local topography, 17
local-global patchwork, 250
locality principle, 762
localization, 1406
    historical -, 223
    of epilepsy focus, 524
    of musical existence, 21
locally
    closed, 1445
    compact, 1464
    finite, 1446

free, 1413
    linear grammar, 660
    Lipschitz, 1452
    ringed space, 1411
    trivial structure, 252
locally almost regular manifolds, category of -, 974
locally compact Hausdorff, 983
Lochhead, Judith, 1264
Locke, John, 1010
locus, Riemann -, 676
logarithmic perception, 550
LoGeoRUBETTE®, 668
logic, 345, 1039, 1067
    absolute, 145
    classical -, 1434
    fuzzy, 336
    harmonic -, 449
    Heyting -, 436, 1460
    inner -, 623
    intuitionistic -, 443, 934, 1434
    musical -, 264
    of orbits, 200
    of toposes, 227
    performance -, 556
    performing -, 774
    predicate -, 436
    topos -, 930
logical, 1430
    algebra, 1433
    connective symbol, 1432
    functor, 946
    motivation, 639
    switch operator, 60
    time, 502
loop, 1375
Lord, John, 189
Loria, Gino, 1048
loudness, 44, 67, 608, 1348
LPS, 595, 621, 1254
Lubet, Alex, vi
Luening, Otto, 250
Lussy, Mathis, 615

M

m^{th} interval succession, 208
M-theory, 1485
M.M., 552, 562
Mälzel, Johannn Nepomuk, 341
Möbius
    bottle, 558
    strip, 476
Möbius strip, 252, 263, 442
Müller, Stefan, v, 1088
Mälzel, Johannn Nepomuk, 552, 562, 571
Mälzel metronome, 28, 1347
Möbius strip, 451, 781

Müller, Stefan, xi, 910
Mac Lane, Saunders, 908, 1035, 1178
Mac OS X, 665, 669
Machado, Antonio, 1004
machine, 1037
    Γ-, 1038
    performance -, 703
    precision, 623
    Turing -, 553
MacLean, Paul, 528
macro, 270
    -event, 74
    germ, 270
macrogesture, 1311
MacroScore, 1106
Mahler, Gustav, 1287
Maiguashca, Mesias, 59, 115
Majithia, Roopen, 886
major, 122, 541
    dichotomy, 519, 543
    mode, 448
    scale, 473, 474
        dominant role of -, 541
    third, 1350
    tonality, 460, 478
        bigeneric -, 449
major-minor problem, 122
makeshift instrument, 1279
making music, 22
Mallarmé, Stéphane, 850
Malt, Mikhail, 811
manifold, 251, 913
    J-, 1062
    almost regular -, 974
    categorical -, 993
    differentiable, 1466
    formant -, 238
    musical -, 241
    of opinions, 828
    semantic -, 241
    tangent -, 1467
    with boundary, 1467
Manin, Yuri, 1177, 1181
Mannone, Maria, vi, vii, 1219
Mansouri, Baya, 1060
map, 1372
    characteristic -, 1428
    coboundary -, 1447
    performance -, 586
    refinement -, 275
    simplicial -, 1445
mapping, gesture-sound -, 1279
maquette, 811
Marceau, Marcel, 865, 875
Marek, Ceslav, 248, 910
marked

counterpoint dichotomy, 518
  class, 518
dichotomy
  autocomplementary -, 518
  class, 518
  rigid -, 518
  strong -, 518
  interval dichotomy, 518
Marquis, Jean-Pierre, 908
Marrou, Henri-Irénée, 1040
Marx, Adolf Bernhard, 496
Maschke, Heinrich, 1395
Mason, Robert, 108, 466
Mason-Mazzola theorem, 109
Mason's theorem, 109
mass-spring, 1346
Massinger, Philip, 679
master, concert -, 626
matching, 761
  of structures, 718
  score-performance -, 761
material
  change of -, 813
  experimental -, 330
  musical -, 811
  of music, 90
  time, 502
Math-motif, 406
mathêmata, 1015
Mathematica®, 769
mathematical
  creativity, 1002
  existence, 145, 328
  gesture, 1002, 1036, 1037, 1040
  intellectuality, 1006
  intuition -, 1027
  model, 464
  morphism, 282
  object, 1021
  overhead, 512
  pulsation, 1002, 1007, 1026, 1033, 1041, 1042, 1058,
    1059, 1064
mathematically equivalent morphisms, 282
mathematics, 5, 160
  creative -, 1033
  history of -, 1040
matrilineal, 628
  scheme, 626
matrix, 898, 1392
  category, 1420
  comparison -, 206
  design -, 724
  initial design -, 724
  interaction -, 709
  of checkboxes, 1110
  product, 1392

Riemann -, 447, 481, 675
  value -, 766
  verse -, 213
matrix-like sketch, 1057
Matterhorn, 151
Mattheson, Johann, 248, 827, 1365
MAX, 115, 210, 441, 789
Max/MSP, 1127
maximal, 313
  meter
    nerve topology, 379
    topology, 268, 378
  structure content, 1362
Maxwell's equations, 973, 1484
mayamalavagaula, 543
Mayer, Günther, 221
Mazur, Barry, 1032
Mazzola, Christina, xii
Mazzola, Guerino, vii, 218, 502, 503, 612, 721, 783, 890,
    900, 1035–1038, 1045, 1163, 1173, 1176, 1295
Mazzola, Guerino, 1181
Mazzola, Patrizio, vi
Mazzola, Silvio, xii
McCullogh, Karl-Erik, 854
McNeill, David, 853, 899, 909, 1180
McPartland, Marian, 883
MDZ71, 239
mean
  performance, 727
  tempo, 727
meaning, 849, 853, 909
  of sound, 241
  paratextual -, 330
  textual -, 330
  topological -, 158
  transformational -, 158
measurable, 1488
measure for complexity, 254
measurement, 28
mechanical
  constraints, 912
  dynamics, 608
  model, 896
  ritard, 896
mechanism
  modulation -, 465
  of conducting, 1291
mediante tuning, 1351
mediation, 774
meet, 1433
mela, 543
melakarta, 543
melismatic improvisation, 1303
MeloRUBETTE®, 744
melodic
  charge, 610

creation, 1077
gesture, 1302
minor, 480, 541
motion, 1305
movement, 1302
variation, 794
melodic minor
scale, 473, 474
tonality, 460
melody, 226, 271
germinal -, 220, 221, 792
retrograde of a -, 114
melody-as-motion, 1307
MeloRUBETTE®, 384, 408, 647, 672
membrane
basilar -, 1356
Reissner's -, 1355
tectorial -, 1357
memory, 527, 528
function
hippocampal -, 528
mental
experiment, 550
organization, 35
time, 547
tone parameters, 67
mental 3D rotation, 872
Merleau-Ponty, Maurice, 854, 860, 867, 871, 1031, 1264
Mersenne, Marin, 1364
MES (Memory Evolutive System), 1060
message, 12, 24
infinite -, 751
passing, 802
messaging, 156
Messiaen
mode, 126
scale, 794
Messiaen, Olivier, 126, 127, 134, 794, 1266
meta-gesture, 1286
meta-object, 802, 811, 814
class, 814
protocol, 813
meta-programming, 801, 813
meta-vocabulary, 199
Metal, 1281
metal in the voice, 1315
metalanguage, 212
metalevel, 17
metamere, 1362
metaphor, 24
physical -, 1288
metasystem, 17
meter, 97, 375–381, 1347
beat -, 97
barline -, 97
local -, 97

method, 156, 802, 1005, 1007
analytical -, 1012
boiling down -, 648
continuous -, 639
dodecaphonic -, 775, 780
group-theoretical -, 198, 205
integration -, 683
of characteristics, 1456
of invention, 1039, 1044
operational -, 28
selection, 805
statistical -, 612, 673
metric, 1449
associated -, 1450
Euclidean -, 229
Minkowski -, 1484
metrical
analysis, 688
component, 267
globalization, 99
grouping, 248
hierarchy, 375
level, 377
linear visualization, 362
profile, 689
quality, 376
rhythm, associated -, 267
similarity, 163, 388
structure (=M), 377
weight, 375, 376, 646
MetroRUBETTE®, 744
metronome, 99
Mälzel -, 28, 1347
MetroRUBETTE®, 376, 670
Mexican hat, 236, 1345
Meyer wavelet, 1346
Meyer-Eppler, Werner, 1353, 1362
mezzoforte, 1349
Michel chromatic, 478
Michel-Angelo, 857
micro
-motif, 646
timing, 220
micro-gesture, 1223
micrologic, 569
microstructure
timing -, 719
microtiming, 1303
middle ear, 1354
middleground, 415
MIDI, 235, 783, 899, 912, 964, 1096, 1116, 1151, 1164
velocity, 913
Mikaleszewski, Kacper, 622
Milmeister, Gérard, 1035, 1088, 1096, 1103
mimesis, 891
cultural -, 874

mimetic
  category, 892
  hypothesis, 1264, 1276
mimic
  gesture, 893
  innervation, 891
mind/body problem, 1264
minimal
  cadential set, 455
  length, 689
Minkowski, Hermann, 310, 1382
Minkowski metric, 1484
minor, 122
  harmonic -, 479
  melodic -, 480, 541
  mode, 448
  natural -, 478
  tonality, 478
mirror, 947, 956, 1268
  gesture, 1270
  neuron, 871, 885, 1264
Mittelgrund, 415
Mitzler, Laurentz, 1365
mixed
  sketch, 1028
  weight, 671
modal
  structure, 314
  synthesis, 1346
mode, 126
  aeolian -, 261
  authentic -, 261
  consonant -, 450
  dissonant -, 450
  dorian -, 261
  ecclesiastical -, 261, 540, 541
  hypoaeolian -, 261
  hypodorian -, 261
  hypoionian -, 261
  hypolocrian -, 261
  hypolydian -, 261
  hypomixolydian -, 261
  hypophrygian -, 261
  ionian -, 261
  locrian -, 261
  lydian -, 261
  Messiaen -, 126
  mixolydian -, 261
  phrygian -, 261
  plagal -, 261
  rhythmic -, 502
  with limited transpositions, 126
model, 1436
  kinetic -, 1306
  mathematical, 464
  mechanical -, 896

notic -, 1306
  physical -, 27
  template fitting -, 1361
modelage, 1003
modeling
  categorical -, 1001, 1003, 1045, 1046
  physical -, 236, 701
models, category of known simple -, 1062
modes of listening, 1281
modification, 1004
  of functional relations, 813
  syntax -, 813
modular
  affine transformation, 786
  composition, 251
modularity dynamical, 666
modulatio, 463
modulation, 459, 463–486, 840, 951, 952
  amplitude -, 236, 833
  degree, 465
  domain, 476
  force, 466, 469
  frequency, 236, 833
  frequency -, 1111, 1342
  just -, 475
  mechanism, 465
  path, 491
  pedal -, 499
  pitch -, 236, 833
  plan, 498, 499
  quantized -, 470
  quantum, 466, 467, 470, 471
  rhythmical -, 473, 501, 503, 794
  theorem, 470
  theory, 1243
  topos-theoretic background of -, 467
  well-tempered -, 469
modulations, limited -, 481
modulator, 73, 236, 470, 490, 491, 494, 1342
modulatory
  architecture, 495
  region, 486
module, 1391
  as basic space type, 59
  cohomology -, 1477
  complex, 288
    constant -, 288
    flasque -, 304
    of A-addressed forms, 289
    representative -, 298
    retracted -, 289
  direct sum -, 1392
  dynamically loadable -, 665
  free -, 1392
  functor, 143
  homology -, 966, 1475

injective -, 1407
  of a commutative local composition, 105
  of global affine functions, 355
  product -, 1392
  projective -, 1406
  semi-simple -, 1394
  shaping -, 665
  simple -, 1394
  structuring -, 665
modules in music, 60
modus ponens, 1434
molecule, 291
  global -, 292
  interpretable -, 292, 316
Molino, Jean, 11, 13, 574
MOLS, 1025
moment, initial -, 574
Monk, Thelonious, 880, 1296
mono, 1420
monochord, 22
monogamic coupling, 632
monoid, 1376
  algebra, 60, 1386
  finite -, 1376
  free -, 1376
  free commutative -, 1376
  homomorphism, 1376
  isomorphism, 1376
  morpheme -, 444
  multigeneric -, 446
  trigeneric -, 444
  word -, 1376
monomorphism, 1420
monotone extraction, 1019
Montaigne, Michel de, 1040
Monteverdi, Claudio, 754
Monti, Georges, 1060
Montiel Hernandez, Mariana, xii, 274
mood, 604
MOP, 813
  graphical -, 813
Morlet wavelet, 1345
morpheme
  harmonic -, 449
  monoid, 444
Morphemfeld, 417
morphic, 617, 1439
morphing, 788
morphism, 162, 908, 941, 1419
  t-fold differentiable tangent -, 552
  t-fold tangent -, 552
  geometric -, 946
  gestural -, 1287
  global -, 246
  local -, 1411
  mathematical -, 282

mathematically equivalent -, 282
  of denotators, 91
  of form semiotics, 1439
  of formed compositions, 291
  of functorial global compositions, 274
  of functorial local compositions, 130
  of gestures, 915
  of global molecules, 292
  of local compositions, 105, 128–132
  of objective global compositions, 274
  of objective local compositions, 129
  of performance cells, 586
  spatial digraph -, 942
  tangent -, 551, 557
morphology, harmonic -, 908
Morris, Robert, 203–205, 211, 314, 316, 409
Morrison, Joseph, 1346
MOSAIC, 1346
mosaic, 310
mother, 596, 619
  primary -, 628, 629
  prime -, 629
  tempo, 562
mother performance, 984
motif, 100, 159, 228, 271
  abstract -, 386
  classification, 187–190
  covering, 384
  Reti's definition of a -, 404
  rhythmic -, 504
  serial, 125
  space, 384
  $\mathbb{Z}$-addressed -, 100
motif i, 1267
motif ii, 1269
motion, 604, 607, 897
  accelerated -, 607
  harmonic -, 415
  melodic -, 1305
  neurophysiological -, 897
  sense of -, 608
  tracking, 1290
  trigger, 607
motivated, 16
motivation, 345
  geometric -, 346, 347
  logical -, 346, 639
motives, enumeration of -, 195
motivic
  analysis, 214, 404
  interpretation, 271, 384
  nerve, 384
  simplex, 385
  weight, 408, 647
  work, 276
  zig-zag, 277, 781

motor action, 608, 897
Motte-Haber, Helga de la, 863, 933
movement, 893
  internal -, 897
  melodic -, 1302
  tensed -, 532
movements, laryngeal -, 1316
Mozart, Wolfgang Amadeus, 189, 377, 490
MSC (Mathematical Subject Classification), 1034
Müller, Stefan, 756, 1441
multigeneric monoid, 446
multigraph, 1375
  finite -, 1375
multimedia object, 362, 370
multiple, 1016
multiple-dispatching, 802
multiplication
  interval -, 512
  scalar -, 1391
multiplicity, 208
  principle, 1013, 1064
multiverse, 1175
Mumford, David, 300
Murenzi wavelet, 1345
music, 3, 7, 9, 13, 22
  absolute -, 774
  alphabet of -, 90
  atonal -, 204
  christianization of -, 891
  composition technology, 464
  concept of -, 21
  critic, 249
   role of -, 752
  criticism, 248, 634
  critique, 751
  definition of -, 5
  deixis, 16
  emotional function of -, 528
  esthetics of -, 323
  fact of -, 10
  film -, 603
  historical dimension of -, 91
  history, 6
  in-time -, 817
  key function of -, 529
  listening, 1353
  material of -, 90
  North Indian raga -, 1305
  psychology, 237, 250
  research, 7
  semiotic perspective of -, 15
  software, 250
  syn- and diachronic development of -, 199
  tape -, 250
  theory, 669
  thinking -, 22

music data, spectral -, 964
music theory
  professional -, 202
  topological -, 1236
musical
  concepts, definition of -, 97
  analysis, 612
  composition, 30
  constraint, 913
  content, 1286
  creativity, 1227
  drama, 753
  embodiment, 1314
  gestalt, 90, 127
  idea, 774
  intellectuality, 1006
  intuition, 202
  logic, 264
  manifold, 241
  material, 811
  onset, 1347
  ontolog, 21
  ontology, 845
  process, 811
  prosody, 220
  reality, 142
  semantics, 134
  sign, 895
  string theory, 910
  taste, 529
  tempo, 28
  time, 892
  topography, 17
  unit, 90
musician
  Hindustani -, 1305
  performing -, 912
musicking, 1301
musicological
  experiment, 30, 31
  ontology, 328
musicology, 3, 13, 669, 719
  cognitive -, 21
  computational -, 21
  historical -, 328
  systematic -, 328
  traditional -, 22, 28
Musikalisches Opfer, 121
Musin, Ilya, 1289
musique concrète, 250
Muzzulini, Daniel, 442, 456, 469, 481

N

n-chain, differentiable -, 975
n-circle, 423
n-cube, 553

singular -, 966
n-dimensional cube, 966
N-formed global composition, 291
n-modular pitch, 205
n-phonic series, all-interval -, 195
N-quotient, 296
Nagasawa, Nobuho, 1276
Nambu-Goto action, 1185, 1484
name, 64
    instrument, 69
    of a denotator, 45
    of a form, 44
names, ordering on -, 76
naming policy, 44, 45, 58
Naradaya-Watson kernel, 709
narration, 773
narrativity, theory of -, 774
Nattiez, Jean-Jacques, 221, 249, 390, 779, 1178
natural, 1422
    decomposition, 707
    distance, 362
    language, 1034
    minor, 478
    transformation, 1422
natural minor tonality, 460
natural sciences, experience in the -, 31
nature
    exterior -, 29
    interior -, 29
nature's performance, 765
navigation, 31, 38
    address -, 140
    conceptual -, 35
    dynamic -, 40
    productive -, 39
    receptive -, 38, 39, 75
    topographical -, 19
    trajectory, 31
    visual -, 361
negation, 346, 1432, 1433
neigborhood, 163
neighborhood, 226, 672, 1443
    elementary -, 403
neo-Riemannian theory, 979, 988, 990
nerve, 776, 1445
    auditory -, 1355
    class -, 283, 309, 322, 988
    discrete -, 254
    functor, 988, 1445
    motivic -, 384
    of a global functorial composition, 282
    of a global objective composition, 253
    weight, 379
        induced -, 379
network, 980
    global -, 990

global Zarlino -, 992
GUI, 1096
homology, 986
    Klumpenhouwer -, 924
    local -, 980, 1253
    non-interpretable global -, 991
Neuhaus, Harry, 622
νεῦμα, 159
Neumann condition, 1483
Neumann, John von, 1494
neumatic notation, 892
neume, 15, 158, 571, 890, 912, 1182, 1293, 1322, 1327
neural pitch processing, 1361
neuron, mirror -, 871, 885, 1264
neuronal oscillator, 606
neurophysiological motion, 897
neurosis semiotic -, 849
neutral, 11, 1342
    analysis, 222, 250
    element, 1376
    level, 211, 956
neutral level, 12, 13
neutralization, 465
Newton, Isaac, 329, 1048
NeXT, 665, 687
NEXTSTEP, 665, 669
nexus, 314
Nicolas, François, 1006
Nietzsche, Friedrich, 850, 1033
nihil ex nihilo, 25
nilpotent, 1396
Node, 1106
Noether, Emmy, 115, 889, 1027, 1087
Noether, Max, 1027
Noh, 13, 342, 632
Noland, Carrie, 1265
Noll, Thomas, xii, 69, 181, 417, 421, 422, 425, 427, 432, 435, 443, 444, 449, 463, 469, 520, 523, 612, 675, 908, 980, 1088, 1377, 1441
non-commutative polynomials, 1386
non-interpretable
    composition, 305, 308
    global
        composition, 990
        network, 991
non-invertible symmetry, 127
non-lexical, 344
non-linear deformation, 640, 681, 733
non-linearity, 1359
non-singular hypergesture, 1247
nonparametric approach, 708
norealworld, 1053
norm, 17, 1450
normal, 1492
    form, 209
        Rahn's -, 209

order, 210
  subgroup, 1379
normalization
  diachronic -, 754
  synchronic -, 754
normative
  analysis, 377
  gesture, 1305
norms equivalent -, 1450
North Indian raga music, 1305
not parallel, 654
notation, 891
  American jazz -, 438
  diastematic -, 1322
  European score -, 67
  lead-sheet -, 438, 441, 571
  neumatic -, 892
  square -, 1329
notched tone space, 1363
note
  alterated, 107
  anchor -, 625
  satellite -, 625
  symbol, 912
note-against-note, 509, 531
nothing, 1067
nothingness, 1068
notic
  analysis, 1306
  model, 1306
notional scenery, 1034, 1046
novelty, absolute -, 1062
number
  complex -, 948
  embedding -, 208
  prime -, 227
  ring, 1112
numbers
  complex -, 1385
  dual -, 508, 1387
  rational -, 1385
  real -, 1385
Núñez, Rafael, 859, 1181

O

object, 156, 802, 1005, 1419
  description, 200
  functional -, 1036
  global -, 245
  mathematical -, 1021
  multimedia -, 362, 370
  prototypical -, 229
  visualization principle, 362
object-oriented programming, 48, 595, 627, 630, 633, 801, 802
objective

closure, 432
gesture, 1042
  global - composition, 252
  local - composition, 90
  local composition, 979
  trace, 101
Objective C, 665, 679
objectlevel, 17
objectystem, 17
observation, 28
OCR, 630
octave, 62, 1350
  coordinate, 1350
  period, 93
octave class, 117
ODE, 652, 683, 1451
Oe, Kenzaburo, 1266
Oettingen, Arthur von, 115, 123, 415, 423, 424, 1350
OFF, 913
off-line algorithm, 761
ON, 912
ON-OFF, 60, 964
ondeggiando, 592
oniontology, 845
onomatopoiesis, 16, 776
onset, 44, 67
  abstract -, 125
  musical -, 1347
  origin, 97
  physical -, 1347
  self-addressed -, 70
  time, 1335
  weight, 99
ontological
  atomism, 24
  coordinate, 10
  dimension, 17
  perspective, 5
  shift, 142
ontology, 9, 142, 149, 152, 328, 1001, 1018
  musical -, 845
  musicological -, 328
  denotator -, 328
  knowledge -, 345
  musical -, 21
  time -, 775
Onuma, Shiro, 1295
opacity, 1108
open
  ball, 1449
  formant set, 238
  functor, 1416
  semiosis, 330
  set, 227, 1443
  source, 666
  covering of a functor, 1416

Open-Editor, 814
OpenMusic, 210, 315, 774, 782, 801–821
openness, 237
operation, 1132
    Boolean -, 784
    flattening -, 75, 270
    graph, 1138
operationalization, 201
operationalized thinking, 161
operator, 598, 616, 619
    Beran -, 744
    agogical -, 720
    articulation -, 591
    artistic -, 1204
    basis-pianola -, 655
    Beran -, 723
    boundary -, 969
    bounded -, 1487
    canonical -, 207
    differential -, 1470
    face -, 968
    field -, 652
    Hamilton -, 1493
    Laplace -, 1471
    Lie -, 638
    Lie type -, 1255
    performance -, 598, 612, 637–661, 1253
    physical -, 616, 651
    pianola -, 659
    prima vista -, 616
    smoothing -, 721
    splitting -, 649
    sub-path -, 483, 1388
    support -, 1336
    symbolic -, 616, 650
    tempo -, 653
    test -, 652
    Todd -, 619
    TTO -, 207
    validation -, 349
opinions, manifold of -, 828
opposite
    category, 1421
    group -, 1378
opposition, 16
optimal path, 675
optimization, local -, 676
Oram, Celeste, 1264, 1274, 1277
orbit, 1378
    set-theoretic -, 1417
    space, 1378
orbits, functor of -, 1417
Orchestervariationen, 115
orchestra instrumentation denotator, 69
orchestral
    gesture, 817

globalization, 555
orchestration, 785
order, 1379, 1381
    normal -, 210
    of a PDE, 1455
ordered
    p-space interval, 207
    pair, 1372
    pc interval, 207
ordering, 362
    alphabetic, 50
    alphabetic -, 36, 38
    encyclopedic, 50
    lexicographic -, 50, 76, 1373
    linear -, 1373
        on a colimit, 77
        on a limit, 77
        on finite subsets, 77
    on
        coefficient rings, 79
        compound (naive) denotators, 51
        compound (naive) forms, 51
        coordinators, 76
        denotators, 75–83
        diagrams, 76
        direct sums, 79
        forms, 75–83
        identifiers, 76
        matrix modules, 79
        Mod, 78–80
        morphisms, 80
        names, 76
        simple forms, 76
        types, 76
        universal construction functors, 77
        $\mathbb{Z}\langle ASCII \rangle$, 79
    partial -, 1373
    powerset -, 51
    principle, 361
        on denotators, 49
    universal -, 39
ordinal, 1420
Oresme, Nicholas, 27, 548
organ of Corti, 1356
organic
    composition principle, 717
    principle, 163
organism
    cellular, 325
    hierarchical -, 249
organization
    degree of -, 718
    mental -, 35
orientation, 7, 262, 509, 1468
    hanging -, 108
    change of -, 509, 515, 531

geographic -, 38
hanging -, 509
local -, 264
ontological -, 9
recursive -, 19
sweeping -, 108, 509
oriented
  contrapuntal interval, 509
  global composition, 291
origin, 267
  of onset, 97
original, 892
OrnaMagic, 780, 787–789
ornament, 592, 786
  pattern, 202
OrnamentOperator, 646
Orsini, Marco, 1293
orthogonality principle, 762
orthonormal decomposition, 10
Orthonormalization, 725
oscillator, 605
  neuronal -, 606
oscillogram, 606
Osgood, Charles, 163
ostinato, 811
ottava battuta, 541
outer
  derivation, 1408
  ear, 1354
  hair
    cell, 1356
  pillar
    cell, 1356
outlet, 805
output, *presto*®-, 784
oval window, 1354, 1358
overhead, mathematical -, 512
overloading, 802
Oxley, Tony, 863

P

p-group, 1380
p-pitch, 206
p-scale, 95
p-space, 206
p-Sylow group, 1380
Pólya, George, 1013, 1039
painting, 151, 783
pair
  ordered -, 1372
  polarized -, 532
  simplicial -, 1447
  topological -, 1474
  Yoneda -, 1437
Palestrina–Fux theory, 540
Pallasmaa, Juhani, 855, 875

Palmer, Richard, 884
paper science, 145
Papez, James, 528
Paré, Ambroise, 29
Para-rhythm, 267
paradigm, 16
  dodecaphonic -, 125
  Fourier -, 232
  general affine -, 134
  gestalt -, 672
  phonological -, 220
παράδειγμα, 158
paradigmatic
  concept, 229
  field, 125
  group, 390
  strategy, 780
  theme, 221, 222, 390
  tool, 789
  transformation equivalence, 212
paradigmatics, uncontrolled -, 165
parallel, 654
  articulation
    field, 567
  crescendo field, 567
  degree, 264
  glissando field, 567
  hierarchy, 591
  hypergesture, 1245
    admissible -, 1250
  not -, 654
  performance
    field, 568
    map, 567
  process, 1139
  space, 590
Parallelklang, 457
parameter
  accessory -, 831
  basis -, 67, 654
  bow -, 833
  cadence -, 454
  color -, 835
  essential -, 831
  exchange, 133
  geometric -, 832
  global technical -, 835
  instrumental -, 832, 835
  local technical -, 835
  pianola -, 67, 654
  primavista -, 594
  space, 357
  stemma, 1257
  system -, 473
  technical -, 236
  vibrato -, 833

parameter exchange, 134
parametric polymorphism, 802
paratextual, 632
    information, 685
    meaning, 330
paratextuality, 349
Parker, William, 863
Parlan, Horace, 875
Parncutt, Richard, 607
parole, 17
part, 246, 250, 273
    dilinear -, 1397
    extraterritorial -, 591
    translation -, 1397
partial, 72, 423
    ordering, 1373
partial differential equation (PDE), 1471
partials, 10
participation value, 525
particle physics, 466
partition, 1372, 1462
    intérieure, 13
partitioning, 210
Pascal, hexagram of -, 1023
Pascal, Pascal, 1023
passing, message -, 802
patch, 802, 811
patchwork, local-global -, 250
path, 1375
    category, 1388, 1458
    closed -, 1375
    harmonic -, 482
    lazy -, 1375
    length, 1375
    modulation -, 491
    optimal -, 675
patrilineal, 628
pattern, 202
pause, 69, 631
    general -, 644
pc, 207
pc-space, 207
pcseg, 207
pcset, 207
    invariant -, 208
PDE, 657, 1455
    quasi-linear -, 1455
PDE (partial differential equation), 1471
Peano axioms, 30
Peck, Robert, 990
pedal
    modulation, 499
    voice, 499
Pederson, Jimmi, 879
peer, 668
Peirce, Benjamin, 1053

Peirce, Charles Saunders, 1053
percept of self-motion, 897
perception, logarithmic -, 550
perceptional pitch concept, 1363
perceptual equivalence, 229
percussion, 220
    event, 503
perfect consonance, 522, 532, 541
performance, 1253
    body, 586
    cell, 585, 1254
    cells
        category of -, 586
        morphism of -, 586
    complexity of -, 548
    culture of -, 623
    daughter -, 984
    definitive -, 895
    development, hierarchy of -, 622
    dynamics of -, 658
    field, 564, 568, 585, 983
        parallel -, 568
        prime mother -, 631
    fields, algorithmic extraction of -, 759
    gestural -, 893
    grammar, 615
        inbuilt -, 753
    hierarchy, 556
    history
        fictitious -, 627
        real -, 627
    infinite -, 549
    initial -, 586, 597
    intermediate -, 622
    logic, 556
    machine, 703
    map, 586
        parallel -, 567
    mean -, 727
    mother -, 984
    nature's -, 765
    of a large orchestra, 626
    operator, 598, 612, 637–661, 1253
    plan, 622
    primavista -, 630
    procedure, 610
    real-time -, 607
    research, computer-assisted -, 629, 701
    score
        global -, 598
        local -, 595, 1254
    structural rationale of -, 325
    synthetic -, 610
    theory, 319, 324, 909, 912
        inverse -, 611, 651, 756
    tradition, 753

vector field, 1253
performance field, time -, 1191
PerformanceRUBETTE®, 583, 652, 654, 679, 733
performative
  body, 1283
  gesture, 1283
performer, 24, 895, 1263
performing
  logic, 774
  musician, 912
perilymph, 1355
period, 415
  fundamental -, 1340
  in the Euler module, 93, 94
  interictal -, 524
  octave -, 93
  of a local meter, 97
  of a Vuza canon, 313
  of a Vuza rhythm, 312
  of duration, 97
  temporal -, 376
periodicity, 708
  higher level -, 99
Perle, George, 203
permutation, 1373
perspective, 24, 150, 152, 324, 466
  change of -, 324
  f-, 275
  of the composer, 246
  ontological -, 5
  variation of -, 151
perspectives, integral of -, 325
Peterson, Oscar, 875, 882
Petsche, Hellmuth, 524, 525
Pfeifer, Rolf, 873
phase, 1340
  portrait, 1454
  spectrum, 238, 1340
phaticity, 212
PHENICX, 1290
phenomenological difference, 756
phenomenology, 1001
  gesture -, 848
phenotype, 782
philosophy, 5
  Chinese -, 1278
  denotator -, 153
  gestural -, 893
  Japanese -, 1266
  of the body, 868
  Yoneda -, 828, 1475
phonatory system
  first part (diaphragm), 1314
  human -, 1314
  second part (larynx), 1314
  third part (resonant space), 1314

phoneme, 234
phonological
  equivalence, 215
  paradigm, 220
  poeticity, 215
photography, 151
phrasal hierarchy, 1311
phrasing, 248
physical
  constraint, 913
  gesture, 911, 1186, 1229, 1289
  gesture curve, 912
  metaphor, 1288
  model, 27
  modeling, 236, 701
  onset, 1347
  operator, 616, 651
  pitch, 1349
  ritard, 896
  sound, 71
  time, 547
  tone parameters, 68
PhysicalOperator, 683
physics, 5
  particle -, 466
physiological correlate, 897
p_i-rank, 1381
pianissimo, 607
pianist, 912
  elementary gesture of the -, 1192
pianist's hand, 911, 912
piano
  hierarchy, 593
  keyboard, 912
Piano concert No.1, 201, 202
pianola
  coordinate, 1347
  deformation, 591, 656
  operator, 659
  parameter, 67, 654
  space, 567, 588, 628
  specialization, 659
piecewise smooth, 1340
Pierce, Alexandra, 889
Pinocchio, 370
pitch, 44, 67
  -class, self-addressed -, 69
  -class set, 203, 207
  absolute -, 577
  alteration, 53, 788
  chamber -, 563, 1349
  class, 93, 116
    segment, 207
    set, 207
  concept, perceptional -, 1363
  concert -, 576

cycle, 207
detector, central -, 1361
difference, 62
distance, 61
   fundamental -, 437
   mathematical -, 61
modulation, 236, 833
   physical -, 1349
processing, neural -, 1361
segment, 207
spaces, 205
symbolic -, 68
pivot, 470
pivotal chord, 463
pixel, 343
Plücker, Julius, 1027
plan, performance -, 622
plane transformation, 786
Plato, 27, 38, 165
platonic ideas, 21
playing, 22
   structural -, 1267
Plomp, Reiner, 1364, 1367
Podrazik, Janusz, 210
Poe, Edgar Allan, 773
poetic Ego, 214, 219
poetical
   function, 16, 116, 212, 241, 248, 774, 781
   functions
      spectrum of -, 218
poeticity, 116, 212
   vertical -, 214
   horizontal -, 214
   phonological, 215
poetics
   timbral -, 241
   verse -, 248
poetology, 211
poiesis, 12, 211
   retrograde -, 14
poietic, 11, 1342
   genealogy, 128
Poincaré, Henri, 1004, 1071, 1079, 1180
point, 147–149, 908, 1371
   generic -, 228
   absorbing -, 432
   accumulation -, 1443
   closed -, 228
   concept, 145
   etymology, 147
   Euler -, 1349
   faithful -, 431
   full, 431
   fully faithful -, 431
   generic -, 269, 1413
   identity, 147

turning -, 465
pointed topological space, 1474
pointer, 23, 38, 146, 849, 1183
   scheme, 48
points, category of -, 981, 1255
Poisson equation, 1189, 1197, 1471
polarity, 519, 527
   at x, 523
   in musical cultures, 543
   profile, 228
polarized pair, 532
politically incorrect, 752
Pollock, Jackson, 860, 1291
Pólya
   enumeration theory, main theorems of -, 191
   theory, 191
   weight function, 191
Pólya, George, 190, 310
polygamic coupling, 632
polygon, 785
polyhedral initial set, 580
polymorphism
   ad-hoc -, 802
   parametric -, 802
polynomials
   commutative -, 1386
   non-commutative -, 1386
polyphony, 826
polyrhythm, 800
polysemy, 108, 164
Popelard, Marie-Dominique, 856
Popper, Karl, 828, 1039
Porphyrean tree, 157
portrait, phase -, 1454
position
   general -, 322
   hand's -, 1192
   privileged -, 520
Posner, Roland, 214, 782
possible, world of the -, 1052
post-serialism, 201
potential
   epileptiform -, 524
   Lagrange -, 1218
   sink -, 608
Pow, denotator, 1113
power
   spectral -, 525
   window, 524
powerset, 1375
   form, 56
   ordering, 51
   type, 47
PR, 377
practising, 633, 634
pre-Hilbert space, 1341

pre-morphism, 338
pre-object, 338
precise conceptualization, 211
precision, 1298
  calculation -, 639
  conceptual -, 32
  human -, 623
  machine -, 623
PrediBase, 665
predicate, 337
  atomic -, 339
  connective, 1435
  deictic, 346
  global -, 454
  logic, 436
  mathematical -, 340, 345
  morphic -, 338
  objective, 338
  primavista -, 341, 632
    European -, 341
    non-European -, 341
  punctual -, 338
  PV -, 341
  relational -, 338
  shifter -, 344
  textual -, 447, 454
  variable, 1435
preferences, 675
prehistory of the string quartet, 826
Prélude op. 28, No. 4, 743
presemiotic, 848, 859, 861, 910, 1180, 1293
presence, 409, 647, 673
presentation, 1030
presheaf, 1422
  continuous -, 982
  gestural -, 962, 1086
pressure
  bow -, 833
  decrease factor, 1337
  variation, 1335
presto®, 41, 115, 201, 202, 219, 220, 239, 437, 525, 576,
    624, 780, 782, 783–789, 791, 1163
  output, 784
Pribram, Karl, 980
Priestley-Chao
  kernel, 744
prima vista operator, 616
primary mother, 628, 629
primavista, 556, 630
  agogics, 629
  dynamics, 629
  parameter, 594
  performance, 617, 630
  predicate, 632
PrimaVista Browser, 1103
PrimavistaOperator, 630

PrimavistaRUBETTE®, 685
prime, 1389
  field, 1386
  form, 210
  mother, 629
    performance field, 631
  number, 227
  spectrum, 227, 239, 925, 1412
  stemma, 628
  vector, 62, 1351
primitive
  element, 1460
  gestural -, 901
  gesture of conductor, 1287
principal component analysis, 739
principal homogeneous *G*-set, 204
principle
  anthropic -, 465, 466, 543
  architectural -, 718
  concatenation -, 133, 512
  dialogical -, 828
  fractal -, 799
  locality -, 762
  multiplicity -, 1013, 1064
  normative -, 378
  object visualization -, 362
  of relevance, 15
  ordering -, 361
  organic composition -, 717
  organic -, 163
  orthogonality -, 762
  packing -, 362
  sonata -, 135
  variation -, 325
  visualization -, 1107
priority, 725, 734
privileged position, 520
problem, 1032
  Cauchy -, 1456
  context -, 674
  hit point -, 580
  mind/body -, 1264
  wild -, 756
procedure
  listening -, 610
  performance -, 610
  rule based -, 615
  rule learning -, 615
  statistical -, 198
  X-ray -, 892
process, 17, 330
  creative -, 1066, 1072
  historical -, 627, 634
  learning -, 556
  musical -, 811
  of conceptualization, 201

parallel -, 1139
unfreezing -, 912
view, 1138
process I, 1267
process II, 1270
product, 1374
ambient space, 104
Cartesian -, 1372
cartesian -, 1424
category, 1422
cellular hierarchy, 590
fiber -, 1387, 1424
group, 1380
matrix -, 1392
module, 1392
of local compositions, 104
of the cells, 587
ring, 1387
semidirect -, 1380
tensor -, 1388
topology, 1444
type, 46
weight function, 191
wreath -, 1380
production, 3, 4
of a musical work, 12
profile, metrical -, 689
program
canonical -, 325
change, 784
programme narratif, 774
programming
constraint -, 774, 801
functional -, 801
language
visual -, 801
object-oriented -, 48, 155, 595, 627, 630, 633, 801, 802
progression
harmonic, 249
chord -, 414
contrapuntal, 249
harmonic -, 128
projecting local composition, 178
projection, 128, 1374
projective
atlas, 296
functions, 297
module, 1406
Prokofiev, Serge, 183
prolongational reduction (=PR), 377
pronoun, 856
proof, 1014
chain, 1434
propagation, sexual -, 627, 628, 637
property
H^*-emergent -, 1062

inheritance -, 395
propositional
formula, 1435
variable, 1432
prosody
Greek -, 1324
musical -, 220
protocol, meta-object -, 813
prototype, 198
prototypical object, 229
pseg, 207
pseudo-metric, 1449
on abstract gestalt space, 394
psychological reality, 548
psychology, 5, 6
cognitive -, 179, 226
gestalt -, 90
music -, 237, 250
psychometrics, 163, 228
psychosomatic disease, 1288
Puckette, Miller, 761
pullback, 1424
pulsation, 1036
mathematical -, 1002, 1007, 1026, 1033, 1041, 1042, 1058, 1059, 1064
rhythmical -, 1326
punctus contra punctum, 931
pure expressivity, 606
pushout, 1424
PVBrowserRUBETTE®, 668
Pythagoras, 436
Pythagorean, 30
school, 97, 340
tonality, 461
tradition, 22, 154, 1364
tuning, 266, 478, 1350
Pythagoreans, 11

Q

quadrant, 1281
quale, 570
qualifier
after -, 815
before -, 815
quality, metrical -, 376
quantifier
existence -, 346
formula, 1435
universal -, 346
quantization, 689
quantized modulation, 470
quantum, modulation -, 466, 467, 470, 471
quantum mechanics (QM), 417, 425, 1485
quartet, string -, 774, 825
quasi-coherent, 1413
quasi-compact, 1445

quasi-homeomorphism, 1414
quasi-linear PDE, 1455
quaternions, 1385
quatuor dialogué, 827
question, open -, 1072
quid, 1204
Quintilian, 856
quiver, 1375
    algebra, 1388
    complete -, 1375
    Riemann -, 482
    Riemann index -, 483
    stemma -, 660
quotient
    category, 1420
    complex, 289
    dominance topology, 231
    group, 1379
    ring, 1386
    topology, 1444

R

radical, 1376
    digraph, 926
    formoid, 928
Radl, Hildegard, 460, 473, 477, 480
Raffael, 154
Raffman, Diana, 22, 570
raga, 543, 1309
Rahaim, Matt, vi, vii
Rahn, John, 203–205, 409
Ramachandran, Vilayanur S., 873, 885
Rameau's cadence, 455
Rameau, Jean-Philippe, 414, 422, 436, 455, 1365
ramification
    mode, 42
    world-sheet -, 1230
Ramstein, Christophe, 901
random, 14
rank, 1392
    torsion-free -, 1381
Raphael, 165
ratiocentricity, 1175, 1177
rational numbers, 1385
rationale, 616
    gestural -, 753
    historical -, 826
Ratner, Leonard, 329
Ratz, Erwin, 466, 497
Ravel, Maurice, 183
RCA, 525
real, 683
    inversion, 124
    numbers, 1385
    performance history, 627
    time, 1189

world, 1044
real-time
    algorithm, 761
    context, 760
    input, 783
    performance, 607
reality, 9, 10, 1018
    historical -, 487
    levels of -, 10
    mental -, 11
    musical -, 142
    physical -, 10
    psychological -, 11, 548
realization, geometric -, 988, 1446
reason, 5
recapitulation, 249, 496
receiver, 212
reception, 3, 4
receptive navigation, 75
recitation tone, 261
recombination, 813
    weight -, 640
reconstruction, 406
recording
    bipolar -, 524
    foramen ovale -, 524
recta, 692
recursive
    classification, 177
    construction, 42
    orientation, 19
    typology, 42, 49
reduced
    diastematic shape type, 387
    strict style, 541
reductio ad absurdum, 1018
reduction, 1400
    curvilinear -, 776
reductionism, 6
Reeves, Hubert, 167
reference
    denotator, 332
    tonality, 448
referentiality, 212
refinement, 1462
    map, 275
reflection, 814
    glide -, 1402
reflexive, 1373
reflexivity, 801
Regener, Eric, 204
region, modulatory -, 486
register, 784, 803
regression analysis, 711, 724, 726
regulae, 1016
regular

denotator, 66–71
division of time, 376
form, 64
representation, 1393
structure, 708
regularity, time -, 99
rehearsal, 556, 609, 612, 633, 634
Reichhardt, Johann Friedrich, 827
reification, 814
Reinhardt, Django, 876
Reissner's membrane, 1355
relation
causal -, 816
commutativity -, 1420
cross-semantical -, 612
equivalence -, 250, 1373
functional -, 1196
K -, 314
Kh -, 314
KI -, 314
temporal -, 816
relative
absolutely -, 1028
automorphism, 1460
delay, 236
dynamics, 685
homotopy, 1447
importance, 648
motivic topology, 400
symbolic dynamics, 68
tempo, 561, 685
topology, 1444
relative local
dynamical sign, 640
tempo, 643
relative punctual
dynamical sign, 640
tempo, 643
relevance, principle of -, 15
religion, Christian -, 847
Rellstab, Ludwig, 18
Remak, Robert, 1396
Remove-Element, 814
renaming, 47
reparametrization, 1185
repetition, 117, 1305
sequential -, 1304
replay, 117
Repp, Bruno, 719, 723, 739, 768
representable
functor, 1423
gesturally -, 962
representation, 1030
adic -, 1389
auditory -, 197
Fourier -, 740, 832

geometric -, 783, 892
linear -, 1393
regular -, 1393
score -, 610
textual -, 776
representative
first -, 180
module complex, 298
reprise, 799
reproduction, identical -, 891
res
cogitans, 1186
extensa, 1186
reset, 60
resolution, 357, 831, 840, 989
cohomology, 355
functor, 294
of a global composition, 294, 323
resolution of the identity, 1494
resolvent set, 1493
response, EEG -, 524
responsibility, collective -, 633
restriction
cellular hierarchy -, 590
of modulators, 498
scalar -, 107, 1392
resultant class, 313
resulting divisor, 313
retard, final -, 607
Reti, Rudolph, 165, 225, 376, 383, 403, 672, 721
Reti-motif, 406
retracted module complex, 289
retraction, 1420
retrograde, 14, 23, 119, 127, 134, 208, 247
address involution, 125
inversion, 61, 120, 134
of a melody, 114
retrogression, 207
reverberation time, 1337
reversed order
score played in -, 119
tape played in -, 120
sound, 121
revolution, experimental -, 29
Reye, Theodor, 1023
RGB, 1115
RH, 525, 629
rhetorical
expression, 570
expressivity, 556
shaping, 556
rhetorics, 827
rhythm, 117, 126, 375–381, 807, 1324
local -, 99, 106
throbbing -, 1271
Vuza -, 312

rhythmic
    germ, 127, 266
    mode, 502
    motif, 504
    scale, 503
rhythmical
    group, 810
    modulation, 473, 501, 503, 794
        theory, 502
    pulsation, 1326
    structure, 794
rhythms, 97
    classification of -, 312
    local -, isomorphism classes of -, 181
Richards, Whitman, 607
richness, semantic -, 570
Richter, Sviatoslav, 622
Riemann
    algebra, 481
    dichotomy, 522
    graph, 677
    harmony, 263
    index quiver, 483
    locus, 676
    matrix, 447, 481, 675
    quiver, 482
    transformation, 315
Riemann Mapping Theorem, 1466
Riemann, Bernhard, 251
Riemann, Hugo, 99, 123, 160, 201, 205, 251, 375, 414,
        417, 436, 447, 449, 463, 469, 481, 484, 509, 523,
        669, 674, 693, 714, 721
Ries, Ferdinand, 825, 827
right
    action, 1378
    adjoint, 1423
    coset, 1379
    ideal, 1386
    inverse, 1378
right-absorbing endomorphism, 432
rigid, 262, 279, 466, 470, 473, 672
    difference shape type, 386
    marked dichotomy, 518
    shape type, 386
rigor, 1039
Riguet, Jacques, 1055
Rilke, Rainer Maria, 850
ring, 1385
    anti-homomorphism, 1385
    endomorphism -, 1391
    homomorphism, 1385
    isomorphism, 1385
    limit -, 1387
    local -, 1395
    number -, 1112
    product -, 1387

quotient -, 1386
    self-injective -, 1407
    simple -, 1386
ringed space, 1411
ritard
    final -, 896
    mechanical -, 896
    physical -, 896
ritardando, 608, 644
Rizzolatti, Giacomo, 873
RMI, 668
Roederer chromatic, 478
Rogers, Carl, 1032
Roland R-8M, 220, 791
role
    exchange, 61
    of a music critic, 752
    of conductor, 1288
Rostand, François, 1043
Rota, Nino, 1292
rotation, 202, 207, 208, 786, 899
    Amuedo's decimal normal -, 210
roughness, 1366
round window, 1358
Rousseau, Jean-Jacques, 248
row-class, 208
Rubato Composer, 947, 1095
RUBATO®, 41, 376, 628, 649, 650, 665–668, 719, 736,
        759
    concept, 665
    Distributed -, 764
    layer, 668
rubato
    Chopin -, 550, 562, 576, 624, 765
    encoding formula, 618
Rubato Composer, 979
rubette, 665, 669–686, 1096
    BigBang -, 947, 1097, 1127, 1157
    for counterpoint, 1097
    for harmony, 1098
    ScorePlay -, 1105, 1107
    Select2D -, 1105
Rufer, Joseph, 125, 134
rule
    based procedure, 615
    contravariant-covariant -, 805
    Fux -, 541
    learning procedure, 615
    preference - (=PR), 377
    well-formedness - (=WFR), 377
rule-based grammar, 615
rules
    differentiation -, 610
    ensemble -, 610
    grouping -, 610
Runge-Kutta-Fehlberg, 652, 683

Ruwet, Nicolas, 222, 249, 779

S

S-duality, 1184, 1484
Sabine's formula, 1338
Sachs, Klaus-Jürgen, 531, 532
Sacks, Oliver, 885
sacred denaturation, 891
Saint-Victor, Hugues de, 846, 862, 909, 1180
Salzer, Friedrich, 414
Sandall, Robert, 884
Sander, Emmanuel, 1022
Sands' algorithm, 310
Sands, Arthur, 310
sapat, 1306
satellite, 370, 1104, 1114
    note, 625
saturation, 434, 1406
    sheaf, 434, 444
Saussure, Ferdinand de, 15, 160, 199, 222, 250, 472, 854,
        856, 867, 1180
Sawada, Hideyuki, 607
SC, 207
scala
    media, 1355
    tympani, 1355
    vestibuli, 1355
scalar, 1391
    multiplication, 61, 1391
    restriction, 107, 1392
ScalarOperator, 683
scale, 94, 442
    12-tempered -, 260
    major -, 473
    melodic minor -, 473, 474
    altered -, 480
    chromatic -, 418
    diatonic -, 543
    harmonic minor -, 472, 474
    harmonical-rhythmical -, 794
    just -, 96, 260
    justest -, 266
    major -, 262, 474
    Messiaen -, 794
    minor
        harmonic -s, 262
        melodic -, 262
    rhythmic -, 503
    whole-tone -, 541
SCALE-FINDER, 210, 441
SCALE-MONITOR, 210, 441
scales, common 12-tempered, 96
Scarlatti, domenico, 1157
scatterplot, 713
scenery, notional -, 1034, 1046
sceptical doubt, 1012

Schönberg, Arnold, 30, 90, 115, 125, 127, 134, 135, 183,
        199, 201, 203, 204, 212, 247, 253, 262, 325, 413,
        422, 442, 463, 465, 466, 502, 775, 780, 1243
Schäfer, Sabine, 687
Schaeffer, Pierre, 250
Schauder basis, 1209, 1486
scheme, 1414
    diagram -, 1420
    Grassmann -, 1416
    matrilineal -, 626
    mental -, 13
    Molino's -, 11
    sonata -, 503
Schenker, Heinrich, 270, 329, 414
scherzo, 791
Schmidt, Erhard, 1396
Schmitt, Jean-Claude, 846, 903, 909, 1180
Schneider, Albrecht, 902
Schoenberg, Arnold, 933
school
    KTH -, 617, 621
    Pythagorean -, 97, 340
    Zurich -, 612
School of Athens, 154, 165
Schopenhauer, Arthur, 748
Schrödinger equation, 1493
Schubert, Franz, 214, 232, 792
Schumann, Robert, 407, 628, 629, 673, 701, 711, 743,
        784, 785, 827, 894, 1260
Schweizer, Albert, 688
science
    cognitive -, 611
    computer -, 156
    contemplative -, 27
    doing -, 28
    embodiment -, 868
    experimental -, 29
    paper -, 145
scientific art, 1045
scientific bankruptcy, 22
score, 11, 13, 60, 341, 783, 891, 899, 909, 912, 1105,
        1263, 1287
    concept, 250, 571, 754, 811
    European - notation, 67
    exterior -, 571
    generic -, 548
    global -, 250, 783
    inner -, 548, 571, 1295
    interior -, 13
    local -, 250, 783
    played in reversed order, 119
    representation, 610
    semantics, 574
    transformation -, 785
score-following, 761
score-performance matching, 761

ScorePlay rubette, 1105, 1107
Scriabin, Alexander, 182, 482, 799
SEA, 384
section, 1412, 1420
  global -, 288, 1424
segment
  class, 206, 208
  pitch -, 207
Seifert-Van Kampen theorem, 929
Select2D rubette, 1105
selection
  axis of -, 116, 212
  field -, 803
  method -, 805
  stepwise forward -, 727
self-addressed
  chord, 185
  contrapuntal intervals, 514
  denotator, 69
  onset, 70
  pitch-class, 69
self-addressed arrow, 514
self-adjoint, 1492
self-injective ring, 1407
self-modulating, 1342
self-motion, percept of -, 897
self-referential, 19, 146
self-similar time structure, 799
semantic
  atom, 442
  charge, 403
    of EEG, 524, 525
  completion, 49
  depth, 383
  differential, 163
  interpretation, 490
  loading, 42
  manifold, 241
  richness, 570
semantics
  functional -, 445
  gestural -, 753
  harmonic -, 436
  incomplete, 83
  musical -, 134
  of weights, 408
  score -, 574
semi-simple module, 1394
semidirect product, 1380
semigroup, 1376
semiosis, 9
  extroversive -, 329
  incomplete -, 330
  introversive -, 329
  open -, 330
  paratextual -, 349

textual, 334
textual -, 348
semiotic, 1265
  component, 854
  constraints, 232
  gesture -, 1183
  marker, visual -, 813
  neurosis, 849
  of $\mathcal{E}$-forms, 1438
semiotical symmetry, 134
semiotics, 5, 14
  of sound classification, 240
semitone, 62
sender, 12, 212
sense of motion, 608
sentence, 1432, 1436
  valid -, 1433
sentic state, 604
sentograph, 895
separable extension, 1460
separating module complex, 296
sequence
  Cauchy -, 1450
  chord -, 486
  contrapuntal -, 531
  Fibonacci -, 340
  interspace -, 192
  of a local composition, 192
sequencer, 789
sequential repetition, 1304
sequentialization, 776
serial
  motif, 125
  technique, 127–128
serial motif
  integrated -, 195
  derived -, 194
serialism, 201
series
  all-interval -, 194, 200
  basic -, 125
  dodecaphonic -, 125, 162, 194, 247, 253, 325
  fundamental -, 115
  (k, n)-, 125, 194
  n-phonic -, 125, 194
  time -, 707
Serre, Jean-Pierre, 1027, 1476
Sessions, Roger, 889, 894, 1182
set, 1371
  Borel -, 1488
  cadential -, 455
  circular -, 66
  class, 207, 208
  closed -, 1443
  complex, 204, 314
    theory, 204

concept, 146
  empty -, 146
  fuzzy -, 163, 1029
  in AST, 203
  index -, 1373
  initial -, 573, 586, 596, 1254
  minimal cadential -, 455
  of operations, 209
  open -, 227, 1443
  pitch-class -, 203
  resolvent -, 1493
  small -, 1420
  source -, 203
  support -, 252
  theory, 249, 1028
SET-SLAVE, 209, 441
set-theoretic orbit, 1417
SETI (Search for ExtraTerrestrial Intelligence), 1175
seventh
  dominant -, 419
  natural -, 423
  subdominant -, 420
  tonic -, 420
sexual propagation, 627, 628, 637
SGC, 208
Shakespeare, William, 637
shape, 405, 1004, 1056
  J-, 1062
  type, 385
shape type
  contrapuntal motion -, 387
  diastematic -, 387
  diastematic index -, 387
  elastic -, 388
  reduced diastematic -, 387
  rigid difference -, 386
  rigid -, 386
  toroidal sequence -, 388
  toroidal -, 388
shapes of the hands, 1310
shaping, 1148
  module, 665
  rhetorical -, 556
  vector, 723
sharp, 109
sheaf, 1415, 1431
  on a base, 1412
  saturation -, 434
sheafification, 1432
shearing, 119, 123, 133, 202, 1402
sheaves, category of -, 1431
Shepard, Roger, 872
Shepp, Archie, 877
Sherman, Robert, 179
shift, 109
  constant -, 109

elementary -, 109
  ontological -, 142
shifter, 574, 578, 856, 892, 895
  esthesic -, 344
  poietic -, 344
Shusterman, Richard, 868
Siebers, Tobin, 880
sieve, 1428
  closed -, 1431
  covering -, 1430
sight reading, 556, 722
sign, 14, 850, 1047
  deictic -, 16
  lexical -, 16
  musical -, 895
  shifter -, 16
signature, 592
  key -, 631
  time -, 631
significant, 14
significate, 14
signification, 14, 15, 337
  process, 14
signs
  of coefficients, 731
  system of -, 5
similarity, 159, 163, 226, 228
  gestural -, 1289
  metrical -, 163, 388
simple
  form, 56
    simplify to a -, 63
  frame, 802
  group, 1379
  module, 1394
  ring, 1386
simple forms, ordering on -, 76
simplex, 596, 988, 1445
  affine -, 1461
  category, 939
  closed -, 1446
  dimension of -, 1445
  motivic -, 385
  singular -, 1447
  standard -, 1447, 1461
simplicial
  cochain complex, 1447
  complex, 779, 1445
  functor, 988
  map, 1445
  metrical weight, 269
  pair, 1447
  weight, 283, 988
simplify to a simple form, 63
Simula, 802
simultaneous local meters, 500

singing, Han -, 1321
singleton interpretation, 275
singular
    n-cube, 966
    cochain, 1447
    homology, 965, 1476
    simplex, 1447
singular homology, hypergestural -, 1236
sink potential, 608
Siron, Jacques, 571
Sirone (Norris Jones), 1176
sister, 660
site, 1430
    Zariski -, 1415
skeletal space, 918, 974
skeleton, 870, 1445
    category, 1421
    global -, 996
    space-time -, 1226
sketch, 1056, 1059
    matrix-like -, 1057
    mixed -, 1028
    of gesture, 1030, 1038
sketches, category of -, 1029
sketchy interpretation, 622
skew field, 1385
skin, 1274
slave tempo -, 624
Slawson, Wayne, 237
slice, 102
    identity -, 275
f-slice, 275
Sloboda, John, 665, 743
slope, global -, 677
slot, 802
slur, 631, 632
SMAC, 612
small
    gestalt, 398
    infinitely -, 569
    set, 1420
smallness, 237
smartphone, 1274
    gesture, 1279
Smith III, Julius O, 1346
smooth, piecewise -, 1340
smoothing
    hierarchical -, 708
    kernel, 708
    kernel -, 708, 721
    operator, 721
SMPTE, 784
SNSF, 612, 665
sober, 1414
    weight, 379
socle, 1396

software
    development, 595
    engineering, 152
    for AST, 209
    music -, 250
solution
    global -, 1454
    local -, 1452
somaethetics, 868
sonata
    form, 249, 496, 791
    principle, 135
    scheme, 503
    theory, 496
sonification, 1111, 1116
sound, 1313, 1335
    classification, 232
    color, 159
    colors, space of -, 237
    conceptualization of -, 14
    design, 1121
    generator, 700
    grouping, 74
    meaning of -, 241
    natural -, 11
    physical -, 71
    pressure level, 1348
    reversed -, 121
    speech, 827
    transformation, 120
Sound Pattern of English (=SPE), 234
sounding analysis, 694
SoundScore, 1107
source
    open -, 666
    set, 203
space, 892, 893, 1067, 1438
    ambient -, 90
    Banach -, 1450, 1486
    basis -, 567, 588, 628
    color -, 832
    compositional -, 204
    concept -, 1034
    contour -, 205
    creative -, 1034
    dual -, 1488
    Euler -, 1349
    finger -, 915
    form -, 55
    fundamental -, 588
    gestural form -, 1087
    hierarchy, 588
    Hilbert -, 1486
    indecomposable -, 588
    instrument -, 597
    lens -, 922

locally ringed -, 1411
motif -, 384
of sound colors, 237
of spectral values, 1317
of vocal gestures, 1315
orbit -, 1378
parallel -, 590
parameter -, 357
pianola -, 567, 588, 628
pre-Hilbert -, 1341
ringed -, 1411
skeletal -, 918, 974
tangent -, 551, 1467
top -, 588
topological -, 1443
vector -, 1392
space-time, 892
skeleton, 1226
span, 520, 672
time -, 70
spanning tree, 1458
spatial
concept, 894
digraph, 914, 940, 1457
morphism, 942
hypergesture, 917
spatialization, 891
SPE, 234
special linear algebra, 1408
specialization, 162, 218, 231, 402, 591, 1026
abstract -, 402
abstract gestalt -, 402
basis -, 656
co-inherited -, 402
gestalt -, 402
Hasse diagram, 220
inherited -, 402
pianola -, 659
topology, 403
specialize, 227
species, 239
spectral
analysis, 524
decomposition, 708
gesture, 1318
music data, 964
participation vector, 524, 525
power, 525
vector, 832
Spectral Theorem, 1494
spectral values, space of -, 1317
spectroid, 908, 925
spectrum, 721
amplitude -, 1340
energy -, 1340
of poetical functions, 218

phase -, 238, 1340
prime -, 227, 239, 925, 1412
speculum mundi, 36
speech, 17
gesture, 1311
sound -, 827
vocalization, 1302
Sperry, Roger, 873
Spicker, Volker, 909
SPL, 1348
spline, 1461
of type $\mathcal{T}$, 1462
split
exact sequence, 1380
local commutative composition, 176
SplitOperator, 682
splitting, 629
operator, 649
spring equation, 1341
SQL, 668
square
Eulerian -, 1025
Latin -, 1024
notation, 1329
Staatliche Hochschule für Musik, 629
Staatliche Hochschule für Musik, 628, 687
stability, gestalt -, 226
stabilizer, 1378
stable concept, 226
staccato, 645
staff, 1322
stalk, 1411
standard
global - composition, 293
atlas, 293
chord, 436
hierarchy, 590
local - composition, 293
of basic musicological concepts, 92
simplex, 1447, 1461
composition, 173
standardized tempo, 723
Stange-Elbe, Joachim, xii, 628, 687
state
emotive -, 1266
sentic -, 604
stationary voice, 499
statistical
approach, 707
method, 612, 673
procedure, 198
Steibelt, Daniel, 120, 134
Steinway, 791
stemma, 556, 612, 619, 621–635, 1253
continuous -, 661
leaves of a -, 628

parameter -, 1257
prime -, 628
quiver, 660
tempo -, 624
theory, 736, 755
tree, 660
stemmata, gesture of -, 1256
stemmatic
  cross-correlation, 634
  grouping, 633
Stendhal, 857
Stendhal syndrome, 1288
stepwise forward selection, 727
stereocilia, 1356
stereotactic depth EEG, 524
Stern, Daniel, 1266
Stimmigkeit, 848
stirrup, 1354
Stockhausen, Karlheinz, 59, 127, 233, 316
Stokes' theorem, 973, 1244, 1248, 1469, 1470
Stokes' theorem for hypergestures, 976
Stolberg, Leopold, 214, 232, 792
Stone, Peter, 115
Stopper, Bernhard, 93
strategy
  experimental -, 694, 701
  gestural -, 1270
  paradigmatic -, 780
  target-driven -, 694, 701
stratum, cognitive -, 869
Straub, Hans, xii, 189, 284, 442, 455, 456, 1517
Straus' zero normal form, 210
Stravinsky, Igor, 1293
strength factor, 610
stretching, 202
  time -, 650
strict
  extension, 443
  style, 541
    reduced -, 541
STRING, 44
string
  empty, 45
  landscape, 1175
  of operations, 209
  quartet, 69, 241, 774, 825
    prehistory of the -, 826
    theory, 825
  theory, 910, 1173, 1175, 1482
    musical -, 910
String Trio op.45, 933
strip
  harmonic -, 253, 262, 442
  Möbius -, 451, 476, 781
strong marked dichotomy, 518

structural
  constant, 1408
  homomorphism, 1385
  playing, 1267
  rationale of performance, 325
structuralist linguistics, 250
structuration, 1026
structure
  formal -, 801
  frame -, 591
  internal -, 254
  interspace -, 192
  limbic -, 524
  local vs. global -, 89
  locally trivial -, 252
  modal -, 314
  of fibers, 756
  regular -, 708
  rhythmical -, 794
  transitional -, 464
structures, matching of -, 718
structures-mères, 1028
structuring module, 665
Stucki, Peter, xii
style, 718
  strict -, 541
sub-complex Kh, 210
sub-path operator, 483, 1388
subbase for a topology, 1443
subcategory, 1421
  address -, 1089
  full -, 1422
  Yoneda -, 1437
subclass, 802
subcomplex, 1445
  full -, 1445
subconscious, 529
subdivision, 622
subdominant, 263, 414, 445, 448
  seventh, 420
subgroup, normal -, 1379
subject, 22
subjective gesture, 1042
subjectivity, 30
subobject, 1428
  classifier, 1428, 1459
  relation, 76
substance, 43
substitution theory, 1363
subtyping, 805
succession, 774
  interval -, 207
successive interval, 526
successively increased alteration, 788
successor
  admitted -, 533

pairing
  allowed -, 531
suchness, 851, 1179
sukoon, 1306
sum
  direct, 63
  disjoint -, 1424
  fiber -, 1424
SUN, 668
Sundberg, Johan, 553, 607, 609, 615, 743, 896, 1182
super-summativity, 167, 226, 271
superclass, 802
supersensitivity, 688
superstring theory, 1485
support, 337
  functor, 257
  of a local composition, 90
  operator, 1336
  set, 252
supporting valence, 1362
surface, integral -, 1456
surgery, concept -, 633
surgical epilepsy therapy, 524
surjective, 1372
suspension, 722
sustain, 1315
Sutera, Salvatore, 1315
svara, 1306
Swan, Richard, 1027
sweeping orientation, 108, 509
Swing, 764
switch, vocabulary -, 240
Sylow
  decomposition, 79, 445, 510
  group, 179
Sylow, Ludwig, 1380
Sylvester, David, 860
symbol
  gestural -, 912
  logical connective -, 1432
  note -, 912
symbolic
  absolute dynamics, 68
  computation, 801
  gesture, 911, 912, 1186, 1228, 1229, 1289
  gesture curve, 912
  kernel, 585
  operator, 616, 650
  pitch, 68
  relative dynamics, 68
Symbolic Composer, 115
SymbolicOperator, 683
symmetric, 1373
  group -, 1378
SYMMETRICA, 311
symmetries

in music, 14, 114–128
musical meaning of -, 132
semantical paradigm for -, 133
symmetry, 91, 98, 113, 162, 1402
  of parameter roles, 127
  breaking, 775
  codification of a -, 128
  contrapuntal -, 532, 1235
  degree of -, 208
  form of a -, 113
  function of a -, 113
  group, 144, 180, 470, 672
    conjugation class of the -, 180
  hidden -, 114
  inner - of C-major, 123
  local -, 533, 534
  non-invertible, 127
  semantical function of -, 113
  semiotical -, 134
  transformation, 249, 250
  underlying -, 129
Synaesthesia Playground, 1274
synaesthetic, 885
synchronic, 16
  axis, 328, 472
  normalization, 754
synonym form, 56
synonymy
  circular -, 64
  type, 47
syntagm, 16
syntagmatic equivalence, 215
syntax modification, 813
Synthesis, 219, 473, 501, 503, 780, 781, 787, 791–799
synthesis, 1338
  modal -, 1346
synthesizer, 1121
synthetic performance, 610
syntonic comma, 97
system
  auditory -, 11
  coefficient -, 1447
  Hess -, 524
  limbic -, 528, 606, 1361
  meta -, 17
  object-, 17
  of signs, 5
    non-linguistic -, 15
  parameter, 473
  vestibular -, 608, 897
  weight -, 632
systematic
  approach, 472
  musicology, 328
  understanding, 826

T

t-fold
  tangent
    composition, 551
    morphism, 552
t-fold differentiable
  tangent morphism, 552
t-gestalt, 391
tönend bewegte Formen, 251, 775
tableau, 1024
tactus, 377
tail, 914, 1457
Takemitsu, Toru, 1263, 1266
Takhtajan, Leon A., 1485
tala, 1309
tangent, 108, 511
  bundle, 1451
  category, 927
  composition, 551
    basis of a -, 551
    global -, 557
  interpretation, 558
  manifold, 1467
  morphism, 551, 557
  space, 551, 1467
    Zariski -, 927, 1415
  tensor, 1468
  torus, 511
tape music, 250
taquin, 1025
target-driven strategy, 694, 701
Tarry, Gaston, 1027
taste
  common -, 752
  musical -, 529
tautology, 1433
Taylor, Cecil, v, 548, 799, 862, 874, 909, 1182, 1291, 1295
Taylor, Yuval, 886
technical parameter, 236
technique
  bodily -, 1280
  cohomological -, 1061
  instrumental -, 832
technology, gesture -, 1283
tectorial membrane, 1357
teleportation, 1180
telling time, 773
tempered, 93
  class chord, 94
  scale space, 96
  tuning, 1350
template fitting model, 1361
tempo, 547, 550, 552, 561, 896
  absolute -, 341, 561, 643
  curve, 202, 220, 562, 607, 623, 723, 784
    adapted -, 576

    deformation of -, 576
  daughter -, 562
  discrete -, 28
  field, 562
  hierarchy, 624
  istesso -, 555
  mean -, 727
  mother -, 562
  musical -, 27, 28
  operator, 653
  relative -, 561, 685
  relative local -, 643
  relative punctual -, 643
  slave, 624
  standardized -, 723
  stemma, 624
  weight -, 653
tempo-intonation field, 565
TempoOperator, 683
temporal
  box, 811
  period, 376
  relation, 816
tenor tone, 261
tensed movement, 532
tension, 414, 647
  contrapuntal -, 531
  global -, 677
  harmonic -, 481, 482
tensor, 1468
  algebra, 1469
  alternating -, 1468
  field, 1468
  product, 1388
    affine -, 1400
  tangent -, 1468
Terhardt, Ernst, 1368
terminal, 1424
terminology, 203
territory, 591
Teschl, Gerhard, 1485
tesselating chord, 309
tesselation, isotypic -, 309
test
  EEG -, 524
  operator, 652
  Turing -, 791
  Wilcoxon -, 526
tetractys, 30, 1364
tetradic interpretation, 276
tetrahedron, 4, 6
text analysis, 609
textual
  abstraction, 362
  content, 1286
  meaning, 330

predicate, 447, 454
representation, 776
semioses
   category of -, 337
textuality, 334–348
texture, four part -, 826
Thalmann, Florian, vi, vii, 979, 1036, 1097
The Topos of Music, 843
theme, 271, 415
   basic -, 202
   paradigmatic -, 221, 222, 390
   Reti's definition of a -, 404
theology, apophatic -, 1004
theorem, 1434
   complement -, 208
   counterpoint -, 534, 538
   Fourier's -, 1340
   Mason's -, 109
   Mason-Mazzola -, 109
   modulation -, 470
   Seifert-Van Kampen -, 929
   Stokes' -, 973, 1244, 1248
theorist, 1263
theory
   American jazz -, 439
   American set -, 979
   catastrophe -, 226, 497
   category -, 908, 1002, 1028
   classification -, 831
   contour -, 271
   counterpoint -, 775, 839
   degree -, 436
   Fourier -, 925
   function -, 437
   Galois -, 1087, 1460
   gestural -, 1182
   gesture -, 893, 1173
   global -, 220
   group -, 212
   homology -, 967
   homotopy -, 1474
   hyperset -, 1094
   landscape -, 1175
   modulation -, 1243
   music -, 669
   neo-Riemannian -, 979, 988, 990
   of ambiguity, 245
   of knowledge, 1005
   of narrativity, 774
   Palestrina-Fux -, 540
   performance -, 319, 324, 909, 912
   rhythmical modulation -, 502
   set -, 249, 1028
   sonata -, 496
   stemma -, 736, 755
   string quartet -, 825

string -, 910, 1173, 1175, 1482
substitution -, 1363
superstring -, 1485
transformational -, 889, 898, 908, 938, 947, 965, 979,
   1119
valence -, 1353
wavelet -, 1344
thesis, 1192, 1287, 1325
   world-antiworld -, 497
Thiele, Bob, 603
thinking, 22
   by doing, 28, 30
   cartesian -, 898, 1182
   music, 22
   operationalized -, 161
thinking music, 22
third, 414
   axis, 96
   chain, 260, 438, 675
      closure, 260
      interpretation, 260
      minimal -, 260
      weak -, 260
   comma, 265
      class, 265
   coordinate, 1350
   degree tonality, 450
   distance, 511
   major -, 62, 1350
   weight, 676
3D vision, 361
threshold
   global -, 675
   local -, 675
throbbing rhythm, 1271
tie, 592
Tierny, Miles, 149, 357
tiling lattice, 426
timbral poetics, 241
time, 4, 338, 892, 1067, 1295, 1326, 1347
   -slice, 250
   -span reduction (=TSR), 377
   complex -, 1186, 1218
   dilatation, 70
   event, 556
   generator, 775
   grouping, 99
   imaginary -, 957, 1189, 1297, 1300
   interval, 70
   layers, 1299
   logical -, 502
   material -, 502
   mental -, 547
   musical -, 892
   onset -, 1335
   ontology, 775

performance field, 1191
physical -, 547
real -, 1189
regularity, 99
reverberation -, 1337
series, 707
signature, 69, 631
span, 70
  reduction, 619
stretching, 650
structure, self-similar -, 799
telling -, 773
told -, 773
unfolding -, 1298
time span, 980
timed co-performance, 1302
timing
micro -, 220
microstructure, 719
Tinctoris, Johannes, 517
Tizol, Juan, 1297
Todd operator, 619
Todd, Neil McAngus, 555, 608, 610, 612, 621, 897, 1182
ToE, 869
told time, 773
tolerance, 642, 681
ToM_CD, vii
Tomasello, Michael, 856
Ton, 507
tonal
ambiguity, 493
function, 249, 263, 447
  value, 447
inversion, 124, 788
tonalities, admitted -, 465
tonality, 249, 263, 414, 436, 447, 453
harmonic minor -, 460
major -, 460, 478
melodic minor -, 460
minor -, 478
natural minor -, 460
Pythagorean -, 461
reference -, 448
third degree -, 450
tone
recitation -, 261
space, notched -, 1363
tenor -, 261
tone parameters
mental -, 67
physical -, 68
tonic, 261, 414, 445, 448
seventh, 420
tonical, 263
Tonort, 507
tonotopy, 1361

tool, paradigmatic -, 789
top space, 588
top-down, 622
topic, 38, 329
topographic cube, 17, 32
topographical navigation, 19
topography, 9, 114
local -, 17
local character of -, 24
musical -, 17
topological
category, 939, 961, 1085, 1464
digraph, 993
functor, 939
meaning, 158
music theory, 1236
pair, 1474
space, 1443
  irreducible -, 1413
topological space, pointed -, 1474
topology, 38, 157, 163, 225
AK_ξ-, 1237
algebraic -, 164, 919
associated -, 1450
base for a -, 1443
coherent -, 1444
coinduced -, 1444
colimit -, 1444
combinatorial -, 254
compact-open -, 914, 937, 1464
discrete -, 1443
dominance -, 231, 402
epsilon -, 397
epsilon gestalt -, 398
extension -, 430
finite cover -, 353
generators for a -, 1463
Grothendieck -, 149, 353, 1430
harmonic -, 443
Hausdorff -, 1445
indiscrete -, 1443
Lawvere–Tierny -, 357
limit -, 1444
maximal meter -, 268, 378
maximal meter nerve -, 379
on gestalt spaces, 395
on motif spaces, 395
product -, 1444
quotient -, 1444
quotient dominance -, 231
relative -, 1444
relative motivic -, 400
specialization -, 403
subbase for a -, 1443
uniform -, 1444
weak -, 1444

Zariski -, 164, 239
topor, 1438
topos, 3, 9, 21, 914, 979, 1029, 1177, 1429
  Boolean -, 1434
  gesture -, 945
  Grothendieck -, 1431
  hyperouranios -, 21
  logic, 435, 930
  of conversation, 826
  Platonic -, 148
topos-theoretic background of modulation, 467
toroidal
  sequence shape type, 388
  shape type, 388
torsion group, 1381
torsion-free rank, 1381
torus, 917
  tangent -, 511
TOS, 605
Toscanini, Arturo, 1292
total, 1372, 1373
touch, 894
Tower, Joan, 1264
Träumerei, 673, 709, 711, 739, 768
trace, objective -, 101
track, 250
tracking, motion -, 1290
tradition, 331
  contrapuntal -, 199, 1367
  performance -, 753
  Pythagorean -, 22, 1364
traditional musicology, 22
transcendence, 21
transcendental element, 1460
transform, Fourier -, 1344
TransforMaster, 789
transformation, 405, 407, 774
  control of -, 200
  gestural -, 1272
  natural -, 1422
  of sound, 120
  plane -, 786
  Riemann -, 315
  score, 785
  symmetry -, 249, 250
transformational
  approach, 204
  invariance, 226, 271
  meaning, 158
  theory, 889, 898, 908, 938, 947, 965, 979, 1119
transit, 1037
transitional structure, 464
transitive, 1373
  action, 1378
  epistemology, 1059
transitivity, 229

translation, 133, 1396
  part, 1397
transposability, 167
transposition, 116, 134, 226, 513
  limited -, 126
transvection, 119, 133
tree, 335
  spanning -, 1458
  stemma -, 660
triad, 90, 414
  augmented -, 262
  diminished -, 262
  major -, 262
  minor, 262
triadic
  degree, 1245
    interpretation, 261
  interpretation, 276, 450, 455, 466
trigeneric monoid, 444
trigger, motion -, 607
trill, 74, 625
True, 1433
Truslit, Alexander, 1293
truth, 1002, 1016
  denotator, 335
TTO operator, 207
tube, Eustachian -, 1355
Tudor, David, 250
tuning, 249
  just -, 1350
  just-tempered -, 1351
  justest -, 460
  mediante -, 1351
  Pythagorean -, 266, 478, 1350
  tempered -, 1350
  12-tempered -, 90
  well-tempered -, 1350
turbidity, 123
Turing
  machine, 553
  test, 791
turning point, 465
12-tempered
  scales, common -, 96
  tuning, 90
two-dimensional
  alteration, 787
TX7, Yamaha -, 525
type, 43, 1438
  casting, 333
  change, 331
  coproduct -, 47
  form -, 55
  gestural form -, 1086
  homotopy -, 1474
  of a cellular hierarchy, 590

powerset -, 47
product -, 46
shape -, 385
synonymy -, 47
types
  of gestures, 1286
  ordering on -, 76
typology
  of forms, 55
  recursive, 49
  recursive -, 42

U

Uhde, Jürgen, 248, 466, 497, 890, 893, 916, 950, 957, 1182
Unbewusstes, 528
uncertainty relation, 245, 425
uncontrolled paradigmatics, 165
underlying symmetry, 129
understanding, 325, 828
  musical works, 324
  systematic -, 826
Underwater, 1283
unfolding, 776
  time, 1298
unfreezing process, 912
unfreezing gestures, 1287
Ungvary, Tamas, 895
unicorned view, 752
uniform topology, 1444
uniformity, 1444
union, 1371
unit, musical -, 90
unitary, 1492
unity, 36, 42, 49
universal ordering, 39
universe, 1175, 1420
  algebraic -, 1029
  of structure, 329
  of topics, 329
unordered
  graph, 1458
  p-space interval, 207
  pc interval, 207
UPIC, 1159
Ursatz, 329
Ussachevsky, Vladimir, 250
Utai, 13, 342

V

vakr, 1306
Valéry, Paul, v, 13, 41, 155, 547, 553, 561, 574, 585, 638, 860, 874, 889, 899, 1042, 1044, 1064, 1177, 1182
valence, 1362
  supporting -, 1362
  theory, 1353

valid sentence, 1433
validation operator, 349
valuation, interpretative -, 13
value
  change, 632
  initial -, 562
  matrix, 766
  participation -, 525
vampire, 947
Vanbremeersch, Jean-Paul, 1060
Varèse, Edgar, 323, 324
variable
  bound, 1436
  causal-final -, 767
  explanatory -, 724
  free, 1436
  inclusion, frequency of -, 732
  individual -, 1435
  predicate -, 1435
  propositional -, 1432
variable address, 53
variation, 405, 407, 507, 787
  melodic -, 794
  of the perspective, 151
  pressure -, 1335
  principle, 325
variational principle
  global -, 1231
  Hamilton's -, 1243, 1481
variations
  calculus of -, 1471
  cycle of -, 791
varieties of sounds, 232
variety instrumental -, 555
varying element, 1023
vector, 1391
  analytical -, 723
  field, 1245, 1451, 1468
  instrumental -, 835
  interval -, 208, 210
  interval-class -, 204
  invariance -, 208
  prime -, 1351
  shaping -, 723
  space, 1392
  spectral -, 832
  spectral participation -, 524, 525
vector field, performance -, 1253
velocity, 608, 784, 1349
  instantaneous -, 27
  bow -, 833
  concept of instantaneous -, 27
  MIDI -, 913
  physical -, 27
verbal description, 622
Vercoe, Barray, 761

Verdier, Jean-Louis, 354
Verillo, Ronald, 607
verse
  matrix, 213
  poetics, 248
vertex, 1375, 1445, 1457
  final -, 660
vertical
  grid vector, 788
  poetical function, 782
  poeticity, 214
vestibular system, 608, 897
Veyne, Paul, 1040
Viète, François, 1033
vibration, gestural -, 893
vibrato, 235, 237, 607, 833
  parameter, 833
Vico, Giovan Battista, 1012
Vieru, Anatol, 211, 314
view, 802
  configuration, 1108
  kernel -, 680
  process -, 1138
  unicorned -, 752
Vigarello, Georges, 874
Villon, François, 116, 198, 199
Vinci, Leonardo da, 856
viola, 825
Viola, Bill, 1265
Violi, Patrizia, 867
violin, 825
  family, 241, 828, 840
  hierarchy, 593
violoncello, 825
virtual acoustics, 701
visual
  navigation, 361
  programming language, 801
  semiotic marker, 813
visualization, 760, 761
  principle, 1107
vitality affect, 1266
vocabulary
  dodecaphonic -, 199
  extension, 40
  switch, 205, 240
vocal
  gesture, 1301, 1313, 1315
  hypergesture, 1316
vocal folds, 1321
vocal gestures, space of -, 1315
vocalist
  Carnatic -, 1301
  Hindustani -, 1301
vocalization, speech -, 1302
Vogel chromatic, 478

Vogel, Martin, 97, 418, 423, 427, 473, 478, 1350, 1499
voice, 509, 1313
  crossing, 509
  instrumental -, 220
  leading, 249
  pedal -, 499
  stationary -, 499
Voisin, Frédéric, 817
Volkswagen Foundation, 665
volume, 188
Vordergrund, 415
vowel, 234
Vuza
  class, 312
  rhythm, 312
Vuza, Dan Tudor, 70, 211, 268, 309, 312
Vygotsky, Lev Semyonovich, 854

W

W, 498
Wagner, Richard, 212, 670
walking, 607, 1304
wall, 1066
  bottom -, 631
Wall, Anthony, 856
Wallis, John, 1007
wallpaper, 1133, 1163
walls, 1072
  extended -, 1072
  open -, 1072
Wanderley, Marcelo, 901, 1182
Ward, Artemus, 615
wave, 1340
  acoustical -, 963
waveguide, 1346
wavelet, 236, 1344
  frame, 236
  Meyer -, 1346
  Morlet -, 1345
  Murenzi -, 1345
  theory, 1344
wavelet-transformed, 1344
weak topology, 1444
Weber-Fechner law, 1348
Webern, Anton von, 124, 125, 127, 163, 204, 246, 325, 711, 752
Wedderburn, Joesph, 1395
wedge, crescendo -, 641
wedge product, 1468
Wegner, Peter, 27
weight, 597, 610, 612, 619, 638, 673
  analytical -, 549, 553, 646
  class -, 283
  combination, 681
  continuous -, 639
  function, 1253

default -, 482
   Pólya -, 191
   product -, 191
harmonic -, 482, 647
induced nerve -, 379
inverted -, 681
metrical -, 375, 376, 646
mixed -, 671
motivic -, 408, 647
nerve -, 379
onset -, 99
profile, 218
recombination, 640
simplicial -, 283, 988
simplicial metrical -, 269
sober -, 379
system, 632
tempo, 653
third -, 676
watcher, 681
weighted graph, 239
Weil, André, 1041
Weissmann, Jody, 1035
well-ordered, 1373
well-tempered
   modulation, 469
   tuning, 1350
Weltgeist, 1174
Wernicke's area, 872
Weyl, Hermann, 161
WFR, 377
whatness, 21, 1179
whereness, 21
White, Andrew, 571
whole, 246, 273
whole-tone scale, 541
Whymper, Edward, 151
Wicinski, A.A., 622
Widmer, Gerhard, 611
Wieland, Renate, 248, 889, 893, 894, 899, 916, 950, 1182
Wieser, Heinz-Gregor, 524
wild problem, 756
Wiles, Andrew, 1179
Wille, Rudolf, 3, 113, 453
window
   oval -, 1354, 1358
   power -, 524
   round -, 1358
Winson, Jonathan, 528
Witten, Edward, 1177, 1485
Wittgenstein, Ludwig, 38, 327, 1178
Wittgenstein, Paul, 883
WLOG, 110
Wolff, Christian, 250
word, 60, 1376

monoid, 1376
work, 11, 13
   identity of a -, 14
   motivic -, 276
   production of a -, 12
world, 460, 497, 498
   imaginary -, 1058
   of the possible, 1052
   real -, 1044
world-antiworld thesis, 497
world-line, 1483
world-sheet, 1185, 1186, 1192, 1471, 1483
   global -, 1224
   ramification, 1230
wrap form morphisms, 331
wrapped as local composition, 91
wreath product, 1380
writing, field -, 803
Wulf, Bill, 32, 667
Wyschnegradsky, Ivan, 93

X

X-ray procedure, 892
Xenakis, Iannis, 30, 211, 1159
XML, 1096

Y

Yamaha, 687, 700, 1342, 1346
   CX5M, 525
   RX5, 220, 791
   TX7, 525
   TX802, 220, 236, 239, 791
Yemenite cantor, 1301
Yoneda
   embedding, 1397, 1423
   pair, 1437
   philosophy, 92, 145, 152, 466, 828, 1475
   subcategory, 1437
Yoneda Lemma, 142, 280, 324, 962, 870, 938, 1006, 1022, 1031, 1034, 1037, 1045, 1056, 1061, 1073, 1179, 1255, 1474
Yoneda, Nobuo, 145, 245, 323, 828, 851, 1179

Z

$\mathbb{Z}$-addressed motives, contrapuntal meaning of -, 100
$\mathbb{Z}$-relation, 210
Zahorka, Oliver, 628, 665, 687, 1088
Zalamea, Fernando, 1037, 1060
Zarca, Bernard, 1028, 1043
Zariski, Oskar, 164
Zariski
   site, 1415
   tangent, 109
      space, 927, 1415
   topology, 164, 239
Zarlino, Gioseffo, 122
Zbikowski, Lawrence, 1265, 1308

Zermelo, Ernst, 1374
zero address, 53
zig-zag, motivic -, 277, 781
Zurechthören, 1353
Zurich school, 612, 979
Zwiebach, Barton, 1482

Printed in the United States
By Bookmasters